60·00

# ENVIRONMENTAL BEHAVIOUR OF PESTICIDES

## AND

## REGULATORY ASPECTS

Developed from a symposium sponsored by
the European Commission within the
framework of COST ACTION 66

Brussels, April 26-29, 1994

Edited by Copin A., Houins G., Pussemier L., Salembier J.F.

© **EUROPEAN STUDY SERVICE**

# Acknowledgments

**We thank for their contribution :**

— The Commission of the European Union and the Cost Action 66

— Fonds de Phytopharmacie - Fytofarmaceutisch Fonds
(Belgian Ministry of Agriculture)

— Phytophar (Belgian Association of Agrochemicals Industry)

— ECPA (European Crop Protection Association)

# PROCEEDINGS OF THE

# 5TH INTERNATIONAL WORKSHOP

# ENVIRONMENTAL BEHAVIOUR OF PESTICIDES

# AND

# REGULATORY ASPECTS

Developed from a symposium sponsored by
the European Commission within the
framework of COST ACTION 66

Brussels, April 26-29, 1994

Edited by Copin A., Houins G., Pussemier L., Salembier J.F.

© 1994 - *EUROPEAN STUDY SERVICE*
Avenue Paola 43 - B-1330 Rixensart - BELGIUM
TEL. : (+32 2) 652 11 84 - FAX : (+32 2) 653 01 80
RCN : 51147 - TVA : BE-659 309 394

ISBN 2-930119-03-9

# SCIENTIFIC COMMITTEE

# ORGANIZING COMMITTEE

# FOREWORD

The 5th International Workshop "Environmental Behaviour of Pesticides and Regulatory Aspects" was organized in Brussels from 26th to 29th April 1994.

This 5th workshop followed a series of international symposia on pesticide behaviour initiated by Paul Jamet in Versailles in 1988. The second workshop was organised in Allicante in 1989 and was followed by two others organised in Munich (1990) and in Rome (1991), respectively.

The organising committee of the Brussels workshop consists of representatives of the Belgian Ministry of Agriculture drawn from the Agricultural Research Center of Gembloux, the Belgian Registration Committee and the Institute for Chemical Research of Tervuren, but also includes representatives of the Faculties of Agricultural Sciences of Gembloux and Louvain-la-Neuve.

We shall now define the spirit in which this 5th International Workshop was developped and the slightly different objectives set.

It was the aim of our organising committee to focus on both scientific and regulatory aspects. Everybody knows that for pesticide registration several criteria need to be considered in the assessment procedure : agronomical aspect (efficiency - selectivity), economic usefulness (treatment cost compared to the cost of the predicted losses at the harvest), toxicological properties (acute toxicity, chronic toxicity, toxicity of residues in food and feed,...) and ecotoxicological effects (behaviour in the various components of the environment and effects on living organisms). This workshop deals only with the behaviour of pesticides in the environment and the assessments needed for registration purposes.

Contrary to a widely spread impression, ecotoxicological problems (behaviour, effects) have been a concern of scientists and legal institutions for many years.

In 1965, Bartha and Pramer presented a design for lab studies on the persistence of pesticides in soils. Helling and Turner's first studies involving an evaluation of the mobility of pesticides in soil date from 1968. In their publication on pesticides (1969), the Council of Europe introduced some first recommendations about environmental studies. In 1973, the Biologische Bundesanstalt für Land und Forstwirschaft (Braunschweig) laid down normalized procedures about leaching and degradation. In 1981, the OECD published guidelines relative to the environment and very recently started work on a special Pesticide Programme.

Besides those initiatives, the F.A.O. created in 1979 the European Cooperative Network on Pesticides with special reference to their impact on the environment. The subnetwork on the pesticide's behaviour in the soil is directed by Mr. Hascoët (INRA France). One of the subnetwork's objectives is to standardize the analytical methods of the behaviour in soil of the pesticides introduced for registration. Meetings were held in France in 1977, in Austria in 1979, in Switzerland in 1981. France hosted the latest in 1984.

This cooperation, initiated by the FAO, has been ongoing through the 4 workshops that preceded the Brussels meeting. The themes evolved around sorption and degradation phenomena, outdoor experiments and water monitoring. The subject of modelisation has barely been discussed.

Thanks to the efforts of the French delegation, the E.C. integrated in 1992 our objectives in those of the European Cooperation in the field of scientific and technical research. This led to the foundation of the Action Cost 66.

Moreover, being aware of the increasing importance of the ecotoxicological criteria in pesticides' registration, rendered in enclosure 6 of the coming directive (91/414), the organizing committee of the 5th Workshop wanted to highlight the oral communications and the posters which emphasize the complementarity between the legal and scientific aspects and provide detailed information useful for the registration procedure.

Consequently, the workshop and the proceedings were divided into 6 different sections. The first concentrates on the **regulatory aspects** and on the role of some important **international organizations** such as EC, OECD, EPPO, etc... Sections 2 and 3 concern the classical problems of **sorption, mobility** and **transformation** (mainly under laboratory conditions) whilst sections 4 and 5 deal with **monitoring** and **outdoor experiments** in the "real" environment. The last section **"mathematical models"** is of particular importance since the models aim to integrate both laboratory and outdoor aspects. Moreover, it seems reasonable to think that the use of models for registration purposes will grow in the future despite much more or less well-founded opposition.

Each section will start with a quite comprehensive description of the State of the Art presented by an eminent personality in that particular field of interest. These proceedings do, of course, include recent research papers presented during the plenary and poster sessions, and also practical discussion points that were submitted during the working group sessions. We hope that the proceedings of this workshop will prove useful for all the people interested in pesticides behaviour and its impact on regulatory procedures.

The organizing Committee

# CONTENTS

Acknowledgments

Organizing and scientific Committees

Foreword

Contents

# SECTION I

# REGULATORY ASPECTS AND ROLE

# OF THE INTERNATIONAL ORGANIZATIONS

Chairman :      A. SCHARPE
European Commission
Directorate-General for Agriculture

# REGULATORY ASPECTS AND ROLE OF THE INTERNATIONAL ORGANISATIONS

## 1.1.  Papers presented by the International Organisations

### 1.1.1.  *The gaps in the regulatory aspects of the environmental guidelines*

## A. Scharpé, L. Smeets

European Commission
Directorate-General for Agriculture

## Introduction

Since January 1993 the European Single Market has been realized. This was only possible by harmonizing the different national regulations in a wide range of sectors. Also for those concerning pesticides used in agriculture it was necessary to harmonize the legislation of the Member States.

Due to differences in data requirements of the various Member States and differences in the methodologies accepted by them for the generation of data, industry often had to generate additional data to address issues that often had already been fully investigated. The additional data thus generated added little to knowledge as to the fate, behaviour and impact of pesticides. Differences in approach to the interpretation of data and in risk assessment could result in different decisions being made.

A Community regime for the authorization of plant protection products was adopted in July 1991 by the Member States and came into force on the 25th of July 1993.

To the extent that certain matters are not covered by the Directive or other Community legislation the Member States will continue to apply further on national provisions.

The Directive provides that the Commission can take certain implementation measures, which when adopted will lead to further harmonization in the Community. In any case the Standing Committee on Plant Health in which the Member States are represented must be involved before any such measure can be adopted.

## Main principles of Directive 91/414/EEC

The regulatory system adopted is a two step procedure, involving approval of active substances at Community level and authorization of formulations for specific uses by the national authorities.

That means that a *Community positive list of active substances* - the Annex I to the Directive - has to be established on the basis of which preparations can be authorized by the Member States.

The positive list of active substances will be established on the basis of a single application, a single dossier, the application of agreed Community criteria, and a single decision of the Commission.

The data requirements for active substances are laid down in Annex II to the Directive.

Inclusion in Annex I may be linked with any appropriate conditions and necessary restrictions on use related to regions, soil types, climates, crops, etc.

Although subject to review at any time if new information becomes available, substances are to be reviewed on a systematic basis 10 years after first inclusion in Annex I.

The *authorization of preparations* has to be granted to the national authorities of the Member States according to common data requirements laid down in Annex III to the Directive, harmonized procedural rules and common criteria. These criteria include that a plant protection product may not be authorized unless it is established that when properly used :

- it is sufficiently effective;
- it is not phytotoxic;
- it does not cause unnecessary suffering and pain to vertebrates to be controlled;
- it has no harmful effect on human or animal health or on groundwater;
- it has no unacceptable influence on the environment (fate and distribution, water contamination, non-target species).

Furthermore it must be possible to determine the active substance(s) in the formulation, to determine its residues. The physical and chemical properties of the preparation must be acceptable.

In order to ensure that Member States apply these criteria in an equivalent manner and at the high level of protection of human and animal health and of the environment sought by the Directive, *"uniform principles"* will be developed.

One of the most essential principles of the regulation with respect to the internal market is the obligation of *mutual recognition* of authorizations granted by another Member State to the extent that the agricultural, plant health and environmental conditions relevant to the use of a product are comparable in the regions concerned.

Conflicts would be resolved, either between the parties concerned or, if not, by a Community procedure and decision.

## Further developments (environmental aspects)

Some of the main measures to be taken for the implementation of the Directive are:

- Development of data requirements (Annexes II and III); detailed provisions for the introductions to both annexes and for the efficacy data have already been published while other parts of the annexes are still under discussion;
- Adoption of Uniform Principles for evaluation and decision making, the Annex VI to the directive; a Commission proposal is under discussion since June 1993.

### Data requirements

The data requirements in Directive 91/414/EEC consist in a list of items for which sufficient information has to be provided. An applicant has always the possibility to justify for the non-submission of certain data.

It is the Commission's intention to define:

— the parameters to decide when particular studies or information are required;

— the parameters to decide when particular studies or information are not required;

— the experimental protocols that have to be used to generate the required data.

This has already been done last year for the efficacy data; the physico-chemical data will be adopted soon.

No further details are available yet for the environmental sections of these annexes. Several working documents are currently under discussion. A tiered approach is followed in most cases.

For laboratory studies internationally agreed protocols are often but not always available. Guidelines of OECD, EPPO and FAO are referred to in the working documents.

Internationally agreed protocols for field studies are not very often available. Expert judgment and discussions between the competent authorities and the applicant form the base to decide on the type of study to be performed.

In the framework of ongoing work in OECD on existing and new test guidelines the European Commission has identified its priorities. These include degradation and leaching studies since guidelines on these items are lacking.

## *Uniform principles*

The uniform principles aim to ensure that all Member States, in making decisions with respect to authorization of plant protection products will apply the requirements of the Directive in an equivalent manner and at the high level of protection of human and animal health and of the environment sought by the Directive.

The Commission proposal for "uniform principles" includes:

— a general introduction which identifies the aims of the uniform principles, the data that have to be taken into account for the evaluation and decision making and the procedures to evaluate data in order to come to a decision;

— principles concerning the evaluation of the available data and concerning the decision making with respect to authorization of plant protection products.

The proposal includes detailed provisions concerning:

— impact on the environment, including its fate in soil, water and air,

— impact on non-target species, including birds and terrestrial vertebrates, aquatic organisms, bees and other beneficial arthropods, earthworms and non-target soil micro-organisms.

This proposal is currently under discussion.

The Commission has developed this proposal on the basis of highest technical expertise available in the Member States. For the environmental impact the EPPO-Council of Europe risk assessment schemes have been taken into account as far as possible.

However, with regard to certain items such as impact of plant protection products on non-target plants and on non-target aquatic flora as well as fate of plant protection products in air, it has appeared that more experience is required before detailed principles on evaluation and decision making can be developed. The Commission will follow the technical and scientific progress in these fields and as soon as possible introduce more detailed rules concerning these areas according to the procedures provided for in the Directive.

In many cases the evaluation implies the use of certain models such as :

— Models to estimate the predicted concentration in soil (PEC's),

— Models to estimate the predicted concentration in groundwater and surface waters,

— Models for predicting degradation and dispersion in air.

The result of these estimations is used to assess inter alia the impact on some non-target species.

At this stage only a general guidance on the models to be used is given in the uniform principles. The models used shall:

—    make a best possible estimation of all relevant processes involved taking into account realistic parameters and assumptions;

—    be reliably validated with measurements carried out under circumstances relevant for the use of the model;

—    be relevant to the conditions in the area of the proposed use.

A lot of work still has to be done:

—    appropriate models do not always exist;

—    if they exist, there is not always enough information available on how to use them;

—    it is not always clear if and how they are validated;

—    their range of validity can be limited.

A group of scientists, regulators and industry intends to coordinate efforts in the area of modelling of the fate of pesticides in the environment. The purpose of the Forum for International Coordination of Pesticide Fate Modelling (FOCUS) is to develop recommendations for the use of models in a regulatory context. The European Commission participates as an observer in this work.

With regard to plant protection products on the market since many years, monitoring data can give very useful information on the fate and distribution in the environment. The role of this information in the authorization process is currently under discussion.

# Conclusion

Many aspects concerning the fate, behaviour and impact of plant protection products on the environment still currently under discussion in the European Union will be discussed during this workshop.

The results of this workshop can therefore be an important input for these discussions and would help to achieve further progress in these fields.

## 1.1.2.   The development of an internationally agreed scheme for environmental risk assessment of plant protection products

# D.G. McNamara

European and Mediterranean Plant Protection Organization,
1, rue Le Nôtre, 75016 Paris, France

## Abstract

In most countries, procedures for the registration of plant protection products include the assessment of the potential risks of environmental damage that might be caused during the use of the products. These risks range from the direct poisoning of wildlife to the presence of chemical residues in ground water or sediments. However, the criteria used to decide on the acceptability of environmental risks, and the ways in which the risks are estimated, vary widely. Many countries have their own well-developed national guidelines and there are several general international guidelines in existence. Each of these takes a slightly different approach to arrive at a decision on the acceptability of a particular product. EPPO and the Council of Europe collaborated with scientists and government expert throughout the Euro-Mediterranean region to establish a consistent and explicit decision-making scheme, by which the logic of assessments could be set down as a step-by-step record of how decisions are reached.

## Introduction

Over the past few years, EPPO (the European and Mediterranean Plant Protection Organization), in association with the Council of Europe, has been associated with a remarkable international collaborative effort to produce a decision-making scheme for the evaluation of the risk of any particular plant protection product to the environment. The result is the Environmental Risk Assessment Scheme (EPPO, 1993). This paper describes the history of this operation, the general structure of the scheme, its present state and future plans.

## Development of an Environmental Risk Assessment Scheme

For a number of years, EPPO has been involved in the harmonization among its member countries of regulations for the registration of plant protection products. However, unlike its role in the other main area of its activities, plant quarantine, EPPO has not been the only international (or indeed, the major) body involved in the harmonization of registration procedures in the European and Mediterranean region. OECD, the Council of Europe and FAO, among others, have all been active in this field, and therefore, EPPO long ago decided that, given its expertise in plant protection, it should concentrate its efforts on the practical evaluation of the efficacy of plant protection products, and to leave other aspects, such as toxicology or residue analysis, to other organizations. To this end, a series of approximately 200 guidelines have been developed over a number of years on the conduct of field trials for efficacy evaluation of plant protection products, each one concerned with a specific pest or type of pest on a specific crop or group of crops (EPPO, 1989).

In 1988, however, the member countries of EPPO expressed concern at the lack of internationally consistent guidelines on how to evaluate the likely effects of the use of a plant protection product on the environment. Certain general international guidelines existed (for example, IUPAC, 1980; OECD, 1981; FAO, 1989; GIFAP, 1990; Council of Europe, 1992) but as they varied to a considerable extent in certain key elements, it was difficult for EPPO member countries to decide which of these should act as the guide for the establishment of national procedures in the registration process. In addition, the member countries of the European Union (EU) (all of whom, incidently, are also EPPO members) were aware that the Commission, in preparing its Directive on the placing of Plant Protection Products on the market, intended to include guidance on ecotoxicological testing as part of the Uniform Principles. The countries recognised that EPPO and the Council of Europe were two bodies with experience in such technical areas, where wide international collaboration and approval could be assured.

The EPPO Working Party on Plant Protection Products (the committee that directs the technical activities of EPPO and which is composed of representatives of the member governments) decided to convene a joint meeting with the Council of Europe to review and evaluate the existing guidelines on the subject of environmental risk assesment and, if necessary, to decide how to develop an internationally acceptable procedure and what form it should take. The Joint Hearing on Ecotoxicology was held in May 1989 in Paris which agreed that a risk assessment scheme could and should be prepared; it proposed the establishment of a joint EPPO/CoE Panel on Environmental Risk Assessment to prepare such a decision-making scheme.

It was agreed that the decision-making scheme should have the characteristics of being a step-wise, logical procedure which would *not* have rigid triggers leading to the production of unnecessary data. It should clarify when tests should be performed with the active ingredient and when with the formulated product, when laboratory tests, semi-field tests and/or field tests should be performed, it should allow the input of expert opinion at various stages of the procedure and during the final evaluation, and should consider the use of post-registration monitoring or surveillance.

The Panel was composed of some 20 experts in environmental testing drawn mainly from government bodies and research institutes in EPPO countries but including also representatives from industry and other relevant international organisations. This Panel decided to subdivide the decision-making scheme into eleven separate risk assessment sub-schemes each one pertaining to a different element of the environment. The elements decided upon were the abiotic parts of the environment - soil, ground water, surface water and air, and the groups of organisms present - aquatic organisms, soil microflora, earthworms, arthropod natural enemies, honeybees, terrestrial vertebrates and non-target plants. In order to prepare each sub-scheme, a separate subgroup of the Panel was created, coordinated by an individual member of the Panel and containing recognised international specialists in this subject-area (Table 1). Some of these sub-groups were closely linked to international groups already in existence, such as the Domsch groups on soil microflora, or IOBC for beneficial arthropods, where a considerable body of knowledge had already been collected. The sub-groups prepared their sub-schemes by correspondence and, if thought necessary by their coordinators, at meetings convened either in association with the Panel or separately. The Panel developed the framework of the decision-making scheme and the general structure of the sub-schemes, and then coordinated the specific activity of the sub-groups during the course of a number of meetings over three years, in Paris, Wageningen, Strasbourg, Slough, Budapest and again Paris.

In April 1993, the first set of chapters of the scheme were published, in English and French, in the EPPO Bulletin (EPPO, 1993). They had previously been approved by the Working Party on Plant Protection Products and by the Council of EPPO at its 41st session in September 1992. The published chapters were the sub-schemes on soil, ground water, surface water, aquatic organisms, earthworms and honeybees, as well as an introductory chapter and a chapter on guidance on identifying aspects of environmental concern. The second set was published in 1994 (EPPO, 1994) and comprised the sub-schemes on soil microflora, beneficial arthropods and terrestrial vertebrates. Further chapters to be completed, in addition to the sub-schemes on air and non-target plants, were on integration of results, validation, glossary, and a list of members of the Panel and the sub-groups. The Council of Europe also intends to publish the complete set of chapters as a separate booklet.

# Aims of the Environmental Risk Assessment Scheme

The assessment scheme aims to :

a)   guide assessors on the questions that should be addressed, and the data that may need to be requested from registrants;

b)   provide information on the test methods and approaches that are suitable in each case;

c)   indicate how the data should be interpreted in a consistent manner, involving expert judgement where appropriate;

d)   produce a reliable assessment of environmental risk, that is suitable to aid risk management.

The assessment procedure may also identify ways in which use could be modified to reduce effects, and therefore has direct implications for risk management. The scheme does not offer fixed criteria on which to decide the acceptability of products and their uses, because that involves other aspects as well as knowledge of the potential environmental risks.

The result is a set of flexible procedures that can be adapted for use in various ways according to the priorities in different countries, yet retain the consistency of a common framework. The scheme is intended to be used by the individual national regulatory authorities. It can also be used by chemical producers during the development of new products in order to predict which data will be needed to answer the environmental concerns of regulatory authorities.

# Structure of the Environmental Risk Assessment Scheme

Each sub-scheme is in the form of a series of questions. These begin with questions about the characteristics (physical and chemical) of the product and its pattern of use (possibilities of reaching the environmental element concerned). The answers to these preliminary questions will direct to other questions further in the sub-scheme concerning data from tests on effects on the particular element. Answers to these questions lead, in turn, to an assessment of the category of risk (negligible, low, medium or high). This is true for all sub-schemes except that on soil which is not considered to be a risk assessment for soil itself, but provides data (transformation rates, change of concentration) to other elements of the environment. An outline diagram accompanies each sub-scheme to give an indication of likely pathways through the decision-making process, but it should be stressed that the diagrams are merely an aid to understanding and are not intended to be used to reach a decision.

The sub-schemes themselves may appear, at first reading, to be complex, because of the total number of questions contained (for example, the sub-scheme on terrestrial vertebrates contains more than ninety steps/questions), but, in fact, the different characteristics of each product being evaluated will result in a different pathway through the scheme, depending on the answers given to the questions along the way. The pathway for each product will be a much simplified version of the whole and will have the additional merit of requiring only *relevant* questions to be answered and, therefore, only requiring a restricted amount of test data to support the claim for registration.

Explanatory notes accompany every sub-scheme in order to provide advice on how to answer certain questions and, occasionally, on how to obtain the necessary data. In addition, there is a section entitled 'analysis of uncertainty' wherein the reader is invited to repeat the passage using different answers to those questions where there might have been uncertainty the first time. For example, where errors of measurement could be suspected, or where assumptions or extrapolation from species to species may have been made.

In addition to limiting the test data to be provided for registration to those that are of true relevance to the type of product and use-pattern of the plant protection product in question, the scheme also has the advantage that the intermediate steps that led to the final registration decision can be recorded and reviewed, if necessary. This is of particular importance at the international level where some countries may share in a partial or complete registration process (for example, within the EU).

## Future plans

From the very beginning of the preparation of the decision-making scheme, it was recognised that a scheme of this complexity and comprehensiveness had not previously been used in practice within a registration process for the evaluation of environmental risk, although parts of it had already been tried out (with success) in a limited number of countries. Therefore, it was decided that a validation process should begin as soon as possible, in which EPPO member countries would be asked to apply the scheme in the evaluation of the environmental risk from a number of 'dummy' products, based on a dossier of data provided by the Panel. The results from this exercise, and other comments offered by the officers who perform the trial evaluation, will be used as the basis for a revision of the entire scheme. Later, it is intended to review the scheme to take account of experience gained through its practical application and of developments in environmental toxicology and chemistry.

## *Conclusion*

Although the scheme will bear the name of EPPO and the Council of Europe, it will represent the product of a considerable amount of careful work by scientists in many countries and the common will of countries in the Europe and the Mediterranean to reach a concensus on standards for ensuring that plant protection products, essential for the production of adequate and healthy food supplies, are not used in a way that might damage our environment.

## *References*

□   [1] Council of Europe (1992). Chapter V. Guidance on environmental phenomena and wildlife data. In *Pesticides* (7th edition). Council of Europe, Strasbourg, France.

□   [2] EPPO (1989). EPPO Guidelines for efficacy evaluation of plant protection products. *Bulletin OEPP/EPPO Bulletin 19,* 183-245.

□   [3] EPPO (1993). Decision-making scheme for the environmental risk assessment of plant protection products. Chapters 1-6, 8 and 10. *Bulletin OEPP/EPPO Bulletin 23,* 1-165.

□   [4] EPPO (1994). Decisionmaking scheme for the environmental risk assessment of plant protection products. Chapters 7, 9 and 11. *Bulletin OEPP/EPPO Bulletin 24,* 1-88.

□   [5] FAO (1989). *Revised Guidelines on Environmental Criteria for the Registration of Pesticides.* FAO, Rome, Italy.

□   [6] GIFAP (1990). *Environmental Criteria for the Registration of Pesticides.* GIFAP Technical Monograph no.3, GIFAP, Brussels, Belgium.

□   [7] IUPAC (1980). Recommended approach to the evaluation of the environmental behaviour of pesticides. *Pure and Applied Chemistry 60,* 901-932.

□   [8] OECD (1981). *Guidelines for Testing of Chemicals.* OECD, Paris, France.

Panel

W.W.M. Brouwer (NL)
E. De Lavaur (FR)
P.W. Greig-Smith (GB)
G. Grolleau (FR)
P. Jamet (FR)
C. Kula (DE)
G. Joermann (DE)
P.J. Lawlor (IE)
J.B.H.J. Linders (NL)
A. Lundgren (SE)

R. Luttik (NL)
P. Matthiessen (GB)
A. Mylymäki (FI)
P. Oomen (NL)
R. Petzold (DE)
H. Schüepp (CH)
E. Seutin (BE)
J. Surjan (HU)
A. Yagüe (Es)

Sub-groups

Soil

J.J.T.I. Boesten (NL)
W.W.M. Brouwer (NL)
F. Copin (BE)
P. Jamet (FR)
D. Riley (GB)
K. Schinkel (DE)
T. Tooby (GB)

Ground Water

L. Bergström (SE)
J.J.T.I. Boesten (NL)
W.W.M. Brouwer (NL)
B. Bügel Mogensen (DK)
A. Del Re (IT)
M.B. Dolan (IE)
M. Galoux (BE)
J. Guth (CH)
A. Hellweg (DK)
P. Jamet (FR)
J.A. Jobsen (NL)
P.H. Nicholls (GB)
K. Schinkel (DE)
A.M.A. van der Linden (NL)
A. Yagüe (ES)
D. Riley (GB)

Surface Water

J. De Greef (NL)
M. Galoux (BE)
J.B.H.J. Linders (NL)
P. Matthiessen (GB)
A. Montiel (FR)
R.R. Stephenson (GB)

Aquatic Organisms

H. Köpp (DE)
P. Kristensen (DK)
P. Matthiessen (GB)
V.K. Miettinen (FI)
R. Stephenson (GB)
C.J. van Leeuwen (NL)

Soil Microorganisms

J. Anderson (DE)
P.C. Brookes (GB)
H. Ehle (DE)
M.P. Greaves (GB)
H. Heinonen-Tanski (FI)
F.J. Lewis (GB)
C. Kula (DE)
P. Perucci (IT)
Z. Pokacka (PI)
W. Reinier van den Berg (NL)
H. Schüepp (CH)
G. Soulas (FR)
L. Torstensson (SE)

Earthworms

P.J. Edwards (GB)
P.W. Greig-Smith (GB)
C. Kula (DE)
E. Steen (SE)
C.A.M. van Gestel (NL)

Arthropod Natural Enemies

N. Carter (GB)
B. de Clerq (BE)
C.A. Dedryver (FR)
S. Hassan (DE)
C. Inglesfield (NL)
U. Heimbach (DE)
P. Oomen (NL)
L. Samsøe-Petersen (DK)

Honey Bees

L.P. Belzunces (FR)
D. Brasse (DE)
P.W. Greig-Smith (GB)
A. Hadhazy (HU)
C. Inglesfield (NL)
F.J. Lewis (GB)
P. Oomen (NL)
J.H. Stevenson (GB)

Terrestrial Vertebrates

P.W. Greig-Smith (GB)
G. Joermann (DE)
P.J. Lawlor (IE)
R. Luttik (NL)
A. Mylymäki (FI)
W. Pflüger (DE)
A. Yagüe (Es)

Air

R. Binner (DE)
W.W.M. Brouwer (NL)
J.P. Garrec (FR)
A.J. Gilbert (GB)
D. Gottschild (DE)
J.A. Guth (CH)
F.A.A.M de Leeuw (NL)
P.C.H. Miller (GB)
P. Robin (FR)
F. van den Berg (NL)

Non-Target Plants

G. Barralis (FR)
V. Breeze (GB)
J.J. Dulka (US)
G. Heidler (DE)
R. Luttik (NL)
W. Pestemer (DE)
F. Rocha, F. (PT)
A.E.G. Tonneijck (NL)
F. Schmider (DE)

**Table 1 :**   Compositions of the EPPO/Council of Europe Panel Environmental Risk Assessment and of the sub-groups on different environmental elements.

## *1.1.3.*      *Development and Revision of OECD Test Guidelines on Environmental Fate of Pesticides*

# Dr. N.J. Grandy

Organisation for Economic Co-operation and Development,
2 rue André Pascal, 75775 Paris, France

## The OECD and its Environment Programme

The Organisation for Economic Co-operation and Development (OECD) is an intergovernmental organisation of 25 democratic nations (Member countries) with advanced market economies[1].

The OECD was established in 1960. Its basic aims are to:

■      achieve the highest sustainable economic growth and employment;

■      promote economic and social welfare throughout the OECD region by co-ordinating the policies of its Member countries;

■      stimulate and harmonise its members' efforts in favour of developing countries.

The OECD is not a regulatory body. It can however take formal actions in the form of Decisions and Recommendations of the Council (i.e., the highest authority of OECD). Council Decisions are legally binding on all the Member countries. Council Recommendations, whilst not legally binding, carry a strong moral obligation.

The OECD is perhaps best known for its work in monitoring and forecasting trends in the world's economies. However, OECD Member countries account for 70% of the world gross domestic product, 70% of world trade, and most of the world production of chemicals. They therefore have a special responsibility with regard to the state of the environment and there has been a strong OECD Environment Programme since 1970. The main activities within the Environment Programme are in pollution prevention and control, environmental health and safety, environment and economics, climate change, waste management, natural resource management and the assessment of environmental performance.

## Chemicals Safety and the OECD

OECD work on chemical safety began in the early 1970's as a result of concern over widespread contamination of the environment and accompanying adverse environmental effects. The initial focus was on specific chemicals, including persistent organochlorine pesticides, polychlorinated biphenyls, mercury, cadmium and lead. However, as thousands of new chemicals and products were entering the world market each year, it was soon recognised that dealing with a few specific chemicals was not sufficient to ensure chemical safety in general. At the same time, many countries began to require, through new legislation, that chemicals be tested and assessed for potential risks before being placed on the market. In response, OECD broadened its activities and took an active role in initiating the development, within its Member countries, of harmonised policies and practical tools for protecting human health and the environment, i.e. the OECD Chemicals Programme. The main activities of the Chemicals Programme are Test Guidelines, Good Laboratory Practice, Hazard Assessment, Harmonisation of Classification and Labelling Systems and Co-operative Activities on Existing Chemicals and Risk Reduction.

---

[1]      The OECD Member countries are : Australia, Austria, Belgium, Canada, Finland, Denmark, Finland, France, Germany, Greece, Iceland, Ireland, Italy, Japan, Luxembourg, Mexico, the Netherlands, New Zealand, Norway, Portugal, Spain, Sweden, Switzerland, Turkey, United Kingdom, United States of America. The Commission of the European Community also participates in the work of the Organisation.

# Pesticides and the OECD

Until recently, the focus of the OECD Chemicals Programme was on new and existing industrial chemicals. This was expanded in May 1992 to include a programme on pesticides, brought about by an increasing interest among OECD Member countries for improved co-operation in the area of pesticide control. The countries recognised that, in spite of the extensive and valuable ongoing activities of a number of international organisations, significant differences still remained in national approaches to pesticide registration. The existence, as at present, of these different pesticide registration procedures with differing data requirements and review processes leads to considerable duplication of effort by regulatory bodies and industry at a time when resources are scarce. Such multiplicity of schemes can also cause barriers to trade. Improved harmonisation of national procedures will increase the possibility of mutual acceptance of data by Member countries, thus enabling a "sharing of the burden" by regulatory authorities and reducing the need for un-necessary and expensive testing by industry.

The OECD Pesticide Programme is overseen by a Pesticide Forum comprised primarily of government regulators, but including representatives from industry and environmental groups. The programme has three major goals:

(i)      to achieve harmonisation of national pesticide assessment and control procedures;

(ii)     to achieve more efficient re-registration of pesticides through Member country co-operation;

(iii)    to promote the reduction of risks from the use of pesticides.

In order to address these goals, the Pesticide Programme includes activities in five areas:

(i)     Test Guidelines;

(ii)    Data Requirements for Registration;

(iii)   Hazard/Risk Assessment;

(iv)    Re-registration; and

(v)     Risk Reduction.

This paper will focus on the work on **Test Guidelines**.

# Test Guideline Activities for Pesticide Testing

## *Survey of Test Guideline requirements*

It was recognised early on in the Pesticide Programme that in relation to the testing of pesticides, certain existing OECD Test Guidelines would need to be revised in order to deal with areas unique to pesticide assessment (e.g. consideration of metabolites). It was also recognised that new Guidelines may need to be developed. Consequently a survey was initiated in May 1992 in which Member countries and GIFAP (on behalf of the pesticide industry) were requested to:

(i)     identify existing Guidelines they felt to be inadequate for pesticide regulatory use and which needed revision;

(ii)    indicate those endpoints for which they believed new OECD Test Guidelines should be developed, giving priorities;

(iii)   identify other international (e.g. FAO, ISO, EPPO) or national test methods and protocols which could be used as a basis for OECD Test Guideline development.

The results of the Survey, in terms of the numbers of revisions and new Guidelines proposed, are shown in Table 1 below.

| Test area | Number of Guideline revisions proposed | Number of new Guidelines (or endpoints) proposed |
|---|---|---|
| Physical-chemical properties and environmental fate | 8 | 45 |
| Ecotoxicology | 9 | 32 |
| Human health effects | 27 | 9 |
| Total | 44 | 86 |

**Table 1 :** Results from the Test Guideline Survey indicating the numbers of revisions and new Guidelines proposed.

Most of the revisions proposed were for human health effects Guidelines, although they were of a general nature (i.e. updating to keep pace with scientific progress). Proposed revisions in the other areas were of a more fundamental nature, indicating that for use with pesticides, additional factors beyond those normally required for general chemicals should be taken into account (e.g. metabolites). Regarding needs for new Guidelines, most proposals were for physical-chemistry/environmental fate and ecotoxicology. This was expected since data requirements in these areas for general chemicals cover many fewer endpoints than do those for pesticides.

At the same time as the Test Guidelines Survey was being carried out, Member countries were also surveyed with respect to their data requirements for the registration of conventional (i.e. chemical) plant protection products. In addition to their data requirements, Member countries were also asked to indicate how frequently the data elements were required (i.e. always, frequently, less frequently or never). It was intended that the results from this survey could serve as a future basis for the identification of agreed common core data sets for pesticide registration and, in addition, provide useful input for the work on Test Guideline revision and development.

## *Priority setting for work on Test Guidelines*

The outcome of the Test Guideline Survey indicated that a considerable amount of work was needed and that priorities had to be set before work could begin. Three Task Forces (physical-chemical properties and environmental fate, ecotoxicology, and human health and occupational exposure) were established in June 1993. Each Task Force comprised of pesticide experts from government and industry (including contract testing laboratories). Individual Task Force members were requested to:

(i)     indicate their priorities (i.e. high, medium, low) for the revision/development of the Guidelines recommended in the Test Guideline Survey;

In indicating their priorities, Task Force members were asked to take into consideration:

—     the frequency with which any particular data requirement is requested for pesticide registration (i.e. from results of the Data Requirements Survey);

—     the priorities proposed by Member countries in the Test Guideline Survey;

—     the availability of existing methods from other fora.

(ii)    indicate the amount of work that would be involved (i.e. small, moderate, large).

Once individual Task Force members had addressed points (i) and (ii) above, each Task Force met in September 1993 to reach consensus on priorities, on the amount of work required and on the way to proceed.

## Task Force priorities for work on environmental fate Test Guidelines

The Annex gives the priorities for work on environmental fate Guidelines agreed by the Task Force. Table 2 summarises the proposals for the work of highest priority.

| Recommended Activities | Workload? | Comments |
|---|---|---|
| 1.  New Guideline for photodegradation in water | Moderate | Ongoing activity |
| 2.  New Guideline for hydrolysis | Small | New activity |
| 3.  Establish Working Group to review methods available and to draft new Guidelines for : <br> - Revision of Guideline 106, Soil Adsorption/Desorption <br> - New Guideline for laboratory soil column studies <br> - New Guideline for soil aerobic/anaerobic metabolism <br> - New Guideline for aerobic/anaerobic degradation in water/sediment systems | <br><br> Small <br> Small <br> Large <br> Large | <br><br> Ongoing activity <br> New activity <br> New activity <br> New activity |
| 4.  Technical workshop on soil selection for tests - both for environmental fate and ecotox studies | | New activity |

**Table 2 :**    Task Force recommendations for work of highest priority for environmental fate

## Activities on environmental fate Test Guidelines

The work on Guidelines suitable for use in pesticide testing is done using the mechanism established within the Test Guidelines Programme (OECD 1993). The central position within the Test Guidelines Programme is held by the National Co-ordinators (one from each Member country and the EC) who oversee the programme and work towards consensus on draft Guidelines. The Task Force recommendations were therefore reviewed by the National Co-ordinators at their meeting in October 1993, who in general agreed with the priorities assigned. With the exception of the proposal for a new Guideline for hydrolysis, all of the activities recommended as being of highest priority in the environmental fate area and shown in Table 2 are either already ongoing or will be initiated during 1994.

*Photodegradation in water.* Three proposals (from Germany/France, the UK and the Netherlands) for a new Guideline on photodegradation in water will be circulated to Member countries for comments. Based on the responses from Member countries, either a single Guideline will be developed, based on the preferred method, or all methods will be included in a Guidance Document.

*Adsorption/desorption.* In April this year, two proposals for Guidelines on adsorption/desorption were circulated to Member countries for comment. The first is a proposal for the revision of Guideline 106, Adsorption/Desorption and is the outcome of an extensive investigation and a ring test within the EC. This work was based on the use of European soils. However, in 1991, the EC Co-ordinating Group on test methods, recommended that a test protocol should be developed which fulfilled the needs of a revised OECD Guideline 106, and most importantly that it be based on the use of soils that are broadly defined, rather than the European reference soils used in the ring test. The main difference in this new draft, compared with the existing Guideline, is that the recommended soils are described by their properties (e.g. organic carbon content, clay content and pH determined in water) rather than by their American classification (i.e. Alfisol, Spodosol, Entisol) and that five rather than three soils are tested.

The second proposal is for a new OECD Test Guideline, i.e. 'Screening Method for the Determination of the Adsorption Coefficient on Soil ($K_{oc}$) using High Performance Liquid Chromatography (HPLC)'. This proposal arose to serve the need within the EC Council Directive 92/32/EEC on the notification of new substances, for information from an adsorption/desorption screening test at the Base Set level.

*Working Group for soil and sediment tests.* A Working Group is being established to develop draft Guidelines for soil column studies and degradation in soil and aquatic sediments (see Table 2). The Working Group will include representation from government and industry. Their task will be to review existing methods available and to draft new proposals for consideration and review by all Member countries as part of the Test Guidelines Programme.

*Technical workshop to discuss soil selection.* During the priority setting process described above, the Task Forces on Physical-Chemistry/Environmental Fate and Ecotoxicology recognised that the selection of soil for use in environmental fate and ecotoxicity studies was a very important issue. A technical workshop will be held in the autumn of 1994 to discuss this issue. In view of work previously done within the EC on soil selection, it is likely that this workshop will be held jointly by OECD and the EC Joint Research Centre.

## *Future Work*

As work proceeds in the areas outlined above, further work on the development of Guideline for environmental fate will be added. New work items will be added based largely on the priorities proposed in 1993 by the Task Force on Physical-Chemical Properties and Environmental Fate.

## *Reference*

□      OECD (1993) Guidance Document for the Development of OECD Guidelines for Testing of Chemicals. *OECD Environment Monographs No. 76*, Paris, 1993.

# ANNEX

## PRIORITIES RECOMMENDED FOR WORK ON TEST GUIDELINES FOR ENVIRONMENTAL FATE

| TEST AREA | SURVEY OF DATA REQUIREMENTS | TEST GUIDELINE SURVEY | Proposed priority | Estimated workload | Comments |
|---|---|---|---|---|---|
| | Requirements listed in survey | Recommendations for revision or development of Guidelines | | | |
| Mobility/leaching in soil | Adsorption/desorption in representative soil types including metabolites & breakdown products | 1. Revise Guideline 106 | HIGH * | Small | Secretariat to take the lead, once CEC proposed draft is received.[1] |
| | Mobility/leaching in representative soil types and mobility of metabolites and breakdown products | 2. New Guideline for column studies (laboratory tests) | HIGH * | Small | Establish Working Group to review and compare the variety of methods currently available and draft a new OECD Test Guideline which would then enter the normal Guideline development process.[1] |
| | | 3. New Guideline for lysimeter studies (semi-field tests) | HIGH | Large | Delay work until late '94/95, when results of research programmes and reviews will be available. |
| | Extent and nature of bound residues in soil | 4. New Guideline for identification/bioavailability of bound residues in soil | - | - | Include identification of bound residues in new Guideline for aerobic and anaerobic metabolism in soil (see 13 and 14). State of the art not yet advanced sufficiently to develop a Guideline to assess bioavailability of bound residues in soil |
| Volatility | Laboratory volatility | 5. New Guideline - volatility from soil | - | - | Combine with new Guideline for aerobic and anaerobic metabolism in soil (see 13 and 14). |
| | | 6. New Guideline - volatility from leaf surfaces | MEDIUM | Moderate | Follow normal Test Guideline development procedure. |
| | | 7. New Guideline - volatility from water | - | - | Combine with new Guideline for aerobic and anaerobic metabolism in water (see 19 and 20). |
| Photodegradation | Photodegradation in water including identification of metabolites and breakdown products | 8. New Guideline - photodegradation in water | HIGH * | Moderate | Secretariat to take lead on this as work has already been agreed by National Co-ordinators. Plan to circulate the various methods available to countries for comment. |
| | Photodegradation on soil including identification of metabolites and breakdown products | 9. New Guideline - photodegradation on soil | HIGH | Large | Ask National Co-ordinators to indicate what methods are available and to indicate if they would like to take the lead. |

| TEST AREA | SURVEY OF DATA REQUIREMENTS | TEST GUIDELINE SURVEY | Proposed priority | Estimated workload | Comments |
|---|---|---|---|---|---|
| | Requirements listed in survey | Recommendations for revision or development of Guidelines | | | |
| Photodegradation (cont) | Rate/route of photochemical degradation in air, identification of breakdown products | 10. New Guideline - photodegradation in air | MEDIUM | Large | As above |
| | no requirement | 11. New Guideline - photo-degradation on leaf surfaces | 0 | | Not needed. Member countries agreed that OECD should not develop methods for residue studies. |
| Degradation/ metabolism in soil | Soil metabolism, aerobic and anaerobic, to determine rate and route of degradation in representative soil types including metabolites and breakdown products | 12. Revise Guideline 304 | 0 | | Guideline not applicable to pesticides. |
| | | 13. New Guideline for aerobic metabolism | HIGH * | Large | Combine aerobic and anaerobic metabolism in soil into one Guideline. Include volatility from soil and the identification of bound residues.[1] |
| | | 14. New Guideline for anaerobic metabolism | | | Recommend Working Group be established to review and compare the variety of methods currently available and draft a new OECD Test Guideline. This draft Guideline would then enter the normal Guideline development process. This Working Group could be the same as that recommended for column studies (see 2 above) |
| Degradation/ metabolism in aquatic systems | Hydrolysis rate including identification of metabolites and breakdown products | 15. Revise Guideline 111 New Guideline needed - leave Guideline 111 | HIGH * | Small | Keep Guideline 111 as screening test. Develop new Guideline for pesticides. |
| | Biodegradation in aquatic systems, aerobic and anaerobic, including identification of breakdown products and metabolites | 16. New Guideline - aerobic biodegradation in water | 0 | - | These are covered by 19 and 20 below. |
| | | 17. New Guideline - anaerobic biodegradation in water | 0 | - | |
| | Adsorption/desorption in water | 18. No recommendation | 0 | - | Use information from the adsorption/desorption studies. |

| TEST AREA | SURVEY OF DATA REQUIREMENTS<br>Requirements listed in survey | TEST GUIDELINE SURVEY<br>Recommendations for revision or development of Guidelines | Proposed priority | Estimated workload | Comments |
|---|---|---|---|---|---|
| Degradation/ metabolism in aquatic systems (cont) | Degradation and distribution in water/sediment system, including identification of breakdown products and metabolites | 19. New Guideline for aerobic degradation in water/sediment | HIGH • | Large | Combine aerobic and anaerobic metabolism in water in same Guideline and include measurement of volatility from water. |
| | | 20. New Guideline for anaerobic degradation in water/sediment | | | Establish Working Group to review and compare the variety of methods currently available and draft new OECD Test Guideline. This Working Group could be the same as that recommended for column studies and soil metabolism (see 2 and 13 & 14) |
| | | 21. Non-specified water/sediment | - | - | Delete |
| Field studies | Terrestrial field dissipation | 28. Develop Guidelines/Guidance for dissipation/fate a) in soil | MEDIUM | Large | Make inventory of guidelines/guidance available and then decide on approach (e.g. Working Group?) |
| | Aquatic (sediment) field dissipation | b) in aquatic environment | MEDIUM | - | As above |
| | Forestry field dissipation | c) forestry | LOW | - | See 3 above |
| | | 29. Lysimeters | - | - | |
| | Higher tier studies - potential to leach & contaminate groundwater | 30. Develop Guideline(s)/Guidance for monitoring | LOW | - | Low priority - post-registration requirement. IUPAC report available end of '93. |
| | Monitoring of transport in air | 31. No recommendation | LOW | - | Research area. |
| | Spray drift field dissipation | 32. Guidance for spray drift | LOW/ MEDIUM | - | Wait for outcome of current US work |
| | Run-off | 33. Guidance for run-off | LOW | - | Await outcome of other work. |
| | no requirement | 34. Guidance - rotational crop | ● | - | Not applicable - concerns residue/efficacy studies. |
| | | 35. Guidance for field volatility | LOW | | |

1 The Task Force recognised the selection of soil for use in environmental fate and behaviour studies as a very important issue and recommended that a workshop be held to try and reach consensus on the approach to be used.

## 1.2.        Special COST Session

### 1.2.1.        *COST a frame for successful European Cooperation in the field of Scientific and Technical Research*

### N.K. Newman

European Commission, Directorate-General for Science, Research and Development (DG XII)

In the mid 1960's, there was a growing realization in the Community of six that the only way to withstand the impending scientific and technical challenge from the U.S.A. and Japan, and to increase the competitiveness of European industry was for research to be coordinated on a European scale and for know-how to be exchanged between the States. This realisation led to the initial idea of creating a framework that was not confined to the six member states of the European Community.

In March 1965, a committee chaired by André MARECHAL began to examine the feasibility of working together with a view to proposing measures for a joint, coordinated policy in the field of scientific and technical research.

The first report, which was submitted to the Council of Ministers on 31 October 1967, proposed the following seven research fields for cooperation at European level :

- Informatics
- Telecommunications
- Transport
- Oceanography
- Materials
- Environmental protection
- Meteorology

In 1969, the European countries which were not members of the Community (namely Austria, Denmark, Finland, Greece, Ireland, Norway, Portugal, Spain, Sweden, Switzerland, Turkey, the United Kingdom and Yugoslavia) were invited to take part in the work of the committee.

The AIGRAIN committee, so named after its new chairman, reported to the Council of Ministers of the European Communities in the same year. On 22 and 23 November 1971, a conference was held in Brussels, attended by the Ministers for Science and Technology of the 19 European countries, at which the first seven agreements at government level were signed. This event is generally considered as the birth of COST.

COST is the acronym for the French equivalent of "European Cooperation in the Field of Scientific and Technical Research" and is distinguished by four particular features :

1.     All member countries of COST can propose research actions at any time (Bottom-up)

2.     Participation in those projects is voluntary

3.     The projects are funded nationally

4.     The cooperation takes the form of "concerted action projects", which is the coordination of national research projects. Each concerted action is administered by a management committee and the secretariat is generally provided by the services of the Commission of the European Communities.

Consequently, the main feature of all COST actions is that they allow a concerted approach to fields of research which are of interest to a minimum number of participants and an interchange of the research results among those taking part. The participants only finance that part of the project for which they are individually responsible, but they have access to all the project results. The other advantage of this type of coordination is that it makes for better management of resources at European level by avoiding duplication of effort. It ensures that the resources allocated to research are used more efficiently and it establishes the necessary network to implement this.

Another feature of COST projects is the large measure of autonomy enjoyed by the countries concerned. Any COST State can take part in a project by signing what is known as a Memorandum of Understanding (MoU). This constitutes the legal basis for the project.

Most COST projects are designed to promote pre-competitive scientific and technical research and any activity which contributes to the attainment of specific objectives or to the opening up of new avenues of research. Consequently, its position is midway between basic research, and technical development.

In contrast to Community research programmes, this form of collaboration does not require an agreed overall research policy. It focuses on specific themes for which there is sufficient interest in the COST member countries.

Again there is no provision for joint funding with this "à la carte" system.

The value added in terms of research comes solely from international collaboration. The informal exchange of research findings between partners and the flexible and unbureaucratic method of management have established themselves increasingly as a tried and tested formula.

The following are examples of particularly successful COST actions :

- The setting up of the European Centre for Medium-Range Weather Forecasts in Reading, UK, under COST action 70.
- The development of advanced high-temperature materials for use in gas turbines and in energy conversion systems - COST 50 and 501.
- The development of electronic traffic aids on major roads (COST 30) and shore-based marine navigation aid systems (COST 301).
- The implementation of COST actions 61, 64, and 68 representing the first concerted efforts in the field of environmental protection in Europe.

In addition, a large number of COST actions, especially in the field of informatics, telecommunications and materials have contributed to the development of the Community's major industry-oriented programmes such as ESPRIT, RACE, EURAM and BRITE but also in new fields of research as environment protection, food technology, chemistry, medicine, biotechnology and forestry.

One might imagine that the development of Community programmes would have put a brake on the cooperation activities within COST, but in fact the opposite proved to be the case. The number of COST projects continued to grow, to the point where, in 1993, 97 actions were underway.

On 21st November 1991 COST celebrated its 20th anniversary. To mark this occasion the Ministerial Conference in Vienna, attended by 23 ministers responsible for Research and Technology, adopted a resolution enlarging the cooperation framework by CZECHOSLOVAKIA, HUNGARY, POLAND as well as ICELAND bringing the COST members states from 19 to 23 European States. The conference underlined COST's dynamism and ability to the changing needs of international cooperation in Science and Technology and to react to the early political developments in the countries of Eastern Europe. On 18th June 1992 the COST Committee of Senior Officials decided to admit SLOVENIA and CROATIA to the COST framework. Since April 1993 the CZECH Republic and SLOVAK Republic replacing the former CZECHOSLOVAKIA are also part of COST. In the frame of the PECO initiative of the Commission of the European Communities to integrate smoothly the new COST members states of Central and Eastern Europe into the COST cooperation by granting research projects.

This shows again the importance given by the Commission of the European Communities to this tried and successfully tested cooperation frame.

COST alongside other forms of European cooperation such as EUREKA, bilateral cooperation with EFTA countries, etc. - will remain a significant and complementary aspect in the creation of a European Science and Technology Community.

Thank you very much for your attention !

*1.2.2.*      *The Cost Action 66 : Fate of Pesticides in the Soil and the*
                     *Environment*

# by P. Jamet [*]

INRA - Direction des Relations Internationales
147, Rue de l'Université 75338 PARIS Cedex 07
( [*] Chairman of the COST Action 66)

## Summary

The soil finally receives most of the plant protection products and plays a part as a main entrance of plant protection products to the environment. Therefore, the behaviour of these chemicals in soil has been increasingly studied for the past twenty-five years. That's the reason it appeared indispensable to elaborate and coordinate an European research program into the **"Fate of Pesticides in the Soil and the Environment"**.

The creation of a network through a COST proposal attempts to realize this scientific goal on the European scale. This COST Action 66 reflects the real to set up a thorough and regular scientific cooperation especially trough the scientific programmes managed by the coordinators of the four working groups (**Degradation and Transformation, Sorption and Mobility, Outdoor Experiments & Monitoring and Mathematical Modelling**) settled during the first Management Committee Meeting. The scientific programmes mostly include applied research objectives since the latter is the ultimate basis for helping in decision making and setting up rules. However, close links with fundamental research through long-term projects are maintained.

In this paper, after a brief history of the COST Action 66, the main and secondary objectives of this COST Action are outlined and the scientific programmes are briefly presented.

## Introduction

### *Environmental Fate of Pesticide*

The soil ecosystem finally receives most of the plant protection products, whatever their pattern of use [7]. Both the fate of these products in the soil and their dispersion in the environment mainly depend on the characteristics and the overall functioning of this ecosystem, and determine the risks of dispersion in the environment (Fig. 1). These risks increase in proportion to the toxicity of the pesticide involved, its application frequency and dosage, and its **persistence and mobility in soil.**

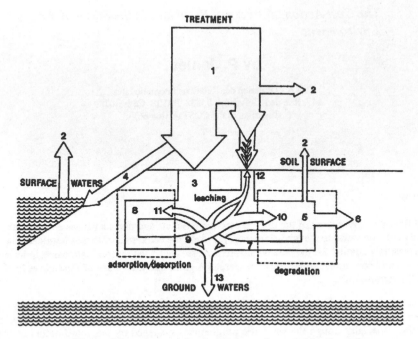

**Fig. 1:** Flow chart of the behaviour of pesticides in soil.

(1) Whatever their patterns of use, most of the pesticides used in agriculture reach the soil during or after treatment. By volatilization (2), pesticides and their metabolites may diffuse into the atmosphere. At the soil surface (3), pesticides can be transferred into the atmosphere (12) by volatilization, degraded by photolysis or transported by run-off (4). In the soil, the pesticide behaviour depends on the simultaneous influence of physical, chemical and biological processes. The soil is an ecosystem which is endowed with high degradation potential (5). The degradation processes produce more or less toxic metabolites, and also mineral compounds such as $H_2O$, $CO_2$, $NH_3$, ... (6). The metabolites can be leached (7), adsorbed by organic or mineral soil components (11) or absorbed by plants (12). When they are adsorbed by soil components (8), pesticides and their metabolites are not leached (13) and not bioavailable. When they desorb, these molecules may be degraded (5), leached (13) or absorbed by plants (12). The soil plays a leading role in the protection of water bodies. It is necessary to protect it from erosion and depletion. (Reproduced with permission from Agro Industry hi-tech. Ref. 7)

## *West European Agrochemical Market*

As a whole, the European community was in 1991 the leading market region [11] with approximately 31.2% share of the total market, followed by North America (26.4%) and the far East (22.5%). Within the European Union the major country markets are represented by France,

Germany, Italy, Spain and the United Kingdom. The West European market declined by 18.1% in 1992 affected by low grower confidence due to Common Agricultural Policy whilst the North American market grew by 3.7%. Fig. 2 depicts the world and West Europe agrochemical market.

Since soil plays such a part as a main entrance of pesticides to the environment, and seing that the West European agrochemical market is one of the most important in the world, it appears to be essential to gather together all the European research teams and indispensable to elaborate and coordinate a European research programme into the **"Fate of pesticides in soil and the environment"**.

## A brief history of the COST Action 66

The European Workshop organized 6 years ago, at the INRA Research Centre of Versailles (16-17 June 1988), gathered 73 scientists and emphasized the necessity of creating a European network for those laboratories which are concerned with the environmental fate of pesticides [4].

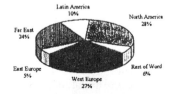
World agrochemical market split by region

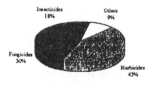
West Europe agrochemical market

**Fig. 2** : Agrochemical overview in 1992 [11].

Awareness of the need for such a network heightened at the 2nd meeting [9]. This meeting was held at Alicante on 30.05/01-02.06.89 and was organized by the Departamento de Quimica Agricola of Alicante University with the participation of over a hundred scientists. On 30.05/01.06.90, some 150 researchers from various countries in Europe and the Mediterranean Basin participated in the 3rd workshop [8] organized at the GSF Institute (Munich). Finally, a 4th workshop was held at Rome on 29-31 May 1991. It was organized by the Istituto per lo Studio del Suelo de Firenze and the University of Rome and attended by over 120 participants [10]. In Rome, it was decided to submit a COST proposal [1]

| Signatories (17) | Date |
|---|---|
| **European Union (8)** | |
| | |
| Belgium | 17-12-92 |
| France | 14-10-92 |
| Germany | 17-12-92 |
| Greece | 20-10-92 |
| Italy | 27-01-93 |
| Netherlands | 14-01-93 |
| Spain | 15-09-93 |
| United Kingdom | 07-10-92 |
| | |
| **EFTA (5)** | |
| | |
| Austria | 09-02-94 |
| Finland | 08-12-92 |
| Norway | 07-10-92 |
| Sweden | 16-11-92 |
| Switzerland | 07-10-92 |
| | |
| **Central Europe (4)** | |
| | |
| Croatia | 18-05-93 |
| Czech Republic | 07-10-92 |
| Hungary | 24-02-94 |
| Slovenia | 29-01-93 |

A COST schema n° 1 was prepared in France [2, 5] and send by the French National COST Coordinator to the New COST Project Group (NPG). This project was discussed in Brussels on November 5, 1991, during the NPG meeting. The technical annex of the Memorandum of Understanding (MoU) was prepared and finalized till June 29, 1992 [3, 6]. Then, the National Authorities began to sign the MoU of this New project (Table 1), namely COST Action 66, in September 1992. Nevertheless, the first Management Committee Meeting (MCM) met for the first time on January 27, 1993, in Brussels. The chairman and vice-chairman were elected, four Working Groups settled and their coordinators appointed (Fig. 3). The 2nd MCM was held in Brussels on June 11, 1993. The third one was held in Paris on November 18-19, 1993. During this meeting the coordinators presented their reports from the Working Groups Meetings (4). More than 80 laboratories are involved in these programmes (Table 2).

**Table 1** : COST Action 66 - Table of signatures (situation on March 1, 1994)

## Objectives

The debates that took place during the fourth workshop have led to clarification of our major objectives for research and co-operation. More details can be found in the MoU.

### *Main Objectives*

— **gather French, then European laboratories which intend to co-operate in investigating the topic "PESTICIDES-SOILS-ENVIRONMENT"** by research on both the consequences of using pesticides in agricultural crops and the ecotoxicological problems resulting from the dispersion of these pollutants in the environment;

— **define a programme for scientific research and exchange** aiming to study the physicochemical and biological processes which control the fate of pesticides in soil and their dispersion in the environment;

— **develop and improve those laboratory and field methods which are essential for predicting and assessing the hazards** involved by qualitative and quantitative transfer of these pollutants to aquatic environments.

| Countries | Laboratories |
|---|---|
| AUSTRIA | 10 |
| BELGIUM | 7 |
| CZECH REPUBLIC | ✶ |
| CROATIA | 5 |
| FINLAND | 6 |
| FRANCE | 10 |
| GERMANY | 5 |
| GREECE | ✶ |
| HUNGARY | ✶ |
| ITALY | 12 |
| NETHERLANDS | 3 |
| NORWAY | 1 |
| SLOVENIA | 3 |
| SPAIN | 6 |
| SWEDEN | 4 |
| SWITZERLAND | 4 |
| UNITED KINGDOM | 7 |
| **TOTAL** | **83** |

**Table 2:** COST Action 66 - Number of participating laboratories. (✶ : inventory in progress)

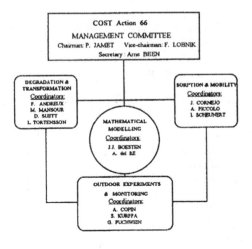

**Fig. 3:** COST Action 66 - Organization chart.

### *Secondary Objectives*

— **To set up an operational network of expertise and means** committed to this research topic, thus making it possible to investigate the diversity of agricultural, soil and climate conditions in Europe.

— **To contribute to defining and bringing into line European regulations** that would include the requirements for optimum equilibrium between the use of toxic products in agriculture and the necessary protection of Man and the Environment.

— **To draw up a yearbook of member laboratories listing their addresses, research potential** for this particular topic, and those features they would preferentially investigate in cooperation with other European partners (Table 2).

## Scientific Programmes

The scientific programmes of the COST Action 66 include the main phenomena for which research approaches are described below (Degradation and transformation; Sorption and mobility). Investigating these phenomena, both in the laboratory and in the natural environment, requires also outdoor experiments and monitoring and expertise for Mathematical Modelling. The main points of these scientific programmes are summarized below.

### *Degradation and Transformation*

(Coordinators: Dr. F. ANDREUX, M. MANSOUR, D. SUETT, L. TORTENSSON)

#### Objectives

— Evaluate current methodology to ensure that future collaborative studies use agreed procedures and conditions.

— Explore the ways in which abiotic processes in soils can influence biological transformations of pesticides to understand the mechanisms of such processes.

— Explore the significances and impact of bound and unextractable residues on total transformation processes.

— Correlate soil microbial activities with soil porosities and degradation kinetics at different stages of the overall transformation and degradation process.

— Develop and test, methods that accommodate the contributions of all these processes.

— Propose validated systems for adoption within the European Union.

#### Scientific results foreseen

— Agree protocols for studying pesticide behaviour under controlled laboratory conditions.

— Use these agreed procedures and conditions to establish the influence of soil, climatic and agricultural variables on the stability and bioavailability of selected relevant pesticides.

— Define and quantify the impact of the biotic and abiotic mechanisms determining the behaviour of these pesticides.

— Identify gaps and deficiencies in existing information and procedures.

### *Sorption and Mobility*

(Coordinators: Dr. J. CORNEJO, A. PICCOLO, I. SCHEUNERT)

#### Soil and adsorption parameters

— Standardization of methods for adsorption coefficients determination and isotherms establishment.

— Identify arable soils in European agricultural areas representative enough to be added to the EUROSOIL system in order to broaden the basis of analytical comparisons.

–    Study of the time dependence of adsorption of pesticides on soils

### Soil organic matter

–    Isolation of well-characterized humic fractions from selected European soils to study specific sorption behaviour of commonly employed pesticides.

–    The influence of soil management practices (addition of sludges, organic wastes, etc ...) on the role of organic matter in controlling pesticides fate.

–    A closer approach to the real soil complex system by taking into account pesticides interaction with synthetically prepared clay-humic aggregates.

–    The role of inorganic and enzymatic catalysis in permanent binding of pesticides on soil organic matter in order to predict future release of pesticides residues in soils.

### Leaching and volatilization

–    Investigate mobility in soil, leaching, volatilization and mineralization of $^{14}$C-labelled pesticides in microcosms

–    Development of new equipments to determine the volatilization of pesticides from soil and plant surfaces or incorporated in the soil should be especially promoted.

–    Standardization of methods to determine the uptake of pesticides in plants.

## *Outdoor experiments and Monitoring*

(Coordinators: A. COPIN, S. KURPA, G. PUCHWEIN)

### Monitoring

–    A set of basic survey data (maps) on pesticide compounds and metabolites

–    Standardized survey data from specific types of habitats comparable between the different geographic areas of Europe

–    Recommendations on the best monitoring strategies to be followed over Europe.

–    Specific data to be used for validating the pesticide models.

### Outdoor experiments

–    A set of basic data from different types of field lysimeters and microlysimeters

–    Detailed data on the fate of pesticide compounds under different farming and environmental conditions

–    Recommendations on the standard methods of various types of field experiments.

–    Specific data to be used for verifying the results of pesticides and for validating the models.

### Monitoring and outdoor experiments

–    Archinfo data files on active ingredients and metabolites monitored in Europe.

## *Mathematical Modelling*

(Coordinators: J.J. BOESTEN, A. del RE)

### Objectives

- To establish a ranked and classified database of available models currently used.

- To develop a protocol for evaluating pesticide fate models including the definition of the different procedures used in pesticide fate modelling, the definition of the input data and test data requirements and the definition of test criteria.

- To identify a small number of data sets that meet the requirements outlined above, and to initiate a coordinated evaluation and comparison of models.

- To define important topics for future research in the field of mathematical modelling.

### Scientific results foreseen

- The exchange of know-how between different scientific groups involved in the mathematical modelling of the fate of pesticide in the soil-crop environment;

- The enhancement of mathematical modelling as a tool to improve scientific insights in the unsteady dynamic processes influencing the fate of pesticides in the soil-crop environment;

- The enhancement of mathematical modelling as a tool to improve the current registration procedures reconciling both ecological and economical constraints;

- The development of objective and standardized procedures for testing pesticide fate models;

- The compilation of high quality datasets which enables the testing of the different pesticide fate models available today.

## *Conclusions*

Setting out a COST research project on the topic *"PESTICIDES-SOILS-ENVIRONMENT"* should allow the above-mentioned objectives to be met, thus leading to precise determination of "how widely the behaviour of pesticides in soil may vary with the agricultural, soil and climate conditions occurring in various European countries" and to assessment of environmental hazards. The answer to such a question should be valuable from both agronomic and economic viewpoints and for any policy about environmental management and the harmonization of the national regulations on agricultural usage of toxic products. Research into this field however requires scientific co-operation of the various COST states since all countries are concerned by this issue. The meetings which have taken place since 1988 bear witness to the growing interest for this subject of scientists and of the research organizations that have already helped us.

Water and soil are two of the most important natural resources which are essential for agriculture. Considering the reform of the Common Agricultural Policy and thereby the necessity of a more efficient but less intensive agriculture, farmers tend to resort to farming practices compatible with the protection of the environment and natural resources.

Research carried out in the laboratories which agree with the elaboration of a European network of scientific co-operation should benefit national and international authorities faced with the question of environmental conservation. The acquired expertise could also be useful for public services and agrochemical industries in the framework of the registration procedure for novel plant protection products.

## Acknowledgments

I would like to express my personal gratitude to my European colleagues, both scientists and members of the Services of the Commission, DG XII and Council COST Secretariat, for their trust and their contribution to finalized the COST Action 66.

## References

☐  [1] **Commission of the European Community (1992)** COST Cooperation: Objectives - Structures - Operations. Published by the Commission of the European Communities. ISBN: 92-826-4371-9.

☐  [2] **COST Secretariat** - Procédures pour l'Introduction de nouvelles propositions COST. Guide de l'usager.

☐  [3] **COST - EFTA Unit (1991)** Preparation of Memoranda of Understanding. Ideas and guidelines for the drafting of technical annexes.

☐  [4] **Jamet P. Ed. (1988)** First International Workshop on the "Methodological Aspects of the Study of Pesticide Behaviour in Soil". INRA - Versailles, June 16-17, 1988. INRA Editions. ISBN: 2-7380-0115-7.

☐  [5] **Jamet P. (1991)** COST: French Proposal for a New Project "Fate of pesticides in the soil and the environment". Schema N 1 - 9 pages.

☐  [6] **Jamet P., Andreux F. (1992)** COST Action 66 "Fate of Pesticides in the Soil and the Environment". Technical Annex of the Memorandum of Understanding. June 29, 1992.

☐  [7] **Jamet P., Deleu R. (1993)** Environmental fate of pesticides: behaviour of pesticides in soil. Agro-Industry hi-tech. Vol 4, N° 3, pp19-21.

☐  [8] **Mansour M. Ed. (1993)** Third International Workshop "Fate and Prediction of Environmental Chemicals in Soils, Plants, and Aquatic Systems". Munich - RFA. (30.05/1.06.1990). Lewis Publishers. ISBN: 0-87371-616-7.

☐  [9] **Navarro J. Ed. (1989)** VII Symposio Internacional de Plaguicidas en Suelos y Plantas. Alicante, 30.05/1-2.06.1989. II Workshop."Pesticides-Soils" ISBN: 84-604-0399-8.

☐  [10] **Piccolo A. Ed. (1992)** Fourth International Workshop "Chemical, Biological and Eco-toxico-logical Behaviour of Pesticides in the Soil Environment". Rome, 29-31 May 1991. Special issue of The Science of the Total Environment (Elsevier Ed.).

☐  [11] **Wood MacKenzie (1993)** Agrochemical Overview 1992 - September 1993.

# 1.3.     Case Studies

## 1.3.1.     *Hazard Assessment of Agro- and Non-agrochemicals in the Netherlands*

## J.B.H.J. Linders & R. Luttik

National Institute for Public Health and Environmental Protection
Toxicology Advisory Centre
P.O. Box 1, NL-3720 BA Bilthoven, The Netherlands.

## Abstract

In recent years an evaluation system was developed in The Netherlands concerning the environmental aspects of agricultural and non-agricultural pesticides in relation to the registration of pesticides, called ESPE. The paper will describe the hazard assessment for environmental behaviour and effects as it is currently carried out in The Netherlands and some policy aspects of the Multi Year Crop Protection Plan. The ESPE-system is completely incorporated in the uniform system for the evaluation of substances (USES) in The Netherlands. USES is thoroughly discussed among ministries, governmental agencies, and industry and therefore commonly accepted as a management tool for estimating the potential risk of substances in general.

## Introduction

Since 1989 and on behalf of the Dutch Ministry of Housing, Physical Planning and Environment, almost all pesticides registered or requested for registration for new substances have been summarized and evaluated at the National Institute for Public Health and Environmental Protection (RIVM), The Netherlands [1] and [2]. Quality requirements and comparisons between substances asked for strict and consistent guidelines and methods for the evaluation process. Therefore, the registration authorities in The Netherlands developed the decision tree approach for the agricultural pesticides and non-agricultural pesticides, as described in [3], [4, 5]. In addition, the evaluation process of pesticides is incorporated in the first official release of the Uniform System for the Evaluation of Substances (USES, 1.0) [6], [7].

## Multi Year Crop Protection Plan

In 1991 the Ministries of Agriculture and Environment in The Netherlands published the Multi Year Crop Protection Plan [8]. The plan states the Dutch policy how to deal with crop protection products in a target setting way. Targets are defined for the period 1990 - 2000, and the targets have to lead to a safe, concurring, and sustainable agricultural practice. The policy strategy of the MYCP can be summarized in the following three main lines: - reduction of the dependence of chemical crop protection, - reduction of the amount of pesticides used, - reduction of the emission of pesticides to the environment.

In addition it was stated that on the basis of environmental criteria for the behaviour and effects of pesticides a substance oriented policy had to be established under the Pesticides Act to abstract certain substances from the Dutch market because of environmental considerations. In other words leading to the banning of undesirable chemical crop protection products. For ground water the EC Directive on drinking water of $0.1\ \mu g.L^{-1}$ for individual substances and $0.5\ \mu g.L^{-1}$ for the sum of all pesticides is taken as a cut-off-value. More research is necessary on the applicability of simulation models and on the validation of these models. As a result from this approach substances causing a threat to ground water are subjected to a banning procedure, leading to a reduction of registered active ingredients. The reduction of the emission of active ingredients to surface water is achieved by improvement of the application equipment and by application measures for risk reduction like treatment free zones, crop free zones along surface water, no allowance for treatment on dry ditch bottoms and ditch slopes. It is foreseen that these measures will cause an emission reduction of c. 90% as an average.

In table 1 an overview is given of the most striking reduction targets in the Dutch Multi Year Crop Protection Plan [8].

Concerning the environment, for the first time clear environmental criteria were set for three items: 1. leaching to ground water, 2. hazard to water organisms, 3. persistency.

In table 2. the criteria are given as they have been established in the MYCP. The MYCP presents a phasing of the policy measures against the worst, in environmental respect, substances in 1995 and applies even more stringent criteria in 2000.

| Type of application/sector/compartment | % in 1995 | % in 2000 |
|---|---|---|
| Soil fumigants | 45 | 68 |
| Soil treatment | 28 | 42 |
| Herbicides | 28 | 40 |
| Insecticides, fungicides, etc. | 25 | 39 |
| Bulb culture | - | 68 |
| Fruit culture | - | 44 |
| All sectors | 37 | 56 |
| Surface water | 70-80 | 90 |

**Table 1 :** Estimated reduction of pesticides in % of total volume applied in The Netherlands.

| Environmental criterion | 1995 | 2000 |
|---|---|---|
| Leaching to ground water : | | |
| - calculated in $\mu g.L^{-1}$ | 10 | 0.1 |
| - measured in $\mu g.L^{-1}$ | 0.1 | 0.1 |
| Hazard to water organisms : | | |
| - fish (PEC/LC50) | 0.1 | 0.1 |
| - crustacea (PEC/L(E)C50) | 1 | 0.1 |
| - algae (PEC/EC50) | 1 | 0.1 |
| Persistency in d | 180 | 60 |

**Table 2 :** Environmental criteria as stated in the MYCP [8].

| Environmental criterion | 1995 | 2000 |
|---|---|---|
| Leaching to ground water : | | |
| - calculated | 40 | 23 |
| - measured | 24 | - |
| Hazard to water organisms | 38 | 22 |
| Persistency | 18 | 46 |

**Table 3 :** Number of substances used in The Netherlands not meeting the criteria from table 2.

The concentration in ground water is calculated using the PESTLA-model [9], while the predicted environmental concentration (PEC) in surface water is calculated with the SLOOT.BOX-model [10], which will be presented in more detail below. Finally, the persistency is based on registration data, in which degradation half-lives (DT50) are given. For all criteria, metabolites are also included.

Applying these criteria to the substances used in The Netherlands at the time the MYCP was presented to the Parliament resulted in table 3, indicating the number of substances not meeting the mentioned criterion. These substance are listed in the MYCP and will be subject to a banning procedure. However, it is possible that according new EU-regulations (91/414/EEC) other measures will be developed.

## Uniform System for the Evaluations of Substances (USES, 1.0)

Coincidently with the reviewing of the registration data of pesticides as described above, a evaluation system was developed to take care for a consistent approach for all the pesticides reviewed. The input to the system was defined by the data asked for in the registration requirements. Additional data was standardized to the Dutch application situation and to generic circumstances. Therefore, comparison of the environmental aspects of the pesticides was possible. As also for new and existing chemicals the need for evaluation of data was obvious, it was decided to join the evaluation systems for all kind of substances in one evaluation system: USES [6]. USES is a tool for rapid, quantitative assessments of the general risk of substances. Risk assessment methods for various categories of substances were integrated, as much as possible, into one assessment scheme. USES is an instrument that can be used by central governments, research institutes, and the chemical industry in decision making. Risk management, however, is outside the scope of USES. USES can be applied to risk assessments and to set priorities for new substances, existing chemicals, plant protection products, and biocides within the scope of The Netherlands' Chemical Substances Act and Pesticides Act. The protection targets are: humans (exposed via the environment and via consumer products), micro-organisms in sewage treatment plants, the aquatic ecosystem, the terrestrial ecosystem, and top predators. In addition and as much as was possible, USES was tailored to perform assessments within the scope of international directives, regulations, and recommendations, such as those set by the European Union and the OECD Chemicals Programme.

USES aims at a quantitative comparison of the results of the effect assessment and the exposure assessment. This comparison is done using the "hazard quotient"; a ratio of the estimated exposure and a suitable effect or no-effect parameter. Hazard quotient estimates are indicators of the likelihood of adverse effects occurring. Priority lists can be generated based on the hazard quotients.

The evaluation of pesticides and biocides is an integrated part of the Uniform System for the Evaluation of Substances. Below the Evaluation System for PEsticides (ESPE), part 1. agricultural pesticides and part 2. non-agricultural pesticides will be dealt with.

### ESPE 1, Agricultural Pesticides

The global exposure scheme of ESPE 1 is presented in Figure 1. Three parts can be clearly distinguished: emission, distribution, and hazard assessment. Emissions are all routes entering the environment, distribution concerns all routes between environmental compartments, and hazard assessment relates to the organisms mentioned. The scheme for non-agricultural pesticides is analogous with respect to the environmental compartments, soil, water, air, and does not change with respect to the organisms at risk, birds, mammals, earthworms, water organisms. The main difference between the two classes of pesticides is their way of entry into the environment. This means that also the transfer functions from one compartment to the other are identical.

The transfer functions per route define the calculation of PECs for water, drinking water, air, soil, and sludge. The hazard assessment part describes the comparison of calculated exposure concentrations with toxicity data for the organisms exposed.

The pesticide applied is emitted into the standard environment, characterizing Dutch conditions. Then, the substance is distributed through the environment and the resulting concentrations in air, water, and soil are calculated: Predicted Environmental Concentrations (PECs). Finally, the comparison is made between the concentration calculated and the effect concentrations measured for the organisms exposed: lethal or effect concentrations $L(E)C_{50}$ or no observed effect concentrations NEC. The ratio PEC/NEC, finally, determines the risk estimate for the organism under concern.

Several aspects of USES and ESPE 1 and 2 are reported elsewhere [3], [4], [5], [6], [7], [9]. Here, only the surface water module SLOOT.BOX will be dealt with.

The concentration in surface water is calculated with the SLOOT.BOX-model, which is described in detail in [10]. The final estimated concentration in surface water is determined by some intrinsic physico-chemical properties, like vapour pressure, solubility, but also base set information, like degradation in standardized laboratory degradation experiments in water/sediment systems. In addition, the application regime is taken into account. By comparing the calculated concentration with the effect concentration, a risk indication for water organisms is made. The emission of pesticides due to drift is mainly based on expert judgement, except the data for application on crops of and 25 cm and the application in fruit trees, which have been actually measured [10]. Therefore, SLOOT.BOX uses validated data on emission of pesticides as input data.

SLOOT.BOX calculates the maximal concentration in the standardized Dutch ditch with a depth of 25 cm after the prescribed mode of application. In Figure 2 an example of a five fold application is shown with an interval of 10 days. This maximum concentration is used for the estimation of the acute exposure of water organisms. The trigger values established in EU Uniform Principles will be used for the decision on further research of registration of the substance. In a formula:

$$\frac{PEC_{ac}}{L(E)C_{50}} = 0.01 \qquad (1)$$

in which :        $PEC_{ac}$ = Predicted acute Environmental Concentration
                  $L(E)C_{50}$ = Lethal or Effect Concentration
                  0.01 = example trigger value.

The most sensitive organism is selected as organism at risk.

For the estimation of the (semi-)chronic concentration exposure concentration the mean concentration over a period equal to the duration of the toxicity test is calculated using the following formula 2 :

$$PEC_{chr} = \frac{C_0 * (1 - e^{-k*t})}{k * t} \qquad (2)$$

in which:

PEC$_{chr}$ = Predicted chronic Environmental Concentration
$C_0$ = maximal concentration after dosage
$k$ = overall rate constant (= (ln2)/DT$_{50}$)
DT$_{50}$ = half-life time of the substance
$t$ = duration of toxicity test.

The chronic PEC is compared to the NOEC-values established in the toxicity tests, according to formula (3), analogous to (1) including an example trigger value.

$$\frac{PEC_{chr}}{NOEC} = 0.1 \qquad (3)$$

Using extrapolation methods the L(E)C$_{50}$- or NOEC-values can be converted to Predicted No-Effect Concentrations (PNEC) or Predicted Concentrations of No Concern (PCNC) taking into account the variability of sensitive species in the actual environment. Again, the most sensitive organism is selected.

Currently, a more sophisticated model is developed in The Netherlands for the prediction of the long term concentration in surface water after pesticide application. This model, that takes also into account drainage, leaching, run-off, and atmospheric deposition will later on be incorporated in the USES-model and will be finished at the end of 1994.

## ESPE 2, Non-Agricultural Pesticides

ESPE 2 has been developed according to the same strategy as ESPE 1. Then an example on the antifouling agent tributyltin is shown. ESPE 2 describes an evaluation system for non-agricultural pesticides [5]. The non-agricultural pesticides are distinguished in six classes: disinfectants, industrial biocides, preservatives, household products, wood preservatives and -protectors, antifouling paints.

Taking into consideration the way of emission into the environment data requirements are set for the registration evaluation. The data will be used in calculation schemes, together with standardized assumptions on the environment, leading to the calculation of PECs. The ratio between the PEC and the MPC (Maximum Permissable Concentration) or sometimes the LC50 or EC50 leads then to a quotient, where a policy decision can be based upon.

Besides the decision trees introduced in the evaluation system for agricultural pesticides [4] two extra trees are developed additionally for beneficial micro-organisms in municipal sewage treatment plants (because an interference with these micro-organisms can lead to an unacceptable contamination of the surface water) and for bats (because remedial treatment of timbers in buildings has been considered a significant cause of mortality to bats).

With the presentation of the second part of the evaluation system ESPE the first step to a methodology for the evaluation of pesticides within the scope of registration has been made. In the system several open ends are still present, because sufficiently accepted methods for all routes are not yet available.

| Variable [unit] | Symbol | Default / Requested | Default Value |
|---|---|---|---|
| **Input :** | | | |
| Number of yachts in yacht-basin [-] | $N_{ship}$ | D | 250 [-] |
| Mean ship deck area [$m^2$] | $DECKAREA_{avg}$ | D | 10 [$m^2$] |
| Water/ship ratio in yacht-basin [-] | $R_{w/s}$ | D | 3 [-] |
| Fraction of ships in water [-] | $F_{ship}$ | D | 0.5 [-] |
| Volume of paint per yacht [$m^3$] | $L_{anti}$ | R/D | 0.002 [$m^3$] |
| Cover of antifouling paint [$m^2.m^{-3}$] | $R_{anti}$ | R/D | 2500 [$m^2.m^{-3}$] |
| Depth of yacht-basin [m] | $DEPTH_{y-b}$ | D | 2.5 [m] |
| Fraction ships in yacht-basin [-] | $F_{s/ns}$ | D | 0.71 [-] |
| Mean flux of compound [$kg.m^{-2}.s^{-1}$] | $FLUX_{anti}$ | R/D | $4.63*10^{-10}$ [$kg.m^{-2}.s^{-1}$] |
| DT50 for advection in the yacht-basin [s] | $DT50_{advec, y-b}$ | D | $4.32*10^{-6}$ [s] |
| Fraction organic carbon in susp. matter [$kg.kg^{-1}$] | $Foc_{susp}$ | D | 0.10 [$kg.kg^{-1}$] |
| Octanol/water partitioning coefficient [-] | $Kow$ | R | [-] |
| Concentration susp. matter in water [$kg.m^{-3}$] | $SUSPCONC_{surf}$ | D | 0.015 [$kg.m^{-3}$] |
| First order degradation rate in water [$s^{-1}$] | $kdeg_{water}$ | R/D | [$s^{-1}$] |
| **Output :** | | | |
| Equilibrium diss. conc. in y-basin water [$kg.m^{-3}$] | $Cwater_{pest, equi}$ | | |

**Table 4 :** Input requirements and output of antifouling calculation.

The largest part of ESPE 2 is incorporated in the USES [6]:

- Biocides in the textile industry
- Biocides in the paper and cardboard industry
- Biocides in the process and cooling-water installations
- Preservatives in the metal industry
- Wood preservatives: creosote impregnation
- Wood preservatives: salt impregnation
- Wood preservatives: drenching and dipping
- Remedial timber treatment in buildings
- Leaching from impregnated wood to surface water
- Antifoulings

As an example of the modelling of non-agricultural pesticides in USES the case of antifouling will be presented in this paper.

The schematic design of the evaluation system in USES for antifoulings is analogous to Figure 1; the emission for antifoulings is only taking place in water and soil. For calculating the hazard/risk of antifoulings in the aquatic environment a middle size yacht-basin is modelled [5], [6]. For this calculation it is necessary to have results of a lixivation study (average flux over a certain period) and information about the Kow and the DT50 in water. See table 4.

For the model calculations performed reference is made to [5]. For the calculation of the equilibrium dissolved concentration in the yacht-basin water the following input is used:

- Mean flux of compound $\qquad$ 2.5 $\mu$g.cm$^{-2}$.d$^{-1}$ $\qquad$ [11]

- First order degradation rate for biodegradation in water  0.041 d$^{-1}$ $\qquad$ [12]

- log Kow $\qquad$ 3.8 $\qquad$ [13]

The calculated concentration in the model yacht-basin is 8.08 $\mu$g/l. This concentration is comparable with measured concentrations in two yacht-basins with stagnant water in the Grevelingen in Zeeland in the Netherlands (Scharendijke 1.8 - 3.4 $\mu$g/l and Brouwershaven 1.4 - 7.2 $\mu$g/l [14]). The antifouling model apparently does not calculate unrealistic high values. The calculated concentrations can be higher than the measured concentrations because the model does not account for periods with strong turbulence (wind) and seepage.

For the hazard/risk assessment of the use of the antifoulant Tributyltin 6 NOECs for freshwater organisms were reported in [14].

The procedure for the derivation of No-Effect Concentrations for the aquatic and terrestrial ecosystem is described in the second article of a series on USES [15]. Because more than 4 NOECs are available for species from different taxonomic groups, the statistical method described in [16] can be used.

The Maximum Permissable Concentration (MPC) calculated for the cation of tributyltin is 0.00896 $\mu$g/l. The hazard quotient (Cwater/MPC) is 902, which means that it can not excluded that the aquatic ecosystem will suffer from the amounts of tributyltin (cation) that can occur in the yacht-basin.

## *Conclusions*

The initial risk assessment of all kinds of substances can be carried out with the decision supporting, computer aided system, called USES. In the system the Dutch methodology for new substances, existing substances and agro- and non-agro-pesticides are presented, giving government, industry and the public insight in the decision making process concerning these substances. The subsystem on pesticides, ESPE 1 for agricultural pesticides and ESPE 2 for non-agricultural pesticides is completely incorporated in USES, 1.0.

Although there are still several gaps in the modelling of processes relevant for decision making, the most important ones for which a methodology was available are included in USES. In near future and if relevant information becomes available, USES will be updated with additional models and scientific developments.

Because USES and therefore ESPE was developed in close cooperation between government, industry and national research institutions in The Netherlands, the support for the system is very broadly present in the country, while no definitive decisions will be based upon the models' outcome. The Dutch government is obliged to discuss the results and to take into account new data. Also according to the Uniform Principles in the EU, an adequate risk assessment has to be carried out. It is thought that USES is a first step in establishing this goal.

Risk assessments for other substances like human and veterinary medicines, food additives, cosmetics, biological products like bacteria, fungi or viruses, etc. can also be incorporated in USES. A suitable emission scenario and relevant distribution route have to be defined.

Consumer exposure to the substances already included in USES is an area that has to be developed further, because only indirect routes are defined.

Other important areas for further development are:

- sediment processes, only used as a sink at the moment,

■   uncertainty analysis, only preliminary modelled for new substances,

# References

□   [1] Canton J.H., J.B.H.J. Linders, R. Luttik, B.J.W.G. Mensink, E. Panman, E.J. van de Plassche, P.M. Sparenburg and J. Tuinstra; *Catch-up operation on old pesticides: an integration*, RIVM report no. 678801002, Bilthoven, 1988, 140 pp.

□   [2] Linders, J.B.H.J.; Jansma, J.W.; Mensink, B.J.W.G. and Otermann, K.; *Pesticides: Benefaction or Pandora's Box? A synopsis of the Environmental Aspects of 244 Pesticides*, RIVM report no. 679101014, Bilthoven, 212 pp (in press).

□   [3] Brouwer, W.W.M.; Vliet, P.J.M. van; Linders, J.B.H.J. Linders; Berg, R. van den *Environmental Risk Assessment for Plant Protection Products: the Dutch Approach*, Risk Assessment, (1993) 4, no3, p 30-35.

□   [4] Emans, H.J.B.; Beek, A.M.; Linders, J.B.H.J. *Evaluation system for Pesticides (ESPE) 1. Agricultural Pesticides*, RIVM, report no. 679101004, Bilthoven, 80 pp.

□   [5] Luttik, R.; Emans, H.J.B.; Poel, P. van der; and Linders, J.B.H.J. (1993) *Evaluation System for Pesticides (ESPE) 2. Non-agricultural pesticides*, RIVM report no. 679102021, Bilthoven, 1993, 60 pp.

□   [6] RIVM; VROM; WVC (1994). *Uniform System for the Evaluation of Substances (USES), version 1.0.*. National Institute for Public Health and Environmental Protection (RIVM), Ministry of Housing, Spatial Planning and the Environment (VROM), Ministry of Welfare, Health and Cultural Affairs (WVC). The Hague, Ministry of Housing, Spatial Planning and the Environment Bilthoven. Distribution No. 11144/150, pp. 345.

□   [7] Linders, J.B.H.J. & Luttik, R.; *ESPE: Risk Assessment for Pesticides in The Netherlands*, Chemosphere, (to be published).

□   [8] Tweede Kamer der Staten-Generaal *Meerjarenplan Gewasbescherming*, 21677, nrs. 3-4, Dutch Government, The Hague, 1992; pp. 298 (in Dutch). ISSN 0921-7371.

□   [9] Boesten, J.J.T.I.; Linden, A.M.A. van der *Modelling the Influence of Sorption and Transformation on Pesticide Leaching and Persistence*, J. Environ. Qual. (1991) 20, 425-435.

□   [10] Linders, J.B.H.J.; Luttik, R.; Knoop, J.M.; Meent, D. van de *Beoordeling van het gedrag van bestrijdingsmiddelen in oppervlaktewater in relatie tot expositie van waterorganismen*, RIVM report no. 678611002, Bilthoven, 1990: pp. 25 (in Dutch).

□   [11] Berendsen, A.M. *Globale schatting van de verontreiniging van de Noordzee door aangroeiwerende verven*. Verfinstituut TNO, (1990) TNO-report nr. 753/1990, pp 9.

□   [12] Watanabe, N.S.Sakei and H.Takatsuki *Examination for degradation paths of butyltin compounds in natural waters*. Wat. Sci. Tech. (1992) 25, 117-124.

□   [13] Laughlin, R.B., H.E. Guard and W.M. Coleman *Tributyltin in seawater: speciation and octanol-water coefficient*. Environ. Sci. Technol. (1986) 20, 201-204.

□   [14] Evers, E.H.G., J.H. van Meerendonk, R. Ritsema, J. Pijnenburg and J.M. Lourens *Butyltinverbindingen. Watersysteemvekenningen, een analyse van de problematiek in het aquatische milieu*. Concept report of RIKZ, (1993) Report DGW-93.015, pp 131.

□   [15] Jager, D.T., T.G.Vermeire, W. Slooff and H. Roelfzema *Uniform system for the evaluation of substances II. Effects assessment*. (1994) (To be published in Chemosphere)

□   [16] Aldenberg, T. and W.Slob *Confidence limits for hazardous concentrations based on logistically distributed NOEC toxicity data*. Ecotoxicol. Environ. Saf. (1993) 25, 48-63.

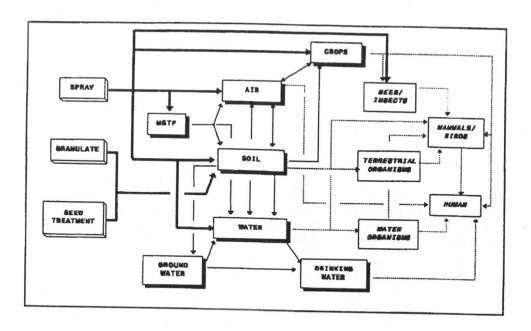

**Figure 1** : Exposure scheme for agricultural pesticides

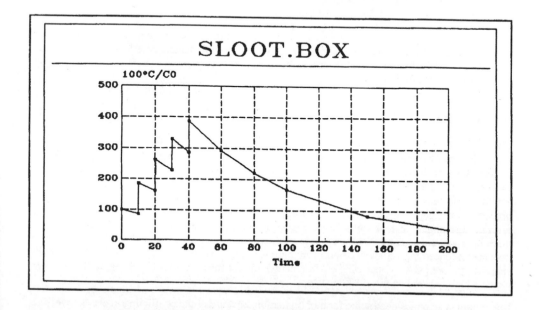

**Figure 2** : Calculation of concentration profile by SLOOT.BOX

*1.3.2.*        *Environmental criteria for pesticides: how to find scientific answers*
            *for political questions?*

# J. Van Wensem & K. Verloop

Technical soil protection committee,
P.O. Box 30947, 2500 GX The Hague, The Netherlands

## Abstract

In the last decade a number of pesticides, which are admitted to the Dutch market according to the Dutch Pesticide Act, were found in concentration levels in the environment that were considered as undesirable. The Dutch Government tried to ban harmful pesticides with three so-called environmental criteria for persistence, leaching to ground water, and acute toxicity for aquatic organisms. This attempt was unsuccessful, since the legal status of the environmental criteria was weak.

The next move was to enforce the legal status by a General Administrative Order (GAO) based on the Pesticide Act. This necessitated further negotiations between the Ministries of Environment and Agriculture, responsible for the GAO. New elements have been added in the form of disclaimers for two criteria: persistence and leaching to ground water.

The political wish to achieve a compromise between environmental and agricultural interests has resulted in criteria that are scientifically difficult to support. For example, the criteria, and their disclaimers, rely heavily on not yet validated mathematical models for pesticides movement and degradation in the environment. The disclaimer concerning persis-tence requires a comparison between predicted levels of pesticides accumulated in the soil after 12 years, and safety levels based on ecotoxicological laboratory experiments. This comparison is scientifically unsound, and can only be considered as a pragmatic solution for a political problem.

The criteria in the present GAO can be used at best to prioritize the banning of the most harmful pesticides. The criteria should not be considered as being "environmental" criteria, as the question of what kind of pesticides are needed in sustainable agriculture has not yet been answered, neither by politicians nor by scientists.

## Introduction

The Technical soil protection committee gives recommandations to the Minister of Environment on scientific subjects, mainly related to the Dutch Soil protection act. All draft General Administrative Orders (GAOs), based on the Soil protection act, are submitted for advice to the committee. Furthermore, draft GAOs with a relationship to soil protection, for example based on the Pesticides act, policy documents on soil protection, and scientific reports that will be used for certain legislation procedures are subjects for advice. The member of the committee are scientists, representative for soil relevant disciplines, such as soil science, soil mechanics, environmental medicine, geohydrology, microbiology, and ecology. With respect to pesticides, the committee has published two advisory documents concerning the development of environmental criteria for pesticides (TCB 1990) and, more specific, the draft GAO concerning environmental criteria for pesticides (TCB 1993). The present paper is mainly based on the 1993 document.

Many human activities need some form of regulation, because of the conflicting interests involved. In the case of pesticides, farmers want to achieve a high agricultural production; industries want to sell pesticides, and the community in general (often represented by the Government) wants to protect the environment against adverse effects of pesticides. Policy making deals with the framing of rules for these human activities. In environmental policy making science is often used as a tool to set limits for processes that are harmful to the environment. Science provides in these cases the yard-stick for: "what is acceptable?"

In The Netherlands we have so-called environmental criteria for pesticides. They concern: persistence of pesticides in soil, leaching of pesticides to ground water, and acute toxicity of pesticides for aquatic life. The environmental criteria are, of course, a result of political compromising; a result of negotiation between different parties. In this paper the attention is focused on the fact that political compromises often lead to (scientifically) complicated solutions, resulting in a vague regulation. Political com- promises ask often more than science can answer.

## History of the Dutch environmental criteria for pesticides

The Dutch Pesticides act came into force in 1962. The act gives rules for the trade in -, and use of pesticides, mainly with respect to reliability of pesticides, and safety of pesticides for humans and (domestic) animals. In 1975 the Pesticides act was extended with a rule concerning the safety of pesticides for the environment. Admission of a pesticide on the market could only be allowed if there was a reasonably certainty that no harmful side effects would occur. A harmful side effect was defined as "damaging soil, water, air, animals or plants in an unacceptable way". In 1989 the "memorandum on environmental criteria" was published by the Ministry of Environment (VROM 1989). The motive for this memorandum was that "the amount and way of pesticide use resulted in general in a spreading of pesticides in the environment that was considered as undesirable". Pesticide use in the Netherlands, including the use of relatively old pesticides, is very high (Figure 1). Measurable concentrations have been reported in ground water, surface water, shallow sea water, and rain water, which not very seldom exceeded guidelines (quality criteria) (Lagas et al. 1989; KNMI/RIVM 1989; Steenwijk van et al. 1992).

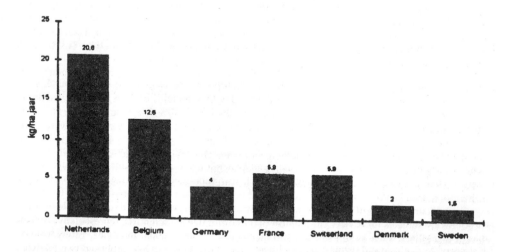

**Figure 1 :** Amounts of pesticides used in several European countries (kg/ha.year) (LNV 1991). The data in this graph have a low reliability, due to different methods used to establish the amount of pesticides used in the different countries.

After the publication of this memorandum the banning of certain pesticides appeared not possible with the 1975 formulation of the Pesticides Act and the criteria mentioned in the memorandum. Undesired and uncontrolled emission/accumulation of pesticides to/in the environment were not considered as "harmful side effects" by the Court, and demonstrating "damage" to the environment caused by a pesticide appeared to be very difficult for the Government. To solve this problem in 1993 an addition was made to the Pesticides act which states: only pesticides are allowed on the market if they meet the requirements given by a GAO concerning rules to prevent affection of the quality of soil, water and air. In 1993 the draft GAO concerning environmental criteria for persistence, leaching to ground water and acute toxicity for aquatic life was published by the Ministry of Environment (VROM 1993). The GAO thus enforces the legal status of the environmental criteria compared to the memorandum.

New pesticides are evaluated at notification. In the Netherlands this is executed by means of an system for hazard assessment for pesticides (Linders & Luttik 1994). Furthermore, old pesticides (pesticides that have been allowed on the Dutch market) are re-evaluated each 10 years with the same system. All pesticides have to meet the environmental criteria required by the GAO at evaluation, otherwise a pesticide is not allowed on the Dutch market. In the case of re-evaluation, the pesticide or certain uses of the pesticide will be banned if the environmental criteria are not met. The decision about admission or banning is made by the Pesticides Admission Board, which is independent from the Government.

## Scientific basis environmental criteria for persistence and leaching

Originally the environmental criteria for persistence in soil and leaching to ground water were based on rather simple principles. These were described in the memorandum of 1989 (VROM 1989). Prevention of accumulation in soil was the basic principle behind the criterion for persistence of a pesticide in soil. The memorandum indicated that a pesticide with a 50% disappearance time (DT50)$\leq$ 60 days was acceptable with respect to accumulation in soil. With respect to leaching to ground water the principle was adopted that ground water should remain suitable for the production of drinking water with normal purification. This meant that ground water should meet the standards from the EC-Directive 80/778/EEC relating to the quality of drinking water intended for human consumption. For the concentration of a pesticide in ground water the standard in the EC-Directive is $\leq$ 0.1 $\mu$g/l (sum of all pesticides $\leq$ 0.5 $\mu$g/l).

In the memorandum the expected accumulation in the top soil and leaching to upper ground water (1-2 m below soil surface) of pesticides was calculated with a model (PESTLA) (Linden van der & Boesten 1989). This model needs two input parameters: the DT50 (50% disappearance time measured in a standardized laboratory test) and the Kom (sorption to organic matter). Calculations were made for a one-year period with one application of the pesticide under "reasonable" bad conditions. It was recognized in general that the model did nothing more than classifying pesticides: the outcome of the model calculations were comparable between pesticides but were not necessarily predicting the real field situation. In this case the model provided a yard-stick for acceptable values for the DT50 and Kom of pesticides.

In the mean time the formulation of the criteria has changed, due to harmonization with the European Union (EU) policy for placing plant protection products on the market and negotiations with the Ministry of Agriculture and several parties with interest. In the draft GAO the criterion for persistence has changed from a DT50 $\leq$ 60 days into a DT50 $\leq$ 90 days (VROM 1993). This change has been made in accordance to the Uniform Principles for the evaluation and authorization of plant protection products, which will be accepted probably by the EU this year (Commission of the European Communities 1993). Also a disclaimer has been added to the criterion, which states that a pesticide with a DT50 > 90 days will not be allowed unless the producer of the pesticide (applicant for admission) is able to prove that the accumulated concentration in the top soil will be lower than a maximum permissible concentration (a safety level), two years after the last application. The criterion for leaching to ground water is still that the concentration of a pesticide in the upper ground water should remain lower than 0.1 µg/l (this is in agreement with the Uniform Principles but subject of discussion in the EU). Also a disclaimer has been added: a pesticide that will occur in higher concentrations than 0.1 µg/l in the upper ground water will not be allowed, unless the producer of the pesticide is able to prove that after four years at ten meters depth the concentration will be lower than 0.1 µg/l (due to degradation in the saturated zone). The Dutch Government will indicate which information is necessary to substantiate the disclaimers.

## Scientific problems

The present criterion for persistence requires knowledge about the accumulation and degradation of a pesticide in the top soil for at least twelve years (a pesticide is admitted for ten years, this period is followed by a two years period of degradation only). The intention of the Dutch policy makers is to give the producers of a pesticide the opportunity to substantiate the disclaimer with model calculations. This means modeling the accumulation of a pesticide in the top soil for a period of twelve years, with realistic assumptions about the application of the pesticide (Figure 2). The model will probably be an extended version of the PESTLA-model. Substantiation of the disclaimer with field data or other models is under discussion. Furthermore, maximum permissible concentrations (MPCs) need to be established for pesticides in soil based on standardized ecotoxicity experiments. In The Netherlands a more or less accepted procedure is available for the derivation of MPCs, but the number of input data needed to derive a reliable MPC is rather high (Slooff 1992; Plassche van de 1994; TCB 1994). Finally, the predicted concentrations in the top soil need to be compared with the MPCs.

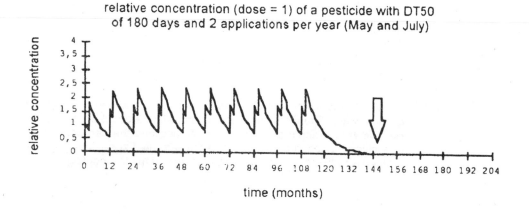

**Figure 2 :**   Schematic representation of the amount of a pesticide present in top soils in a period of twelve years. The disclaimer for the criterion for persistence in soils requires a comparison of the concentration at t = 144 months (or the equilibrium concentration) with a maximum permissible concentration (TCB 1993).

The present criterion for leaching to ground water requires modeling: (1) leaching to upper ground water, (2) leaching to deeper ground water (saturated zone), and (3) breakdown of pesticides in the saturated zone over a 4-years period. The first process has been described with the PESTLA-model. The outcome of the PESTLA-model is considered as being relative; pesticides are compared to each other. The Dutch Government decided in 1991 that the PESTLA-model should be validated and optimized in such a way that the model calculations will predict absolute concentrations in the upper ground water (LNV 1991). As a consequence the PESTLA-model is validated at the moment by the National Institute of Public Health and Environmental Protection (RIVM). Suitable data sets appear to be very scarce and validation for a period longer than one year is not possible. Most experts involved in the validation still have the opinion that the use of the PESTLA-model as a tool to compare pesticides at a relative basis should be preferred above the use of the PESTLA-model as a predictor of absolute (real) concentrations in the field (Wijland 1994). Leaching to upper ground water for a period of one year is a simple process compared to leaching to deeper ground water and degradation of pesticides in the saturated zone over a four years period (Figure 3). It might be expected that extension of the PESTLA-model again leads to a model that compares pesticides at a relative basis. By adding the disclaimer to the criterion for leaching to ground water, again the problem arises how to deal with this "relative" versus "absolute" problem in the hazard assessment system.

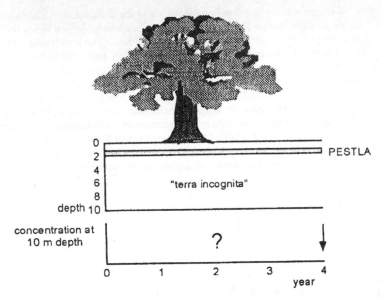

**Figure 3 :**   Schematic representation of the disclaimer for the criterion for leaching to ground water. The disclaimer requires knowledge about the transport a pesticides to a depth of 10 m below surface and the degradation of a pesticide in a period of four years in the saturated zone. The grey area marked with "PESTLA" indicates the present status of modeling.

## Questions and solutions

The addition of the disclaimers to the criteria for persistence in soil and leaching to ground water appears to introduce considerable (scientific) problems. Knowledge about long term processes in soil and ground water is lacking or scarce. Models that substantiate the disclaimers will at best lead to comparison of pesticides at a relative basis. Thus model calculations may easily lead to different conclusions than field data for a pesticide in a relevant soil. It is also imaginable that different models will be developed.

An other question is how to deal with bound residues in soil. In general there is no consensus about the risk of soil bound residues for soil organisms. The disclaimer for persistence in soil requires comparison of the (calculated or measured) concentrations in top soils with maximum permissible concentrations (MPCs). MPCs are to be derived from (sub)chronic laboratory tests with soil organisms. In these tests the amounts of active ingredients, metabolites and bound residues in will be very different from these amounts in top soils after twelve years (Figure 4) (TCB 1993). What will be the relevance of this comparison?

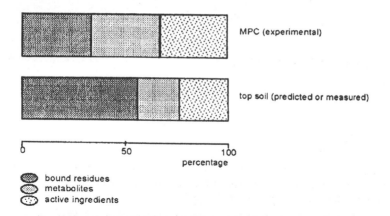

bound residues
metabolites
active ingredients

**Figure 4 :**   Hypothetical distribution (in percentages) of the amount of a pesticide over three fractions: bound residues, metabolites, and active ingredients. The distribution will differ between experiments and top soils. The toxicity of the fraction will differ, which makes the comparison of concentrations of a pesticide in ecotoxicological experiments, expressed as an maximum permissible concentration, and the concentration predicted or measured in top soils, unreliable (TCB 1993).

In an optimistic view we may expect that all these problems and questions will be solved sooner or later. The GAO, however, is planned to come into force on January 1, 1995. This means that in the next future the solutions will lack scientific reliability. We will have to deal with "most sensible" solutions, as model validation has not taken place yet. It is even questionable that validation of the models will be possible; model makers did often not intend to make "real-life" models. And specific field data (time, space, conditions) will be opposed to generic rules. The past experience has learned that this all may affect the juridical strength of the environmental criteria. The criteria in the present GAO can be used at best to prioritize the banning of the most harmful pesticides. The criteria should not be considered as being "environmental" criteria, as the question of what kind of pesticides are needed in sustainable agriculture has not yet been answered, neither by politicians nor by scientists.

## *Conclusions*

1.   Simple environmental protection goals have evolved in complicated rules due to compromising.

2.   Complicated rules result in:

     —    very high (unrealistic) demands for scientific solutions;
     —    uncertainty with respect to protection of the environment;
     —    uncertainty with respect to the juridical strength of the environmental criteria.

3.   Policy makers should be more aware of the scientific possibilities.

4.   Scientists should try to indicate the "general" applicability and limitations of their work.

5.   More integration is needed between science and policy making.

## Acknowledgments

We thank our colleagues at the secretariat of the Technical Soil Protection Committee for support and discussion.

## References

□    Commission of the European Communities, 1993. Proposal for a Council Directive establishing Annex VI of Directive 91/414/EEC concerning the placing of plant protection products on the market. Brussels, 20 April 1993.

□    KNMI/RIVM, 1989. Netherlands Precipitation Chemistry Network. Monitoring Results 1988. KNMI publ. 156-11, De Bilt, RIVM report no. 228703 012, Bilthoven.

□    Lagas, P., B.Verdam & H.L.J. van Maaren, 1989. Veldonderzoek Bestrijdingsmiddelen, rapportage van de 4e bemonstering. RIVM rapport nr. 728473 003, Bilthoven.

□    Linden, A.M.A. van der & J.J.T.I. Boesten, 1989. Berekening van de mate van uitspoeling en accumulatie van bestrijdingsmiddelen als functie van hun sorptiecoëfficiënt en omzettingssnelheid in bouwvoormateriaal. RIVM rapport nr. 728800 003, Bilthoven.

□    Linders, J.B.H.J. & R. Luttik, 1994. Hazard assessment of agro- and non-agrochemi- cals in the Netherlands. Proceedings 5th International Workshop on Environmental Behaviour of Pesticides and Regulatory Aspects, Brussels, April 26-29, 1994 (in press).

□    LNV, 1991. Meerjarenplan Gewasbescherming. Tweede kamer, vergaderjaar 1990-1991, 21 677, nrs. 3-4.

□    Plassche, E.J. van de, 1994. Towards integrated environmental quality objectives for several compounds with a risk for secondary poisoning. RIVM report no. 679101 012, Bilthoven.

□    RIVM, 1988. Zorgen voor morgen. Nationale Milieuverkenningen 1985-2010. Rijks- instituut voor Volksgezondheid en Milieuhygiëne, Bilthoven.

□    Slooff, W., 1992. RIVM Guidance Document. Ecotoxicological effect assessment: deriving maximum tolerable concentrations (MTC) from single species toxicity data. RIVM report no. 719102 018, Bilthoven.

□    Steenwijk, J.M. van, J.M. Lourens, J.H. van Meerendonk, A.J.W. Phernambucq & H.L. Barreveld, 1992. Speuren naar sporen I. RIZA nr. 92.057, Lelystad, DGW nr. 92.040, Den Haag.

□    TCB, 1990. Advies bodembescherming en bestrijdingsmiddelen. Technische commissie bodembescherming, A89/05, Leidschendam.

□    TCB, 1993. Advies Besluit milieutoelatingseisen bestrijdingsmiddelen. Technische commissie bodembescherming, A05(1993), Den Haag.

□    TCB, 1994. Advies project Integrale Normstelling Stoffen, deel B. Technische commissie bodembescherming, Den Haag (in preparation).

□    VROM, 1989. Notitie Milieucriteria ten aanzien van stoffen ter bescherming van bodem en grondwater. Tweede kamer, vergaderjaar 1988-1989, 21 012, nr. 1.

□    VROM, 1993. Ontwerp-Besluit milieutoelatingseisen bestrijdingsmiddelen. Versie van 19 juli 1993. Ministerie van VROM, Den Haag.

□    Wijland, R. 1994. Internal report on the PESTLA validation workshops held on September 8, 1993 and February 22, 1994. Technical Soil Protection Committee, The Hague.

# SECTION II

# SORPTION

Chairman :     A. PICCOLO
               (Univ. Napoli, Italy)

# SORPTION

## 2.1. Introductory presentation

### 2.1.1. *Retention and Bioavailability of Pesticides in Soil*

## R. Calvet and E. Barriuso

INA-PG, INRA, 78850 Thiverval-Grignon, France

### Introduction

Retention phenomenon (sorption and incorporation/irreversible retention) has been recognized as one of the major processes determining the fate of pesticides in natural systems. Many studies have shown their influence on the behaviour of pesticides in soil, including toxic effects against target and non target organisms, transformations (biological and non biological) and transport processes (volatilization, runnoff and leaching). The relationship between pesticide bioavailability and retention has been appeared of primary importance while being not yet fully described and understood. As a matter of fact, it is not possible to give a general relationship between pesticide bioavailability and retention. The purpose of this paper is to recall some experimental facts and to draw attention to important questions regarding this relationship.

A compound is bioavailable when it can be absorbed by a living organism in a given medium. This is a property which represents a potential to be absorbed, that is, to be easely transported to an organism or to be in its immediat vicinity. From this point of view, a condition to be fulfilled by a chemical species is to be mobile either in the gaz phase or in the liquid phase, making pesticide bioavailability strongly dependent on retention/release phenomena. For a compound, the bioavailable amount can only be defined for a given medium, a given living organism, a given set of conditions and for a given process (Calvet, 1989). It is important to distinguish the potential bio-availability of a compound from the bioavailable amount of that compound. The later can only be known through a process mediated by a living organism. For example, the bioavailable amount of an herbicide estimated through a phyto-toxic effect is not the same as that which would be determined from a degradation experiment.

Pesticide bioavailability can be assessed through toxic effects against living organisms (plants, soil microflora, fauna ....), biodegradation and health hazards for man *via* consumption of contaminated plants, animals or water. Only, bioavailability to soil living organisms will be considered below.

### Bioavailabilty of Sorbed Molecules

The primary effect of retention is to lower the pesticide concentration of the soil solution and of the soil atmosphere in contact with the solid phase. Retention by the solid phase is mainly due to colloïd constituents and involves two broad groups of phenomena : sorption and incorporation/irreversible retention.

Sorption can be chronologically divided in two steps. The first step is an interface phenomenon, that is, adsorption on colloïdal surfaces. In the upper layer of the soil, the adsorbing surface is essentially organic due to coating of mineral surfaces by organic polymers. The second step is a diffusion into organic molecular aggregates and into the more or less altered plant tissues. Consequently, sorption of non-ionic molecules (adsorption and intra-particle diffusion) can be mainly described on the basis of interactions with the organic matter but has nothing to do with a water organic/solvent partition which is not necessary to explain sorption experimental data. This two steps description of sorption is compatible with published data, particularly kinetic characteristics.

To be bioavailable, sorbed molecules must be released at a rate which allows a sufficient amount to be absorbed. Sorbed molecules in the soil constituents (essentially the organic matter) are less available than sorbed molecules on the surface of constituents (that is adsorbed molecules) because they have to diffuse out the solid phase before being able to desorb.

Molecules in solution are immediatly bioavailable during a short period of time (8-10 days) as shown by Stalder and Pestemer (1980) for several herbicides. For longer periods, bioavailable amount depends probably on transport phenomena to the living organisms as was early stated by Bailey and White (1964). This is the most simple situation as compared to those of sorbed and unextractable molecules.

Concerning the plant roots, the overall observed effect is a decreasing bioavailability with increasing sorption (Bailey and White, 1964 ; Calvet et al, 1980 ; Sims et al, 1991 ; Gaillardon et al, 1992). It is well known, for example, that the rate of application of herbicides must be increased as the soil organic matter content increases. Figure 1 illustrates the effect of sorption on atrazine phytotoxicity to oat plants, the growth inhibition-dose relationship being dependent on the sorption properties of the soils. The same kind of effect can also be observed on figure 2 where atrazine toxicity to green algae in suspension decreases as the sorption coefficient increases. Another example is given by Günther et al (1989) who have compared the availability to plants of several sulfonylurea herbicides. They have found that bioavailability in two soils having approximately the same clay content strongly depends on the organic matter content. The general trend of bioavailability-sorption relationship evaluated through phytotoxic effect is quite clear. However, some questions remain as shown in figure 3 where results obtained with various soils are reported : the bioavailability of diuron is identical in soils with different sorption properties (group A) and is different in other soils which have the same sorption properties (group B). Three sets of data show that microbial degradation is affected by sorption of pesticide molecules.

Some correlations have been observed between degradation rate and soil properties. This is the case of the clay content, an increase of which entails a decrease of the degradation rate of atrazine (Soulas, 1975). Walker et al (1983) have reported a positive correlation between the half-life of simazine determined in laboratory experiments and soil properties such as organic carbon, clay contents and pH. Increase of sorption capacity with increasing clay and organic carbon content may explain the observations but it should be kept in mind that differences between degradation in various soils may be due to differences between microbial populations. Another way to study the effect of sorption on microbial degradation is to run experiments with amended media with organic matter. Addition of organic matter to a soil may affect the sorption of pesticide as well as microbial populations because it is both a sorbent and a nitrogen/carbon substrate making such experiments difficult to be easely interpreted (Hance, 1970). The third set of data is provided by modeling studies. Microbial degradation in soil is generally better described when adsorption/desorption and diffusion are introduced in models (Sims et al, 1991; Scow and Hutson, 1992 ; Duffy et al, 1993).

## Bioavailability of Bound Residues

A more or less important fraction of retained molecules cannot be released from the solid phase. This has been observed in many instances with various pesticide/soil systems and leads experimentally to retention hysteresis, partial and time-variability of residue extraction by water and organic solvents.

Many experiments made with radiolabelled molecules show that the extraction with organic solvents is never complete. Unextractable residues have been called "bound residues" and have been reviewed by Khan (1980), Calderbank (1989), Bertin and Schiavon (1989) and by Barriuso et al (1991). Several authors have reported experimental data showing clearly that bound residues are constituted both by parent products and degradation products. Parent products have been found to represent various fractions of bound residues : 18% for pyridil carbamate, more than 50 % for prometryn, 10% for atrazine and traces for several other pesticides such as picloram, triallate and methylparathion (cited by Calderbank, 1989). It has been observed that the extractability decreases with aging because phenomena responsible for incorporation are probably more efficient with an increasing residence time in soil (figure 4). Irreversible retention of pesticides is not yet fully understood. Several hypothesis have been put forwards to explain their formation and properties but they have not been completely verified. Entrapping in humic polymers aggregates as suggested by Wershaw and Goldberg (1972), Khan (1973) and more recently by Singh et al (1989) has been advanced as a possible explanation for hysteresis, compatible with humic substance structure (Wershaw, 1986). Hydrophobic molecules may be trapped inside the most hydophobic part of humic aggregates and then, prevented from a release with aqueous and organic solvents. Another process, probably a major one leading to an irreversible retention, is the chemical binding of pesticide molecules to the organic matter (Bollag and Loll, 1983). Reactions are due to microorganisms and plant enzymes and can be considered as some kind of degradation. Involved reactions appear to be the same as those responsible for humic substance formation (Bollag et al, 1992 ; Scheunert et al, 1992). Another process to be refered to is the biological incorporation. This retention phenomenon is due to pesticide absorption by plants and soil micro-organisms. Some fraction of this absorbed pesticide may return to the soil either through the crop residues or dead micro-organisms and thus may be incorporated in the soil organic matter.

Bioavailability of bound residues can be analyzed through toxic effects and biodegradation studies. Bollag et al (1992) observed that phenolic compounds were not toxic for *Rhizoctonia practicola* when the culture medium contains extracellular laccase. They attributed this observation to the ability of the enzyme to transform or to polymerize the parent compound and then to decrease its toxicity. Scribner et al (1992) shown that aged simazine residues were less available for plant uptake and microbial degradation.

From a general point of view, bound residues have probably a very low bioavailability for plants and soil organisms, if they have anyone. They are retained through covalent linkages and by entrapping in organic matter aggregates and they can only be released with drastic treatments such as chemical extraction or pyrolysis. It is certainly unlikely to find natural phenomena in soil with such strong effects so that slow release of bound residues is merely probable. This may favor a complete degradation and theoritically prevent from any harmfull accumulation. However, some doubt still remains about delayed hazards. Several observations show that bound residues can be recovered quite a long time after a pesticide application. Scheunert et al (1992) have been able to extract 4-chloroaniline which represents 15% of [14]C buturon applied 16 years before. Also, it has been shown that small amounts of bound residues can be absorbed by plant roots (Kloskowski and Führ, 1987 ; Andreux et al, 1993). Moreover, retention by organic matter does not necessarily lead to immobilization of the pesticide or its degradation products. When retention takes place on hydrosoluble organic polymers, pesticide water solubility can be enhanced (Chiou et al, 1986 ; Barriuso et al, 1992) and transport may be favored (Ballard, 1971; Scheunert et al, 1992).This may contribute to increase the bioavailabilty of some compounds for living organisms, particularly in aquatic media.

## Evaluation and Prediction of Pesticide Bioavailability in Soil

The complexity of involved phenomena makes difficult any evaluation and prediction of the pesticide bioavailability in soil. Today, only some qualitative statments are possible, based on the knowledge of sorption coefficients and on release characteristics with water and organic solvents.

### *The sorption coefficient*

In spite of the lag of quantitative relationship between sorption and bioavailability, sorption coefficient values can be used to assess qualitatively the biological behaviour of a pesticide in soil. Basically, it is assumed that bioavailability is a decreasing function of the sorption coefficient. Consequently, any factor having an influence on sorption may have an influence on bioavailability.

Sorption coefficients can be either directly determined or estimated. When they are determined, the main difficulty is due to non-equilibrium. Brusseau et al (1991) have described the rate limiting processes according to two general classes : transport and sorption related processes. Transport limited sorption results from heterogeneous flow domains in relation to macroscopic heterogeneities (aggregates, pores of various sizes). Sorption non-equilibrium may be due to chemical non-equilibrium or to rate limited mass transfers. Chemical non-equilibrium is caused by rate-limited interactions between sorbate and sorbent and is probably unimportant for hydrophobic organic compounds sorption but not for polar compounds. Three diffusive processes may be responsible for rate limited mass transfers : film diffusion, retarded intra-particule diffusion and intra-sorbent diffusion. This last phenomenon concerns the diffusion of hydrophobic organic compounds into the organic matter particles or aggregates and polar molecule diffusion into mineral particules.

Sorption coefficients may be estimated through relations with water solubility and soil organic carbon content. This implies that sorption can be described by linear isotherms. Such isotherms are frequently reported but they are not specific of particular sorbents or sorbates (Calvet, 1993). Thus, it is not possible to relate univocally a linear isotherm to a particular sorbent/sorbate system or to a given particular mechanism as was previously pointed out by Mingelgrin and Gerstl (1983).

According to the literature, sorption of pesticides depends on the electrical charge and the ionizability, on the hydrophobicity and the polarity and on the geometrical and topological characteristics of the molecules (Sabljic, 1984 ; Woodburn et al, 1992). So very broad relations may be suggested between these coefficients and pesticide bioavailability in soils. It has been observed that molecule affinity for water has a deep influence on sorption of non-ionic molecules and has been extensively discussed (Karickhoff, 1981 ; Chiou et al, 1983 ; Hasset and Banwart, 1989). According to many observations, the normalized sorption coefficient (Kom, on an organic matter content basis, and Koc, on an organic carbon content basis) is a decreasing function of the water solubility and an increasing function of the water/octanol partition coefficient. These coefficients may be used to classify pesticides according to their bioavailability in a given soil. Nevertheless, it is only an approximate procedure.

Many soil characteristics are related to pesticide sorption and their relative importance depends on the polarity of molecules (Barriuso and Calvet, 1992). Clay and organic matter contents are two important factors. However, if clays are mainly responsible of cation adsorption and may play a role in clay reached soils and subsoils, the soil organic matter is essential for sorption of non-ionic organic compounds. Sorption coefficient values are generally correlated with the soil organic carbon content essentially when a large domain of organic carbon content is considered. This correlation is not so well defined when the organic carbon content is less than 3% which corresponds to the majority of cropped soils.

In opposition to what it was hopped, Koc is generally quite variable with coefficient of variation often greater than 100% (Gerstl, 1992). This coefficient becomes no longer relevant when the OC fraction/clay fraction ratio is low (Ainsworth et al, 1989). For ionizable compounds, pH is of primarily importance. Adsorption increases with decreasing pH for weak acids and weak bases showing a maximum for some compounds.The shape of this variation depends on the apparent pKa of the molecule (Feldkamp and White, 1979) and on the variation of the surface electric charge. One may think that bioavailability of ionized molecules would follow the same variation as a function of pH.

Bioavailability of pesticides in soils depends certainly on molecular mechanisms of interactions between molecules and the soil solid phase. Taking into account energetic characteristics of sorption isotherms, one can suggest that bioavailability may decrease according to the following order : S-isotherms>C-isotherms>L-isotherms>H-isotherms (nomenclature of adsorption iso-therms is that proposed by Giles et al, 1960). However, this remains to be studied more completely.

### *Release characteristics*

To be bioavailable, a molecule must be in the fluid phase so that release is a key phenomenon. Thus it may be usefull to use release characteristics to evaluate bioavailability of pesticides.

It is well known that retention is not completely reversible causing release isotherms to be different from retention isotherms. Strictly speaking, sorption is a reversible process and includes adsorption, desorption and diffusion in and out the solid phase. Several explanations have been proposed to account for an hysteretic behaviour. Some are related to experimental conditions which lead to an underestimation of the equilibrium concentration of the liquid phase : artifacts due to the method, failure to establish equilibrium ; in this last case, it is better to speak of resistant forms of pesticides. Other potential causes for hysteresis correspond to a real incorporation into the solid phase : modifications of molecule/surface interactions, irreversible retention into the solid phase (mainly the organic matter) leading to an entrapping of the molecules, chemical reactions between the pesticide molecules and the solid phase which is equivalent to a degrada-tion phenomenon. Some of these causes have been experimentally verified. This is the case of degradation, of entrapping and of chemical reactions (Bollag et al, 1992 ; Andreux et al, 1993). It is worthnoting that hysteresis can diplay a range of intensity from reversibility to the absence of any release, corresponding to an equivalent range of bioavailability. Here again, the statment can only be qualitative.

At a given time, some molecules cannot be released in water but can be extracted by an organic solvent, others remaining in the solid phase as bound residues. It is impossible to predict from extracted amounts the fate of such molecules and particularly their future bioavailability.

## Concluding Remarks

Some data on relationships between retention and bioavailability of pesticides are reported in the literature. These are essentially qualitative. Experimental results show, sometimes clearly, that bioavailability is closely related to sorption characteristics, soil and molecule properties. However, these relationships are not yet completely described and understood. Our knowledge about this topic will be probably improved by a better energetic and kinetic description of release phenomena. Nevertheless, it is essential to draw attention to the great importance of biological experiments. They remain certainly the most efficient way to proceed since bioavailability can only be assessed through effects on living organisms. A way for future researches could be to relate energetic and kinetic characteristics of retention to bioavailability of pesticide in soils.

## *References*

❑　Ainsworth C.C., Zachara J.M. and Smith S.C. (1989). Carbazole sorption by surface and subsurface materials: Influence of sorbent and solvent properties. *Soil Sci. Soc. Am. J.*, 53, 1391-1401.

❑　Andreux F., Scheunert I., Adrian P. and Schiavon M. (1993). The binding of pesticide residues to natural organic matter, their movement, and their bioavailability. in Fate and prediction of environmental chemicals in soils, plants and aquatic systems. Mohammed Mansour (Ed.), Lewis Publishers.

▫   Bailey G.W. and White J.L. (1964). Review of adsorption and desorption of organic pesticides by soil colloids, with implications concerning pesticide bioactivity. *Agri. Food Chem.*, 12, 324-382.

▫   Ballard T.M. (1971). Role of humic carrier substances in DDT movement through forest soil. *Soil Sci. Soc. Amer. Proc.*, 35, 145-147.

▫   Barriuso E., Schiavon M., Andreux F. and Portal J.M.(1991). Intér+t et limitation des méthodes de séparation des micro-polluants organiques des sols. *Science du Sol*, 29, 301-320.

▫   Barriuso E. and Calvet R. (1992). Soil type and herbicide adsorption. *Intern. J. Environ. Anal. Chem.*, 46, 117-128.

▫   Barriuso E., Baer U. and Calvet R. (1992). Dissolved organic matter and adsorption-désorption of Dimefuron, Atrazine, and Carbetamide by soils. *J. Environ. Qual.*, 21, 359-367.

▫   Barriuso E. and Koskinen W. (1994). Incorporation of atrazin non extractable (bound) residues in soil size fractions. *Soil Sci. Soc. Amer. J.* (in press).

▫   Bertin G. and Schiavon M. (1989). Les résidus non extractibles de produits phytosanitaires dans les sols. *Agronomie*, 9, 117-124.

▫   Bollag J.M. and Loll M.J. (1983). Incorporation of xenobiotics into soil humus. *Experiencia* 39, 1221-1231.

▫   Bollag J.M., Myers C.J. and Minard R.D. (1992). Biological and chemical interactions of pesticides with soil organic matter. *Sci. Total Environ.*, 123/124, 205-217.

▫   Bouchard D.C., Enfield C.G. and Piwoni M.D. (1989). Transport process involving organic chemicals. In: Reaction and movement of organic chemicals in soils (B.L. Sawhney & K. Brown, ed.), SSSA, special publication n°22.

▫   Brusseau M.L., Jessup E., Suresh P. and Rao C. (1991). Non equilibrium sorption of organic chemicals: Elucidation of rate-limiting processes. *Environ. Sci. Technol.* 25, 134-142.

▫   Calderbank A. (1989). The occurence and significance of bound pesticide residues in soil. *Rev. Environ. Contam. Toxicol.*, 108, 69-103.

▫   Calvet R. (1989). Adsorption of organic chemicals in soils. *Environ. Health Persp.*, 83, 145-177.

▫   Calvet R. (1993). Comments on the evaluation of the sorption coeeficient for modeling the fate of pesticides in soils. In: Del Re A.A.M., E. Capri, S.P. Evans, P. Natali and M. Trevisan (Eds.). IX Symposium Pesticide Chemistry - Mobility and degradation of xenobiotics, 11-13 October, Piacenza, Italy.

▫   Calvet R., Tercé M. et Arvieu J.C. (1980). Adsorption des pesticides par les sols et leurs constituants. IV : conséquences des phénomènes d'adsorption, *Ann. Agro.*, 31, 385-411.

▫   Chiou C.T., Porter P.E. and Schmedding D.W. (1983). Partition Equilibria of nonionic organic compounds between soil organic matter and water . *Environ. Sci. Technol.*, 17, 227-231.

▫   Chiou C.T., Malcolm R.L., Brinton T.I. and Kile D.E. (1986). Water solubility enhancement of some organic pollutants and pesticides by dissolved humic and fulvic acids. *Environ. Sci. Technol.*, 20, 502-508.

▫   Duffy M.J., Carski T.H and Hanafey M.K. (1993). Conceptually and experimentally coupling sulfonylurea herbicide sorption and degradation in soil. In: Del Re A.A.M., E. Capri, S.P. Evans, P. Natali and M. Trevisan (Eds.). IX Symposium Pesticide Chemistry - Mobility and degradation of xenobiotics, 11-13 October, Piacenza, Italy.

▫   Feldkamp J.R. and White J.L. (1979). Acid-base equilibria in clay suspensions. *J. Colloid Inter. Sci.*, 69, 97-106.

☐ Gaillardon P., Barriuso E. and Calvet R. (1992). La disponibilité des herbicides dans le sol. ANPP - 15ème Conf. Columa, Journées Intern. sur la Lutte contre les mauvaises herbes, Versailles, 2,3,4 Décembre 1992.

☐ Gaudry J.C., Gaillardon P. and Calvet R. (1982). Etude de l'influence de l'adsorption sur la phytotoxicité de l'atrazine et de l'isoproturon dans le sol. (publication interne).

☐ Gerstl Z. (1990). Estimation of organic chemical sorption by soils. *J. Contam. Hydrol.*, 6, 357-375.

☐ Giles C.H., MacEwan T.H., Nakhwa S.N. and Smith D. (1960). Studies in adsorption. Part XI. A system of classification of solution adsorption isotherms and its use in diagnosis of adsorption mechanisms and in measurement of specific surface areas of solids. *J. Chem. Soc.* 3, 3973-3993.

☐ Günther P., Rahman A. and Pestemer W. (1989). Quantitative bioassays for determining residues and availability to plants of sulphonylurea herbicides. *Weed Res.*, 29, 141-146.

☐ Hance R.J. (1970). Influence of sorption on the decomposition ofpesticides. In Sorption and transport processes in soils. *Soc. Chem. Industry Monograph*, 37, 92-104.

☐ Harris C. and Sheet J.J. (1964). Influence of soil properties on adsorption and phytotoxicity of CIPC, diuron and simazine. *Weeds*, 215-219.

☐ Hasset J.J. and Banwart W.L. (1989). The sorption of Nonpolar Organics by Soils and Sediments. Reactions and Movement of Organic Chemicals in Soils. *SSSA* Special Publication n°22.

☐ Karickhoff S.W. (1981). Semi-empirical estimation of sorption of hydrophobic pollutants on natural sediments and soil. *Chemosphere*, 10, 833-846.

☐ Khan S.U. (1973). Interaction of humic acid with chlorinated phenoxyacetic and benzoic acids. *Environ. letters* 4, 141-148.

☐ Khan S.U. (1980). Bound pesticide residues in soil and plants. *Residue Review*, 84, Springer Verlag, 1-25.

☐ Kloskowski R. and Führ F. (1987). Aged and bound herbicide residues in soil and their bioavailability. Part 1: Uptake of aged and non-extractable (bound) [3-$^{14}$C] metamitron residues by sugar beets. *J. Environ. Sci. Health*, B22, 509-535.

☐ Lefebvre-Drouet E., Rousseau M.F. and Calvet R. (1988). Observations sur la biodisponibilité de l'atrazine pour des algues vertes (chlorelles) en suspension. *Chemosphere,* 17, 243-246.

☐ Mingelgrin and Gerstl Z. (1983). Reevaluation of partitioning as a mechanism of nonionic chemicals adsorption in soils. *J. Environ. Qual.*, 12, 1-11.

☐ Sabljic A. (1984). Predictions of the nature and strength of soil sorption of organic pollutants by molecular topology. *J. Agric. Food Chem.*, 32, 243-246.

☐ Scheunert I., Mansour M. and Andreux F. (1992). Binding of organic pollutants to soil organic matter. Int. *J. Environ. Anal. Chem.*, 46, 189-199.

☐ Scow K.M. and Hutson J. (1992). Effect of diffusion and sorption on the kinetics of biodegradation: Theoretical considerations. *Soil Sci. Soc. Am. J.*, 56, 128-134.

☐ Scribner S.L., Benzing T.R., Shoabaiu Sun and Boyd S.A. (1992). Desorption and bioavailability of aged simazine residues in soil from a continuous corn field. *J. Environ. Qual.*, 21, 115-120.

☐ Sims G.K., Radosevich M., He X.T. and Traina S.J. (1991). The effects of sorption on the bioavailability of pesticides. In Biodegradation: Natural and Synthetic Naturals. Edited by W.B. Betts, Springer Verlag, 120-137.

◻  Singh R., Gerritse R.G. and Aylmore L.A.G. (1989). Adsorption-Desorption behaviour of selected pesticides in some Western Australian Soils. *Aust J. Soil Res.*, 28, 227-43.

◻  Soulas G. (1975). Influence du taux d'argile sur la persistance del'atrazine dans le sol. 8ème Conférence du COLUMA tome 1,3-10.

◻  Stalder L. and Pestemer W. (1980). Availability to plants of herbicide residues in soil. Part I: A rapid method for estimating potentially available residues of herbicides. *Weed Research*, 20, 341-347.

◻  Walker et al. EWRS Herbicide-Soil working group. (1983). Collaborative experiment on simazine persistence in soil. *Weed Res.*, 23, 373-385.

◻  Wershaw R. L. (1986). A new model for humic materials and their interactions with hydrophobic organic chemicals in soil-water or sediment-water systems. In: D.L. Macalady (Ed.), Transport and Transformations of Organic Contaminants. *J. Contam. Hydrol.*, 1, 29-45.

◻  Wershaw R.L. and Goldberg M.C. (1972). Interaction of organic pesticides with natural organic polyelectrolytes. *Am. Chem. Suc. Adv. Chem.*, 121, 149-158.

◻  Woodburn K.B., Delfino J.J. and Rao P.S.C. (1992). Retention of hydrophobic solutes on reverse-phase liquid chromatography supports: Correlation with solute topology and hydrophobicity indices. *Chemosphere*, 24, 1037-1046.

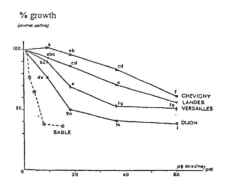

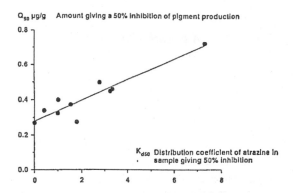

**Figure 1 :** Toxicity to oat plants *(Avena sativa)* of atrazine in soils having different sorption properties. The order of decreasing sorption is : Chevigny>Landes>Versailles>Dijon>sable (Gaudry et al, 1982)

**Figure 2 :** Effect of sorption on the amount of atrazine leading to 50 % reduction of pigment production by green algae, *Chlorella pyrenoïdosa* (Lefebvre-Drouet et al, 1988)

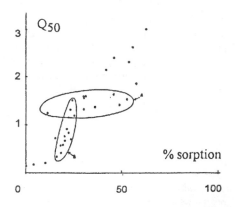

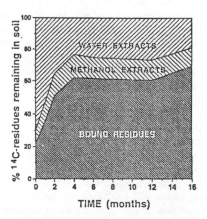

**Figure 3 :** Relation between the sorbed amount of diuron and the amount Q50 able to reduce the plant production to 50 % (Harris and Sheet, 1964)

**Figure 4 :** Time variation of 14C-atrazine residues extracted by water and methanol (Barriuso and Koskinen, 1994)

## 2.2.　　　Platform Presentations

*2.2.1.*　　　*Adsorption Mechanisms of s-Triazine and Bipyridylium Herbicides on Soil Humic Acids*

## N. Senesi, V. D'Orazio and T. M. Miano

Istituto di Chimica Agraria, Università di Bari, Bari, Italy

## (Full text not received)

*The main objectives of this work were : (a) to characterize comparatively the humic acids (HA) isolated from three soils sampled in the same hops field in Bavaria, the first under the hops plants (A), the second between the plant rows (B), and the third in a nearby not cultivated control site (C); and (b) to evaluate comparatively the adsorption mechanisms and the types of binding occuring in the interaction of these HAs with atrazine and simazine and diquat and paraquat. The three soils were supplied by GSF-Munchen in the framework of an EEC-funded STEP-Research Project. The HAs ($HA_A$, $HA_B$, $HA_C$) were isolated according to the conventional procedures of sodium hydroxide-sodium pyrophosphate extraction, precipitation by HCl to pH $\leq$ 2, and mild purification. The HA-herbicide interaction products were prepared by shaking 60 mg HA in 100 ml of 0.5 mM herbicide solution for 24 h at room conditions, separation, retreatment, repeated $H_2O$ washings, and freeze-drying. Both the unreacted HAs and their interaction products with each herbicide were characterized for elemental and functional group composition, and by Fourier-transform infrared (FT-IR), fluorescence and electron spin resonance (ESR) spectroscopies. The $HA_A$ sample showed a little higher H, N and S% and C/N ratio and a little lower 0%, C/H and O/C ratios, phenolic OH and total acidity contents, $E_4/E_6$ ratio and free radical concentration, with respect to the $HA_B$ and $HA_C$ samples. The FT-IR and fluorescence spectra did not show relevant differences. These results indicated that the hops cultivation affected only secondarily the structural and functional chemical properties of HA in these soils. The three HAs also exhibited a similar behaviour in their interaction with each herbicide examined. All the interaction products showed an increase in the N % with respect to the unreacted HA, thus indicating the water-stable association of HA with the N-containing herbicide molecules. In any case, FT-IR spectra of the HA-herbicide interaction products showed an increase of the relative absorption intensity of aromatic and/or aliphatic groups and carboxylate groups and a decreased intensity of bands of carboxylic and, secondarily, phenolic OH groups, with respect to the unreacted HAs. An appreciable shift of diagnostic out-of-plane C-H vibration modes of the bipyridylium rings was observed in FT-IR spectra of diquat and paraquat adsorbed onto HA. A small decrease in fluorescence intensity and a shift of the emission maximum to a longer wavelength were observed for HA-atrazine and HA-simazine interaction products, with respect to unreacted HAs. A shift of the fluorescence emission maximum to a shorter wavelength was measured for HA-diquat interactions. Fluorescence excitation and synchronous-scan spectra of HAs remained almost unchanged after the interaction with any herbicide. An increase in organic free radical concentration of about 50 % of the original HA concentration was measured in all HA-herbicide interaction products. The results of spectroscopic analyses indicated that the main binding mechanisms occurring in the interaction involved the formation of ionic and charge-transfer bonds and, possibly, hydrogen bonds, with a secondary participation of fluorescent groups of HAs.*

## *Acknowledgement*

This research-work has been supported by an EEC-STEP Project grant, contract CT90-0025 (TSTS).

2.2.2.        *Association of Dichlorprop and Atrazine with Soil Organic Carbon*

# G. Riise & M. Nandrup Pettersen

Laboratory of analytical chemistry, Agricultural University of Norway,
PO Box 5026, N-1432 Ås, Norway

## Abstract

Different experimental approaches have been used in order to study sorption and desorption of [$^{14}$C]dichlorprop and [$^{14}$C]atrazine in Norwegian soil types containing different amounts of organic carbon at different pH values. Except for a silty loam soil, there was generally a good relationship between $K_{oc}$ and $K_{ow}$ coefficients for dichlorprop when accounting for pH values in the system (log $K_{oc} = 0.5 \log K_{ow} + 1.6$). Similar relationships were not obtained for atrazine, as the sorption seemed to be influenced both by the properties of soil organic carbon and other soil components.

Distribution coefficients ($K_d$ values) for dichlorprop indicate that the sorption and desorption process are not completely reversible or that desorption kinetics are slower than for sorption kinetics. Comparing $K_d$ values for the sorption and desorption of atrazine, it seemed that atrazine sorbed to organic rich soils was more readily desorbed than atrazine sorbed to a silt loam soil.

The percentage amount of dichlorprop and atrazine retained in soils after extraction with different leaching agents were highly related to content of soil organic carbon. Column experiments with silt loam showed that a large fraction of dichlorprop and only a small fraction of atrazine leached through the silt loam columns during the addition of artifical rainwater. An additional amount of atrazine was mobilized from the columns during the addition of $NH_4OAc$.

## Introduction

Soil organic carbon is often considered to be one of the single most important factors for the sorption of several pesticides in soil. Regardless of the source of organic carbon, partitioning coefficients for nonionic pesticides are often converted to $K_{oc}$ values based on the percentage weight of organic carbon in soil. For nonionic pesticides, relationships between $K_{oc}$ values and the hydrophobicity ($K_{ow}$) of the pesticides are well established (Briggs 1973, Karickhoff et al. 1979).

For ionizable pesticides, pH of the soil-water system influences the dissociation of the pesticides and thereby their sorption behaviour in soil. The weak acid dichlorprop (pKa app. 3) is for instance rather mobile at pH values where its anionic form predominates (Riise and Salbu 1992). Due to the low pKa value of the weak base atrazine (pKa app. 1.7), this herbicide is often considered to be neutral within normal pH ranges in soil.

In addition to sorption capacity, the strength of binding is important for the transport and biological uptake of pesticides retained in soil. When predicting the potential risk for pesticides to reach non target sites, information concerning leaching of pesticides in different soil types is essential.

The aim of this paper was to study the influence of soil organic carbon and pH on sorption and desorption processes of dichlorprop and atrazine in Norwegian soil types. In order to study the leachability of dichlorprop and atrazine, different extraction agents were applied in sequence (artificial rainwater, $NH_4$-acetate) in both batch and column experiments. [$^3$H]water is used as a tracer for the movement of water in the column experiments.

## Experimental

**Soil samples** were taken from the plough layer of cultivated soils (0-12 cm), air dried and passed through a 2-mm sieve before use. Organic carbon, determined in a LECO EC-12 (752-100) Carbon Determinator with an IR detector, varied from 30.4 to 3.3 %. Absorbance was measured at 465 and 665 nm ($E_4/E_6$ ratio) on humic material extracted into 0.05 M $NaHCO_3$. The organic rich soils (soil I and II) were characterized by higher $E_4/E_6$ ratio, cation exchange capacity and lower pH compared to the silt loam and loam soil (soil III and IV) (Table 1).

|                              | Soil I    | Soil II   | Soil III  | Soil IV |
|------------------------------|-----------|-----------|-----------|---------|
| Texture                      | org. rich | org. rich | silt loam | loam    |
| CEC mequiv 100 $g^{-1}$      | 99        | 55        | 23        | 18      |
| pH (0.01 M $CaCl_2$)         | 4.4       | 4.1       | 5.2       | 5.0     |
| ($H_2O$)                     | 4.6       | 4.3       | 5.5       | 5.3     |
| Base sat. (%)                | 49        | 34        | 65        | 49      |
| Organic C (%)                | 30.4      | 14.6      | 5.8       | 3.3     |
| E4/E6                        | 9.5       | 9.4       | 6.7       | 5.9     |

**Table 1 :** Physical and chemical properties of soils used in the experiments

The following **pesticides** were used in the experiments : [*U-phenyl ring* [14]C]dichlorprop (2-2,4-dichlorophenoxy)propionic acid) with a specific activity of 55.01 µCi/mg (Agrolinz 1989) and [*triazine*-2-[14]C]atrazine (2-chloro-4-ethylamino-6-isopropylamino-1,3,5-triazine) with a specific activity of 55.5 µCi/mg (CIBA-Geigy AS).

### *Sorption/desorption vs pH*

Centrifuge tubes of pollyallomer quality, containing 0.5 g soil and 5 ml 0.01 M $CaCl_2$ spiked with [[14]C]pesticide, were adjusted to different pH values by adding dilute HCl or NaOH. The suspensions were shaken for 24 h by an automatic rotating shaker. Prior to separation, the suspension was centrifuged for 30 min at 11000 x g. Desorption experiments were carried out on soil residues (not pH adjusted) from the sorption experiments by adding 5 ml 0.01 M $CaCl_2$ adjusted to different pH values by dilute HCl/NaOH. Sorption, µg pesticide per g soil ($C_s$), was estimated from the difference between the initial concentration ($C_o$) and the final concentration ($C_w$) in the solution after the equilibration. Single distribution coefficients ($K_d$) are given by the ratio : $C_s/C_w$. After removing 1 ml for scintillation counting, pH was measured in the soil suspension with an Orion model SA pH meter. The experiments were carried out in triplicates.

### *Extraction agents applied in sequence*

Batch experiments : Desorption of pesticides sorbed to soil was carried out by first extracting the soil residues with distilled water for 0.5 h. and secondly with 1 M Na-acetate (pH 7) for 2 h. The ratio between soil and liquid was 1:10. The solid and liquid phase were separated by centrifugation for 30 min at 11000 x g between each step.

Column experiments : 15 kBq [[3]H]water and 6.4 kBq [[14]C]pesticide were applied to the top of water saturated soil columns (i.d. 2.7 cm, I. 10 cm) containing silty loam from Mørdre (E-Norway) (Soil III). Artificial rainwater was continuously added by means of a peristaltic pump. After the dichlorprop and atrazine columns had received 300 ml and 250 ml, respectively (corresponding to approximately 530 and 430 mm precipitation), the artificial rainwater was replaced by 1 M $NH_4$-acetate. 15 kBq [[3]H]water was added immediatly prior to the shift in leaching solution in order to follow the transport of NH4OAc through the columns. Measurements of $^3$H and $^{14}$C were conducted on 1 ml aliquotes from the leachates.

### *Octanol-water extraction*

Water samples (0.2 mM KCl) adjusted to different pH values with dilute HCl/NaOH were spiked with [$^{14}$C]pesticides. Flasks containing 50 ml of the water samples and 5 ml of 1-octanol (PS-97 % purity, KEBO Lab) were shaken for 16 h and stored for 1-2 h prior to the separation of the two phases. The $^{14}$C activity was measured in the two phases.

### *$^{14}$C and $^3$H measurements*

The activities of $^{14}$C and $^3$H were measured with a Packard Tri-Carb 4530 liquid scintillation counter. The ratio between sample and scintillation cocktail was 1:10.

## Results and discussions

The hydrophobicity of ionizable pesticides can be rather dependent on pH. The weak acid dichlorprop has low $K_{ow}$ coefficients at pH values higher than 5 where its anionic form predominates. When dichlorprop is converted to its neutral form at decreasing pH values, the result is a rapid increase in hydrophobicity (Fig. 1). Compared to dichlorprop, the hydrophobicity of the weak base atrazine is less dependent on pH within normal pH ranges in soil. Based on $K_{ow}$ coefficients, nonspecific hydrophobic interactions with soil organic matter should be more important for atrazine than dichlorprop at pH values higher than pH 4.

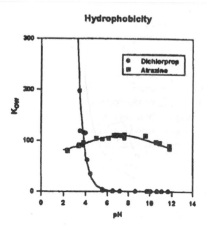

**Figure 1 :** $K_{ow}$ partitioning coefficients of dichlorprop and atrazine vs pH.

Soil to soil variations with respect to the different soil types capacity to sorb dichlorprop is reduced by converting $K_d$ coefficients to $K_{oc}$ coefficients (Fig. 2). Accounting for pH values in the system, the following relationship has been estimated for dichlorprop: $\log K_{oc} = 0.5 \log K_{ow} + 1.6$ (r2 = 0.986) (Riise and Salbu 1992). In general, the silt loam soil has a higher capacity to sorb dichlorprop compared to the other soil types, indicating that other soil properties than percentage weight of soil organic carbon can also influence the sorption behaviour (e.g. Fe, Mn oxides).

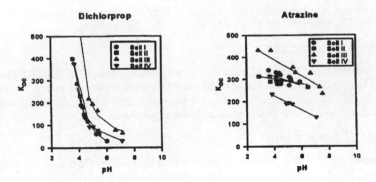

**Figure 2 :** $K_{oc}$ coefficients of dichlorprop and atrazine vs pH in different soil types

For atrazine, the relationships between $K_{oc}$ coefficients and pH show large soil to soil variations (Fig. 2). The organic rich soils (Soils I and II) with high $E_4/E_6$ ratios, have generally higher and less pH dependent $K_{oc}$ coefficients compared to the loam soil (soil IV). High $E_4/E_6$ ratios of humic material extracted into 0.05 M $NaHCO_3$ indicates the presence of smaller molecules which contain less C but more O, COOH groups, and total acidity than larger molecule with low $E_4/E_6$ ratios (Chen et al. 1977). This may indicate that the properties of the soil organic carbon influence the sorption pattern of atrazine, and soil organic carbon in peat areas are more reactive to atrazine compared to soil organic carbon in the loam soil. However, this theory does not explain the high sorption of atrazine for the silt loam soil, especially at low pH values. This can either be explained by the presence of soil organic carbon that efficiently sorb atrazine, or the presence of other soil components that are responsible for the sorption pattern.

The sorption and desorption curves for dichlorprop are not overlapping for either of the soils. The deviations are most significant for organic rich soils (soil I and II) at high pH values (Fig. 3). This can indicate that the desorption kinetics are slower than the sorption kinetics or that the sorption and desorption processes are not completely reversible. Similar results were obtained for atrazine in the silt loam soil. In the organic rich soils, however, atrazine seemed to be more readily desorbable (Fig. 3). Despite these results, the amount of atrazine retained after the desorption was higher than for dichlorprop in similar soil types.

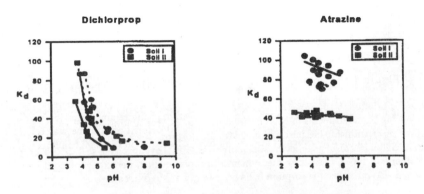

**Figure 3 :** Sorption (solid curves) and desorption (dotted curves) of dichlorprop and atrazine vs pH

The percentage amounts of dichlorprop and atrazine retained after the extractions with distilled water and NH4OAc were dependent on content of soil organic carbon (Fig. 4). In the soil with the highest amount of organic carbon (30.4 % org. C), 40 % and 70 % of dichlorprop and atrazine were retained, respectively. A larger fraction of dichlorprop, compared to that for atrazine, was extracted by the use of NH4OAc in the organic rich soils, indicating that a relatively large fraction of dichlorprop is "exchangeable" or can be easily mobilized by changes in ionic strength.

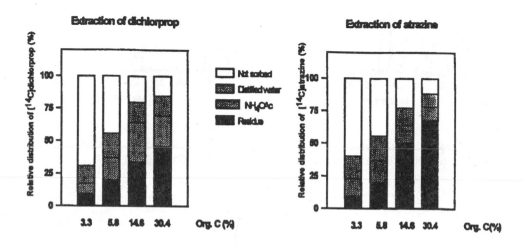

**Figure 4 :** Sequential extraction of dichlorprop and atrazine from different soil types

Results from the column experiments with the silt loam soil showed that the major fraction of dichlorprop was leached during the addition of artificial rainwater (81 %), and only 13 % was retained in the column at the end of the experiments (Fig. 5).

Atrazine showed a different leaching pattern compared to dichlorprop (Fig. 5). Only a small amount of atrazine leached through the column during the addition of artificial rainwater (4 %). Furthermore, the atrazine peak appeared at the same time as the rainwater peak. An additional amount of atrazine was mobilized during the addition of NH4OAc, first a small peak at the same time as the NH4OAc peak, and then an long tail. This can indicate that a fraction of the retained atrazine was associated with exchangeable sites in soils. However, as much as 80 percent of atrazine was retained in the column at the end of the experiment. Thus, the mobility of atrazine in silt loam columns is rather low.

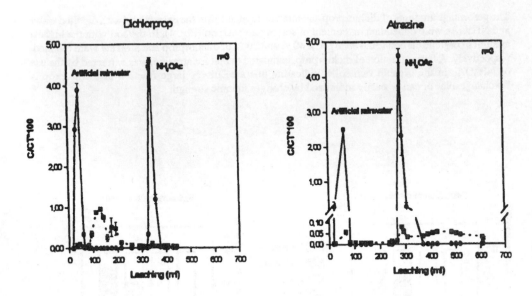

**Figure 5 :** Sequential leaching of dichlorprop and atrazine through silt loam columns (soil III), [³H]water solid lines and [¹⁴C]pesticides dotted lines (n=3)

## Acknowledgments

G.R. thanks the The Norwegian Research Council for Science and Humanities for a Research Fellowship and personell at the Laboratory for analytical chemistry and at the Norwegian Plant Protection Institute for all help during the survey.

## References

☐   Agrolinz 1989. Information from the producer.

☐   Briggs, G.G., 1973. A simple relationship between soil adsorption of organic chemicals and their octanol/water partitioning coefficients. *Proceedings 7th British Insecticides and Fungicide Conference*, 1: 83-86.

☐   Chen, Y., N. Senesi and M. Schnitzer, 1977. Information provided on humic substances by $E_4/E_6$ ratios. *Soil Sci. Soc. Am. J.*, 41: 352-358.

☐   CIBA-Geigy AS. Information from the producer.

☐   Karickhoff, S.W., D.S. Brown and T.A. Scott, 1979. Sorption of hydrophobic pollutants on natural sediments. *Water Res.*, 13: 241-248.

☐   Riise, G. and B. Salbu, 1992. Mobility of dichlorprop in the soil-water system as a function of different environmental factors. I. *A batch experiment. Sci. Tot. Environ.* 123/124: 399-409.

## 2.2.3.    Influence of Repeated Atrazine Treatments on the Adsorption of the Pesticide on a Brown Soil

# C. Bernhard-Bitaud [1,2], M. Schiavon [1], F. Andreux [2,3]

[1]   Laboratoire de Phytopharmacie, I.N.P.L.- E.N.S.A.I.A., 2, av. de la Forêt de Haye
      54505 Vandoeuvre Cedex, France

[2]   C.N.R.S. Centre de Pédologie Biologique, 17, rue Notre-Dame des Pauvres,
      B.P.5, 54501 Vandoeuvre Cedex, France

[3]   Centre des Sciences de la Terre, Université de Bourgogne, 6, bd Gabriel
      21000 Dijon, France

## Abstract

Two experiments have been performed in order to evaluate the influence of pluriannual atrazine treatments on adsorption properties of a brown soil from the Lorraine plain, eastern France. The study soil was formerly developed under a permanent grassland, and had been cultivated with maize since 1989.

The first experiment under actual field conditions showed that atrazine treatments modified the adsorption capacity of the soil. A strong correlation between the soil organic matter (SOM) content and the adsorption of atrazine was observed. The decrease of SOM content with cropping time was emphasized in the absence of weeds, as a result of the specific effect of the herbicide. Not only the amount of atrazine adsorbed, but also the adsorption kinetics, were affected by the treatment.

In the second experiment, adsorption assays were carried out on samples from plots with treated and untreated soils, and with controlled SOM content. No effect of the herbicide could be demonstrated after one year, whereas the influence of the SOM quality was questioned.

## Key words

*Adsorption, atrazine, kinetics, maize, soil organic matter.*

## Introduction

The herbicide atrazine (2-chloro 4-ethylamino 6-isopropylamino 1-3-5-s-triazine) has been used for weed control in maize cropping since 1959. There is an abundant literature showing that its fate in soil depends mainly on soil sorbing capacity. As maize cropping and atrazine treatment are often conducted repeatedly on the same soil from one year to the next, it was interesting to investigate the effects of previous atrazine treatments on further adsorption of this herbicide on a soil from a maize field. The present communication discusses results of adsorption measurements on soils from a field experiment and from a semi-controlled plot combining the effects of soil organic matter (SOM) and atrazine treatment.

## Materials and methods

### *Field study*

The study soil was a brown soil from the Lorraine region (France) (Florentin, 1978), which had been cultivated with maize since 1989. One area had been cultivated without chemical weed-control and the other area with atrazine

treatment. Soils with silty clay and silty clay loam texture predominated in the untreated and treated areas, respectively. The two parts of the field had similar SOM content (Table 1).

| | % Clay | % Silt | % Sand | % C * | % N * | C/N * | pH (H2O) * | pH (KCl) * |
|---|---|---|---|---|---|---|---|---|
| Untreated | 50 | 49 | 1 | 4.8 ± 0.6 | 0.49 ± 0.06 | 9.7 ± 0.2 | 5.2 ± 0.2 | 4.5 ± 0.2 |
| Treated | 35 | 60 | 5 | 4.7 ± 0.4 | 0.51 ± 0.05 | 9.2 ± 0.1 | 5.5 ± 0.2 | 5.0 ± 0.3 |

\* Mean values of five samples.

**Table 1** : Physical and chemical characteristics of the brown soil in 1989

The surface soil (0-0.20 m) was sampled in March 1990 and March 1992, before the maize was sown. In the two areas, five samples were collected along a transect, at about 10 m from each other. Each sample was then air-dried, sieved at 2 mm, and analysed. The carbon (C) and nitrogen (N) content was measured by combustion in a Carlo-Erba MOD 1106 analyser. In each area, a particle-size fractionation at 50 μm under water was carried on three selected samples. The coarse fraction (50-2000 μm) was considered to contain the non humified SOM, and the fine fraction was enriched in humified material. The humification degree Tc was measured as the percentage of the total organic C present in the 0 - 50 μm fraction (Andreux et al., 1980).

The measurement of $^{13}C$ isotope abundance made it possible to calculate $C_{C4}$ the proportion of soil C derived from residues of plants with a C4-type photsynthesis in a given soil (Balesdent et al., 1987). In this case, the C4 plants are mainly represented by the maize crop, but also is the *Amaranthus retroflexus*, which is one of the weeds identified in the untreated area. $C_{C4}$ is expressed in g of C from C4 plants per 100 g of soil.

### *Semi-controlled experiment*

Surface soil (0-0.20 m) from the untreated area was collected in July 1992 and stored in duplicate bottomless-containers (0.50 m long, wide and deep) placed in soil in the neighbourhood of the field. Four different treatments were applied to the soil in these containers :

(a)    Chemical treatment with atrazine (1,5 kg ha$^{-1}$);

(b)    Hand weeding;

(c)    Atrazine treatment and incorporation of air-dried weeds (800 g m$^{-2}$)

(d)    No treatment (spontaneous weeds).

Samples from each container were collected in July 1992 and July 1993, air-dried, and sieved to 2 mm.

### *Atrazine adsorption studies*

Adsorption kinetics were carried out on samples from the two areas, using a batch equilibration method. Average samples from each area were prepared, by mixing equal weights of the five samples, air-dried and sieved at 2 mm. They were crushed, passed through a 50 μm sieve, then mixed at constant temperature of 20°C and under constant shaking with a 0.01 M $CaCl^2$ solution of $^{14}$C-ring labelled atrazine, using a soil/solution ratio of 1 : 5 (w/v). The atrazine concentration in the solution was 15 mg $L^{-1}$. At 0, 10, 30 mn, 1, 3, 7, 24, 48 h, three aliquotes of the sus-

pension were sampled, centrifuged at 4000 $g$, and the radioactivity present in the supernatant was measured by liquid scintillation counting, using a TRICARB 460 CD.

Experimental values were adjusted to an hyperbolic model (Jamet *et al.*, 1985) :

$$Q_{ads}(t) = \frac{Q_{max} \cdot t}{k + t} \qquad \text{(Eq. 1)}$$

The values of $Q_{max}$ and k were evaluated by the Gauss-Newton method from SAS-STAT software (Tab. 2). Equilibrium was considered to be reached when $Q_{ads} \geq 99\%$ of $Q_{max}$. The values of $Q_{eq}$ and $t_{eq}$ were calculated with the model.

The kinetic study revealed that the amount of atrazine adsorbed after 24-hour-shaking, $Q_{ads}$ (24h), was a satisfactory approximation of $Q_{max}$. This value was measured as described for the kinetic study on the samples from the semi-controlled experiment.

# Results

## *Soils from the field experiment*

The analyses of the SOM presented on Table 2 showed that (i) the treated area was poorer in C at each date than the untreated one, and the soils collected in 1992 were poorer than those collected in 1990 in each area; (ii) the proportion of C from C4 plants increased with time, representing from 4 to 8% of the total C content, but a very low difference was observed at each date between the two areas; (iii) the C/N ratio seemed to be higher in the untreated area than in the treated one, and higher in 1992 than in 1990, but with very low differences; (iv) the humification degree increased in the two areas, but in a more significative way in the untreated one.

|      |           | % C | % C<sub>C4 plants</sub> | C/N | Tc |
|------|-----------|------|------------------|------|------|
| 1990 | Untreated | 4.8 ± 0.5 | 0.21 | 9.8 ± 0.2 | 44 % ± 2 |
| 1990 | Treated   | 4.2 ± 0.6 | 0.26 | 9.5 ± 0.3 | 52 % ± 6 |
| 1992 | Untreated | 4.3 ± 0.3 | 0.31 | 9.9 ± 0.2 | 59 % ± 1 |
| 1992 | Treated   | 4.1 ± 0.2 | 0.34 | 9.8 ± 0.3 | 54 % ± 1 |

**Table 2 :** Changes in the characteristics of soil organic matter
in the surface layer of the brown soil

The changes with time of the quantity of atrazine adsorbed (Qads) on the soil samples collected in 1990 and 1992 in the two parts of the field experiment are shown in Figure 1.

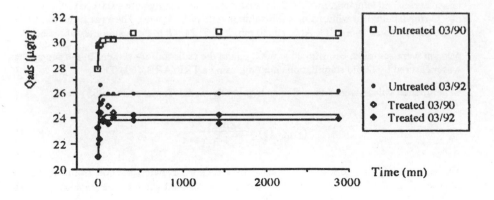

**Figure 1 :** Changes on the adsorption kinetics of atrazine on the brown soil.

For all soil samples, the equilibrium was reached after a short phase of fast adsorption. The values of the parameters $Q_{eq}$ and $t_{eq}$ varied from 23.7 µg/g in the treated area in 1992 to 30.0 µg/g in the untreated area in 1990, and from 5 mn in the treated area in 1990 to 17 mn in the treated area in 1992, respectively (Table 3).

|      |           | Qads (24 h) | Qmax | k    | Qeq  | teq (mns) |
|------|-----------|-------------|------|------|------|-----------|
| 1990 | Untreated | 30.8        | 30.3 | 0.09 | 30.0 | 9         |
| 1990 | Treated   | 24.2        | 24.2 | 0.04 | 24.0 | 5         |
| 1992 | Untreated | 25.9        | 25.9 | 0.18 | 25.6 | 15        |
| 1992 | Treated   | 23.5        | 23.9 | 0.14 | 23.7 | 17        |

**Table 3 :** Measured adsorption parameter (Qads), adjustment parameters (Qmax and k)
to the hyperbolic model, and calculated equilibrium parameters (Qeq and teq),
in the case of soils from the field experiment.

In each part of the field, the equilibrium adsorption values were higher in 1990 than in 1992, and at each date, were higher in the untreated than in the treated area. Moreover, the shapes of the curves were different and showed that the asymptote was reached faster in the soils collected in 1990 than in those collected in 1992.

*Soils from the semi-controlled experiment*

In general, little difference could be observed between the different assays. At the begining of the experiment, the soil C content was about 10% lower than in the soils from the field experiment. Only in the soils with spontaneous or added weeds, the C content increased with time in a significant way (95%).

In the presence of weeds, especially when no atrazine was added, a slight decrease in the adsorption of the pesticide appeared (Table 4). In the soils without weeds, no statistically significant changes of C content was observed. The adsorption of atrazine showed a tendency to decrease, especially when atrazine was added.

| Treatment | | July 1992 | | July 1993 | |
|---|---|---|---|---|---|
| | | % C | Qads (24h) | % C | Qads (24h) |
| Atrazine, no weeds | (a1)* | 4.4 | 27 | 4.2 | 26 |
| | (a2) | 4.3 | 28 | 4.2 | 25 |
| No atrazine, no weeds | (b1) | 4.2 | 29 | 4.3 | 30 |
| | (b2) | 4.3 | 26 | 4.1 | 24 |
| Atrazine + weeds | (c1) | 4.2 | 29 | 4.3 | 29 |
| | (c2) | 4.4 | 27 | 4.6 | 26 |
| No atrazine + weeds | (d1) | 4.3 | 26 | 4.3 | 24 |
| | (d2) | 4.1 | 28 | 4.4 | 26 |

\* The two lines correspond to two different containers.

**Table 4** : Carbon content and adsorption parameter Qads (24h) measured from the semi-controlled device.

## Discussion

It is now well known that the adsorption of atrazine on soil is correlated to the SOM content (e.g. Bailey and White, 1970 ; Barriuso and Calvet, 1992). The linear correlation observed in the present study between C content and amount of atrazine adsorbed after 24 hours ($r^2 = 0.993$), was in close agreement with this general rule.

The lower adsorption values observed in the samples collected in the same area of the field in a two year interval can be related to the decrease of C content with cropping time. Such a decrease has been reported by several authors during the first years of cropping in soils formerly developed under grass-land (Martell and Paul, 1974 ; Tiessen et al., 1982 ; Mann, 1986). The higher C content in the area which did not receive atrazine can be explained by either higher returns from maize, or lower mineralisation of SOM, or returns from weeds, or a contribution of all these factors.

$\delta^{13}$C measurements (Table 2) made it possible to establish that returns from C4 plants represented a low proportion of soil C after two years, and that the difference between the two areas regarding to the returns from these plants was far below the difference on total C content. Therefore, the C4 plants, and mainly the maize, were not responsible for the difference in C content between the treated and untreated areas.

It was observed that the treated area was under a more loamy soil than other areas of the field. Loamy soils are known to be more affected by SOM decrease under cultivation than clay soils (Tiessen and Stewart, 1983). Representative samples of the clayey area showed a C content of 3.6 % under treatement, and of about 4.2 % without treatement, which allowed to consider that the variability in soil textural properties could not be responsible for the differences in C content. Nevertheless, a stimulative effect of atrazine on the mineralizing microflora cannot be excluded; but results from literature are contradictory.

In these conditions, returns from weeds (excepted perhaps *Amaranthus retroflexus*) were necessarily suspected to have an important role in the change of C content. Maize and weed samples collected in September 1992 on three 1 m$^2$ -spots showed that weeds represented between 17 and 45 % (fresh weight) of the total aerial plant material in the untreated area, but only from 2 to 9 % in the treated area (Table 5). Moreover, the untreated area was hoed during the summer, which allowed a second growth of weeds after a first incorporation of organic matter in soil. Such a difference may therefore explain the difference in total C content between the two areas.

| | Not treated area | | | Treated area | | |
|---|---|---|---|---|---|---|
| Sample | 1 | 2 | 3 | 1 | 2 | 3 |
| % fresh weight | 17 | 45 | 19 | 9 | 2 | 5 |

**Table 5 :** Weed percentage (fresh weight basis) of total plants collected in three different spots of the study areas.

As the weeding practice induced lower returns of organic matter in soil, and a faster decrease of C content, a subsequent decrease of the adsorption potential measured by $Q_{eq}$ also occured in the treated area with respect to the untreated area.

The $t_{eq}$ value, which is representative of the soil reactivity, was lower in the soils collected in 1990 than in those collected in 1992. This could not be explained by the difference in C content since the soil of the treated area collected in 1990 and the soil of the untreated area collected in 1992 had similar C content but different $t_{eq}$ values. Other authors have shown that the different components of SOM have different affinities to atrazine (Dunigan and Mac Intosh, 1971; Schiavon et al., 1990). The incidence of the humification degree was stressed by Walker and Crawford (1968) and Paya-Perez et al. (1992) and more recently by Dousset et al. (1994). In the present study, the humification degree increased in both parts of the field between 1990 and 1992, probably as a result of cultivation (Tiessen and Stewart, 1983 ; Dalal and Mayer, 1986). In 1990, the humification degree was higher in the treated than in the untreated area, whereas the opposite occured in 1992. Such a change was probably in relation with the growth of weeds which represented a material of low lignin content and high content of easily degradable material. Thus, weeding would induce a slower increase of the humification degree. The fact that $t_{eq}$ increased strongly when the humification degree increased suggests that SOM fractions of lower humification degree could be especially active during the first phase of the herbicide adsorption.

This semi-controlled experiment was designed to investigate whether the previous atrazine treatment of soil could influence the adsorption of this herbicide during further treatments. The lack of specific effect of the previous atrazine treatment could be explained by the recent start of the experiment. Similarily, it was noticed that the added organic matter from the weeds had no positive effect on atrazine adsorption, probably because weeds were incorporated only five months (in February) before the second sampling. They were not enough decomposed to exhibit any affinity for the herbicide. On the contrary, a slight decrease of adsorption was noticed, suggesting a hindrance of reactive sites or a higher mobility of atrazine sorbed on disolved organic matter, due to the presence of this recently incorporated organic matter.

## *Conclusion*

The field experiment conducted in this study showed that the main effect of previous atrazine treatment was the lower amount of organic returns due to weeding. From the first year, the presence of herbicide emphasized the decrease of SOM content due to cultivation. Such a decrease of SOM content resulted in a decrease of the adsorption coefficient of atrazine. In the treated area, an increasing mobility of atrazine was liable to take place, with higher pollution risks for underground waters. However, the strong difference observed in 1990 between the treated and the untreated areas became narrower in

1992, suggesting that further changes can be expected. Moreover, if increasing humification degree is confirmed as slowing down atrazine adsorption, any parameter affecting humification can be expected to affect the fate of atrazine in soil.

## Acknowledgment

The authors thank J.-M. PORTAL and B. GERARD from the Centre de Pédologie Biologique (C.N.R.S.) for carbon analysis, H.-P. GUIMONT from the Domaine Experimental de la Bouzule (E.N.S.A.I.A.) for managing field experiments, and P. JAME from the Service Central d'Analyses (C.N.R.S.) for $\delta^{13}$C measurements.

## References

□ ANDREUX F., BRUCKERT S., CORREA A., SOUCHIER B., 1980. Sur une méthode de fractionnement physique et chimique des agrégats des sols : origines possibles de la matière organique des fractions obtenues. *C. R. Acad. Sc. Paris* (291 D), pp. 381 - 384

□ BAILEY G. W., WHITE J.L., 1970. Factors influencing the adsorption, desorption and movement of pesticides in soil. *Residue Reviews* (32), pp. 29 - 92

□ BALESDENT J., MARIOTTI A., GUILLET B., 1987. Natural $^{13}$C abundance as a tracer for studies of soil organic matter dynamics. *Soil Biol. Biochem.* (19, 1), pp. 25-30

□ BARRIUSO E., CALVET R., 1992. Soil type and herbicides adsorption. *Intern. J. Environ. Anal. Chem.* (46), pp.117-128

□ DALAL R.C., MAYER R.J., 1986. Long term trends in fertility of soils under continuous cultivation and cereal cropping in Southern Queensland. *Aust. J. Soil Res.* (24), pp. 265-300

□ DOUSSET S., MOUVET C., SCHIAVON M., 1994. Sorption of terbuthylazine and atrazine in relation to the physico-chemical properties of three soils. *Chemosphere* (28, 3), pp.467-476

□ DUNIGAN E.P., MAC INTOSH T.H., 1971. Atrazine-soil organic matter interactions. *Weed Science* (19, 3), pp. 279-282

□ FLORENTIN L., 1978. Carte des sols du domaine experimental de la Bouzule. ENSAIA, unpublished document

□ JAMET P., THOISY J.-C., LAREDO C., 1985. Etude et modélisation de la cinétique d'adsorption des pesticides dans le sol. *Les Colloques de l'INRA* (31), pp. 135-146

□ Mac GLAMERY M.D. , SLIFE F.W., 1966. The adsorption of atrazine as affected by pH, temperature, and concentration. *Weeds* (14, 3), pp. 237-239

□ MANN L.K., 1986. Changes in soil carbon storage after cultivation. *Soil Science* (142, 5) pp. 279 - 288

□ MARTELL Y.A., PAUL E.A., 1974. Effects of cultivation on the organic matter of grassland soils as determined by fractionation and radiocarbon dating. *Can. J. Soil Sci.* (54), pp. 419 - 426

□ PAYA-PEREZ A.B., CORTES A., SALA M.N., LARSEN B., 1992. Organic matter fractions controlling the sorption of atrazine in sandy soils. *Chemosphere* (25, 6), pp. 887-898

□ SCHIAVON M., BARRIUSO E., PORTAL J.M., ANDREUX F., BASTIDE J., COSTE C., MILLET A., 1990. Etude du devenir de deux substances organiques utilisées dans les sols, l'une massivement (l'atrazine) et l'autre à l'état de trace (le metsulfuron-métyl), à l'aide de molécules marquées au $^{14}$C. Ministère de l'Environnement, SRETIE/MERE, 7219, Opération 237 01 87 40131, 75 p.

❑   TIESSEN H., STEWART J.W.B., 1983. Particle-size fractions and their use in studies of soil organic matter, part II : Cultivation effects on organic matter composition in size fractions. *Soil Sci. Soc. Am. J.* (47), pp. 509-514

❑   TIESSEN H., STEWART J.W.B., BETTANY J.R., 1982. Cultivation effects on the amount and concentration of carbon, nitrogen and phosphorus in grassland soils. *Agron. J.* (74, 5), pp. 831 - 835

❑   WALKER A., CRAWFORD D.V., 1968. The role of organic matter in adsorption of the triazine herbicides by soil. *IAEA Symposium on Isotopes and Radiation in Soil-Organic-Matter studies*, Vienna, pp. 91-103

## 2.2.4.    An Availability Test of Pesticides from Soils : Comparison with their migration potentiality through soil layer chromatography

# R. Deleu [1], A. Hormatallah [2], A. Copin [1] & Ph. Drèze [1]

[1]    Faculté des Sciences Agronomiques de Gembloux, Belgium

[2]    Institut Agronomique et Vétérinaire Hassan II, Morocco

The role of the adsorption phenomena on the behaviour of a pesticide in the environment (Leaching, volatilization, run-off, metabolism, bioaccumulation, ...) and on the efficiency of its action explains the interest shown by many researchers to evaluate, in  laboratory conditions, the nature and importance of this adsorption and its reversibility (desorption).

Determining the parameters of the Freundlich or derivative equations remains the most utilized method (Barthelemy et al. 1988, Barriuso et al 1991, Calvet 1989, Green and Karickhoff 1990, Jamet et al. 1988). The OECD normalized this method under guideline 106. It cannot be applied to products with a low water solubility and its precision is limited when either adsorption or desorption is weak: several authors (Green et al 1980) state getting an acceptable intensity estimation and reversibility of this process through small soil columns. In both cases, the used material, especially the invested time and the difficulty in automatizing manipulation remain a hindrance.  The technique of soil layer chromatography based on Helling and Turner's work (1968) seems to alleviate part of the inconveniences linked to both previous techniques.  It is rapid and allows comparison, in homogeneous conditions, of several components of a same soil layer (Jamet and Thoisy-Dur 1986). Its use however requires the usage of [14]C radiolabelled pesticides and is also inefficient when used in sandy soils.  In those conditions, we believe it to be important to develop a new technique : the "Availability Test" (Deleu et al 1988).  Such a test does not require radioactive molecules, like the chromatography on soil layer.  The method is fast because it associates, in one manipulation, the adsorption and the desorption. Opposed to isotherms, the desorption phase is unique.  This technique permits humid soils. The results obtained in the lab are valid based on field samples and analyzed through extraction of organic solvent (Total residue) and different volumes of CaCl2 (available residue).

## Availability test

### Test description

The availability test is conducted by extracting a soil mass spiked with pesticide, through different volumes of CaCl$_2$ 5.10$^{-3}$M. The analysis of the different solutions allows the evaluation of the pesticide concentrations  for several soil/solution ratios.  The different stages for the availability value use atrazine as model.

Four initial concentrations are applied to the soil: 0.43 - 0.94 - 4.51  and 10.61 ppm (µg/g). X being the percentage of dry matter in the different soil suspensions  and Y being the pesticide concentration in soil solution, it is possible to relate an equation type $Y = A\, e^{BX}$ for the four predefined concentrations. When using atrazine, those equations are:

$$0.43 \text{ ppm}: \ Y = 0.03513*e^{0.03635\,X}$$
$$0.94 \text{ ppm}: \ Y = 0.07869*e^{0.0375\,X}$$
$$4.51 \text{ ppm}: \ Y = 0.41248*e^{0.03694\,X}$$
$$10.61 \text{ ppm}: \ Y = 8.8793*e^{0.03704\,X}$$

The coefficients A and B of those relations come each from the initial concentration. They are defined through atrazine by:

$$A = 0.00906 + 0.08296\ X$$
$$B = 0.03654 * X^{0.00624}$$

where X is the total concentration of atrazine on the soil (total residue).

By replacing the values of A and B in the general equation, it is possible to calculate and draw the curves shown in figure 1.

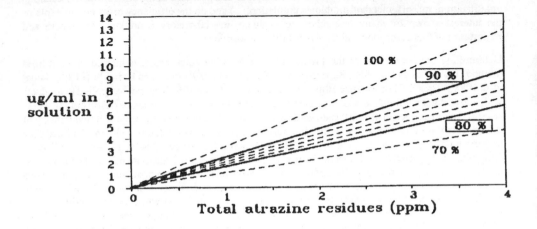

**Figure 1** : Relationships between total residues in soil (ppm)
% dry soil and atrazine concentrations in solution (µg/ml)

By means of this chart, it is possible knowing the atrazine concentration in soil (total residue obtained by chemical analysis) and the percentage of soil humidity, to evaluate the herbicide concentration level in the soil solution, potentially available for the transfer (run-off, leaching) for the vegetal and telluric organisms (metabolism, bioaccumulation).

To confirm the lab results and verify the reality of the availability model, we applied the test to natural field samples with known total residue compounds through chemical analysis. We extracted isoproturon and chlortoluron soil samples with different volumes of CaCl2. In those parcels, with any herbicide, the obtained relations for the field samples are less than 10% off from the surveyed expected areas. For isoproturon "aged" residues (54 days after application), the predicted model continues reproducing the plain field reality. This shows that the availability test translates properly the adsorption and desorption phenomena of the compounds with soil colloids.

The availability integrating the adsorption and desorption phenomena also depends on the percentage of organic C in the soil and, in the specific case of acid pesticides (sulfonylurea, imidazolinones, ...), is influenced by the soil's pH.

## *Influence of the organic C percentage in soils*

The availability test has been used on 5 soils (silt loam) with organic C ranging from 0.6 to 3.6 % (0.58 - 1.00 - 1.17 - 1.50 and 3.55). The pH of those soils stays relatively close (between 6.8 and 7.7)

With isoproturon, relationships have been established for four initial herbicide concentrations (0.03 - 0.435 - 4.035 and 16.035 µg/g) between the percentage of organic C in the 5 soil types and the pesticide solution concentration. Figure 2 shows the relationship for the 16.035 ppm concentration. However, the availability of weak acid molecules (with pKa ranging between 3 and 6 like the triasulfuron, the metsulfuron-methyl, the tribenuron-methyl, the chlorsulfuron and the imazapyr) is high on those same 5 soils. It ranges between 70 and 100% of the initial quantity, but is not influenced by the organic C percentage.

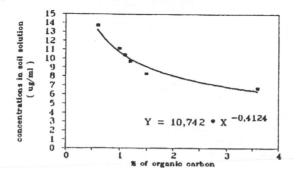

**Figure 2 :** Availability of isoproturon - Relationship with % organic C -
initial concentration 16.035 ppm

## *Influence of the pH in soils*

A pesticide with a higher rate of pKa than the pH in soil exists mainly in a neutral form. This form enhances its adsorption on the negative soil colloids. This aspect contains a certain reversibility, their chemical remanence and biological persistence continue through time. Their ecotoxicological behaviour (migration, dispersion, ...) becomes fundamentally modified compared to the non adsorbed anionic form.

Three soils demonstrated the pH influence on the availability of triasulfuron (pKa = 4.5) and imazapyr (pKa = 3.6). Those soils, with organic C almost at 1, have pH values of 4.2 - 5.5 and 6.8.

The imazapyr quantity fixed on soil with pH equal to 4.2 is more important (50% average depending on the soil/solution ratio) than on the other 2 soils (20% average). Up to 50% of the triasulfuron (pKa = 4.5) is withheld in the first soil. In the soil with a pH = 5.5, the value ranges between 20 and 30% of the initial quantity, while in the last soil, no adsorption of herbicide occurs (between 0 and 5 %). The influence of the soil pH on the availability of the acid components is therefore confirmed (Barriuso and Calvet 1992).

Applying the availability test on 14 pesticides allowed us to draw figure 3. It shows the percentage of pesticide in soil solution for an initial concentration of 0.1 µg/g with a soil humidity of 20%. The chosen soil has a 6.8 pH and 1% of organic C.

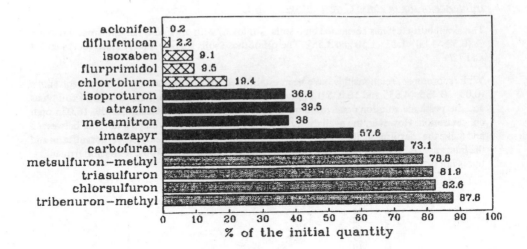

**Figure 3 :** Availability of 14 pesticides (in %) for an initial concentration
of 0.1 ppm and 20 % of humidity

This figure demonstrates that the pesticide availability is not necessarily linked to their solubility in water. For example, the metamitron, with a 1800 ppm solubility, only shows an availability of 38%, while the triasulfuron, with 1500 ppm solubility reaches 82%.

The atrazine and metamitron, with a solubility of 30 ppm for the former and 1800 ppm for the latter, do not differ in availability. This parameter integrates efficiently the intensities of adsorption, desorption and its hysteres phenomenon.

## _Comparison between chromatography on soil layer and availability test_

Soil thin layer chromatographies have been executed on 125 μm sieved soil with a technique inspired from Helling and Turner (1968). The pesticide's Rf (X) have been compared to their availability value, i.e. the percentage of pesticide in solution (Y). A positive correlation between those 2 parameters exists in each case (see figure 4 for the case with soil of pH 6.8 and 1% of organic C). This significant correlation proves the value of the availability test as an indicator for the moving potential of a pesticide.

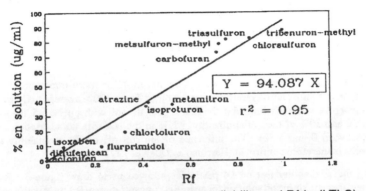

**Figure 4 :** Relationship between availability and Rf (soil TLC)
for 0.1 ppm and 20 % of humidity

# Conclusions

This availability test makes it possible :

1.  to evaluate quickly and simply the pesticide concentration in the soil solution. It depends, together with the basic soil characteristics, on the total pesticide residue and the soil humidity.

2.  to exclude from the experiment constraints highlighted by the thin soil layers, soil columns or adsorption isotherms.

3.  to make an objective estimate of the pollution risks of ground and surface waters.

4.  to know the components concentration in the soil solution, with a phytotoxic test, thereby removing the agronomist's concerns when evaluating a pesticide's toxicity in a culture.

# Acknowledgments

The authors thank the IRSIA - Brussels for subsidising this research.

They would also like to thank Rhone-Poulenc, Dow Elanco, Ciba-Geigy and Dupont de Nemours for providing them with the radiolabelled pesticides which simplified this study a great deal.

# Bibliography

□   Barthélemy J.P., Baudhuin Th., Bourdeaud'Huy J.L., Copin A., Deleu R., Hoyoux D. (1988)
    Détermination des isothermes d'adsorption et de désorption des pesticides.
    In : *Methodological aspects of the study of pesticide behaviour in soil*. INRA Versailles -
    191-194

□   Barriuso E., Andreux F., Schiavon M., Portal J.M. ( 1991 )
    Intérêts et limitations des méthodes de séparation des micropolluants organiques des sols -
    *Science du Sol* Vol 29, 310-320.

□   Barriuso E., Calvet R. (1992)
    Soil type and herbicides adsorption
    *Intern. J. Environ. Anal. Chem.* Vol. 46 117-128

□   Deleu R., Copin A., Barriuso E., Salemblier J-F. ( 1988 )
    Test pratique de détermination des phénomènes d'adsorption-désorption des pesticides d'un
    sol. - In : *Methodological aspects of the study of pesticide behaviour in soil*. INRA Versailles,
    50-55.

□   Green R.E., Karickhoff S.M. ( 1990 )
    Sorption Estimates for Modeling - In: *Pesticides in the soil environment* - Soil Science Society
    of America Chap.4, 79-101.

□   Green R.E., Davidson J.M., Biggar J.W. (1980)
    An assessment of methods for determining adsorption-desorption of organic chemicals.
    In: *Agrochemicals in soils* - Pergamon Press - 73-82

□   Helling C.S., Turner B.C. ( 1968 )
    Pesticide mobility : determination by soil thin-layer chromatography - *Science* 162, 562-563.

❑   Jamet P., Thoisy-Dur J-Ch. ( 1986 )
     Evaluation et comparaison de la mobilité de différents pesticides par chromatographie sur
     couche mince de sol.

❑   Jamet P., Copin A., Deleu R., Thoisy-Dur J.Ch., D; Hoyoux, Carletti G. ( 1988 )
     Aspects méthodologiques de l'étude de l'adsorption et de la désorption des herbicides dans le
     sol. - *Proc. EWRS Symp.* *"Factors affecting herbicidal activity and selectivity"* pp 263-268.

2.2.5.     *Bioavailability of 2,4-D in the soil. A modeling approach.*

## B. Lagacherie and G. Soulas
### Technical assistance : Nadine Rouard

INRA, Laboratoire de microbiologie des sols
17 rue Sully, 21034 DIJON-cedex France

## Introduction

The environmental fate of pesticides is mainly determined by their capacity to be sorbed onto soil constituents as well as their susceptibility to be degraded. These two processes are often studied independently but their real impact need to be examined from a more integrated point of view taking into account interactions between the biological and physico-chemical phenomena. As a rule, it has been shown that adsorption on soil colloids controls biological availability of the pesticide molecule and limits its degradation by the soil microflora (Ogram et al., 1985; Sprankle et al., 1975 ; Moyer et al., 1972; Weber and Coble, 1968; Kerhoas and Thoisy-Dur, 1988). However, accelerated breakdown has been found in soils with a high content of organic matter that may promote activity of the soil microbial degraders or may be directly responsible for chemical hydrolysis of the pesticide. Moreover, bioavailability may also directly affect the physiological route by which a pesticide is degraded. This is happening when metabolizing and cometabolizing microorganisms are concurrently acting upon the chemical. An increased concentration in the soil solution gives a selective advantage to those of the microorganisms that are able to use the pesticide as sole source of carbon and energy. So the balance between cometabolism and metabolism may be affected by all physical processes that modify the distribution of the pesticide between the different phases of the soil and its fraction amenable for biodegradation. The aim of this work was to take advantage of the modelling approach to test some mechanistic hypothesis concerning the main physical or physico-chemical processes that may be implicated in the determination of the bioavailability of a chemical in the soil.

## Materials and methods

We defined the bioavailable part of a pesticide as the fraction that can be found as a solute in the soil water present in pores of sufficient size to allow penetration by soil microorganisms. From a technical point of view, we started from the consideration that this bioavailable fraction is correlated to, and could be tentatively estimated by, the amount of the pesticide that is recovered after short-term extraction with water.

### *Theoretical considerations*

Basically, two main physical processes have been considered to affect the biological availability of pesticides: diffusion and adsorption. The diffusion process causes translocation of the pesticide from the macropores open to the microorganisms to micropores where microbial life is excluded. If $C_M$ and $C_m$ are the concentrations of the pesticide respectively in the macro- and the microporosity and k a rate constant for the passage of material between the two pore systems, following equation applies:

$$\frac{dC_M}{dt} = k(C_M - C_m)$$

(1)

Corresponding models are referred to as "two-region" models by analogy with the concept developped by van Genuchtem and Wierenga (1978) to study movement of water and solutes in non-homogeneous soils.

Moreover, exchange of material between the liquid and the solid phases has been assumed to be driven by the difference between the actual and the maximum surface concentration at equilibrium and the transfer of the pesticidal molecule between the liquid and the solid phase was assumed to conform to simple non-equilibrium linear adsorption isotherm; if S is the surface concentration of the pesticide (in mg kg$^{-1}$), $K_d$ the partition coefficient, a the rate constant for the adsorption process, r the bulk density of soil and q the volumetric water content, then:

$$\frac{dC_M}{dt} = -\alpha \frac{\rho}{\theta} (K_D C_M - S) \qquad (2)$$

From the definitions of $\rho$ and $\theta$ equation (2) may be written as:

$$\frac{dC_M}{dt} = -\alpha\left(\frac{K_D}{H}C_M - S\right) \qquad (3)$$

where H is the gravimetric water content of the soil. In some instances, the sorptive capacity of the soil appears to be adequately described by considering two different adsorption sites that differ in their kinetic properties. These so-called "two-site" models were first presented by Lapidus and Amundsen (1952) and further developped by different authors.

In this work, three different models, based on different mechanistic hypotheses have been successively tested. Their respective flow diagrams are given on figure 1. The first model is the most simple and based on the assumption that recovery of the pesticide from the microporosity during the time of the extraction is rapid as compared to recovery of adsorbed material. For that reason no distinction was introduced between macro- and microporosity. Moreover, we have postulated that all adsorption sites were kinetically equivalent. Under such hypotheses the model is:

$$\frac{dR}{dt} = -\alpha\left(\frac{1+K_D}{H}R - 100\right) \qquad (4)$$

where R is the percentage of extractable radioactivity.

We derived the second model by further assuming that two types of adsorption sites could be distinguished on consideration of their kinetic properties: for a fraction $f_a$ of these sites adsorption equilibrium is reached very quickly while it takes a longer time for the other sites to come to an equilibrium. With that distinction model 2 takes the form :

$$\frac{dR}{dt} = -\alpha_1\left[\frac{K_D^1 f_a}{H}R - R_1\right] - \alpha_2\left[\frac{(1-f_a)K_D^2 R}{H} - R_2\right]$$

$$\frac{dR_1}{dt} = \alpha_1\left[\frac{K_D^1 f_a}{H}R - R_1\right]$$

$$\frac{dR_2}{dt} = \alpha_2\left[\frac{(1-f_a)K_D^2 R}{H} - R_2\right]$$

where $R$, $R_1$, $R_2$ are respectively the percentages of radioactivity found in the soil solution, on pseudo-equilibrium and on kinetic sites of adsorption. All other parameters have been previously defined.

Evidence that adsorption in the soil was often biphasic gives support to the assumption of some heterogeneity in the kinetic properties of adsorption sites. But another hypothesis may provide an alternative explanation to this observation. It may be postulated that only a fraction $f_a$ of the adsorption sites is directly accessible from the larger pores, the remaining sites being located in, and accessed through, the smaller pores. On these sites adsorption is under control of a diffusion process driven by the difference in the pesticide concentration of the water phases present in each pore volume. This hypothesis was tested with model 3 for which the following system of ordinary differential equations applies:

$$\frac{dR}{dt} = -k \left[ \frac{1-f}{f} R - r \right] - \alpha_1 \left[ \frac{f_a K_D^1}{fH} R - R_1 \right]$$

$$\frac{dR_1}{dt} = \alpha_1 \left[ \frac{f_a K_D^1}{fH} R - R_1 \right]$$

$$\frac{dr}{dt} = k \left[ \frac{1-f}{f} R - r \right] - \alpha_2 \left[ \frac{(1-f_a) K_D^2}{(1-f) H} r - r_1 \right]$$

$$\frac{dr_1}{dt} = \alpha_2 \left[ \frac{(1-f_a) K_D^2}{(1-f)H} r - r_1 \right]$$

where $R$ is the percentage of radioactivity found in the soil solution present in the larger pores, $R_1$ the percentage of radioactivity on those of the adsorption sites that are located in these macropores. $r$ and $r_1$ are the corresponding percentages with reference to the microporosity. Among other parameters, $f$ is a partition coefficient that is related to the distribution of the soil solution between the two pores volumes. The value of this partition coefficient was evaluated in a preliminary experiment carried out in exactly the same conditions as those that will be described therafter, [14]C-2,4-D being replaced by [36]Cl. It was estimated to 0.95. As a further simplifying assumption we have also postulated that all adsorption sites were kinetically and energetically equivalent. In other words the same set of parameters were chosen for adsorption in both pore spaces.

## _Soils and chemicals_

The experimental part of this study was made as described by Soulas (1992). A sieved (1-2 mm) silty clay soil with the following characteristics: 33% clay, 52% silt, 15% sand, 1.18% organic C, 0.14% total N, and 1.3% $CaCO_3$ was used in all experiments. The pH in water was 7.8, the CEC 20.8 meq 100 $g^{-1}$, and the water holding capacity (WHC, determined at 1000 g) 25%. When necessary, the soil was partially air-dried to a water content of about 12-14% (W/W, on an oven dry basis) making subsequent sieving easier. Finally, soil aggregates were wetted to 72% of the WHC.

Carboxyl-[14]C 2,4-dichlorophenoxyacetic acid (specific activity 329 MBq $mmol^{-1}$) (Amersham International Ltd) was dissolved with solid non-radioactive analytical grade chemical in phosphate buffer (M/20, pH 7) after evaporation of the solvent of the stock radioactive solutions.

Portions of 5 g (oven dry basis) of moist soil aggregates were sterilised by three successive chloroform fumigations for 16 h with an interval of one week in between the fumigations. A total of 210 samples was so prepared. A week after the last fumigation these samples received 100 ml of appropriate solutions of radioactive material to give the soil a final water content of

80% of the W.H.C. and concentrations of the chemical of 0.064, 0.16, 0.32, 0.96, 1.60, 4.10, and 6.40 mg kg$^{-1}$. The amounts of initially added radioactivity were in a range varying from 0.65 to 0.43 kBq per sample. After treatment soil samples were placed in 250 ml glass bottles where two additional plastic scintillation flasks were also present, one with 10 ml of water and the other with 5 ml of a 0.2N solution of sodium hydroxide. Incubation jars were closed and stored in the dark in climatic chambers at 20±1°C for the duration of the experiment. After 0.018, 0.6, 1, 2, 4, 7, 11, 15, 28, and 42 days of exposure triplicate samples from the different groups of concentration were taken for extraction on a reciprocal shaker (300 rpm) with 50 ml of a borate buffer (H$_3$BO$_3$: 2 g l$^{-1}$, Na$_2$B$_4$O$_7$,7H$_2$O: 14 g l$^{-1}$; pH: 9) in 150 ml centrifuge tubes. After centrifugation at 6000 g 5 ml of the supernatant were analysed for radioactivity by scintillation counting. Correspondingly, the sodium hydroxide of the CO$_2$ traps was analysed for radioactivity by scintillation counting to detect possible failure of the sterilisation procedure that would result in reactivating mineralisation of the pesticide.

The time variations of the extractable radioactivity were analysed using a modelling approach. The different models were fitted using non-linear regression analysis with a computer program, NL, an extension of the S$^{TM}$ library (Becker et al., 1988) developed by the Department of Biometry (INRA). Parameter estimation was based on a least squares procedure using the Marquardt algorithm.

# Results and discussion

The main results are summarized on figure 1 with, on the left hand side, a graph of the flow chart of each model and, on the right hand side, the corresponding plots comparing experimental results and simulated curves for the same set of data. Table 1 gives the best fit estimates of the parameters and an evaluation of the goodness of fit ($\sigma^2$)for each model.

*Model 1.* With this model, estimations of the parameters $\alpha$ and K$_D$ have been performed in two different conditions. In a first run, the initial percentage of the radioactivity present in the liquid phase, R$_0$, was fixed to its theoretical value, i.e. 100%, while estimating other parameters. A second identification was performed with R$_0$ being considered as a parameter to be estimated. Plot A of figure 1 and table 1 clearly show that the goodness of fit was greatly improved when the R$_0$ value was adjusted. For the different test concentrations we have found R0 values ranging from 89 % of the initially added radioactivity at 0.06 mg kg$^{-1}$ to 100% at 6.4 mg kg$^{-1}$. From this first trial, it clearly appears that at the lower concentrations there is an immediate consistent removal of radioactivity from the soil solution model 1 may not account for.

| | Application rate of 2,4-D (mg kg$^{-1}$ soil) | | | | | | |
|---|---|---|---|---|---|---|---|
| | 0.064 | 0.160 | 0.320 | 0.960 | 1.920 | 4.096 | 6.400 |
| **Model 1** | | | (fixed $R_0$) | | | | |
| $\alpha$ | 0.347 | 0.301 | 0.510 | 0.483 | 0.241 | 0.417 | 0.199 |
| $K_D$ | 0.121 | 0.125 | 0.103 | 0.093 | 0.071 | 0.066 | 0.071 |
| $\sigma^2$ | 30.04 | 31.17 | 24.67 | 37.72 | 10.56 | 17.67 | 10.08 |
| | | | (floating $R_0$) | | | | |
| $\alpha$ | 0.181 | 0.165 | 0.245 | 0.217 | 0.143 | 0.208 | 0.148 |
| $K_D$ | 0.128 | 0.131 | 0.108 | 0.098 | 0.074 | 0.068 | 0.073 |
| $\sigma^2$ | 10.06 | 8.32 | 7.24 | 15.45 | 4.38 | 6.61 | 7.57 |
| **Model 2** | | | | | | | |
| $f_a$ | 0.190 | 0.119 | 0.219 | 0.280 | 0.320 | 0.515 | 0.180 |
| $\alpha_1$ | 19.000 | 22.383 | 21.527 | 29.527 | 33.162 | 32.482 | 7.324 |
| $K_{D1}$ | 0.125 | 0.204 | 0.116 | 0.098 | 0.040 | 0.035 | 0.014 |
| $\alpha_2$ | 0.203 | 0.186 | 0.277 | 0.239 | 0.154 | 0.214 | 0.165 |
| $K_{D2}$ | 0.128 | 0.121 | 0.105 | 0.099 | 0.091 | 0.104 | 0.107 |
| $\sigma^2$ | 9.28 | 7.9 | 6.85 | 15.48 | 4.46 | 6.76 | 7.97 |
| **Model 3** | | | | | | | |
| $f_a$ | 0.500 | 0.500 | 0.600 | 0.600 | 0.600 | 0.600 | 0.900 |
| $k$ | 0.741 | 0.767 | 0.694 | 0.712 | 0.391 | 0.493 | 5.464 |
| $\alpha$ | 5.569 | 5.781 | 6.034 | 8.557 | 0.492 | 7.285 | 0.137 |
| $K_D$ | 0.104 | 0.105 | 0.087 | 0.076 | 0.058 | 0.049 | 0.058 |
| $\sigma^2$ | 15.81 | 16.30 | 10.71 | 21.57 | 9.60 | 11.11 | 6.67 |

**Table 1** : Bets fit estimates of the parameters for the different models

*Model 2*. Such a "two-site" model gives the best fit to our data. This can be appreciated either visually from the comparison between observed and simulated percentages of soluble radioactivity (plot B, figure 1) or statistically from the value $\sigma^2$ (table 1). Parameter $f_a$ was estimated by trial and error; table 1 gives the values that minimize $\sigma^2$ while estimating other parameters. Best fit estimates confirm the possible existence of two categories of adsorption sites: the rate constants $\alpha_1$ and $\alpha_2$ differ by more than two orders of magnitude. Distribution coefficients for both sites show different trends with changes in the pesticide concentration: it decreases with increasing concentration for the "pseudo-equilibrium" sites while it remains about constant for "kinetic" sites.

*Model 3*. In two occasions (1.92 mg kg$^{-1}$ and 6.4 mg kg$^{-1}$) best fit estimates do not fall in the range of values found for other concentrations. This probably reflects sensitivity of the model to the quality of the data gathered in the few hours at the beginning of the experiment. From purely statistical considerations, this "two-region" model compares less favourably to experimental data leading to the conclusion that underlying hypotheses are less realistic than those of model 2. From a mechanistic point of view such a model probably needs a reexamination in another environmental context favoring high quality data concentrated during the first hours of experimentation.

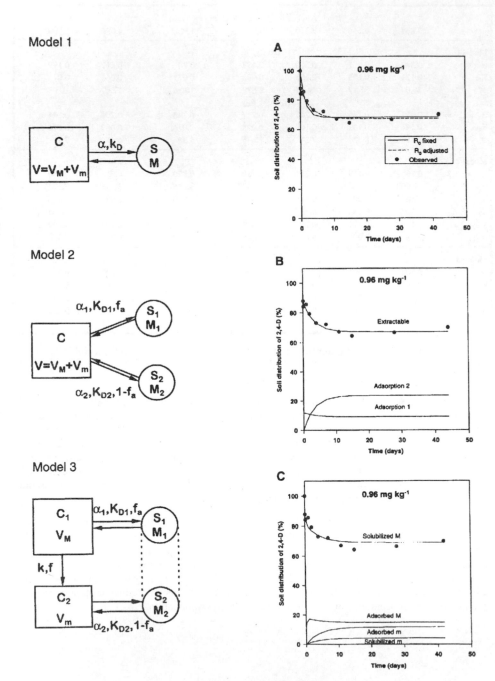

**Figure 1 :** Flow diagrams of the different models and example of comparison
of observed and simulated extractable radioactivity

# Acknowledgements

The authors wish to thank Dr. S. Huet, INRA, Biometry laboratory, Jouy-en-Josas, for her assistance in using the non-linear estimation NL software package

# References

☐ Kerhoas L. and Jeanne-Chantal Thoisy-Dur. 1988. Relation entre adsorption et persistance d'un pesticide dans le sol. Etude du comportement de la fluméquine. In *"Methodological aspects of the study of pesticide behaviour in soil"* P. Jamet ed. pp. 195-200.

☐ Lapidus L. and N.R. Amundsen. 1952. Mathematics of adsorption in beds. The effect of longitudinal dispersion in ion-exchange and chromatographic columns. *J. Phys. Chem.*, 56, 984-988.

☐ Moyer J.R., R.J. Hance and C.E. McKone. 1972. The effect of adsorbents on the rate of degradation of herbicides incubated with soil. *Soil Biol. Biochem.*, 4, 307-311.

☐ Ogram A.V., R.E. Jessup, L.T. Ou and P.S.C. Rao. 1985. Effects of sorption on biological degradation rates of (2,4-dichlorophenoxy)acetic acid in soils. *Appl. Environ. Microbiol.*, 49, 582-587.

☐ Soulas G. 1992. Biological availability of pesticides in soil. 2,4-D and glyphosate as test cases. p.219-224. In J.P.E. Anderson, D.J. Arnold, F. Lewis and L. Torstensson (ed.) *Proc. Int. Symp. on Environmental Aspects of Pesticide Microbiology*, Sigtuna, Sweden, 17-21 August 1992. Univ. Press, Uppsala, Sweden.

☐ Sprankle P., W.F. Megitt and D. Penner. 1975. Adsorption, mobility and microbial degradation of glyphosate in the soil. *Weed Sci.*, 23, 229-234.

☐ Van Genuchtem M. Th. and P.J. Wierenga.1976. Mass transfer studies in sorbing porous media. I. Analytical solutions. *Soil Sci. Soc. Am. J.*,40, 473-480.

☐ Weber J.B. and H.D. Coble. 1968. Microbial decomposition of diquat adsorbed on montmorillonite and kaolinite clays. *J. Agric Food Chem.*, 16, 475-478.

### 2.2.6.        *Adsorption and Mobility of Bentazone in Soils*

# G. Dios Cancela; E. Romero Taboada; F. Sánchez Rasero;
# A. Peña Heras and C. de la Colina González

Estación Experimental del Zaidín, (Consejo Superior de Investigaciones Científicas),
C/ Profesor Albareda, 1, Granada (España).

## Introduction

Bentazone [3-isopropyl- 1H- 2,1,3,-benzothiadiazin- (4)3H-one 2,2 dioxide] is widely used for treatment of broad-leaves weeds in differents crops. As a consequence of this widespread use, bentazone residue levels about $1.0 \, \mu g \, l^{-1}$ have been detected in deep groundwater in western Europe (Leistra 1989). However, Stoller et al (1975) and Kördel et al (1991), from 1 meter lysimeters, found that bentazone does not tend to accumulate in soil and does not move to deeper soil zones and consequently pointed out that volatilization and/or mineralization should play an important role in the fate of this herbicide from the soil system.

Because of this contradiction, data on the environmental fate of bentazone in different soils are required in order to determine the potential risk of this herbicide to reach groundwater. In this paper, adsorption and leaching experiments of this herbicide in different soils are described.

## Material and Methods

### *Material*

Ten soils from Doñana National Park, Huelva (Spain) were studied and their properties shown in table I. The samples were from A-horizon, previously passed by 2 mm sieve.

Organic compounds.- Bentazone as an analytical standard of known purity, was supplied by BASF. AG. Its solubility in water, at 20°C, is about 500mg/l.

### *Adsorption measurements*

The adsorption of bentazone at 15°C was measured. Bentazone solutions at 300, 200, 100 and 50 mg $L^{-1}$ were used. Aqueous suspensions of the samples were prepared by adding 20 cm$^3$ of the different bentazone solutions to 1 g. of adsorbent in 50 cm$^3$ tubes. The tubes were mechanically shaken (end-over-end) for 24 hours in a thermostatic chamber (kinetic experiments indicated that adsorption of bentazone by the samples reached an apparent equilibrium after 4 h). The suspensions were then centrifuged at 1720 g for 20 min and bentazone was determined in the supernatant phase.

| Soils | Phyllosilicates % | Smectite % | c.c.c. meq/100 g | M.O. % | N % | C/N | pH |
|-------|-------------------|------------|------------------|--------|------|------|------|
| $P_1$ | 87 | 20 | 24.7 | 2.23 | 0.18 | 7.2 | 8.17 |
| $P_2$ | 25 | 14 | 8.8 | 1.38 | 0.08 | 17.2 | 7.99 |
| $P_3$ | 47 | 30 | 21.4 | 1.31 | 0.11 | 11.9 | 7.69 |
| $P_4$ | 21 | 12 | 9.0 | 0.61 | 0.10 | 6.1 | 5.94 |
| $P_5$ | 44 | 38 | 27.9 | 1.45 | 0.19 | 7.6 | 7.96 |
| $P_7$ | 45 | 45 | 28.8 | 2.45 | 0.21 | 11.7 | 7.73 |
| $P_8$ | 37 | 18 | 13.3 | 1.74 | 0.14 | 12.4 | 8.02 |
| $P_9$ | 23 | 6 | 3.5 | 1.10 | 0.07 | 15.7 | 5.96 |
| $P_{10}$ | 77 | 39 | 24.2 | 0.52 | 0.08 | 6.5 | 8.13 |
| $P_{11}$ | 31 | 26 | 14.5 | 0.85 | 0.10 | 8.5 | 7.48 |

**Table 1** : Properties of the Soils

## *Leaching experiments with soil columns*

Break-through-curves (BTC) for two soil materials ($P_{10}$ and $P_2$) with bentazone were obtained experimentally from soil column experiments.

The cylinders for the columns consisted of 3 mm polyacrilate, with 9.3 cm i. d. and 20 cm long. The end of the cylinder was provided with a sand filter (5 cm thick) graded from coarse sand (around 0.2 cm) at the bottom to fine sand (around 0.02 cm) on the top. The bottom of the cylinder was also provided with an outlet tube for the collection of the percolated water. Column packing was done by filling the cylinder containing the sand filter with dry soil. A layer of 1 cm of coarse sand was placed on the top of the columns.

A rainfall simulator, fed with water from a variable peristaltic pump (Autoclude, model VL, England), was placed on the top of the cylinders. Chloride-ion was used as a tracer to characterize water flow in the soil columns. After a few days of percolation, when steady-state water flow ocurred in the column, 50 ml of $CaCl_2$ in water at 0.1 mM/ml were dropped uniformily on the column. The irrigation period in the $P_{10}$ soil-column, which showed slower flow than $P_2$ was 40 ml per day, while the $P_2$ soil-column was irrigated 8 hours per day with the rainfall simulator. The chloride efluents from both columns were collected every day and every hour or half an hour, respectively and stored at 4°C until its analysis.

After the collection of the Cl- ion from the column, bentazone in water (10 ml) was dropped uniformily on the soil- columns ($P_{10}$ and $P_2$). The amounts of bentazone applied were 3.13 mg for $P_{10}$ soil-column and 4.79 mg for $P_2$ soil-column.

A survey of the Characteristic of the leaching experiment with the soil columns is given in table III.

The transport of water-tracer Chloride-ion in the soil-columns was simulated with a computed model based in the PESTLA model described by Boesten and Van der Linden (1991), considering no adsorption and no transpoformation of chloride-ion and a steady-state water flow at constant soil water content. Break-Throught Curves (BTC) for Cl- and bentazone were evaluated by fitting the calculated BTC to the experimnental result.

*Chemical analysis*

Samples containing bentazone were analysed using Liquid Chromatography following the procedure described by Sánchez Rasero et al. (1992).

Chloride-ion concentration in the eluates was measured using a selective chloride electrode and a pH/mv meter (CRISON). Chloride-ion samples were diluted 1:10 in deionized water.

# Results and Discussion

## *Adsorption*

Adsorption isotherms of bentazone by different soils fit well, in every case, the Freundlich equation (figure 1). The values of K and $1/n$ , as well as the correlation coefficients for every soil, are shown in table I. The $1/n$ values for $P_3$, $P_7$, $P_8$, $P_{10}$ and $P_{11}$ soils are below one, which indicates convex isotherms or from the L type in the Giles classification (1960). The shape of the isotherms suggest that for these soils the number of useful adsorption sites tend to a limit. The $P_1$, $P_2$ and $P_5$ soils give lineal isotherms ($1/n \approx 1$), while for $P_4$ and $P_9$ soils the $1/n$ values are above one, which indicates concave isotherms or from S type. At low equilibrium concentrations bentazone seems to adsorb poorly. When Ce increases there is a significant increment of bentazone amount adsorbed. This could be explained by the lower pH values of the latter two soils, so when pesticide concentration increases it could be partially protonated and adsorbed by ion exchange.

As bentazone is not adsorbed by the soils, it will therefore move easily through them, being able to reach rivers and groundwater.

## *Mobility*

The Freundlich adsorption coefficients measured in the present study together with Ldis values determined by the chloride curve fitting in the $P_{10}$ and $P_2$ soil columns, were applied to the simulation model to fit the experimental leaching of bentazone for both soil types, (Do in the simulation was estimated to be 2.8 $cm^2$/h from Zhong et al.,1986).

Results of chloride-ion leaching through $P_{10}$ soil are shown in figure 2. Cl- ion is present at high concentrations already in the first eluates, reaching its maximun concentration after 8 days. There is a steep concentration decrease after the maximun, which slows down at the end.

Figure 2 shows the results obtained for the fitting of the experimental data with the theoretical model using Ldis = 2 cm. The experimental chloride ion curve fits well with an uniform flow through the column, except for the first eluates/first days, in which the ion concentration is higher than expected. The point for maximum concentration agrees for both the experimental and theoretical data, but only the chloride concentration at the beginning of the curve appear at shorter times than in the simulated. Figure 3 shows the leaching of bentazone in soil column $P_{10}$.

The first bentazone concentration in the eluate appears latter than that of the tracer, reaching a maximum pesticide concentration 2 days later. The comparison of the experimental curve with the theoretical model for Ldis = 2 cm shows a faster flow at the beginning and a higher concentration of bentazone at the end of the curve. The best fitting is obtained with Ldis = 1 cm for the first part of the curve, showing a dispersion at the end of the column, as happened with the dispersion length 2 cm. This lack fit at the end of the curve could be due to the possible different adsorption of bentazone between the batch and the column experiments. The mass balance for bentazone shows that this compound is totally eluted from the column.

Notice that both assays for chloride-ion and bentazone were carried out one after the other, to avoid problems in the HPLC system with the presence of salts. This event may explain the differences observed, due to a decrease in mechanical dispersion of the columns with time, especially taking into account the smectite content of the soil employed.

Figure 4 shows the leaching of chloride ion through $P_2$ soil column. In this case, the water flow velocity is faster than in the $P_{10}$ soil-column. The chloride-ion was first determined in the eluate two hours after the tracer application, reaching the maximun concentration after 5.5 hours. The chloride ion curve does not fit well with a uniform flow through the column using Ldis 1 or 0.5 cm. In both fits, the maximun in the theoretical curves appears earlier than in the experimental one. Maybe the theoretical curve with Ldis 0.5 cm fits better at the beginning and at the end of the experimental curve. Figure 5 shows the results obtained for bentazone for the fitting of the experimental data with the theoretical model using Ldis = 0.5 cm. As can be observed the fitting is much better than with chloride ion for the maximum concentration. The lower experimental concentration of bentazone at the end of the curve could be explain by the same reason as in the $P_{10}$ soil-column.

## Conclusions

Bentazone is poorly adsorbed by the ten soils studied. According with this, bentazone will probably pass through soils reaching easily surface and groundwaters.

The main difference between the two soils studied is the water flow velocity. This different behaviour could be explained by their smectite content affecting soil structure and porosity.

The model applied to explain the leaching of bentazone in both cases is suitable for the description of the movement of a pesticide like bentazone in soil with this characteristics.

To improve the simulation of the leaching of bentazone in soil-columns, the tracer and the solute must be apply at the same time. By other hand, the adsorption coefficient must be measured in the same experimental condition.

## References

□    J.J.T.I. Boesten and A.M.A. Van der Linden. "Modeling the influence of adsorption and transformation on pesticides leaching and persistence". *J. Environ. Qual.* 20, (1991), 425-435.

□    W. Kördel, M. Herrchen and R.T. Hamm. Lysimeter experiments on bentazon. *Chemosphere*, vol. 23 n1 (1991) 83-97.

□    M. Leistra and J.J.T.I. Boesten. Pesticide of groundwater in Western Europe. *Agriculture, Ecosystems and Environment*, 26 (1989) 369-384.

□    F. Sánchez Rasero, M.E. Báez and C.G. Dios. Liquid chromatographic analysis of bentazone in the presence of soil and some of its normal constituents, with photodiode array detection. *The Science of the Total Environment*, 123/124 (1992) 57-61.

□    E. W. Stoller, L.M. Wax, L.C. Harderlie and F.W. Slife. Bentazon leaching in four Illinois soils. *J. Agric. Food Chem.* 23, n4 (1975) 682-684.

□    W. Z. Zhong, A. T. Lemley and R. J. Wagenet. Quantifying pesticide  adsorption and degradation during transport through soil to groundwater. In: *Evaluation of pesticides in ground water*. Willa Y. Garner; Richard C. Honeycutt, and Herbert N. Nigg, Ed.ACS Symposium series No. 315 (1986) 61-67.

| Soils | K (mg/kg) | 1/n | r |
|:-----:|:---------:|:---:|:-:|
| $P_1$ | 0.160 | 1.02 | 0.999 |
| $P_2$ | 0.028 | 1.11 | 0.995 |
| $P_3$ | 1.03 | 0.77 | 0.986 |
| $P_4$ | 0.0044 | 1.45 | 0.990 |
| $P_5$ | 0.131 | 1.06 | 0.991 |
| $P_7$ | 0.583 | 0.85 | 0.972 |
| $P_8$ | 3.47 | 0.60 | 0.992 |
| $P_9$ | 0.05 | 1.42 | 0.985 |
| $P_{10}$ | 0.188 | 0.82 | 0.999 |
| $P_{11}$ | 0.039 | 0.85 | 0.991 |

**Table II :** Freundlich constant (K and 1/n) and correlation coefficients (r) for the adsorption of bentazone by soils

| Soil-column | $P_{10}$ | $P_2$ |
|:------------|:--------:|:-----:|
| Column length (cm) | 10 | 9.5 |
| Surface area of the column ($cm^2$) | 6.97 | 6.97 |
| Soil bulk density ($g/cm^3$) | 1.086 | 1.230 |
| Volume fraction of water ($cm^3/cm^3$) | 0.58 | 0.48 |
| Water flux density | 0.529 ($cm^3/cm^2$ d) | 1.083 ($cm^3/cm^2$ h) |

**Table III :** Characteristic of the leaching experiment with the soil-columns $P_{10}$ and $P_2$

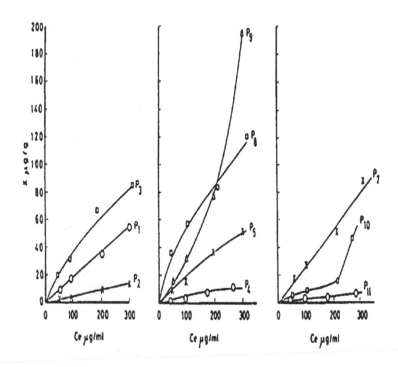

Fig. 1. - Adsorption isotherms of bentazone in ten soils at 15'C

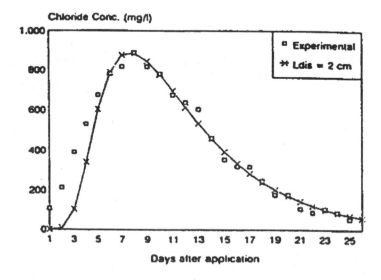

Fig. 2. - Experimental (square point) and theoretical (solid line) chloride
concentration in the effluents from $P_{10}$ soil-column

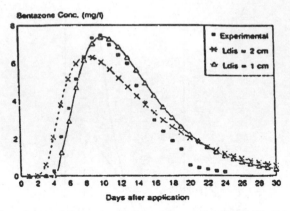

Fig. 3. - Experimental (square point) and theoretical (solid line) bentazone
concentration in the effluents from $P_{10}$ soil-column

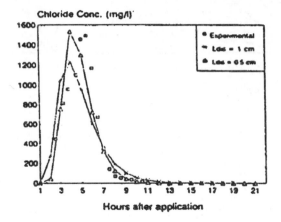

Fig. 4. - Experimental (square point) and theoretical (solid line) chloride
concentration in the effluents from $P_2$ soil-column

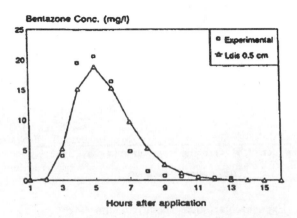

Fig. 5. - Experimental (square point) and theoretical (solid line) bentazone
concentration in the effluents from $P_2$ soil-column

## 2.2.7. Study on the Availability of Isoproturon and its Degradation Products in Soil

# P. Gaillardon

Station de Phytopharmacie, INRA
Route de Saint Cyr, 78026 Versailles Cedex

## Abstract

Changes in the distribution coefficient (Kd) of isoproturon between a clay loam soil and the soil solution was determined during a 6 week incubation period, in laboratory conditions, at various temperatures and herbicide doses (3 or 0.6 $\mu$g g$^{-1}$ at 4°C and 3 $\mu$g g$^{-1}$ at 11 and 18°C ), and constant soil moisture (40 % w/w). At 4°C, degradation of isoproturon was very low, Kd value raised for about 3 weeks and was higher at low dose. At 18°C, degradation was rapid and Kd value increased continuously. Long term adsorption and nonlinear adsorption isotherm are shown to be primarily responsible for changes in isoproturon availability during degradation in soil. Isoproturon was degraded to two major metabolites : monomethyl isoproturon that was mainly adsorbed by soil and an unidentified $X_5$ metabolite (possibly hydroxy di-desmethyl isoproturon) that was mainly in soil solution. Decreasing temperature reduced isoproturon degradation and monomethyl isoproturon accumulation whereas $X_5$ metabolite accumulation could be favoured.

## Introduction

The availability of herbicides in soil is of major importance in controlling biological activity, transport and degradation. Many studies provide evidence that availability decreases during incubation. They involve soil solution and simple water extract analysis (Walker, 1987; Pignatello and Huang, 1991; Gaillardon and Sabar, 1994) or desorption isotherm determination (Barriuso et al., 1992). In previous work, we observed that the distribution coefficient of isoproturon between solid and liquid phases of a clay loam soil increased during incubation at 18°C. This was assumed to be due to long term adsorption of herbicide by soil, nonlinear adsorption isotherm which may favour adsorption of low concentrations induced by degradation and adsorption-desorption hysteresis. Because biodegradation of isoproturon depends on temperature, low temperatures are likely to prevent degradation and to provide suitable conditions for studying long term adsorption. An experiment was thus conducted to measure changes in the concentrations of isoproturon and its degradation products in soil and soil solution during incubation at various temperatures and herbicide doses in order to determine the duration of adsorption, the influence of herbicide dose and their involvement in the availability of this herbicide. Furthermore, analysis of degradation products may provide information about their availability in soil and the effect of temperature on their accumulation.

## Materials and Methods

Screen (0.5 - 2 mm) air dried clay loam soil samples (20 g containing 1.1 ml water) were treated with 1.5 ml of aqueous solution of $^{14}$C-carbonyl-isoproturon to give a 3 $\mu$g g$^{-1}$ concentration (on dry soil basis) with 20.3 KBq. Each sample was mixed and placed in a beaker as four consecutive layers watered to 40 % (w/w). Three glass microfibre filters (Whatman GF/A) were inserted between the central layers .

Samples were incubated at 4, 11 and 18°C and analysed one day after treatment, refered to as 0 week in figures, and then weekly or every two weeks for 6 weeks, as described previously (Gaillardon and Sabar, 1994). Briefly, radioactivity concentration in soil solution was determined in the central filter by weighing and liquid scintillation counting. Qualitative analysis of soil solution was performed by TLC of the radioactive compounds recovered from the external filters. Amounts of isoproturon and its metabolites in soil solution were calculated. Soil samples were extracted once by water and then by methanol. Extracts were analysed by TLC and amounts of isoproturon and its metabolites in soil were calculated. Unextracted radioactivity was measured by combustion of dried residual soil. At 18°C released $^{14}CO_2$ was trapped in 0.1 N NaOH for determination.

In order to evaluate the effect of herbicide concentration on adsorption, a similar experiment was conducted with soil samples treated with low isoproturon dose (0.6 µg g$^{-1}$) at 4°C to minimize degradation.

## Results and Discussion

When the high isoproturon dose (3 µg g$^{-1}$) was incubated in soil at 4°C, most of the applied radioactivity could be extracted at any time and unextracted radioactivity remained below 2 %, as shown in figure 1 which represents data for 0 and 6 week incubation time. By contrast, at 11 and 18°C, extracted radioactivity decreased from 99 to 70 and 41 % and unextracted radioactivity increased from 1 to 19 and 31 % within 6 weeks, respectively. Up to 20 % of the applied radioactivity was evolved as $^{14}CO_2$ at 18°C. At the highest two temperatures, the recovered radioactivity progressively decreased to 90 %.

At the high dose, degradation of isoproturon in soil was observed at every temperature. However, temperature greatly influenced disappearance of this herbicide. At 4°C, degradation started 2 weeks after treatment and herbicide amount decreased to 93% of that applied within 6 weeks (Fig. 2). At 11 and 18°C, degradation was faster and herbicide amounts decreased to 48 and 20 %. In these conditions, herbicide amounts in soil solution also declined from 25 to 14 , 8 and 3%.

The distribution coefficient (Kd) of isoproturon between soil and soil solution, calculated from the above data, increased during incubation (Fig. 3). At 4 and 11°C, it raised for 2 to 3 weeks and then it showed little changes. Because of limited degradation and almost total recovery of the applied $^{14}C$ in soil extracts at the lowest temperature, this increase may be mainly attributed to long term adsorption. In that condition, adsorption continued for more than two weeks but precise determination of equilibrium attainment requieres degradation is totally prevented. At 18°C, Kd value showed a more continuous increase. Changes in the first part of the curve may be also due to long term adsorption. However, because of rapid degradation of isoproturon, higher adsorption of low herbicide concentrations and/or adsorption-desorption hysteresis might explain changes in the second part of the curve.

When isoproturon was applied to soil at 3 or 0.6 µg g$^{-1}$ and incubated at 4°C, degradation expressed as percent was quite similar (data not shown). Whatever the dose, Kd values increased during incubation but they were higher at low dose (Fig. 4). This clearly demonstrated that isoproturon was more adsorbed at low concentration. Thus nonlinear adsorption isotherm may be responsible for Kd changes during incubation at high temperature. However, adsorption-desorption hysteresis cannot be totally reject and if so, causes of this phenomenon would remain to be explained.

Qualitative study of degradation of isoproturon applied at high dose showed  that at the highest temperature, amount of radioactive degradation products found in soil rapidly increased up to 22 % within 3 weeks and then it showed little changes although herbicide disappearance continued (Fig. 5). At 11°C, this amount increased more slowly but it reached the same final level although isoproturon degradation was reduced. At the lowest temperature, accumulation was greatly reduced to less than 5 % of the applied radioactivity. At 18°C, amount of degradation products found in soil solution increased for two weeks up to a maximum value around 8 % before decreasing. In general, it represented a minor part of the corresponding amount in soil. In contrast, at 11°C, this amount continuously increased up to higher values and accounted for half of amount in soil. At the lowest temperature, it slowly increased but it constituted the major part of amount in soil.

At least 8 radioactive degradation products could be detected in soil by thin layer chromatography. The major compounds were monomethyl isoproturon and an unidentified compound referred to as $X_5$ metabolite which might be the hydroxy di-desmethyl derivate of isoproturon (Fig. 6). In soil, amounts of monomethyl isoproturon increased with incubation time up to 1.5 , 12 and 16 % of the applied radioactivity at 4 , 11 and 18°C, respectively, so that decreasing temperature reduced accumulation. Whatever the temperature, low levels were always found in soil solution ( 1.5 %), showing that this compound was mainly adsorbed by soil. In contrast, amounts of $X_5$ metabolite in soil and in soil solution were practically the same, and hence this compound was not adsorbed by soil. At 18°C , amounts rapidly increased for one week then they dropped suggesting subsequent transformation of this compound. In general, they were low as compared to that of monomethyl isoproturon in soil. At 11°C , the $X_5$ metabolite accumulated to a higher level, suggesting that transformation might have been delayed. In that case, amounts were closed to that of monomethyl isoproturon at least for the first weeks. At 4°C, accumulation was reduced but amounts were similar to that of monomethyl isoproturon. Thus contrary to monomethyl isoproturon, decreasing temperature may favour accumulation of the $X_5$ metabolite and low temperatures seem less effective in reducing accumulation of this compound. These effects combined with presence of the $X_5$ metabolite in soil solution may explain the relatively high level of degradation products found in soil and soil solution at the intermediate temperature and their relative abundance in soil solution at low temperature.

## Conclusion

The main result of this study is to show that a long term adsorption phenomenon takes place in soil for more than two weeks and is responsible for an increase in the distribution coefficient of isoproturon between soil and soil solution during incubation at any temperature. A further increase in Kd value is observed when degradation decreases herbicide concentration in soil. This is primarily due to nonlinear adsorption isotherm. In soil, isoproturon undergoes transformation into several degradation products which may be mainly adsorbed by soil such as monomethyl isoproturon or conversely contained in soil solution such as the $X_5$ metabolite. Temperature has quantitative and qualitative effects on compound accumulation. Decreasing temperature reduces isoproturon degradation and monomethyl isoproturon accumulation in soil but accumulation of $X_5$ metabolite may be favoured. Accordingly, unexpected high levels of degradation products may be observed in soil and soil solution at temperatures around 11°C and if small amounts are present at low temperature (around 4°C) they are mainly contained in soil solution.

## References

◻   BARRIUSO E., KOSKINEN W. and SORENSEN B. (1992). Modification of atrazine desorption during field incubation experiments. *The Science of the Total Environment.* 123-124, 333-344

◻   GAILLARDON P. and SABAR M. (1994). Changes in the concentrations of isoproturon and its degradation products in soil and soil solution during incubation at two temperatures. *Weed Research* (in press)

◻   PIGNATELLO J.J. and HUANG L.Q. (1991). Sorption reversibility of atrazine and metolachlor residues in field soil samples. *Journal of Environmental Quality*, 20, 222-228

◻   WALKER A. (1987). Evaluation of a simulation model for prediction of herbicide movement and persistence in soil. *Weed Research*, 27, 143-152

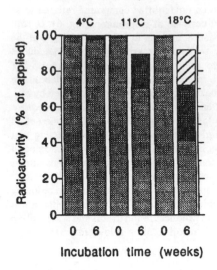

Fig. 1 : Distribution of the radioactivity recovered after soil treatment with 3 µg g$^{-1}$ of isoproturon and incubation at 3 temperatures : extracted ▦ , unextracted ■ and $^{14}CO_2$ ▨

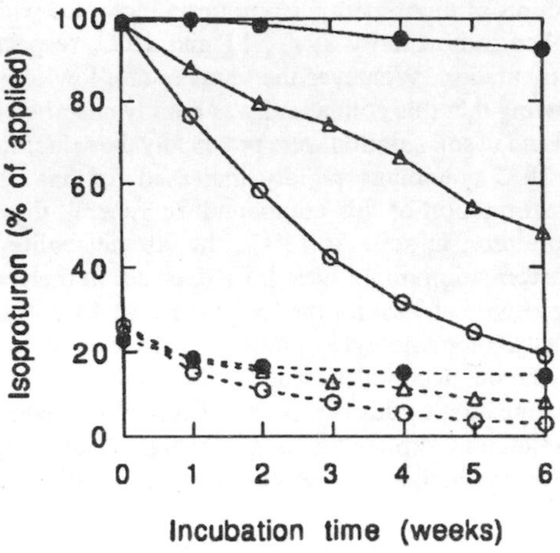

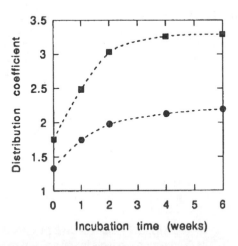

Fig. 2 : Amounts of isoproturon in soil (full lines) and soil solution (broken lines) after treatment at 3 µg g-1 and incubation at 4 ( ● ), 11 ( △ ) and 18° C ( ○ )

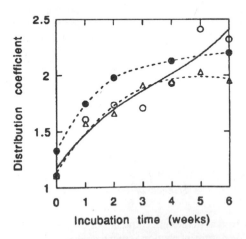

Fig. 3 : Values of the distribution coefficient of isoproturon between soil and soil solution after treatment at 3 µg g$^{-1}$ and incubation at 4 ( ● ) 11 ( △ ) and 18° C ( ○ )

Fig. 4 : Influence of initial dose on Kd values. Isoproturon was applied at 3 ( ● ) or 0.6 ( ■ ) µg g$^{-1}$ and soil was incubated at 4° C

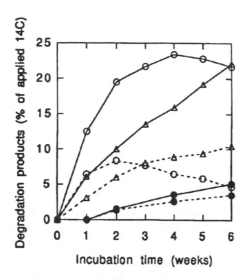

Fig. 5 : Amounts of degradation products of isoproturon in soil (full lines) and soil solution (broken lines) after treatment at 3μg g⁻¹ and incubation at 4 ( ● ), 11 ( △ ) and 18°C ( ○ )

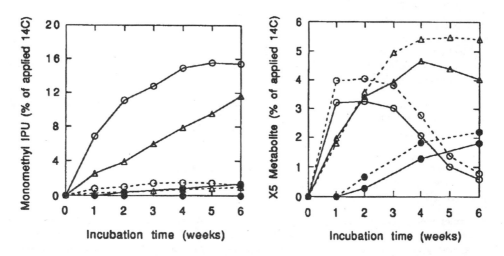

Fig. 6 : Amounts of the main two metabolites of isoproturon in soil (full lines) and soil solution (broken lines) after treatment at 3μg g⁻¹ and incubation at 4 ( ● ), 11 ( △ ) and 18°C ( ○ )

*2.2.8.*     *Identification, Collection, Preparation and Characterisation of EURO-Soils*

# W. Karcher, G. Kuhnt*, M. Herrmann** and H. Muntau

Environment Institute, Joint Research Centre Ispra,
Commission of the European Communities

\*     Department of Geography, University of Kiel
\*\*   German Federal Environmental Agency, Berlin

## Abstract

Following a Commission Communication to the Council and the European Parliament, the European Chemicals Bureau (ECB) was established within the Environment Institute of Joint Research Centre Ispra with effect of 1st January, 1993 (O.J. n°C1, p.3;5/1/93).

The principal task of the Bureau is to carry out and coordinate the scientific/technical work which is needed for the implementation of EC legislation (directives, regulations) in the area of chemical control. In this context, the Bureau has to coordinate also the development, updating and adaptation to technical progress of the experimental testing methods which have to be applied to determine the properties of hazardous or dangerous chemicals.

This work is performed in coordination with OECD in order to lay down testing methods which can be used worldwide and are accepted also outside the EC. Therefore, the Commission decided to adapt the OECD Test Guideline 106 'Adsorption/ Desorption' for European soil testing in cooperation with the German Federal Environmental Agency and the University of Kiel, which identified five frequent soil types of the European Community and their most representative sites of occurrence.

Harmonisation of soil investigation within the European Community as well as testing and notification for new chemicals to be distributed on the EC market require a source of standard soils for adjusting regional differences in soil analysis and classification procedures as well as testing material. This leads to a much better comparability of scientific findings and test results and therefore increases the value of individually obtained data.

However, essential prerequisites for these standard soils are that the material obtained and distributed reflect the physical-chemical properties of the most common soils of the European Community and that the specimens are taken at regionally representative sampling sites.

Therefore, the Commission of the European Communities together with the German Federal Environmental Agency, the EEC Joint Research Centre in Ispra/Italy and the University of Kiel launched a project on the identification, collection, treatment, preparation and characterisation of EC-representative soils.

The selection procedure was especially developed to guarantee that the soil samples taken are representative with respect to soil characteristics, frequency distribution and spatial patterns. The multi-level approach applied includes digital map evaluation, statistical verification on the basis of metric soil profile data, field research, soil sampling and treatment.

The soils, collected in Greece (Orthic Rendzina), Italy (Vertic Cambisol), Wales (Dystric Cambisol), France (Orthic Luvisol) and Germany (Orthic Podzol), have been treated at Ispra according to the rules in force for reference material preparation.

All together, 200 kg of each soil were prepared showing a particle < 2 mm and 250 g-aliquotes were distributed to a series of laboratories.

The material collected which is very well analysed and characterised reflects the wide variety of the pedological and sorption-controlling properties of the most wide-spread European soils and can therefore be used as reference material for sorption testing of chemicals and other pedological investigations.

Since 1990 the material gained much popularity as 'European Reference Soils' ('EURO-Soils') for a wide range of biological and chemical investigations and this growing interest prompted the Environment Institute to increase its stock and collect further 200 kg for future use.

# Introduction

With respect to the harmonisation of testing and notification procedures for new chemicals to be distributed on the EC market it is the basic assumption that data which were measured and submitted to one competent authority in one Member Country are valid and should fulfil the requirements for all other EEC Member States (DIRECTIVE 79/831/EEC). Most properties which have to be determined according to OECD Test Guidelines (OECD 1981) are inherent data of the chemical (e.g. vapour pressure, melting point) or their test media (e.g. water, air or solvents) are available in equal or comparable quality in all Member States (KUHNT & MUNTAU 1992a).

The OECD Test Guideline 106, however, belongs to those guidelines requiring environmental specimens. It has been developed for determining the potential mobility of chemicals in soil. Considering the fact, that there is a variety of more than 300 principally different kinds of soil within the EC realm, harmonisation of soil investigation within the European Community as well as testing and notification for new chemicals to be distributed on the EC market require a source of standard soils for adjusting regional differences in soil analysis and classification procedures as well as testing material. This leads to a much better comparability of scientific findings and test results and increases the value of individually obtained data.

Therefore the Commission of the European Communities together with the German Federal Environmental Agency and the EEC Joint Research Centre in Ispra/Italy launched a project on the identification, collection, treatment, preparation and characterisation of EC-representative reference soils (BRÜMMER et al. 1987, KUHNT et al. 1991). In the following, the various steps of the project as well as the main characteristics of the EURO-Soils will be briefly described.

# Identification of soil types and sampling sites

From its very nature, the geographical analysis of spatial distribution implies that sampling procedures must always tackle the elementary problem of obtaining representative specimens. That is, samples must be representative in the sense that their properties reflect those of a whole set of cases with a measurable degree of accuracy (FRÄNZLE 1984). In view of the wide variability of the EC soil cover only a careful and systematic study of the particular distribution functions or associations can ascertain the representativity of the reference samples. Therefore, a multi-level approach was developed to identify EC-representative soil types and sampling sites (KUHNT 1989a, 1993)

The first step is the evaluation of small-scale maps in order to define the typical and most frequent soils of the European Community. For this purpose the "Soil Map of the European Communities 1:1 Mio." had been digitised, because only maps yield the areal information necessary to calculate frequency distributions. Since soil properties are also dependent on other factors such as vegetation, land use or climate, the respective data were collected and converted in a way that they can form additional levels of information. This implies the possibility to more precisely define the environmental boundary conditions under which certain soil types mostly occur.

The second step is the nearest-neighbourhood analysis which leads to the determination of regionally representative sampling sites. This geo-statistical procedure is able to unfold the most typical neighbourhood relationships and association patterns of the soil types as they appear on the map. From various investigations we found that most of the relationships are genetically founded and determined by specific geological, geomorphological and climatic conditions. Due to this reason the neighbourhood analysis is a capable tool for the identification of regional representative sampling sites (KUHNT 1989b).

In a third step a comprehensive study of the literature and the evaluation of large-scale maps as well as metric soil profile data collected and stored in a data bank are necessary to ascertain if the thus defined representative soils adequately reflect the wide variability of the whole soil inventory from an ecochemical point of view. This means that the soils selected have to cover the wide range of sorption-controlling properties mainly occurring in the soils of the European Community.

By combining the total of information gained, five EC reference soils and the corresponding sampling locations were identified (cf. Table 1).

Cambisols, Luvisols, Podzols and Rendzinas are the most frequent soils of the European Community, covering about 70 per cent of the area. The dominant soil moisture regimes within the EC realm are udic and xeric, and the dominant temperature regimes are mesic and thermic. Cambisols are distributed all over the Community, therefore two representatives of this group under different climatic conditions are considered. Luvisols mostly occur in North and Central Europe, Rendzinas are frequent in the South, and Podzols are typical soils of the North. Accordingly, the sampling sites are located within the climatic zones where these soils predominantly occur. From a detailed analysis of maps of the natural vegetation and land use in the European Community it was found that pasture and meadow, arable ground, coniferous forest and broad-leaved trees must be taken into account to assure representativity in this case as well. As a consequence, during field work special attention was also paid to these requirements. The synopsis of the geological situation demonstrates that a reasonable diversity in the parent material of soil formation was also achieved.

All together, the combination of different soil types on alternate parent material under varying climatic conditions and numerous types of vegetation forms the best preconditions for obtaining reference material that is either representative for the EC territory or that differs with respect to its sorptive properties.

| EC SOIL MAP 1:1 Mio. | CAMBISOLS | | LUVISOLS | PODZOLS | RENDZINAS |
|---|---|---|---|---|---|
| FAO SOIL MAP OF EUROPE | Brown Forest Soils p.p. | Brown Mediterranean Soils p.p. | Gray-brown Podzolic Soils | Podzolized Soils | Rendzinas |
| FREQUENCY (%) | 44.7 | 44.7 | 15.7 | 6.7 | 5.0 |
| FAO SOIL UNIT | Dystric Cambisol | Vertic Cambisol | Orthic Luvisol | Orthic Podzol | Orthic Rendzina |
| SOIL CLIMATE | | | | | |
| Moisture Regime | udic | xeric | udic | udic | xeric |
| Temperature Regime | mesic | thermic | mesic | mesic | thermic |
| VEGETATION / LAND USE | Pasture | Meadow | Arable Ground | Coniferous Forest | Broad-leaved Trees Scrub |
| GEOLOGY / PARENT MATERIAL | Till Glacial drift | Marine Deposits | Loess | Fluvioglacial Sediments | Lacustrine Deposits |
| REPRESENTATIVE SAMPLING LOCATION | Radyr Wales | Aliminusa Sicily | Rots Normandy | Gudow Schl.-Holst. | Souli Peloponnesos |
| EC MEMBER STATE | Great Britain | Italy | France | Germany | Greece |

**Table 1:** Main characteristics of EC-representative soils and sampling sites

## Site exploration and sampling procedures

Step four of the multi-level approach consists in the verification of the theoretical investigations by visual inspection in the field, including site exploration and geological and pedological mapping, to finally locate discrete soil profiles where samples are taken and analysed to determine the validity of the selection.

As a matter of fact, even large-scale pedological or geological and geomorpho- logical maps only reflect natural conditions in a more or less generalised way. Therefore a careful and systematic evaluation of the specific situation within the areas identified as being representative must be performed in the field in order to exactly determine the optimum location for the sampling of specimens.

Since the formation and development of different soil types are highly dependent on topography as well as on geology and geomorphology, the first step in site exploration consists in a detailed survey of these parameters. Moreover, the spatial configuration and appearance of vegetation and land use patterns must be carefully analysed in order to determine the degree of alteration in soil quality due to these factors.

With respect to the main task of the sampling campaigns, special attention was paid to adequately defining the optimum sampling point within the areas specified by computer analysis. This involved a detailed analysis of numerous soil cores taken by drilling equipment, the estimation or measurement of the main sorption controlling parameters, a topographical, geomorphological, geological and hydrological survey and a detailed evaluation and documentation of the land use situation (KUHNT & MUNTAU 1992b).

To determine the appropriate site for sampling, the area taken into consideration was analysed by means of grid sampling with an auger. The drill cores serve to determine the predominant soil types and the associated soils. Each drill core was scrutinised with respect to the above mentioned soil properties. The routine field methods for these parameters are important because a pedological evaluation of the area must be accomplished by finding the optimum profile, i.e. the profile must reflect the typical soil constituents of this soil type and the predominant spatial and pedological association patterns of the region.

With the help of routinely used field methods, the main sorption controlling parameters were determined or estimated for each core. After determining the final location of the sampling site, it is characterised according to the criteria topographical position, altitude, relief, inclination, exposition, climate, vegetation and land use.

For the pedological documentation of the profile and its diagnostic horizons, a pit including all genetic soil horizons was dug, prepared at the location and samples were taken for closer characterisation. Finally, after carefully removing vegetational cover and litter, approx. 200 kg of topsoil material was sampled, filled into barrels and transported to the labs of the Environment Institute, JRC Ispra for further treatment.

## Treatment and characterisation of reference material

According to the OECD Test Guideline 106, the batch experiments have to be performed using air-dried fine soil. To obtain this, the samples have been dried in air-conditioned labs for 3 to 4 months. Before sieving, the aggregates were crushed by hand because using mechanical hammers or ball mills influences the test results in an unpredictable way. The sieved soil material (< 2 mm) was then filled into large homogenisation drums and after one week of continuous rotation, filled into 250 g bottles. To prevent microbial degradation during the tests, the material has been sterilised by radiation. One part of the samples were used for performing an intercomparison test on adsorption/desorption, the remainder is stored under dark and cool conditions at the EEC Joint Research Centre in Ispra/Italy.

Homogeneity testing was carried out by performing numerous sorption experiments using atrazine as a test substance. Ten different samples from one bottle were taken to determine the intra-bottle variation. Correspondingly, the inter-bottle variation is defined as the standard deviation of test results using material from ten different bottles. Considering the fact that soil is a highly heterogeneous medium the values shown in Figure 1 have to be regarded as satisfactory (KUHNT 1992).

Table 2 and Figure 2 show the main pedological and physical-chemical character- istics of the five EURO-Soils as well as the results of sorption experiments with atrazine, lindane and 2,4-D. Due to the lack of space the data can not be interpreted in detail. However, the five EURO-Soils show completely different composition and properties and thus reflect the wide variability of European soils. The soil samples which cover the required standards for reference material can be used to validate any results of sorption experiments.

This may form an EC-wide basis for valid interpretations, quantitative comparison of test results and extrapolation purposes.

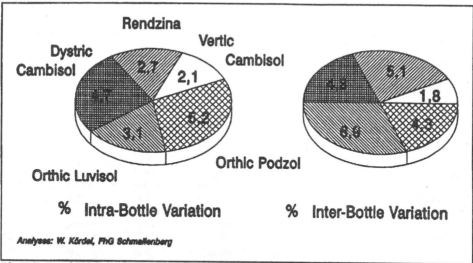

Fig. 1: Variation of sorption capacities of the EURO-Soils: Standard deviation of k' values for atrazine (% of mean values)

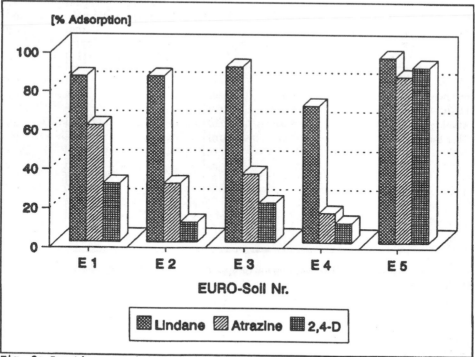

Fig. 2: Sorption capacities of the five EURO-Soils for different chemicals

|  | EURO-SOIL 1 | EURO-SOIL 2 | EURO-SOIL 3 | EURO-SOIL 4 | EURO-SOIL 5 |
|---|---|---|---|---|---|
| FAO-SOIL UNIT | Vertic Cambisol | Orthic Rendzina | Dystric Cambisol | Orthic Luvisol | Orthic Podzol |
| EC-MEMBER STATE | Italy | Greece | Great Britain | France | Germany |
| GRAIN SIZE % |  |  |  |  |  |
| Sand total | 3.3 | 13.4 | 46.4 | 4.1 | 81.6 |
| coarse + medium | 2.0 | 4.4 | 23.1 | 1.1 | 64.8 |
| fine | 1.3 | 9.0 | 23.3 | 3.0 | 16.8 |
| Silt total | 21.9 | 64.1 | 36.8 | 75.7 | 12.7 |
| coarse | 4.0 | 21.3 | 19.4 | 52.2 | 7.4 |
| medium | 9.7 | 23.1 | 11.6 | 19.4 | 4.3 |
| fine | 8.2 | 19.7 | 5.8 | 4.1 | 1.0 |
| Clay | 75.0 | 22.6 | 17.0 | 20.3 | 6.0 |
| pH value ($H_2O$) | 5.9 | 8.0 | 5.8 | 7.0 | 4.6 |
| pH value (KCl) | 5.1 | 7.5 | 5.2 | 6.5 | 3.4 |
| total carbon % | 1.5 | 10.9 | 3.7 | 1.7 | 10.9 |
| $CaCO_3$ % | 0.0 | 60.45 | 0.0 | 0.0 | 0.0 |
| organic carbon % | 1.30 | 3.70 | 3.45 | 1.55 | 9.23 |
| organic matter % | 2.65 | 6.40 | 6.44 | 2.86 | 15.92 |
| N total % | 0.17 | 0.20 | 0.26 | 0.16 | 0.30 |
| C/N ratio | 7.65 | 18.50 | 13.27 | 9.69 | 30.77 |
| P total % | 0.15 | 0.15 | 0.38 | 0.29 | 0.21 |
| CEC mval/100g) | 29.9 | 28.3 | 18.3 | 17.5 | 32.7 |
| Fe total % | 3.71 | 0.99 | 1.44 | 1.15 | 0.10 |
| Fe amorph. % | 0.32 | 0.02 | 0.48 | 0.19 | 0.06 |
| FE HCl-sol. % | 0.18 | < 0.01 | 0.22 | 0.15 | 0.01 |
| Al amorph. % | 0.06 | 0.02 | 0.16 | 0.08 | 0.10 |
| Al HCl-sol. % | 0.08 | < 0.01 | 0.17 | 0.16 | 0.09 |
| $SiO_2$ % | 56.22 | 21.60 | 68.45 | 68.63 | 71.57 |
| $Al_2O_3$ % | 23.92 | 8.66 | 11.92 | 12.07 | 3.85 |
| CaO % | 0.41 | 30.62 | 0.20 | 0.71 | < 0.02 |
| $K_2O$ % | 1.85 | 1.27 | 1.59 | 1.84 | 0.63 |
| $Fe_2O_3$ % | 10.76 | 1.66 | 4.14 | 2.71 | < 0.05 |
| MgO % | 1.12 | 1.82 | 1.19 | 1.11 | 0.65 |
| $TiO_2$ % | 0.99 | 0.25 | 0.65 | 0.72 | 0.36 |

**Table 2 :** Main pedological characteristics of the EURO-Soils

# References

◻ BRÜMMER, G., FRÄNZLE, O., KUHNT, G., KUKOWSKI, H. & VETTER, L. (1987): Fortschreibung der OECD-Prüfrichtlinie Adsorption/Desorption im Hinblick auf die Übernahme in Anhang V der EG-Richtlinie 79/831: Auswahl repräsentativer Böden im EG-Bereich und Abstufung der Testkonzeption nach Aussagekraft und Kosten.- Umweltforschungsplan des Bundesministers für Umwelt, Naturschutz und Reaktorsicherheit, Forschungsbericht 106 02 045, Kiel.

◻ DIRECTIVE 79/831/EEC: Official Journal No L 259 of 15.10.79, p 10.

◻ FRÄNZLE, O. (1984): Regionally Representative Sampling. In: LEWIS, R. A., STEIN, N., LEWIS, C. W. (Eds.): Environmental Specimen Banking and Monitoring as Related to Banking. Boston, The Hague, Dordrecht, Lancaster, 164-179.

◻ KUHNT, G. (1989a): Selection of Representative Soil Samples for Testing the Sorption Behaviour of Chemicals.- in: JAMET, P. (Ed.): Methodological Aspects of the Study of Pesticide Behaviour in Soil.- Paris, 151-158.

◻ KUHNT, G. (1989b): Die groräumige Vergesellschaftung von Böden. Rechnerge-stützte Erfassung pedogenetischer Zusammenhänge, dargestellt am Beispiel der Bundesrepublik Deutschland.- ERDKUNDE 43, 170-179.

◻ KUHNT, G. (1992): Leitlinien und Grundzüge eines europäischen Referenzbodensy-stems zur Chemikalienprüfung.- In: KUHNT, G. & ZÖLITZ-MÖLLER, R. (Hrsg.): Beiträge zur Geo0kologie aus Forschung, Praxis und Lehre.- Kieler Geographische Schriften 85, Kiel, 275-293.

◻ KUHNT, G. (1993): The EURO-Soil Concept as a Basis for Chemicals Testing and Pesticide Research.- In: MANSOUR, M. (Ed.): Fate and Prediction of Environmental Chemicals in Soils, Plants and Aquatic Systems. Boca Raton, Ann Arbor, London, Tokyo, 83-93.

◻ KUHNT, G. & MUNTAU, H. (1992a): European Reference Soils for Sorption Testing.-Fresenius Envir. Bull 1, 589-594.

◻ KUHNT, G. & MUNTAU, H. (Eds. 1992b): EURO-SOILS - Identification, Collection, Treatment, Characterization.- Commission of the European Communities, Joint Research Centre, Ispra/Italy.

◻ KUHNT, G., HERTLING, T., SCHMOTZ, W. & VETTER, L. (1991): Auswahl von Referenzböden für die Chemikalienprüfung im EG-Bereich.- Forschungsbericht 106 02 058 im Umweltforschungsplan des Bundesministers für Umwelt, Naturschutz und Reaktorsicherheit, Kiel.

◻ OECD (1981) Guidelines for Testing of Chemicals.- Paris.

2.2.9.        *Column and Batch Experiments on Atrazine Sorption on Three Soils from the Centre Region, France*

# C. Mouvet [1], S. Dousset [1,2,3] & M. Schiavon [2]

[1]    BRGM, Geochemistry and Physicochemistry Dept.

[2]    ENSAIA, Phytopharmacie

[3]    Post-graduate fellowship from the Région Centre

## Introduction

Monitoring networks in France have revealed the presence of atrazine in both surface and groundwater supplies at levels sometimes in excess of the CEC threshold value of 0.1 µg/L. Values above the WHO guideline value of 2.0 µg/L can be observed in some surface waters (Ministère de la Solidarité, de la Santé et de la Protection Sociale, 1990). The Centre Region of France, southwest of Paris, is an area of intensive agriculture with a significant proportion of maize for which atrazine is the most widely used pesticide with a dose of 1.5 a.i. per ha. The filtering capacity of a given soil in a given region for a given pesticide depends on a combination of the soil's intrinsic physico-chemical characteristics and the prevailing climatic and agronomic conditions. We therefore decided to study in the laboratory the interaction of atrazine with three soils representative of the Centre Region. Field work carried out at the same time is presented elsewhere (Dousset, 1994).

## Materials and Methods

### *Treatment of soil samples*

Soils were sampled at a depth of 0 to 30 cm, air-dried, homogenised and passed through a 2.0 mm sieve. Soil dispersion before fractionation without removal of carbonates was performed according to the Na-resin method of Rouiller et al. (1972). Organic carbon was determined by a combustion method at 940°C using a Carmhograph 12 Wösthoff after removing the carbonates with HCl 1 N. The mineral composition of clays was determined by X-ray diffraction using a Philips PW 1710 diffractometer. The organic matter studies included a granulometric fractionation of soil aggregates in an aqueous medium to allow the separation of a < 50 µm fraction corresponding to the humified compounds (Bruckert et al., 1978) and a > 50 µm fraction corresponding to the less humified organic matter. After removing the carbonates, the humic compounds were extracted from the < 50 µm fraction with a pyrophosphate-NaOH solution (1%; 0.1N). These extracts were acidified with 2N HCl to pH 1.5 and allowed to stand at 4°C for 12 h. Soluble fulvic acids (FA) were then separated from the coagulate humic acids (HA) by centrifugation.

### *Batch experiments*

Concentrations of atrazine ranging from 0.08 to 20 mg/L (approximately 170 Bq/mL) were prepared with distilled water. Soils were ground and passed through a 50 µm sieve. The soil/solution ratio was 1/5. The polypropylene tubes were rotated for 48 hours at 20 ± 1°C to achieve equilibrium. The suspensions were then centrifuged for 30

min at 4000 g in a thermostated centrifuge. Three replicates were made for each sample. The herbicide concentration in solution was measured by liquid scintillation counting. The sorption isotherms were described with the Freundlich equation :

$$x/m = K_f\, C_{eq}{}^{n} \tag{1}$$

with :       $x/m$ = amount of herbicide sorbed (mg/kg)
             $C_{eq}$ = equilibrium herbicide concentration (mg/L)
             $K_f$ = empirical constant (L/kg)
             $n$ = empirical constant (dimensionless).

### Column experiments

Soils (fraction > 1 mm) were packed into High Density Polyethylene columns, 1.5 cm in diameter and 10 cm in length. The bed length was adjusted to 6.0 cm. The columns were connected to a multi-channel peristaltic pump and fraction collectors. Effluent breakthrough curves (BTC's) for bromide and atrazine were measured under steady saturated water flow conditions with a step-input boundary condition at an average pore-water velocity of 5.65 cm/h. The influent concentration for Br and atrazine were 20 and 1 mg/L, respectively. Br and atrazine concentrations in the column effluent fractions were measured by ionic chromatography and HPLC, respectively.

## Results

The main physical and chemical characteristics of the three soils are given in table 1. Significant differences appear in the granulometry and in the organic carbon and $CaCO_3$ contents.

| Soil type | pH (H₂0) | Sand % | Silt % | Clay % | Org.C % | CaCO₃ % |
|---|---|---|---|---|---|---|
| High clay | 8.0 | 24.5 | 13.0 | 62.5 | 1.08 | 3.2 |
| Loamy clay | 8.2 | 3.6 | 64.7 | 31.7 | 1.11 | 1.9 |
| Calcareous clay | 8.0 | 29.2 | 19.5 | 51.3 | 1.50 | 26.4 |

**Table 1** : Selected physical and chemical characteristics of the soils

The sorption isotherms of atrazine on the three soils is illustrated in figure 1, while values of $K_f$ and $n$ are given in table 2.

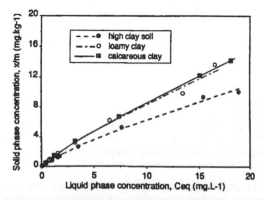

**Figure 1** : Sorption isotherm of atrazine by the three soils with a 48 h equilibration time. Circles and squares are experimental data. Curves are fitted to Freundlich equation.

| Soil type | Kf | n | $r^2$ |
|---|---|---|---|
| High clay | 0.88 - 1.05 | 0.78 - 0.84 | 0.997 |
| Loamy clay | 1.03 - 1.41 | 0.78 - 0.90 | 0.990 |
| Calcareous clay | 1.16 - 1.24 | 0.84 - 0.87 | 0.999 |

**Table 2 :** Parameters of the Freundlich equation describing the adsorption isotherm of atrazine by the three soils. Numbers are the 95% confidence intervals.

The same amount of atrazine was adsorbed by the loamy clay and calcareous clay soils in spite of their different clay (31.7 and 51.3%, respectively) and organic carbon contents (1.11 and 1.50%, respectively). Furthermore, despite similar organic carbon contents, the high clay soil (1.08%) and the loamy clay soil (1.11%) adsorbed different amounts of the herbicides. This does not agree with the generally observed correlation between adsorption and organic carbon or clay contents (Liu et al., 1971; Walker and Crawford, 1968).

Preferential sorption of atrazine on smectites compared to illite or kaolinite has been shown (Bailey et al., 1968). The highest smectite content measured here was, nevertheless, for the high clay soil, which has the lowest sorption capacity (Dousset et al., 1994). The contribution of the clay fraction to the sorption of atrazine appears therefore rather limited, contradicting the observations of Walker and Crawford (1968) for soils with an organic carbon content of less than 3 or 4%.

Numerous studies have shown that the sorption of non-ionic organic chemicals in soils is strongly dependent on the organic content of the soils (e.g., Singh et al., 1990). This has lead to the well known normalisation of sorption to the soil organic carbon (OC) content, $K_{oc} = 100 \, K_f / OC$. The $K_{oc}$ values and the carbon distribution in the humic fraction of the three soils are given in table 3.

| Soil type | Koc | Degree of humidification | Degree of polymerization |
|---|---|---|---|
| High clay | 88.9 | 68.3 | 1.4 |
| Loamy clay | 109.9 | 85.9 | 1.8 |
| Calcareous clay | 80.0 | 73.1 | 2.1 |
| *Degree of humification = C from (fulvic + humic acids + humin)/C total* | | | |
| *Degree of polymerization = C from fulvic acids/C from humic acids* | | | |

**Table 3 :** $K_{oc}$ and carbon distribution in the humic fraction of the three soils.

The $K_{oc}$ of the loamy clay soil is the highest of the three. This may be explained by the fact that this soil has the highest amount of humified matter. The $K_{oc}$ of the calcareous clay soil, lowest of the three, might be explained by its lower degree of humification (73.1%) and polymerisation (2.1). The lower $K_{oc}$ of the high clay soil may be linked to its rather poorly humified organic matter.

Results of the column experiments are illustrated in figures 2 to 4.

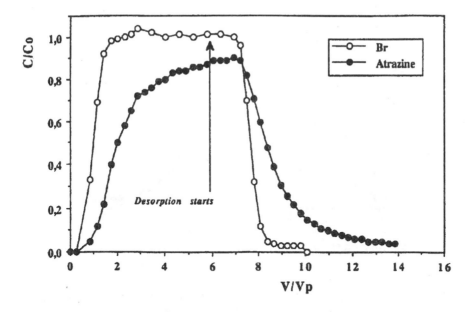

**Figure 2 :** BTC of Br and atrazine for the high clay soil.

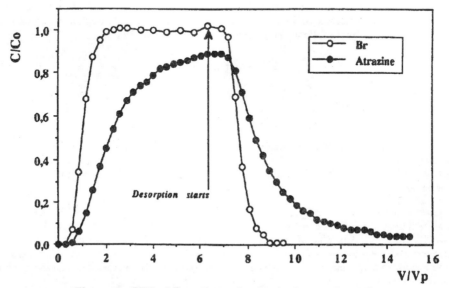

**Figure 3 :** BTC of Br and atrazine for the loamy clay soil.

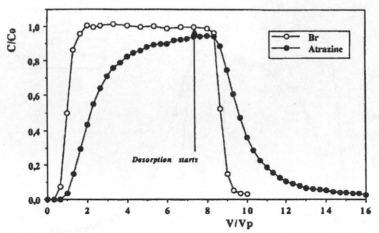

**Figure 4** : BTC of Br and atrazine for the calcareous clay soil.

BTC's that are symmetrical and sigmoidal in shape, such as those for Br in all three soils, suggest the absence of physical nonequilibrium processes in the porous medium. The BTC's for atrazine are asymmetrical, with profound tailing in approach to C/Co = 1. Tailing or increased dispersion is typical for BTC's obtained under nonequilibrium conditions. The fact that only the BTC's for the sorbing solute exhibit increased dispersion or asymmetry suggests that the mechanism responsible for nonequilibrium is related to the sorption process (Nkedi-Kizza et al., 1989).

Various methods can be used to calculate the leaching retardation factors (Rf). One method involves sorption isotherm data (method 1), while others (e.g. method 2) involve analysing the column BTC data. Dual calculation enables a comparison of sorption behaviour measured under static conditions (batch isotherms) with that characterised under dynamic flowing conditions (BTC data). The retardation factor (Rf) is a measure of the mobility of a solute being eluted through a soil column. For method 1, Rf is given by the following equation (Nkedi-Kizza et al., 1987) :

$$Rf = 1 + \rho \, Kd/\theta \qquad\qquad (2)$$

with   $Kd$ = sorption coefficient $(cm^3/g)$
       $\rho$ = bulk density $(g/cm^3)$
       $\theta$ = volumetric water content $(cm^3/cm^3)$.

In method 2, the value of Rf is set equal to the number of pore volumes required for the effluent pesticide concentration to reach 0.5 Co. This method is based on the assumption of symmetrical sigmoidal BTC's with the implicit assumption that equilibrium conditions prevail during pesticide leaching through the soil column (Nkedi-Kizza et al., 1987).

Rf values computed from the isotherm data and the values from the BTC at C = 0.5 Co are shown in table 4. All the values of Rf computed from the isotherm data are greater than those obtained from the BTC's at C = 0.5 Co. The differences in $K_f$ values in the various soils observed in the batch experiments are not found in the Rf values. Nonequilibrium during the column experiments, probably due to a too high pore water velocity, may explain both of these differences between the batch and the column results.

| Soil type | Rf from the BTC at C/Co = 0.5 | Rf computed from the batch data (see*) | Kf, batch sorption coefficient |
|---|---|---|---|
| High clay | 2.2 - 2.3 | 2.5 - 2.8 | 0.88 - 1.05 |
| Loamy clay | 2.0 - 2.1 | 2.8 - 3.4 | 1.03 - 1.41 |
| Calcareous clay | 2.2 - 2.4 | 3.1 - 3.3 | 1.16 - 1.24 |
| * Rf = 1 + Kf* bulk density/porosity | | | |

**Table 4 :** Comparison between the Rf values from the column BTC's and those computed from the batch isotherm data, together with the sorption coefficient Kf values from the batch isotherms.

## Conclusion

Batch experiments have revealed significant differences between soils with respect to their sorption capacity of atrazine. For similar organic carbon contents, as is the case for the loamy clay and high clay soils, the degree of humification of the organic matter plays an important role in sorption. Column experiments yielded BTC's indicative of nonequilibrium processes, most likely linked to the sorption processes. The resulting retardation factors were smaller than those calculated from the batch data and did not vary with the soil sequence as did those of the static sorption coefficients. The main advantages and disadvantages of the batch and the column approaches are summarised in table 5.

Batch experiments enable rapid comparisons of soils and pesticides and provide soil adsorption coefficients which can be used for modelling. The fact that the interactions linked to the water movement cannot be taken into account is a major drawback of this approach. Column experiments, in which the contact between the soil and the water is based on the water movement, are therefore complementary to the batch procedure.

| Parameter | Batch | Column |
|---|---|---|
| Ease of set-up | + | - |
| Standardized protocol | + | - |
| Solid/ water ratios (range ; ease of change) | + | - |
| Soil types (variety) | + | - |
| Ease of interpretation | + | - |
| Similarity with reality | - | + |
| Non saturated conditions | - | + |
| Consolidated solids | - | + |
| Hydrodynamic component | - | + |
| Kinetics studies | + | + |

**Table 5 :** Advantages and disadvantages of the batch and column methodological approaches.

# References

▢   Bruckert, S., Andreux, F., Correa, A., Ambouta, K.J.M. and Souchier, B., 1978. Fractionnement des agrégats appliqués à l'analyse des complexes organo-minéraux des sols. *Note
technique n°22*. Centre de Pédobiologie Biologique, CNRS, Nancy.

▢   Dousset, S. 1994. Evaluation des potentialités de lessivage de la terbuthylazine et de l'atrazine
dans trois sols de la Région Centre. *PhD thesis*, 161 p. INPL/ENSAIA, Nancy.

▢   Dousset, S., Mouvet, C. and Schiavon, M., 1994. Sorption of terbuthylazine and atrazine in
relation to the physico-chemical properties of three soils. *Chemosphere*, 28, 3, 467-476.

▢   Liu, L.C., Cibes-Viade, H. and Koo, F.K.S., 1971. Adsorption of atarzine and terbacil by soils.
*J. Agri. Univ. Puerto-Rico* LV, 4, 451-460.

▢   Ministère de la Solidarité, de la Santé et de la Protection Sociale, 1990. Teneur en triazine des
eaux destinées à la consommation humaine. DSG/PGE/1.D, n°717, 9p., 1 Place de Fontenoy,
F 75350 Paris 07.

▢   Nkedi-Kizza, P., Brusseau, M.L., Rao, P.S. and Hornsby, A.G., 1989. Nonequilibrium sorption
during displacement of hydrophobic organic chemicals and 45Ca through soil columns with
aqueous and mixed solvents. *Environ. Sci. Technol.*, 23, 7, 814-820.

▢   Nkedi-Kizza, P., Rao, P.S. and Hornsby, A.G., 1987. Influence of organic cosolvents on
leaching of hydrophobic organic chemicals through soils. *Environ. Sci. Technol.*, 21, 11,
1107-1111.

▢   Rouiller, J., Burtin, G. and Souchier, B. 1972. La dispersion des sols dans l'analyse granu-
lométrique des sols. Méthode utilisant les résines échangeuses d'ions. *Bull. E.N.S.A.I.A.*, XIV,
193-205.

▢   Singh, G., Spencer, W.F., Cliath and Van Genucthen, M. Th., 1990. Sorption behaviour of
s-triazine and thiocarbamate herbicides on soils. *J. Environ.Qual.*, 19, 520-525.

▢   Walker, A. and Crawford, D.V., 1968. The role of organic matter in adsorption of triazine
herbicides by soils. *I.A.E.A. Symposium on the use of isotopes and radiations in soil organic
matter studies*. SM 106/19, 91-107.

# Acknowledgements

The financial contribution of the Conseil Régional de la Région Centre, through a post-graduate
fellowship to S. DOUSSET, was greatly appreciated. This is BRGM contribution n°94026.

## 2.2.10.    *Affinity and Mobility of Organophosphorus Nematicides in Soils and Plants*

# M. Ammati[1], M. Mansour[2], M. El M'Rabet[1] and K. Kacimi[3]

[1]    I.A.V. Hassan II, Rabat-Instituts - Maroc

[2]    GSF Oekochemic, Freising-Attaching - Germany

[3]    Dept. de Chimie Organ., Fac. des Sciences, Rabat - Maroc

### (Full text not received)

*Isazofos and fenamiphos are soil applied nematicides, widely used on crops conducted under plastic houses; Undisturbed soil columns (1 m length x 12 cm diameter) were used to assess the mobility and distribution of these pesticides in soil profiles, water leachates and banana seedlings. Each column was under irrigation regime equivalent to agricultural practice (150 ml/day/column) and supplied by peristaltic pumps. The amount of active ingredients applied corresponded those applied by the growers (1250 mg a.i./nematicide). Fenamiphos and its oxidation toxic metabolites (sulfone and sulfoxide), isazophos was monitored for 13 weeks; in the water leachates collected daily under each column; in the soil profile for 5 weeks and in the roots and shoots for 3 weeks. Quantification and identification of different active ingredients were done with GLC, equipped with and NPD Detector. For fenamiphos and its toxic metabolites, 19.7, 6.6 and 3.4 % of the initial concentration were observed respectively in soil profile, water leachates and plant material (roots and shoots). In water leachates the maximum concentration was observed after four weeks and was evaluated at 47.5 ppm. Fenamiphos and its toxic metabolites attained 65.6 ppm in the roots and 428 ppm in the shoots. The soil adsorption, $K_{oc}$ for fenamiphos and its toxic metabolites combined, varied within the soil profile and was of 90 in the first top 20 cm and increased to 230 in the lower layer (> 80 cm). For isazofos, 37.15, 1.4 and 0.1 % of the initial concentration were respectively observed in the soil profile, the water leachates and the plant material. The maximum concentration in the water leachates was 4.4 ppm after six weeks. The concentrations of isazofos attained 2.7 and 6.7 ppm respectively in the roots and the shoots. The soil adsorption $K_{oc}$ relatively high and varied from 2190 in the first top 20 cm to 4320 at 40 cm depth. Isazofos was not detectable below 40 cm depths. This study showed the high mobility of fenamiphos and its toxic metabolites compared to isazofos. The soil adsorption $K_{oc}$ and the mobility of different active ingredients in plant material and soil are nicely related. If this relationship can be extended to other classes of compounds, this experimental design could be used to predict soil adsorption and mobility of nematicides in plants. Similar results were observed in the banana fields.*

## 2.2.11.    Mobility of Phenylurea Herbicides in Soil Columns

### C. Brouard, J.-P. Taglioni, J. Fournier

Laboratoire de chimie bioorganique, Centre Régional d'Etude
des Produits Agropharmaceutiques (CREPA),
8 rue Becquerel, Angers-Technopole, 49070 BEAUCOUZE.

## Introduction

The migration of pesticides in soils is of extreme importance owing to eventual contamination of groundwater supplies. This migration mainly depends on the amount of water moving through the soil and to the extend of pesticide sorption on soil. Many environmental factors affect the migration of herbicides in soil.

The purpose of this study was to investigate the influence of the herbicide chemical structure on the adsorption and on the mobility through a soil column of phenylurea herbicides.

We have studied the mobility of five phenylurea herbicides : metobromuron, chlortoluron, isoproturon, chloroxuron and difenoxuron. As expected, the movement of herbicides in soils has been shown to be depended on their physical and chemical properties.

We will examine, in the laboratory, the difference between the mobilities of these ureas through a soil column.

## Structure and Chromatographic Analysis of Urea

### *Structure (1.2)*

This study concerns five herbicides of the phenylurea family. The name and molecular formulae are given below.

ISOPROTURON

METOBROMURON

CHLOROXURON

CHLORTOLURON

DIFENOXURON

*Analysis of phenylurea herbicides by liquid chromatography*

A High Performance Liquid Chromatography procedure is used to analyze phenylurea herbicides in soil extracts.

*- H.P.L.C instrumentation* :

The instrumental system consisted of a Waters 600 Multisolvent injector, a Waters 991 photodiode array detector and a Waters 5200 Printer plotter.

The column used was a 25 cm * 4.6 mm I.D column packed with ODS-bonded, 5 μm spherical silica (Lichrospher RP-18, Merck).

Operating conditions were as follows : mobile phase consisted of methanol - water mixture (60 : 40 v/v) ; flow rate, 1 ml/min (pressure = 2700 Psi) ; injection volume, 25 μl.

| | Metobromuron | Isoproturon | Chlortoluron | Chloroxuron | Difenoxuron |
|---|---|---|---|---|---|
| λ (mm) | 244,2 | 241,6 | 242,4 | 245,5 | 246,8 |
| TR (min) | 13,10 | 13,64 | 12,36 | 33,65 | 14,95 |

**Table 1** lists the retention time and the optimum absorption wavelengths

# Mobility of Urea in Soil Columns (3,4,5,6,7)

## *Materials*

Studies were performed using a 30 cm * 1.5 cm I.D column (Bioblock), fitted with a pump, flow rate 3 ml/hour (Gilson - Grosseron).

## *Characteristics of the soil*

The soil which has been selected for this study is the soil of Grignon kindly supplied by Professeur Calvet.

| Clay g/kg | Coarse silt g/kg | Fine silt g/kg | Fine sand g/kg | Coarse sand g/kg | % C | % N | pH (eau) | CEC cmol/kg |
|---|---|---|---|---|---|---|---|---|
| 249 | 472 | 215 | 54 | 10 | 0,60 | 0,078 | 8,1 | 12,3 |

## *Procedure*

Column was packed with 10 grams of soil and was satured with demineralized water. The column was eluted with an aqueous solution of urea (1 ug/ml), with a flow rate of 3 ml/hour. Elution was stopped when the effluent concentration approached or became identical to the input concentration : the column was then eluted with demineralized water (pH = 6.1).

Column effluent was sampled every thirty minutes and analyzed for urea concentration by H.P.L.C procedure.

## *Results*

The curves of the measured concentrations for isoproturon, chlortoluron, metobromuron, chloroxuron and difenoxuron in the soil column eluent are displayed on figure 1.

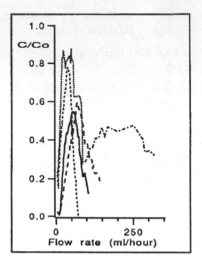

**Figure 1**

The eluent volume needed for an urea to obtain in the effluent 50 % of the inlet concentration (C/Co = 0.5) is used as a direct estimation of its retardation factor.

|  | Metobromuron | Chlortoluron | Isoproturon | Difenoxuron | Chloroxuron |
|---|---|---|---|---|---|
| Retardation factor | 45,2 | 21,4 | 12,5 | 63,1 | -- |

Because of strong holding of chloroxuron by soil, we couldn't determine the retardation factor; the adsorption was so long that the value C/Co = 0.5 cannot be reached even after 300 ml.

From reported retardation factor values for the five ureas, it can been seen that they followed the order :

cloroxuron > difenoxuron > metobromuron > chlortoluron > isoproturon

If we compare molecular formulae of these ureas, only chloroxuron and difenoxuron present a phenoxy group in their structure. This chemical group can be thought to be partly responsible of the strong holding of chloroxuron and difenoxuron by the soil components.

In a previous study, Freudlich isotherms have been determined in our laboratory for each urea on Grignon soil ; similar results had been noticed with these obtained in the mobility studies. The Freundlich coefficient K was very important for chloroxuron and it was lesser for difenoxuron.

## Mobility of Isoproturon in Soil Columns

### *Effect of the soil mass (column height) on the movement of isoproturon*

Figure 2 shows the curves of isoproturon concentration in the effluent versus elution volume as influenced by the mass of Grignon soil used in the column.

5, 8 10 and 20 grams of soil were mixed with respectively 5, 8, 10 and 20 grams of Fontainebleau sand.

In this study, the column was eluted with an aqueous solution of calcium chloride 0.01 M to be the ionic force of soil solution reproduced.

| Quantity of soil (g) | 5 | 8 | 10 | 20 |
|---|---|---|---|---|
| Height (cm) | 4,9 | 6,0 | 8,1 | 16,3 |
| Retardation factor | 9,33 | 14,66 | 14,66 | 32,00 |

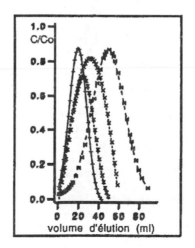

**Figure 2**

We observed that a ratio of 1.65 in the heights of soil columns between 5 and 10 grams results of approximatively 1.6 in the ratio of eluent volumes needed to reach C/Co = 0.5, and that a ratio of 2.01 in the heights of soil columns between 10 and 20 grams results of approximatively 2.18. The light difference between these two ratio may be attributed in part to the higher dispersion that occured in the long column (20 grams).

### *Effect of the ionic force on the movement of isoproturon*

For this study, the column was filled with a mixture of 10 grams of soil and 10 grams of Fontainebleau sand. The columns were eluted with aqueous solutions of calcium chloride of different concentrations (0.01 ; 0.1 ; 1 M). The effect of ionic force on the mobility of isoproturon is illustrated in figure 3. We observed that the retardation factor gains a limit with the ionic force increasing.

|     | H₂O | CaCl₂ 0,01 M | CaCl2 0,1 M | CaCl₂ 1 M |
|-----|-----|-------------|-------------|-----------|
| $\mu$ | 0 | 0,03 | 0,3 | 3 |
| Rf | 23,2 | 20,0 | 16,8 | 15,8 |

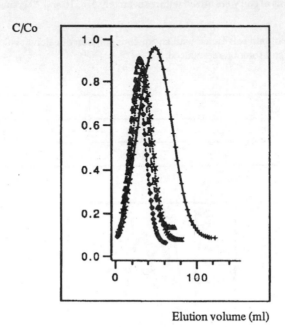

Elution volume (ml)

**Figure 3**

As it can be seen from figure 4, the mobility of isoproturon increases with the ionic force.

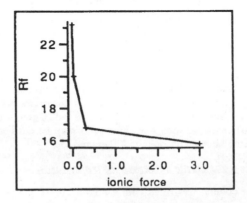

**Figure 4**

# Conclusion

The purpose of this study was to compare the mobility of phenylurea herbicides through a soil column.

Among the five ureas, we have observed that chloroxuron and difenoxuron were strongly held by the soil and so their mobilities slow down compared with the other ureas.

Isoproturon, chlortoluron and metobromuron are more mobile in the soil than chloroxuron and difenoxuron. This implies that chloroxuron and difenoxuron are held to a greater extend by the soil collo5ds than the other ureas. The similarity in molecular chemical formulae of these two compounds would suggest that phenoxy group plays a leading place in the holding phenomena.

But other factors are involved in the differential mobilities of these five ureas in soils. The most obvious reason for these observed results is their solubility in water. As expected, this study shows that the retardation factors increase when the water solubility of ureas is lower. Since soil holding and mobility are related to competition between adsorption on soil components and soil water dissolution, this may explain the reduced mobilities of chloroxuron and difenoxuron.

# References

□    [1] The agrochemicals Handbook, The Roy. Chem. Soc. ed., Nottingham, 1993.

□    [2] Index Phytosanitaire, ACTA, Paris 1992.

□    [3] R. Calvet, I.N.R.A., I.N.A-PG, Centre de Grignon, "Evaluation des coefficients d'adsorption et prédiction de la mobilité des pesticides dans les sols", *Methodological aspects of the study of pesticide behaviour in soil*, INRA Versailles, P. Jamet éd., June 16-17, 1988.

□    [4] Garnson Sposito, "The Chemistry of soils", Oxford Univeersity Press, 1989.

□    [5] S.-Q. Zheng, J.-F. Cooper, P. Fontanel, Laboratory of analytical Chemistry, Faculty of Pharmacy, 34060 Montpellier, France, "Movement of pendimethalin in soil of the south of France", *Environ. Contam., Toxicol.*, 50:492-498, 1993.

□    [6] Wondimagegnehu Mersie and Chester L. Foy, "Adsorptio, Desorption and Chlorsulfuron in soils", *J. Agri. Food Chem.*, 34, 89-92, 1986.

□    [7] Hugh J. Beckie and Robert B. McKercher, Department of Soil Science, University of Saskatchewan, Saskatoon, Canada S7N OWO, "Mobility of two sulfonylurea herbicides in soil", *J. Agri. food Chem.*, 38, 310-315, 1990.

# Acknowledgement

These studies were financed by ADEME and the Ministère de l'agriculture.

# 2.3.        Poster Presentations

## 2.3.1.        *Mobility of Diazinon in Soils Modified by Addition of Organic Matter*

# [1]M. Arienzo, [2]M. Sanchez-Camazano, [2]M. J. Sánchez-Martín, [2]T. Crisanto

[1]    Dipartimento di Scienze Chimico-Agrarie, Universitá di Napoli "Federico II",
       Via Universitá 100-I, Portici, Italy
[2]    Instituto de Recursos Naturales y Agrobiologia. C.S.I.C. Salamanca. Spain

## Introduction

One of the major anthropogenic supplies of organic matter to soils is in the form of organic fertilizers or amendments, whether solid or liquid; it is thus surprising that little research has so far been conducted on their potential influence on soil pesticide transport in soil. Zsolnay (1992) studied the influence of an organic fertilizer on atrazine mobility, and Guo et al. (1991; 1993) investigated the effect of carbon-rich residues on atrazine and alachlor mobility. No study on the influence of exogenous organic matter on insecticide mobility has to date been reported.

In this work, the influence of two organic amendments and a liquid humic amendment on diazinon mobility was studied by leaching soil packed columns and using $^{14}$C-labelled diazinon. The effects of a carbon-rich organic compound of known structure on the pesticide mobility was also examined.

Diazinon (O,O-diethyl-O-2-isopropyl-6-methylpyrimidin-4-yl phosphorothioate) is a non-systemic insecticide commonly applied to horticultural, sugar can and cotton plant crops. Several authors have detected it in ground waters (Hallbeg, 1989) and others have studied its soil persistence relative to other pesticides (Sattar, 1990). In previous work on the adsorption and mobility of diazinon in various soils, the authors (Arienzo et al., 1994) found the organic matter content to be very important in both processes.

## Materials and Methods

The soils used were a chromic cambisol (soil) and a cambic arenosol (soil 2). Table 1 shows some selected characteristics of both.

$^{14}$C-labelled diazinon of specific activity 185 MBq/mg was purchased from International Isotope (Munich, Germany). Unlabelled diazinon was supplied by Riedel de Haen (Hannover, Germany). Its solubility in water is 40 mg/l.

Two peats commonly employed as agricultural amendments, namely Torfsicosa (TP), (Plantaflor Humus Veraufs GmbH, Vechia, Germany) containing 90% organic matter, and Grun Garant (GGP) (Deutsche Torfgesellschoft GmbH, Saterland Scharrel, Germany), containing 60% organic matter, were used. A liquid humic amendment, viz. Huminag (HH) (Braker Laboratories, S.L., Valencia, Spain) containing 17% humic and fulvic acids was also employed. The carbon-rich organic compound of known chemical structure was hexadecyltrimethylammonium bromide (HDTMA) (Fluka AG, Busch, Switzerland). The organic carbon content of the amendments was as follows: 52.2% for TP, 34.7% for GGP, 22.1% for HH and 62.7% for HDTMA.

| Soil | Texture | pH | Organic matter % | Sand % | Silt % | Clay % | Clay a mineralogy[a] |
|------|---------|-----|------------------|--------|--------|--------|-----------------------|
| 1 | Clayed sand | 4.6 | 1.04 | 73.8 | 12.1 | 14.1 | I, K, S |
| 2 | Sandy | 6.1 | 0.47 | 86.4 | 2.9 | 10.7 | S, I, K |

[a] In order of abundance: I, Illite; S, Smectite; K, Kaolinite

**Table 1** : Selected characteristics of the soils

Soil column leaching was performed as follows: columns were 40 cm long X 5 cm inner diameter; dose of amendment applied was 4 t/ Ha as total carbon; five ml of 200 µg/ml diazinon in ethanol (5 kg/Ha) of 36 kBq/ml specific activity were added at the top of each column; the columns were then leached with 200 ml (10.2 cm) of water on a daily basis for 12 days; column leachates were monitored daily for their insecticide content by using 1 ml of sample to measure the activity (DPM) and after draining for some time, the columns were cut breadthwise at 5-cm intervals. $^{14}$C-diazinon was determined in soil and leachate using a Harvey OX-500 biological oxidizer and a Beckman LS 1800 scintillation counter. Each experiment was performed in duplicate and two columns of each soil to which no external organic matter was added were also analysed for comparison.

## Results and Discussion

Modifying soil 1 with HDTMA decreased the cumulative amount of diazinon in the leachates from 49.51% in the control soil to 18.33% (Table 2). The decrease produced by the three organic amendments was smaller (to 41.85-46.25%).

The diazinon peak concentration in the percolation curves was 5.80% for the control column and only 2.30% for the HDTMA-modified soil column (Table 3). Those for the soils modified with the other three types of amendment were 5.12% (GGP), 5.44% (TP) and 5.54% (HH), respectively. The peak of the percolation curve appeared at a water drained volume of 737 ml for the control soil and 744, 769 and 711 ml for the HH-, GGP- and TP-modified soil, respectively. On the other hand, the peak for the HDTMA-modified soil appeared at 385 ml. In any case, the percolation curve for the HDTMA-treated soil did not exhibit a well-defined maximum, but rather several small, separate peaks (the diazinon concentration varied over a narrow range, viz. 1.75-2.30%, along the percolation curve).

The control column retained 25.15% of the amount of diazinon applied (Table 2), whereas that containing HDTMA-modified soil retained as much as 70.48%. The amounts retained by the soils treated with the other three amendments were very similar to one another, as were the amounts of pesticide in their respective leachates.

As regards partitioning of retained diazinon in the column, most of the residual pesticide (50%) was held in the first segment of the HDTMA-modified soil. The pesticide also built up in the first segment of the TP- and GGP-treated soil columns. On the other hand, residual pesticide in the soil treated with the liquid amendment did not accumulate in any particular segment, but partitioned evenly across the column.

The mobility of diazinon in modified soil 2 was very similar to that in soil 1 (Table 2 and 3). However, there were some, small differences probably arising from textural dissimilarities between the two soils; thus, while soil 1 was essentially clayed sand, soil 2 was basically sandy.

| Segment | Control | Soil treatment | | | |
|---|---|---|---|---|---|
| | | Torfsicosa TP | Grün Garant GGP | Humimag HH | HDTMA |
| **Soil 1** | | | | | |
| 1 | 2.50 | 13.80 | 11.51 | 4.96 | 51.99 |
| 2 | 2.70 | 2.95 | 3.34 | 5.80 | 5.47 |
| 3 | 5.20 | 3.21 | 3.76 | 5.83 | 3.44 |
| 4 | 5.00 | 3.11 | 5.44 | 5.42 | 3.47 |
| 5 | 5.10 | 2.94 | 4.07 | 5.87 | 2.94 |
| 6 | 4.80 | 4.57 | 5.51 | 5.99 | 3.14 |
| Soil | 25.15 | 30.57 | 33.63 | 33.86 | 70.48 |
| Water | 49.51 | 41.85 | 46.25 | 44.98 | 18.83 |
| Total | 74.66 | 72.42 | 79.88 | 78.88 | 89.31 |
| **Soil 2** | | | | | |
| 1 | 3.14 | 10.82 | 10.03 | 3.12 | 47.60 |
| 2 | 3.39 | 2.63 | 4.00 | 2.92 | 5.86 |
| 3 | 3.30 | 3.11 | 4.04 | 2.79 | 4.55 |
| 4 | 2.81 | 3.15 | 3.54 | 3.38 | 1.72 |
| 5 | 3.95 | 3.34 | 3.54 | 2.91 | 1.61 |
| 6 | 3.45 | 4.33 | 4.43 | 4.42 | 2.10 |
| Soil | 20.04 | 27.37 | 29.57 | 19.51 | 63.44 |
| Water | 57.77 | 42.27 | 46.44 | 53.15 | 23.70 |
| Total | 74.81 | 69.64 | 76.01 | 72.66 | 87.14 |

**Table 2 :** Amounts of diazinon retained and leached (% of applied) in the columns of soils modified with organic matter

The cumulative amount of diazinon in the leachate from the control column of soil 2 was 54.77% of the amount applied and that retained in the column accounted for 20.04% (the two values are slightly higher and lower, respectively, than those for soil 1). The amounts of diazinon retained by the columns treated with HDTMA and the three organic amendments were also somewhat smaller, while the cumulative amounts were larger relative to soil 1. Diazinon peak concentrations ranged from 7.07% for the control soil to 3.14% for the HDTMA-modified soil. The percolation curve for the column treated with the liquid organic amendment exhibited both the maximum concentration (6.57%) peak at 768 ml and another peak (4.95%) at a lower volume (395 ml).

As regards diazinon partitioning in the columns, about 50% of the amount applied to the soil was retained by the first segment in the HDTMA-treated column. The pesticide also accumulated preferentially in the first segment of the columns modified with the solid organic amendments, but evenly throughout that modified with the liquid amendment.

The above results show that organic materials reduces the mobility of diazinon in soils. Of the four amendments studied, HDTMA is the most efficient for hindering leaching of the pesticide. On the other hand, the other three organic amendments are similarly efficient, even though there is a slight difference in this respect between the two peats and the liquid organic amendment.

| Treatment | Soil 1 | | Soil 2 | |
|---|---|---|---|---|
| | Volumen ml | Peak concentration % | Volumen ml | Peak concentration % |
| Control | 737 | 5.80 | 763 | 7.07 |
| HDTMA | 385 | 2.30 | 396 | 3.14 |
| GGP | 744 | 5.54 | 395 | 4.95 |
| | | | 768 | 6.57 |
| TP | 769 | 5.12 | 757 | 5.65 |
| HH | 771 | 5.44 | 690 | 5.94 |

**Table 3 :** $^{14}$C-Diazinon peak concentration (% of applied) and leachate volumens (ml) for the soils modified with organic matter

The results obtained show that the mobility of diazinon in the two soils modified with organic amendments is influenced by: 1) hydrophobic nature of diazinon, 2) carbon content of the amendments, 3) soil texture, 4) hydrophobic nature of humic and fulvic acids (active organic components of TP, GGP, and HH), 5) highly hydrophobic character of HDTMA, 6) presence of dissolved organic matter in HH (liquid amendment).

## Aknowledgements

This work was financially supported by the Spanish "Comisión Interministerial de Ciencia y Tecnologia " as part of Project Nat. 91-0616. The authors wish to thank L. F. Lorenzo and A. Nuñez for their kind technical assistance.

## References

□ Hallberg, G.R., 1989. *Agric. Ecosys. Environ.*, 26: 299-367.

□ Arienzo, M., Crisanto, T., Sànchez-Martìn, M.J. and Sànchez-Camazano, M., 1994. *J. Agric. Food Chem.* (in press).

□ Guo, L., Bicki, T.J. and Hinesly, T.D., 1991. *Environ. Toxicol. Chem.*, 10:1273-1282.

□ Guo, L., Bicki, T.J., Felsot, A.S. and Hinesly, T.D., 1993. *J. Environ. Qual.*, 22: 186-194.

□ Sattar, M.A., 1990. *Chemosphere*, 20: 387-396.

□ Zsolnay, A., 1992. *Chemosphere*, 24: 663-669.

2.3.2.      Role of Soil Fractions in Retention and Stabilisation of
            Pesticides in Soils

# E. Barriuso, P. Benoit and V. Bergheaud

I.N.R.A., Unité de Science du Sol, 78850 Thiverval-Grignon, France

## Introduction

Chemical fractionation of soil organic matter, mainly using alkaline solvents, has been used to study
the interactions between soil organic compounds and pesticides (Andreux et al., 1991). The distribution of pesticide bound residues in fulvic and humic acids and humin, has been published for many
pesticides and soils (Khan, 1980; Bertin and Schiavon, 1989). Physical soil fractionation methods are
less employed in soil-pesticides studies (Barriuso et al., 1991). They allow to separate soil organic
fractions according to physical properties and to their association with soil mineral constituents
(Turchenek and Oades, 1979; Andreux et al., 1991). Soil physical fractionation separates fresh or less
humified organic matter from the humified organic matter. The former is localized in the coarsest
fractions while the latter is in the finest fractions, usually smaller than 50 or 20 μm. Generally, organic
compounds localized in 20-2 μm fractions are more stable, less biodegradable and have a larger
aromaticity; whereas, the organic compounds in < 2 μm fractions are more labile, have more aliphatic
constituents and include more biochemical residues from microbial biomass (Catroux and Schnitzer,
1987; Christensen, 1992).

These different nature and properties of soil organic constituents can provoke different pesticide fate,
mainly through pesticide retention processes (Nkedi-Kizza et al., 1983; Dios Cancela et al., 1989).
Sorption can be the first step of pesticide residues stabilization, and pesticide sorbed can evolve to
unextractable forms, as bound residues. The aim of this paper is to present some application of soil
fractionation techniques with the soil fractions characterisation in relation to availability to sorb
pesticide, then to form stabilised residues as bound residues. As illustration, results from an *Inceptisol*
soil and the atrazine are presented.

## Materials and Methods

Soil used was a Typic eutrochrept, from an experimental corn field, with a silt loam texture, an organic
C content of $14.0\,g\,kg^{-1}$, and a pH of 6.4. Soil was spiked in laboratory with $^{14}C$-atrazine, and incubated
under outdoor conditions. The incubation conditions were described in a earlier study (Barriuso et al.,
1991).

Soil samples were fractionated after 6 months of incubation. First, water-soluble $^{14}C$-residues were
exhaustively extracted by successive shaking with 0.01 M $CaCl_2$ water solution. Then, the soil samples
were size fractionated: two fractions (200 and 200-50 μm) were separated by sieving; three by
successive sedimentation (50-20, 20-2, and 2-0.2 μm); and one (< 0.2 μm) by centrifugation after
flocculation with 0.01 M $CaCl_2$. Each isolated fraction was exhaustively extracted with methanol.
Then, the soil size fractions containing the atrazine non-extractable (bound) residues were air-dried
at 40°C. Humic compounds were extracted by shaking each soil size fraction in a water alkaline
mixture of 0.1 M NaOH and 0.1 M $Na_4P_2O_7$; that allowed isolated the humin, fulvic and humic acids
for each fraction.

Additional experiences were done to evaluate the mineralization rate of bound residues of each
fraction. Aliquots of the soil size fractions were mixed with a fresh soil and incubated for 100 days
in hermetic jars at 28°C and 80% of water retention capacity of soil-fraction mixtures. $^{14}CO_2$ evolved
was trapped in 0.5 M NaOH.

Soil without [14]C-atrazine was submitted to the same fractionation procedure, and fractions, recovered by dry-freezing, were used for batch atrazine adsorption measurements. An aliquot of each fraction was suspended in a given volume of a solution of 10 mg l[-1] [14]C-atrazine in 0.01 M $CaCl_2$; the solution/weight of fraction was 5/1, and kept constant for all fractions. After 24 h equilibration by shaking at 25°C, supernatant was recovered by centrifugation and their radioactivity measured. Kd and Koc for each fraction was calculated as classically.

The total organic C in the fractions was determined by dry combustion (Elemental Analyser CN 1500, Carlo Erba Instruments). The radioactivity in liquid samples was measured with a Kontron V liquid scintillation counter with quenching correction by external standardization. The total radioactivity of solid samples was determined by trapping the [14]$CO_2$ produced after sample combustion with a Packard Sample Oxidizer B306 and counting the radioactivity as previously described.

## Results and Discussion

Distribution of soil C into soil size fractions is shown in Table 1. 19% of the soil organic matter was found in the fractions coarser than 20 µm, representing the less humified organic matter. The C content of soil size fractions increased when grain size decreased, and the highest was in the clay fractions.

| | Organic C distribution | | Atrazine adsorption characteristics | | | |
|---|---|---|---|---|---|---|
| | Fraction gC kg[-1] | Soil % soil C | Kd 1 kg[-1] | Koc 1 kg[-1] | (x/m) * mg kg[-1] | % (x/m) % total |
| Whole soil | 13.7 | 100 | 0.81 | 59 | 7.34 | 100 |
| > 200 µm | 7.4 | 5 | 0.63 | 86 | 0.52 | 5 |
| 200-50 µm | 10.7 | 7 | 1.13 | 106 | 0.83 | 7 |
| 50-20 µm | 3.7 | 7 | 0.30 | 82 | 0.72 | 7 |
| 20-2 µm | 14.5 | 42 | 0.96 | 66 | 3.31 | 44 |
| 2-0.2 µm | 28.1 | 29 | 1.31 | 47 | 1.49 | 30 |
| < 0.2 µm | 27.7 | 7 | 1.83 | 66 | 0.46 | 7 |

*Amount adsorbed on fractions expressed in relation to weight contribution of each fraction to wheight of whole soil.

**Table 1 :** Carbon content of soil size fractions, soil carbon distribution into fractions, and atrazine adsorption parameters for the different fractions

This C content in fractions increased with atrazine adsorption (Table 1). That is a general result because of the strong relation between C content and adsorption: adsorption increased when the particle size decreased because of increasing C content of soil particle-size separated (Nkedi-Kizza et al., 1983; Dios Cancela et al., 1989). Special adsorption behaviour was found in the finest fraction (< 0.2 µm), which C content was similar to this of 2-0.2 µm fraction whereas, atrazine adsorption was higher.

Koc values allow to point out that less humified organic matter in the coarsest soil size fractions had higher adsorption capacity than humified organic matter in the finest fractions (Table 1). The decrease in the capacity to adsorb atrazine with respect to particle size was apparently in contradiction with the increase in specific surface area. These results indicated that specific surface area of humified organic compounds may have been overestimated, as pointed out by Chiou et al. (1990). It is possible that the organic surface of the finest size fractions was not accessible, and reactive groups not available. Increase of specific surface when particle size decreases can result in organic matter consuming most reactive groups to establish interactions with mineral sites. Low specific surface area of minerals in coarse fractions would have organic matter with free reactive groups capable of binding.

A important result was the additivity of atrazine adsorption on separate fractions as compared with adsorption on total soil (Table 1). That could indicate that during soil adsorption measurements in batch all surfaces are accessible to atrazine, without hiding in aggregates structures disruptables by the fractionation procedure.

Stabilization of atrazine residues as soil bound residues was a rapid process and nonextractable atrazine residues were formed immediately after atrazine application (Barriuso and Koskinen, 1994). The proportion of non extractable residues is dependent of soil C content (Khan, 1980; Bertin and Schiavon, 1989; Calderbank, 1989; Barriuso et al. 1991). Equilibrium with residues in compartments defined through their extractability in water and methanol took place in soils. Increasing atrazine residence time in soil decreased the availability of residues extracted in water, and increased bound residues. After 6 months, distribution of $^{14}$C-residues remaining in soil was: 28 % as water soluble, 3 % as methanol soluble, and 64 % as bound residues (which represented 19 % of $^{14}$C-atrazine initially applied.

Fresh less humified organic matter localized in size fractions coarser than 50 μm had the largest capacity to form atrazine bound residues (Table 2). The organic matter capacity for atrazine stabilization as bound residues decreased as particle size decreased (Fig. 1). Nature of organic compounds in relation to size fractions could explain this result; however, decreased accessibility of organic compounds in the finest fractions of aggregate structures could not be discarded.

Bound residues of whole soil were mainly localized in humified fractions, especially in the two silt fractions 20-5 and 5-2 μm (Fig. 2). The bound residues distribution in relation to the organic matter content shows that the less humified organic matter immobilized 3 to 4 times more atrazine residues than the well humified organic matter of the smallest fractions. It could be expected that the humified organic matter would present the greatest reactivity toward atrazine, mainly because of its large specific surface area (Broersma and Lavkulich, 1980).

|              | mg $^{14}$C-atrazine-equivalent kg$^{-1}$ fraction | | | | |
|--------------|-------|-------|-------------|-------------|----------------|
|              | Total | Humin | Humic Acids | Fulvic Acids | % in whole soil |
| > 200 μm     | 1.44  | 0.88  | 0.14        | 0.42        | 12             |
| 200-50 μm    | 2.74  | 1.94  | 0.24        | 0.56        | 8              |
| 50-20 μm     | 0.18  | 0.01  | 0.05        | 0.12        | 4              |
| 20-5 μm      | 1.38  | 0.99  | 0.17        | 0.22        | 32             |
| 5-2 μm       | 1.75  | 1.04  | 0.17        | 0.54        | 17             |
| 2-0.2 μm     | 1.35  | 0.73  | 0.01        | 0.61        | 24             |
| < 0.2 μm     | 1.23  | 0.53  | 0.04        | 0.66        | 5              |

**Table 2 :** $^{14}$C-atrazine equivalent content in non extractable (bound) residues in each soil size fraction and their distribution into humic and fulvic acids, and humin

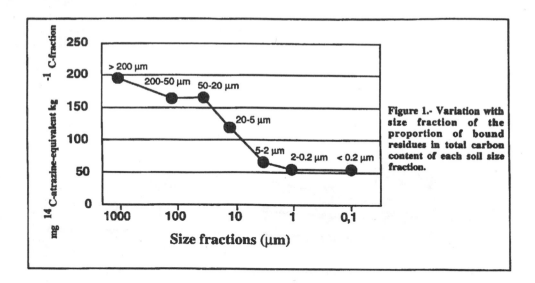

Figure 1.- **Variation with size fraction of the proportion of bound residues in total carbon content of each soil size fraction.**

However, bound residues distribution could be indicative of different immobilization mechanisms depending on humification degree of soil organic matter, or of different accessibility of the reactive groups of organic compounds according to their association with a particle size fraction. It was hypothesized that bound residues would result from bioaccumulation by soil microorganisms, especially fungi, localized in the coarsest fractions (Barriuso et al., 1991). These microorganisms can absorb and degrade large amounts of atrazine (Percich and Lockwood, 1978). The large amount of dealkyl derivatives found in methanol extract in 200-50 μm fraction agreed with this possibility (Barriuso and Koskinen, 1994).

After alkaline extraction, the proportion of [14]C-residues bound to the humin was the highest in non humidified fractions. In the humified fractions (< 20 μm), the proportion of [14]C-residues bound to the humin decreased, while that bound to the fulvic acids increased with decreasing of the fraction size (Table 2). Fulvic acids appear to play an active role in the initial rapid immobilization of bound residues (Schiavon et al., 1977; Bertin et al., 1991).

Fig. 2 shows the mineralization kinetics of bound residues of different soil size fractions. Mineralization rates strongly depended of the fraction size, and they progressively increased when the fraction size decreased. After 100 days of incubation, nearly 8 % of [14]C- residues bound to the finest fraction (< 0.2 μm) were mineralized. In contrast, less than 1% of bound residues of the coarsest fraction (> 200 μm) were evolved as $CO_2$. These incubation test allowed to evaluate the bound residues availability and their potential biological remobilization. We could hypothesize that bound residues associated with less humified materials would undergo similar pathways as other organic compounds in these humified materials during the humification process. However, it did not seem so from mineralization rates of carbon in the non humified fractions (results not shown).

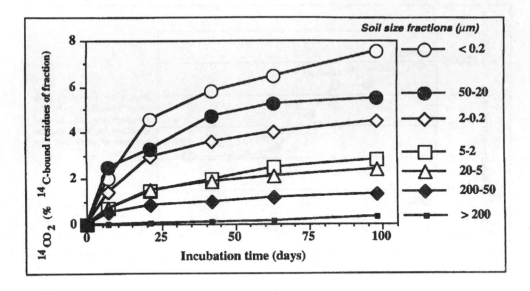

**Figure 2 :** Mineralization kinetics of bound residues of isolated soil size fractions incubated in soil.

In summary, a combination of physical and chemical soil fractionation methods could be used in environmental research especially to characterize the retention properties of organic matter and to evaluate the proportion of bound residues susceptible of mobilization. Bound residues associated with incompletely decomposed plant or animal materials in the coarsest fractions could be released during their decomposition or humification. On the other hand, bound residues associated with dispersible fine clay fractions or with soluble humic fractions, such fulvic acids, would present a high mobilization potential, if leaching of particles or soluble organic matter occur.

## Acknowledgements

Authors thanks M. Schiavon for doing the field incubation and Th. Choné for C analysis.

## References

❑    Andreux, F., M. Schiavon, G. Bertin, J.M. Portal, and E. Barriuso, 1991. The usefullness of humus fractionation methods in studies about the behaviour of polluatants in soils. *Toxicol. Environ. Chem.* 31/32:29-38.

❑    Barriuso, E., and W. Koskinen, 1994. Incorporation kinetic of atrazine non-extractable (bound)residues in soil size fractions under field conditions. *Soil Sci. Soc. Am. J.*, in press.

❑    Barriuso, E., M. Schiavon, F. Andreux, and J.M. Portal. 1991. Localization of atrazine non-extractable (bound) residues in soil size fractions. *Chemosphere* 22:1131-1140.

❑    Bertin, G., and M. Schiavon. 1989. Les résidus non extractibles de produits phytosanitaires dans les sols. *Agronomie* 9:117-124.

◻    Broersma, K., and L.M. Lavkulich. 1980. Organic matter distribution with particle-size in surface horizons of some sombric soils in Vancouver Island. *Can. J. Soil Sci.* 60:583-586.

◻    Calderbank A. 1989. The occurrence and significance of bound pesticide residues in soil. *Rev. Environ. Contam. Toxico.* 108:71-103.

◻    Catroux, G., and M. Schnitzer. 1987. Chemical, spectroscopic, and biological characteristics of the organic matter in particle size fractions separated from an Aquoll. *Soil Sci. Soc. Am. J.* 51:1200-1207.

◻    Chiou, C.T., J.F. Lee, and S.A. Boyd. 1990. The surface area of soil organic matter. *Environ. Sci. Technol.* 24:1164-1166.

◻    Christensen, B.T. 1992. Physical fractionation of soil and organic matter in primary particle size and density separates. *Adv. Soil Sci.* 20:1-90.

◻    Dios Cancela, G., J.A. Guillen Alfaro, and S. Gonzalez Garcia. 1989. Adsorción de clorprofan (CIPC) por suelos. II. Adsorción por las distintas fracciones granulométricas. *An. Edafol. Agrobiol.* 48:23-37.

◻    Khan, S.U. 1980. Plant uptake of unextracted (bound) residues from an organic soil treated with prometryn. *J. Agric. Food Chem.* 28:1096-1098.

◻    Nkedi-Kizza, P., P.S.C. Rao, and J.W. Johnson. 1983. Adsorption of diuron and 2,4,5-T on soil particle-size separates. *J. Environ. Qual.* 12:195-197.

◻    Percich, J.A., and J.L. Lockwood. 1978. Interaction of atrazine with soil microorganisms: population changes and accumulation. *Can. J. Microbiol.* 24:1145-1152.

◻    Turchenek, L.W., and J.M. Oades. 1978. Fractionation of organo-mineral complexex by sedimentation and density techniques. *Geoderma* 21:311-343.

2.3.3.     *Adsorption and Mobility of Metolachlor in Surface Horizons of Soils with Low Organic Matter Contents*

# T. Crisanto, M. Sánchez-Camazano, M. J. Sánchez-Martín

Instituto de Recursos Naturales y Agrobiología, C.S.I.C. Salamanca, Spain

## Introduction

The application of herbicides to soils to combat weeds forms an integral part of modern techniques of agricultural production. Study of the processes of adsorption and mobility of herbicides is considered to be of special interest, the former owing to their influence in the magnitude and effects of other processes (volatilization, degradation ...) and the latter because it allows the evaluation of the possibility of contamination of ground water.

Metolachlor is a herbicide from the chloroacetanilide group that has been used widely since 1980 for the control of weeds in soybean, corn and many other crops. Its presence in underground water has been detected by different authors at concentrations ranging from 0.08 to 680 µg/l (Chesters et al., 1989). The literature contains several references to its behaviour in soils that highlight the effect of organic matter on its efficacy, adsorption and mobility. These works have generally been carried out on small groups of soils, some of which include soils with high organic matter contents or with fractions of soil organic matter.

However, no references have been found on works addressing the effect of soil properties on the adsorption and mobility of this herbicide using a large number of soils, all of them with low organic matter contents.

Metolachlor [2-chloro-N-(2 ethyl-6-methylphenyl)-N-(2-metoxy-1-methyl ethyl)-acetamide] is widely used for different types of crops grown on soils with organic matter contents of less than 2.6 % in irrigated zones of the province of Salamanca (West-Central Spain). It was therefore thought of interest to study the adsorption and mobility of this herbicide in 33 soils from the area. The study on adsorption was carried out by adsorption isotherms and on mobility by soil thin layer chromatography (soil TLC).

## Materials and Methods

**Soils :** Selected characteristics of the soils are included in Table 1.

**Pesticide :** Metolachlor. Unlabelled metolachlor and $^{14}$C-metolachlor specific activity 1.94 MBq/mg) was obtained from CIBA-GEIGY (USA). Solubility in water : 530 mg/L.

**Adsorption Isotherms :** Soil/Solution : 1g/4mL. Pesticide concentrations : 4, 6, 9, 12, 15, 18 mg/L and solution activity 17 Bq/mL. Temperature : 20 C (thermostated chamber). Time treatment : 48 h with intermitting shaking. Quantitative determination of pesticide : $^{14}$C-activity (Beckman LS 1800 Liquid Scintillation Counter).
Adsorption capacity parameters : K and Kd (Freundlich adsorption equation).

**Soil Thin Layer Chromatography :** Plates : 5x20 cm. Soil/Solution : 1/2. Plates thickness : 500 µm. Volume spotted 5 µl $^{14}$C-pesticide solution (200 Bq). Mobility parameter : Rf determined by Lineal Analizer, Berthold TLC Tracemaster 20.

| Soil | Soit texture | pH | O.M. % | Sand % | Silt % | Clay % | *) Clay Mineralogy |
|------|------|------|------|------|------|------|------|
| 1 | Loamy sand | 5.6 | 0.69 | 81.20 | 8.20 | 10.60 | I, K, V |
| 2 | Loamy sand | 5.0 | 0.80 | 83.30 | 6.00 | 10.70 | I, K,V |
| 3 | Sandy clay loam | 7.0 | 1.56 | 61.40 | 14.30 | 24.30 | I, K, M |
| 4 | Loamy sand | 7.0 | 1.51 | 73.40 | 12.70 | 13.90 | I, K, V |
| 5 | Loamy sand | 6.0 | 1.44 | 75.40 | 11.30 | 13.30 | I, K, M |
| 6 | Loamy sand | 5.9 | 0.94 | 78.40 | 10.00 | 11.60 | I, K, V |
| 7 | Loamy sand | 6.1 | 1.30 | 74.80 | 13.70 | 11.50 | I, K |
| 8 | Sandy loam | 7.2 | 1.22 | 61.30 | 18.85 | 19.85 | I, K, V |
| 9 | Loamy sand | 6.6 | 1.52 | 73.30 | 14.25 | 13.45 | I, K, V |
| 10 | Loamy sand | 5.3 | 1.11 | 83.15 | 5.95 | 10.90 | I, K, V |
| 11 | Sandy loam | 7.5 | 0.77 | 65.30 | 15.50 | 18.20 | I, K, M |
| 12 | Loamy sand | 4.6 | 0.59 | 85.00 | 6.20 | 8.80 | I, K, M |
| 13 | Loamy sand | 7.7 | 0.85 | 72.60 | 13.60 | 11.80 | I, K |
| 14 | Loamy sand | 6.1 | 1.50 | 78.20 | 9.70 | 12.10 | I, K |
| 15 | Sandy clay loam | 7.2 | 0.78 | 66.00 | 12.70 | 20.75 | I, K, M |
| 16 | Sandy clay loam | 7.6 | 2.04 | 56.00 | 21.35 | 22.20 | I, K, M |
| 17 | Sandy clay loam | 5.9 | 0.91 | 57.65 | 16.90 | 25.45 | I, K, M |
| 18 | Loamy sand | 4.8 | 0.49 | 67.85 | 12.15 | 20.00 | I, K, M |
| 19 | Loamy sand | 5.6 | 0.43 | 82.60 | 8.10 | 9.30 | I, K, M |
| 20 | Loamy sand | 5.1 | 0.66 | 78.10 | 9.30 | 12.60 | I, K, M |
| 21 | Loamy sand | 5.2 | 0.69 | 81.80 | 9.70 | 8.50 | I, K, M |
| 22 | Loamy sand | 6.1 | 0.73 | 77.10 | 8.60 | 14.30 | I, K, M |
| 23 | Loamy sand | 7.0 | 0.73 | 86.80 | 3.70 | 9.50 | I, K, M |
| 24 | Loamy sand | 7.0 | 0.87 | 82.70 | 5.80 | 11.50 | I, K, M |
| 25 | Loamy sand | 6.1 | 1.20 | 76.80 | 10.25 | 12.95 | I, K, M |
| 26 | Loamy sand | 6.9 | 0.63 | 73.50 | 10.75 | 15.75 | I, K, M |
| 27 | Loamy sand | 4.4 | 0.67 | 84.70 | 8.10 | 7.20 | I, K |
| 28 | Loamy sand | 5.8 | 1.01 | 71.35 | 11.50 | 17.15 | I, K |
| 29 | Sandy loam | 5.0 | 1.01 | 68.60 | 12.00 | 19.40 | I, K, M |
| 30 | Sandy loam | 6.1 | 1.00 | 69.90 | 12.55 | 17.55 | I, K, M |
| 31 | Loamy sand | 5.0 | 0.97 | 72.50 | 6.80 | 20.70 | I, K, M |
| 32 | Loam | 5.4 | 2.59 | 75.80 | 45.70 | 18.50 | I, K, C |
| 33 | Loamy sand | 5.6 | 1.92 | 71.55 | 18.10 | 10.35 | I, K, C |

*) I, Illite; K, Kaolinite; M, Montmorillonite; V, Vermiculite; C, Chlorite

**Table 1** : Selected Characteristics of the Soils

## Results and Discussion

The adsorption isotherms of metolachlor for the 33 soils were obtained. These correspond to type L or type S according to the Giles et al. classification (1960). In general, no relationship can be seen between the initial curvature of the isotherms and soil composition. However, it is seen that the initial slope of the curves is greater for the soils with higher organic matter contents.

The adsorption isotherms fit the Freundlich equation. The Freundlich constant (K), distribution coefficients (Kd) and Kom were used to compare the adsorption of metolachlor by the different soils (Table 2).

| Soil | K | n | Kd | log $K_{om}$ | Rf |
|------|------|------|------|------|------|
| 1 | 0.66 | 1.40 | 0.33 | 1.98 | 0.55 |
| 2 | 1.16 | 1.69 | 0.44 | 2.16 | 0.63 |
| 3 | 1.52 | 1.03 | 1.37 | 1.99 | 0.35 |
| 4 | 1.83 | 1.82 | 1.62 | 2.08 | 0.47 |
| 5 | 2.00 | 1.10 | 1.64 | 2.14 | 0.41 |
| 6 | 0.39 | 0.87 | 0.54 | 1.62 | 0.48 |
| 7 | 0.88 | 1.05 | 0.73 | 1.83 | 0.44 |
| 8 | 1.55 | 1.40 | 0.79 | 2.10 | 0.30 |
| 9 | 1.74 | 1.41 | 0.82 | 2.06 | 0.50 |
| 10 | 1.24 | 1.23 | 0.80 | 2.05 | 0.45 |
| 11 | 0.31 | 0.81 | 0.50 | 1.60 | 0.55 |
| 12 | 0.88 | 1.04 | 0.81 | 2.17 | 0.55 |
| 13 | 0.78 | 1.80 | 0.25 | 1.96 | 0.48 |
| 14 | 0.51 | 1.00 | 0.50 | 1.53 | 0.47 |
| 15 | 0.41 | 1.00 | 0.38 | 1.72 | 0.42 |
| 16 | 1.34 | 1.00 | 1.32 | 1.82 | 0.38 |
| 17 | 1.31 | 1.13 | 1.01 | 2.16 | 0.44 |
| 18 | 0.15 | 0.74 | 0.31 | 1.49 | 0.65 |
| 19 | 0.36 | 0.93 | 0.42 | 1.92 | 0.66 |
| 20 | 0.60 | 1.07 | 0.52 | 1.96 | 0.58 |
| 21 | 0.64 | 0.96 | 0.68 | 1.97 | 0.71 |
| 22 | 0.07 | 0.63 | 0.31 | 0.98 | 0.73 |
| 23 | 1.03 | 1.63 | 0.38 | 2.15 | 0.65 |
| 24 | 0.64 | 1.16 | 0.45 | 1.87 | 0.68 |
| 25 | 0.42 | 0.96 | 0.44 | 1.54 | 0.64 |
| 26 | 0.39 | 1.16 | 0.27 | 1.79 | 0.54 |
| 27 | 0.72 | 1.40 | 0.34 | 2.03 | 0.63 |
| 28 | 1.14 | 1.51 | 0.51 | 2.05 | 0.43 |
| 29 | 1.10 | 1.46 | 0.53 | 2.04 | 0.51 |
| 30 | 0.86 | 1.37 | 0.39 | 1.93 | 0.34 |
| 31 | 1.26 | 1.47 | 0.60 | 2.11 | 0.49 |
| 32 | 1.54 | 1.05 | 1.38 | 1.77 | 0.36 |
| 33 | 1.20 | 1.18 | 0.69 | 1.80 | 0.45 |
| LSD (0.05 %) | 0.58 | | 0.18 | | 0.14 |

**Table 2 :** Constants of the Freundlich adsorption equations (K and n),
distribution coeficients (Kd), log $K_{om}$ and Rf values

With a view to determining the influence of soil properties on the adsorption of metolachlor by a statistical approach, simple correlations between the constants of adsorption and mobility and soil properties were obtained.

| Variable | K | $K_d$ | Rf |
|----------|---|-------|-----|
| K | -- | -- | -0.604[a] |
| $K_d$ | -- | -- | -0.536[a] |
| O.M. % | 0.632[a] | 0.698[a] | -0.625[a] |
| Clay % | 0.233 | 0.310 | -0.543[a] |
| Clay + Silt | 0.351[b] | 0.482[a] | -0.632[a] |
| Clay + Silt + O.M. | 0.368[b] | 0.499[a] | -0.641[a] |

[a] Significant at 0.01 to 0.001 level

[b] Significant at 0.05 to 0.01 level

**Table 3 :** Simple correlations coefficients (r) between Freundlich constant K, distribution coefficient ($K_d$) and Rf and soil characteristics

The results show that both the content in organic fraction and the content in the inorganic fraction of soils are very important parameters in the adsorption and mobility of metolachlor in soils with low organic matter contents.

The simple correlation coefficients between the K constant and soil properties reveal the existence of a very significant correlation ($p < 0.01$) between the organic matter content and a significant correlation ($p < 0.05$) between K and the contents in silt + clay and organic matter + silt + clay.

The simple correlation coefficients between the Rf values and soil properties indicate the existence of a very significant negative correlation ($p < 0.01$) between the Rf values and the contents of organic matter, clay, silt + clay and organic matter + silt + clay.

According to the classification of Helling and Turner (1968), metolachlor can be considered to be moderately mobile in 78.8 % and mobile in 21.2 % of the soils studied.

## References

□ Chesters, G.; Simsiman, G.V.; Levy, J.; Alhajjar, B.J.; Fathulla, R.N.; Harkin, J.M. *Rev. Environ. Contam. Toxicol.* 110, 1-74, 1989.

□ Giles, C.H.; Mac Evan, T.H.; Nakhava, S.N.; Smith, *D. J. Chem. Soc.* 786, 3973-3993, 1960.

□ Helling, C.S.; Turner, B.C. *Science* 162, 562-563, 1968.

2.3.4.     *Adsorption Studies of Chlorophenols from Aqueous Solutions into*
           *Al-Pillared Clays and Mesoporous Aluminum Phosphates*

# T.G.Danis, T.A.Albanis, D.E.Petrakis and P.J.Pomonis

Department of Chemistry, University of Ioannina, Ioannina 451 10, Greece

## Abstract

Adsorption and removal of chlorophenols studied in aqueous suspensions of Al-pillared montmorillonite (Al-Mont) and mesoporous Alumina-Aluminum Phosphates (AAP). For concentrations between 25 to 300 μg/L, Al-pillared montmorillonite adsorbs 26.3% of 2,4-dichlorophenol, 75.6% of 2,4,6-trichlorophenol and 95.2% of pentachlorophenol. The mixture mesoporous Alumina-Aluminum Phosphates the adsorption increases with the ratio of P/Al. The mixture AAP, P/Al = 0.6 adsorbs 14.8% of 2,4-dichlorophenol, 27.1% of 2,4,6-trichlorophenol and 58.3% of pentachlorophenol.

## Introduction

Among the different pollutants of aquatic ecosystems, phenols, especially the chlorinated ones, are considered of urgent priority since they are harmful to organisms even at ppb levels [1,2]. Pentachlorophenol and some tetrachlorophenols are used, primarily as wood preservatives or fungicides [3,4]. Residues of chlorophenols have been found wordwide in soil, water and air samples, in food products, in human and animal tisssues and body fluids. [3,4].

The removal of such compounds at such low levels consists a difficult problem. Among the methods employed are either destructive oxidation, often assisted by light [5] or adsorption into porous solids such as active carbon [6,7] and clays natural or pillared ones [8,9]. Nevertheless such adsorption studies have often taken place at ppm levels which are three orders of magnidut above the ppb levels imposed as acceptable limits by EC. So we considered worth to check the adsorption capacity of pillared clays at these low levels. Besides we thought that a comparison of such microporous materials with mesoporous Alumina/Aluminum Phosphate (AAP) solids [10] might be of interest as indicative of the mechanisms controlling the process.

## Experimental procedure

Aluminum pillared montmorillonite (Al-Mont) of specific surface area $130m^2.g^{-1}$ (BET) and range of $d_{001}$ spacing from 12-18 A when fired at 500°C in air, was used as adsorbent. The mesoporous Alumina-Aluminum Phosphates were prepared as reported in [10] and had a P/Al ratio equal to 0.3, 0.4, and 0.6 with surface areas of $150m2.g-1$ (BET) and pores with rp 80-100A. The adsorption studies took place by mixing 0.3g of the solid in 25 ml of triply distilled water, into which an initial concentration of 2,4-dichlorophenol, 2,4,6-trichlorophenol and pentachlorophenol between 25 to 300 ppb was put. The suspension was left under slow agitation for 24 hrs. Then it was precipitated in a centrifuge (6000 rpm) and the supernatant was analyzed after acetylation in a G.C. equipped with E.C.D. From the results the adsorption isotherms were easily constructed (Figures 1 and 2). The desorption studies took place by putting back into the centrifuged slurry another portion of 25ml of water and reaching again equilibrium under slow agitation, followed by centrifugation and G.C. analysis as above. In the AAP samples desorption was possible in water. On the contrary for Al-pillared Montmorillonite no desorption took place in aqueous solution and the cited desorption isotherms were taken in acetone (table 1).

# Results and Discussion

Amounts of chorophenols remaining in equilibrium concentration, adsorbed, desorbed with water and acetone and remaining adsorbed or degradated are summarized in Table 1 and 2. The remaining free amount of chlorophenols was found from the equilibrium concentration in aqueous solutions. The desorbed amount was found from extractions of solid remainder with water and acetone. The differences of the sum of desorbed quantities from adsorbed initial amount of chlorophenols was considered to represent the ramaining adsorbed or catalytic degradated amount.

From the results depicted in figure 1, it can be easily seen that Al-pillared montmorillonite is a very good adsorbent of much superior capacity as compared to AAP. The desorption isotherm from Al-Montmorillonite shows a significant hysteresis as compared to adsorption. It should be noticed that desorption of PCP and TrCP was not possible from this material into water suspension and the observed desorption was observed in acetone. In conclusion it could be said that Al-pillared montmorillonite is a very good adsorbent of PCP and TrCP. The limits of its adsorbance capacity as well as its selectivity for different molecules are currently examined.

The adsorption results of studied chlorophenols shown that the adsorption of these compounds inreases as the number of chlorines in their molecules increases (figure 3). This phenomenon due to the increament of lipophilic character of chlorophenols and decrement of their solubility as the number of chlorines increase.

| Compound | Equilibrium Conc. (%) | Adsorption (%) | Desorption in water (%) | Desorption in acetone (%) | Remain Adsorbed (%) |
|---|---|---|---|---|---|
| PCP | 4,80 | 95,20 | 7,42 | 79,86 | 7,92 |
| TrCP | 24,38 | 75,62 | 13,09 | 55,89 | 6,64 |
| DiCP | 73,69 | 26,31 | 15,15 | 8,54 | 2,62 |

**Table1 :** Amounts of chlorophenols in equilibrium concentration, adsorbed, desorbed with water and acetone and remain adsorbed or degraded in aqueous suspensions of Al-pillared montmorillonite (n=3)

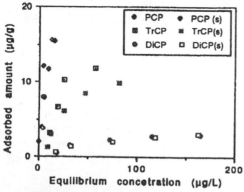

**Figure 1 :** Adsorption of PCP, TrCP and DiCP separately and simultaneously on Al-pillared montorillonite

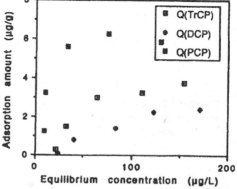

**Figure 2 :** Adsorption of PCP, TrCP and DiCP on alumina-aluminum phosphates (AAP)

In alumina-aluminum phosphate materials it seems that the adsorption capacity increases with the addition of phosphorous, which increases the surface acidity, but never reaches the capacity of Al-pillared montmorillonite. In these materials the adsorptive capacity increased as the ratio of P/Al increases. A change of the ratio of P/Al from 3:7 to 6:4 increases the adsorption amount of PCP from 2.8 to 67.9%, for initial concentration 100 μg/L.

| Compound | Equilibrium Conc. (%) | Adsorption (%) | Desorption in water (%) | Desorption in acetone (%) | Remain Adsorbed (%) |
|----------|----------------------|----------------|-------------------------|---------------------------|---------------------|
| PCP  | 41,65 | 58,35 | 30,30 | 12,90 | 15,15 |
| TrCP | 72,89 | 27,11 | 14,36 | 4,64  | 8,11  |
| DCP  | 85,23 | 14,77 | 6,25  | 1,96  | 6,56  |

**Table 2 :** Amounts of chlorophenols in equilibrium concentration, adsorbed, desorbed with water and acetone and remain adsorbed or degraded in aqueous suspensions of alumina aluminum phosphate, P/Al=0.6 (n=3)

The elimination of chlorophenols in AAp suspensions is increased as compatre to the Al-pillared montmorillonite suspensions (Table 2 and figure 4). We consider that this fact shows the catalytic degradation of chlorphenols by AAP materials and this degradation is significant in case of penta-chlorophenol (15.5%). It seems the acidic surface of AAP has an ability to degradate the adsorbed chlorophenol molecules.

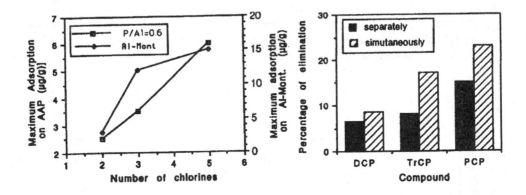

**Figure 3 :** The influence of Cl atoms number of chlorophenols in their adsorption by Al-pillared

**Figure 4 :** Elimination of PCP, TrCP and DiCP separately and simultaneously on allumina-aluminium phosphate, P/Al==0.6

# References

□   [1] P.M. Chapman, G.M.Romberg and G.A.Vigers, *J.Water Pollution Control Federation*, 54 (1982) 292.

□   [2] M.Kubata, N.Shindo and K.Munakata, *J.Agric.Chem.Soc. Japan*, 44(1970) 169.

□   [3] J.H. Exxon, Vet. Hum. *Toxicol*. 26(6): 508-519, 1984.

□   [4] A. Bevenue, H. Beckman. Pentachlorophenol: A discussion of its properties and its occurence as a residue in human and animal tissues. *Residue Reviews* 19: 83-128, 1967.

□   [5] N.-M.Hermann, C.Guillard and P.Pichat, 1st European Workshop Meeting on "Environmental Industrial Catalysis", Editors P.Ruiz, F.C.Thyrion and B.Delmon, Preprints p.13 (1992) and refs wherein.

□   [6] M.C.Bonnet, B.Welte and A.Montiel, *Wat.Res*. 26(1992)1673.

□   [7] I. Abe, K. Hayashi, M. Kitagawa and T. Hirashima, *Bull. Chem. Soc. Jpn*. 56(1983)1002.

□   [8] Zhouglua Hu and E.F.Vansant, 1st Europ. Workshop Meeting on "Environmental Industrial Catalysis", Editors, P.Ruiz, F.C.Thyrion and B.Delmon, Preprints p.253 (1992) and refs wherein.

□   [9] L.J.Michot and T.J.Pinnavaia, *Clays and Clay Minerals*, 39 (1991) 634.

□   [10] D.E.Petrakis, P.J.Pomonis and A.T.Sdoukos, *J.Chem.Soc. Faraday Trans*, 87(1991) 1439.

## 2.3.5.    *Mobility of Triazine Metabolites in the Soil*

# L. Donati, J. Keizer, P. Bottoni, R. Scenati and E. Funari

Istituto Superiore di Sanità - Laboratorio di Igiene Ambientale -
Viale Regina Elena, 299 - 00161 Rome, Italy

## Introduction

For non-ionic pesticides, the organic carbon/water partition coefficient (KOC) is considered to be a very good index of mobility for most soils [1, 2, 3].

In a previous study we determined the KOC of atrazine, hexazinone and terbutylazine and two degradation products of triazines that can have some environmental relevance: deethylatrazine (DEA) and deisopropylatrazine (DIA) [4]. We have developed this study and in this paper we present the KOC determined for deethyldeisopropylatrazine (DEDIA), which is a metabolite common to many triazines, deethylterbutylazine (DET) which is a metabolite of terbutylazine and 3-(4-hydroxycyclohexyl)-6-(methylamino)-1-methyl-1,3,5-triazin-2,4(1H,3H)-dione which is a metabolite of hexazinone.

In the previous study, the KOC of the substances examined were derived both from their KD values determined from sorption isotherms and by calculating the KOC from a HPLC method. The comparison between the KOC values obtained with the HPLC method and those from KD showed that they were generally similar and the ratios between their average values were in most cases within 2. Even though this comparison of course cannot be considered exhaustive as to the validity of the HPLC method for determining the KOC values for non-ionic pesticides it seems reliable at least for the triazines and this is the reason why in this study we have used only this method.

## Materials and Methods

The following compounds were used as reference substances for the elaboration of the regression lines with the HPLC method: benzene, naphtalene, xylene, anthracene, phenanthrene, toluene, ametryne, metolachlor, prometryne, metribuzin, propoxur, chlorpyrifos-ethyl, chlorpyrifos-methyl, alachlor, cyanazine, diazinon, fenitrothion, molinate, simazine and trifluralin. Table 1 shows the man KOC values of these substances calculated from the available data.

KOC values of the substances examined were measured essentially by applying published HPLC methods [5, 6, 7]. KOC values of these substances were derived from the regression lines elaborated using the 21 reference substances listed above. The mean KOC values of these substances were correlated with their relative capacity factors (k) determined in the following chromatographic conditions: 1) column: C-18, 25 cm; solvents: 85% methanol, 15% water; 2) column: C-18, 25 cm; solvents: 70% methanol, 30% water; 3) column: C-8, 15 cm; solvents: 60% methanol, 40% water. The retention time (TR) was determined in triplicate for each substance and each condition. The time representing the dead volume of the system (T0) was determined using uracil as a reference substance. The capacity factor was calculated as: (TR-T0)/T0.

Once the capacity factors of the examined substances were determined, their KOC values were obtained using the regression lines. The KOC of DEDIA was calculated using the three conditions mentioned above whereas for the other two substances examined, the KOC was determined only in the condition 2.

# Results

Table 2 shows the KOC values for the three metabolites examined together, with those of their parent compounds and the other metabolites previously determined.

Where a comparison is possible it turns out that the three experimental conditions applied do not strongly influence these values.

These data show that all the metabolites studied have low KOC values, always lower than their parent compounds.

| SUBSTANCE | MEAN ± SD | DATA NUMBER | REFERENCES |
|---|---|---|---|
| alachlor | 203 ± 71 | 5 | (1,2,3,4,5) |
| ametryne | 386 ± 6 | 2 | (1,2) |
| anthracene | 21000 ± 5000 | 2 | (6,7) |
| atrazine | 154 ± 26 | 6 | (1,2,3,4,8,9) |
| benzene | 69 ± 43 | 5 | (6,7,10,11) |
| chlorpyrifos-ethyl | 7274 ± 5421 | 7 | (12,14,1,3,13) |
| chlorpyrifos-methyl | 3310 | 1 | (1) |
| cyanazine | 188 ± 8 | 4 | (1,3,4,8) |
| diazinon | 643 ± 558 | 2 | (3,9) |
| fenitrothion | 424 ± 170 | 2 | (12) |
| metolachlor | 163 ± 45 | 3 | (2,3,4) |
| metribuzin | 90 ± 52 | 3 | (1,3,15) |
| molinate | 85 ± 5 | 2 | (12) |
| naphtalene | 883 ± 277 | 4 | (6,7,11,16) |
| phenanthrene | 18000 ± 4546 | 3 | (6,7,11) |
| prometryne | 700 ± 83 | 3 | (1,3,9) |
| propoxur | 47 ± 6 | 2 | (12) |
| simazine | 189 ± 56 | 6 | (1,3,5,8,9,17) |
| toluene | 135 ± 78 | 4 | (10,11,18) |
| trifluralin | 6417 ± 4386 | 7 | (12,14,1,3,5,19) |
| xylene | 507 ± 288 | 2 | (11,20) |

**Table 1 :** Mean Values of Koc for the reference substances
calculated from Literature Data

| Substances | Condition 1 | Condition 2 | Condition 3 | Mean ± SD |
|---|---|---|---|---|
| Atrazine | 105 | 111 | 184 | 133 ± 44 |
| DEA | 39 | 38 | 52 | 43 ± 8 |
| DIA | 31 | 28 | 39 | 33 ± 6 |
| DEDIA | 18 | 19 | 22 | 20 ± 2 |
| Terbutylazine | 186 | 199 | 323 | 236 ± 62 |
| DET | | 69 | | |
| Hexazinone | 38 | 62 | 64 | 55 ± 14 |
| Hexazinone metabolite | | 17 | | |

**Table 2 :** Koc values calculated from the HPLC-capacity factor at the 3 different experimental conditions

## Discussion and Conclusions

From this study it turns out that the three triazine metabolites examined have very low KOC values, which indicate their tendency to move easily through the soil.

If we consider these substances together with their parent compounds and the other metabolites examined in the previous study we can conclude that hexazinone and all the triazine metabolites have KOC values lower than that of atrazine, so showing their high mobility in the soil.

As a consequence of the shortage of data on their soil degradation half-time at present it is not possible to formulate any prediction on leaching potential of the metabolites examined.

## References

□   [1] Kanazawa, J. 1989. Relationship between the soil sorption constants for pesticides and their physicochemical properties. *Environ. Toxicol. Chem.* 8: 477-484.

□   [2] Karickhoff, S.W., Brown, D.S. and Scott, T.A. 1979. Sorption of hydrophobic pollutants on natural sediments. *Water Res.* 13: 241-248.

□   [3] Pionke, H.B. and De Angelis, R.J. 1980. Methods for distributing pesticide loss in field runoff between the solution and adsorbed phases. In: Creams and Field Scale Model for Chemical Runoff and Erosion from Agricultural Management Systems. *U.S. Dep. Agric. Conservation Research Report 26.* pp. 607-643.

□   [4] Donati, L., Keizer, J., Bottoni, P., Scenati, R. and Funari, E. 1994. KOC estimation of deethylatrazine, deisopropylatrazine, hexazinone and terbutylazine by reversed phase chromatography and sorption isotherms. *Toxicol. Environ. Chem.* (in press).

□   [5] McCall, P.J., Swann, R.L., Laskowski, D.A., Unger, S.M., Vrona, S.A. and Dishburger, H.J. 1980. Estimation of chemical mobility in soil from liquid chromatographic retention times. *Bull. Environ. Contam. Toxicol.* 24: 190-195.

□  [6] Szabo, G., Prosser, S.L. and Bulman, R.A. 1990. Prediction of the adsorption coefficient (Koc) for soil by a chemically immobilized humic acid column using RP-HPLC. *Chemosphere* 21 (6): 729-739.

□  [7] Hodson, J. and Williams, N.A. 1988. The estimation of the adsorption coefficient (Koc) for soils by High Performance Liquid Chromatography. *Chemosphere* 17 (1): 67-77.

## References of Table 1

□  1.Kenaga, E.E. 1980. Predicted bioconcentration factors and soil sorption coefficients of pesticides and other chemicals. *Ecotoxicol. Environ. Safety* 4: 26-38.

□  2.Donigian, A.S. Jr. and Carsel, R.F. 1987. Modeling the impact of conservation tillage on pesticide concentration in ground and surface waters. *Environ. Toxicol. Chem.* 6: 241-250.

□  3.Johnson, B. 1989. Setting revised specific numerical values. *California Department of Food and Agriculture.* October 1989. EH 89/13.

□  4.Soil Conservation Service. 1980. Water Quality Workshop: Integrating water quality and quantity into conservation planning. *SCS National Technical Center*, Ft. Worth, TX.

□  5.Wilkerson, M.R. and Kim, K.D. 1986. The pesticide contamination prevention act: setting specific numerical values. *California Department of Food and Agriculture.* December 1, 1986.

□  6.Karickhoff, S.W., Brown, D.S. and Scott, T.A. 1979. Sorption of hydrophobic pollutants on natural sediments. *Water Res.* 13: 241-248.

□  7.Karickhoff, S.W. 1981. Semi.empirical estimation of sorption of hydrophobic pollutants on natural sediments and soils. *Chemosphere* 10(8): 833-846.

□  8.Brown, D.S. and Flagg, E.W. 1981. Empirical prediction of organic pollutants sorption in natural sediments. *J. Environ. Qual.* 10(3): 382-386.

□  9.Rao, P.S.C. and Davidson, J.M. 1980. Estimation of pesticide retention and transformation parameters required in nonpoint source pollution model. In: *Environmental Impact of Nonpoint Pollution.* M.R. Overcash and J.M. Davidson (Eds.) Ann Arbor Science Publishers INC, Ann Arbor, MI. pp. 23-67.

□  10.HSDB, Hazardous Substances Data Bank, Bethesda MD, National Library of Medicine.

□  11.Vowless, P.D. and Mantoura, R.F.C. 1987. Sediment-water partition coefficients and HPLC retention factors of aromatic hydrocarbons. *Chemosphere* 16(1): 109-116.

□  12.Kanazawa, J. 1989. Relationship between the soil sorption constants for pesticides and their physicochemical properties. *Environ. Toxicol. Chem.* 8: 477-484.

□  13.Felsot. A.S. and Dahm, P.A. 1979. Sorption of organophosphorus and carbamate insecticides by soil. *J. Agric. Food Chem.* 27: 557-563.

□  14.McCall, P.J., Swann, R.L., Laskowski, D.A., Unger, S.M., Vrona, S.A: and Dishburger, H.J. 1980. Estimation of chemical mobility in soil from liquid chromatographic retention times. *Bull. Environ. Contam. Toxicol.* 24: 190-195.

□  15.Jury, W.A., Focht, D.D. and Farmer, W.J. 1987. Evaluation of pesticide groundwater pollution potential from standard indices of soil-chemical adsorption and biodegradation. *J. Environ. Qual.* 16: 422-428.

□  16.Lokke, H. 1984. Sorption of selected organic pollutants in Danish soils. *Ecotoxicol. Environ. Safety* 8: 395-409.

□      17.Glotfelty, D.E., Taylor, A.W., Isensee, A.R., Jersey, J. and Glenn, S. 1984. Atrazine and simazine movement in the Wye River estuary. *J. Environ. Qual.* 13: 115-121.

□      18.Garbarini, D.R. and Lion, L.W. 1986. Paritioning equilibria of volatile pollutants in three-phase systems. *Environ. Sci. Technol.* 20: 1263-1269.

□      19.Grover, R., Bauting, J.D. and Morse, P.M. 1979. Adsorption and bioactivitu of diallate triallate and trifluralin. *Weed. Res.* 19:(6): 363-369.

□      20.Daniels, S.L., Hoerger, F.D. and Moolenaar, R.J. 1985. Environmental exposure assessment experience under the Toxic Substances Control Act. Environ. *Toxicol. Chem.* 4: 107-117.

2.3.6.    *Sorption of Chlorophenols in Mineral Sorbents*

# Z. Fröbe, S. Fingler, E. Maric and V. Drevenkar

Institute for Medical Research and Occupational Health,
2 Ksaverska St., 41000 Zagreb, CROATIA

## Abstract

To investigate the sorption of hydrophobic but ionizable polar compounds by inorganic sorbents, a series of model experiments was carried out in which three chlorophenols, 2,4,6-tri-chlorophenol (TCP), 2,3,4,6-tetra-chlorophenol (TeCP) and pentachlorophenol (PCP), were sorbed by four different mineral sorbents, quartz, calcite, kaolinite and montmorillonite. The results clearly showed the existence of sorption interaction. Freundlich isotherm coefficients $K_f$ and $1/n$ were calculated for all the three compounds and compared for all the sorbents tested. The $K_f$ values and isotherm non-linearities which were observed in almost all experiments indicated that the sorption intensity in the mineral surface could be compared to that in the organic phase.

## Introduction

The transport, persistence and biological effects of a chemical in soils and aquatic systems strongly depend on its tendency to be sorbed by natural sorbents (1). Most of the sorption mechanisms proposed for non-ionic compounds involve "hydrophobic interactions" based on the dominance of the organic phase attached to a sorbent surface. The role of the inorganic mineral matrix in the sorption process is generally underestimated, although the interaction of solute with mineral surface has been indicated (2-4). This hydrophilic contribution to sorption may be significant or even dominate the overall interaction in the case of high sorbate polarity and/or low organic matter content of the sorbents (1).

Chlorophenols belong to a class of polarizable hydrophobic compounds suitable for testing the ability of mineral sorbents having no organic matter to sorb ionizable organic solutes. At typical ambient pH values chlorophenols, in particular the highly chlorinated TeCP and PCP, are present in the aqueous solution partly as phenolate anions. Being weak organic acids, but becoming increasingly acidic with an increase in the chlorine substitution (Table 1), phenols with more chlorine atoms are more likely to be ionized at environmental pH values. The sorption behaviour of such ionizable but still highly hydrophobic compounds should therefore comprise both molecular and ionic forms.

Mineral sorption of some phenolic compounds has been reported in literature, and it was correlated with the oxide content in natural soil materials (5). Binding of phenols to clay minerals was suggested to proceed by oxidative polymerization occurring on metalion (Fe, Al, Ca, Na)-exchanged montmorillonites (6). Smectite clay, material having montmorillonite, was found to exhibit high sorptive capabilities for PCP after saturation with organic cations (7).

| Sorbates | | TCP | TeCP | PCP |
|---|---|---|---|---|
| Acidity (pK$_a$) | | 6.15 | 5.40 | 5.25 |
| Hydrophobicity (log K$_{ow}$) | | 3.72 | 4.42 | 5.24 |
| Sorbents | Quartz | Calcite | Kaolinite | Montmorillonite |
| Mean particle size (μm) | 307.8 | 59.5 | 4.2 | 19.2 |
| SSA (m$^2$ g$^{-1}$) | 0.03 | 0.20 | 11.70 | 270.00 |
| Acidity in water suspension (pH) | 7.5 | 7.2 | 7.0 | 5.3 |

Kow=octanolwater partition coefficient

**Table 1 :** Physico-chemical characteristics of chlorophenols and mineral sorbents

## Experimental

### *Chemicals*

2,4,6Trichlorophenol (TCP), 2,3,4,6tetrachlorophenol (TeCP) and pentachlorophenol (PCP) were purchased from H.P. Chem. Service, West Chester, PA, USA. Quartz, calcite, kaolinite and montmorillonite were all products of Aldrich, USA. Mineral sorbents were characterized with respect to their granulometric composition and specific surface area (SSA). The mean particle size and SSA values of mineral fractions used in sorption experiments are presented in Table 1.

### *Sorption experiments*

Portions of 0.1 g of sorbents, previously saturated with deionized water, were equilibrated with 11 mL solutions of chlorophenol mixture in 0.01 M CaCl$_2$, with concentrations ranging from 0.02 to 0.4 uM for each compound. The suspensions were agitated overnight with a mechanical shaker. Concentrations of nonsorbed solutes in 5 mL aliquots of clear supernatant were analysed by gas chromatography after accumulation from water sample by C$_{18}$ reverse phase adsorption and acetylation with acetic anhydride (8). The sorption results were evaluated with respect to Freundlich sorption isotherm.

## Results and Discussion

In spite of the fact that water phase acidity influences the extent of ionization of polar and easily ionizable compounds like chlorophenols, and hence is likely to influence their sorption behaviour as well, no attempt was made to control aqueous pH values by addition of a buffer system which would maintain the same acidity in all sorption experiments. Although this would have contributed to the uniformity of sorption results, the experimental conditions were deliberately chosen to mimic natural conditions as much as possible, allowing sorption to occur under nondisturbed conditions.

| | TCP | | TeCP | | PCP | |
|---|---|---|---|---|---|---|
| | Kf | 1/n | Kf | 1/n | Kf | 1/n |
| Quartz | 1.280 | 0.177 | 2.173 | 0.434 | 10.367 | 0.920 |
| Calcite | 5.210 | 0.359 | 6.373 | 0.467 | 13.955 | 0.784 |
| Kaolinite | 5.856 | 0.751 | 2.203 | 0.340 | 0.463 | 0.133 |
| Montmorillonite | 5.220 | 0.412 | 6.392 | 0.401 | 26.883 | 0.505 |

**Table 2 :** Freundlich sorption coefficients for sorption of chlorophenols in mineral sorbents, obtained by fitting experimental data to Freundlich sorption isotherm: $x/m = K_f.C_e^{1/n}$.

As shown in Table 1 all of the sorbent suspensions are neutral to slightly alkaline, except that of montmorillonite. According to the $pK_a$ values of TCP, TeCP and PCP, the ionized forms are estimated to dominate in most of the sorption experiments. This assumption is based on the rule that for dominance of phenolates, aqueous acidity should be a pH unit higher than the $pK_a$ value of the investigated chlorophenol, a condition being fulfilled for the investigated compounds in all sorbents but montmorillonite. Therefore, non-ionized forms of chlorophenols should be taken into account only when sorption in montmorillonite is being considered.

For the mineral sorbents having no organic matter, SSA is anticipated to become a sorbent characteristic most likely to reflect sorption capacity. Four mineral sorbents which have been chosen to test this prediction, cover a wide range of SSA values, from only 0.03 m$^2$ g$^{-1}$ for quartz to as much as 270 m$^2$ g$^{-1}$ for montmorillonite. The latter sorbent is especially superior in this respect as a clay mineral characterized by a very large SSA, which should considerably improve its sorbing ability.

The results of all sorption experiments obtained by fitting experimental data sets to Freundlich sorption isotherm are summarized in Table 2. The analysis of Freundlich sorption coefficients not only shows the existence of sorption interaction, but also proves that the intensity of such hydrophilic sorption in mineral sorbents since no hydrophobic partition is possible in the total absence of organic matter is comparable to the intensity of hydrophobic sorption observed for the same sorbates in natural sorbents having a high organic matter content (4). However, except in kaolinite, the magnitudes of sorption coefficients $K_f$ for the three chlorophenols tested, follow in general the solubility dependence, being lowest for the least hydrophobic TCP and highest for the most hydrophobic PCP. This indicates that the hydrophobicity of sorbate still plays an important role in the sorption process. This finding is more likely due to the incompatibility of the increasingly hydrophobic sorbates with water as a solvent, rather than being a consequence of their site specific compatibility with mineral surface where sorption takes place. Kaolinite shows an oposite trend, sorbing TCP with the highest intensity, while the sorption of PCP is the lowest of all.

The Freundlich sorption coefficients 1/n, reflecting the isotherm nonlinearities, are considerably lower than unity in almost all sorbents. The change in the isotherm slope is considered to reflect the change of sorbent capacity in the concentration range under study and the lack of isotherm linearities suggests that the sorbent surface approaches saturation. Sorption coefficients 1/n are therefore expected to correlate with sorbents SSA. However, only for TCP sorption such a correlation is observed, again with the exception of kaolinite, indicating a decrease of isotherm non-linearities in sorbents with increasing SSA. In general, 1/n values are random for all the three compounds regardless of their speciation and indicate a significant deviation from linearity for both non-ionized species in montmorillonite and correspondent phenolates in the other three sorbents.

The observed sorption behaviour of chlorophenols proves a strong "hydrophilic interaction" with mineral surface. However, contrary to hydrophobic partition which seems to be dominated by organic matter content, the sorption results obtained so far with purely inorganic sorbents do not allow a priori estimation of sorption behaviour utilizing commonly measured physical properties of sorbates and sorbents.

# References

◻    [1] S.W. Karickhoff, J. Hydraulic Eng., 110(1984)707.

◻    [2] Z. Fröbe, S. Fingler, V. Drevenkar and M. Juracic, accepted for publication in Sci. Tot. Environ.

◻    [3] M.C. Hermosin, I. Roldan and J. Cornejo, Sci.Tot.Environ., 123/124(1992)109.

◻    [4] E. Morillo and C. Maqueda, Sci.Tot.Environ., 123/124(1992)133.

◻    [5] J. Artiola-Fortuny and W.H. Fuller, Soil Sci., 133(1982)18.

◻    [6] Y. Soma and M. Soma, Environ. Health Perspect., 83(1989)205.

◻    [7] J.A. Smith, P.J. Witkowski and C.T. Chiou, Rev. Environ. Contam. Toxicol., 103(1988)127.

◻    [8] S. Fingler, V. Drevenkar and Z. Vasilic, Mikrochim. Acta [Wien] II(1987)163.

## 2.3.7.    *Adsorption and Mobility of Diuron, Atrazine and MCPA on a Peat-amended Soil*

# E. González-Pradas, M. Villafranca-Sánchez, M. Fernández-Pérez and M. Socías Viciana

Departamento de Química Inorgánica. Universidad de Almería.
04071-ALMERIA (SPAIN).

## Summary

The adsorption processes of diuron, atrazine and MCPA by a calcareous soil from Spain after organic matter amendment (from 0.40% to 10.30%) with a commercial peat have been studied by using batch experiments. The $K_f$ (Freundlich parameter) values range between 2.17-34.28 mg kg$^{-1}$ (diuron), 0.94-5.47 mg kg$^{-1}$ (atrazine) and 0.24-2.21 mg kg$^{-1}$ (MCPA). From the correlation between $K_f$ and the organic carbon content of the samples, the $K_{oc}$ parameters have also been calculated. Leaching of diuron, atrazine and MCPA has also been studied in the two soil samples containing the highest and the lowest percentages of organic matter by using soil columns in laboratory conditions. According to both the $K_{oc}$ values and the data from the leaching experiments, diuron was much less mobile than atrazine and MCPA.

## Key Words

Adsorption, Leaching, Herbicides, Calcareous soil, Soil amendment

## Introduction

Reported incidents on pesticide contamination of the environment or on adverse environmental effects from pesticide use can often be traced to improper application or inappropriate practices. Either a lack of knowledge or a disregard for the sensitive nature of the environment has been the root of many pesticide contamination problems. The basic problem for most cases has been a lack of understanding of the processes affecting the behaviour and fate of pesticides from the point they enter the environment to the point at which they would affect the target organisms (1).

Since concerns are mostly associated with the presence of pesticides in the soil environment, it is essential that the processes affecting the transport of pesticide in the soil be understood before any cause-and-effect relationship can be established. The fate of a pesticide in the soil environment is governed by the retention, transformation, and transport processes, and the interaction of these processes (1, 2). In addition to the variety of processes involved in determining pesticide fate, many factors can affect the kinetics of the processes. The chemical structure of the pesticide governs the reactivity of the chemical, such as its efficacy in controlling pests and its mobility and degradability. Properties of the chemical, such as its solubility in water, vapor pressure, and polarity, also affect its behaviour in the environment (3). Likewise, soil properties such as its organic matter and clay contents, pH, Eh, and ion exchange capacity affect the behaviour of the chemical in the environment (1, 2).

Given that diuron, atrazine and MCPA are three widely applied herbicides for general weed control, and regarding that organic matter amendment of soils is a very usual practice (4), we have considered it useful to study the sorption processes and mobilities of diuron, atrazine and MCPA by a calcareous soil from Almería (a Mediterranean region in Southeastern Spain), after organic carbon amendment, from 0.40 % to 10.30 %, by using a commercial peat. The objective of this paper is to study on the one hand, the retention processes of the herbicides diuron, atrazine and MCPA on the organic matter-amended calcareous soil by using sorption isotherms, and on the other hand, the movement of these chemicals through the soil by using soil columns in laboratory conditions.

We claim to evaluate the potential for groundwater contamination of the three herbicides by calculating the adsorption parameters and measuring the amounts of each chemical leached as a function of increasing volumes of water added.

## Materials and Methods

Analytically pure diuron, atrazine and MCPA were used as adsorbables in this study. The soil chosen for this study was a calcareous soil (Camborthids) from Almeria; air-dried less than 2-mm samples were analysed by standard methods (5). Some relevant properties of the soil are: pH=8.1; Organic Matter content (OM)=0.40%; Clay content=8%; Cation Exchange Capacity (CEC)=2.1 meq/100g. A more complete discussion about the characteristics of this soil is given in a previous paper (6); this soil will be referred in the text as original soil, and will be labelled as T-0. The organic matter amendment of the soil was carried out by thoroughly mixing 50 g of the original soil with a varied amount (0.6-6 g) of a commercial peat containing 91.8% of OM. The OM range of the resulting amended soils (from T-1 to T-6) was (1.42% to 10.30%).

The adsorption experiments were carried out by shaking (at 25°C) 3 g of soil and 25 ml of a 0.02 M $CaCl_2$ aqueous solutions of varied herbicide concentrations (2-14 mg $L^{-1}$). After shaking the concentrations of diuron, atrazine and MCPA were determined by high-perfomance liquid chromatography (HPLC) at 250 nm, 222 nm and 228 nm, respectively.

The leaching experiments were carried out by using PVC columns (4.4 x 30 cm) filled with the soils T-0 and T-6. The columns were added, respectively, with 3 mg each of diuron, or atrazine, or MCPA. Each product was added in the form of concentrated aqueous solution to the top of the column and incorporated into the soil surface layer. Every day an aliquot (500 ml) of distilled water was added to each soil column by a controlled system which allowed a constant water flux of 0.35 ml $min^{-1}$. Water addition was stopped and the experiment terminated when the concentration of the chemical in the effluent decreased to a value less than 1% of total amount added.

## Results and Discussion

Figure 1 shows the adsorption isotherms of diuron, atrazine and MCPA on the soil samples studied. The adsorption isotherms were compared using the $K_f$ parameter of the Freundlich's adsorption equation. The $K_f$ values for the three herbicides and the soil samples studied increase from 2.17 mg $kg^{-1}$ (diuron), 0.94 mg $kg^{-1}$ (atrazine) and 0.24 mg $kg^{-1}$ (MCPA), for the original soil (containing 0.40% OM), up to 34.28 mg $kg^{-1}$ (diuron), 5.47 mg $kg^{-1}$ (atrazine) and 2.21 mg $kg^{-1}$ (MCPA), for the T-6 sample (containing 10.30% OM).

From the correlation analysis between the $K_f$ values and the organic carbon content of the soil samples, the $K_{oc}$ values (adsorption constant per kilogram of the organic carbon in soil) were calculated for the three herbicides; these values being respectively 651.0 mg $kg^{-1}$, 93.7 mg $kg^{-1}$ and 48.4 mg $kg^{-1}$ for diuron, atrazine and MCPA. The values of $K_{oc}$ suggest that diuron has the highest potential for adsorption on these amended soils, this potential being intermediate for atrazine and the lowest one corresponding to MCPA. The following equation has been established for the correlation between $K_{oc}$ and the partition coefficient octanol-water ($K_{ow}$):

$$\log K_{oc} = 2.53 \ 10^{-3} K_{ow} + 1.66 \qquad (r=0.999; p=0.05)$$

In relation to the leaching experiments, Figure 2 (A: for the T-0 -original soil- and B: for the T-6 soil) shows the percentages of recovery in the effluents (calculated on the basis of the total amount of herbicide added) for diuron, atrazine and MCPA as a function of increasing volumes of water added as aliquots of 500 ml each. For the original soil, 91.62%, 49.48% and 33.02% of herbicide were respectively recovered after 500 ml, 7500 ml and 8500 ml of water had leached the soil columns treated with MCPA, atrazine and diuron. For the T-6 soil, 49.35%, and 38.45% of herbicide were respectively recovered after 1000 ml and 9000 ml of water had leached the soil columns treated with MCPA and atrazine. The experiment involving the leaching of diuron was finished after 10000 ml of water had leached the soil column, due to after 8500 ml the percentage recovered reached a constant value.

Considering the above and taking into account that both $K_{oc}$ values and data from leaching experiments are generally used for predicting the fate of pesticides in the environment, the results obtained in this paper could be useful in order to know the fate of the herbicides diuron, atrazine and MCPA in environmental areas (such as the Mediterranean Southeastern Spain), in which, due to the poor organic matter content, organic matter amendment of soils is a very usual practise. So, considering the different lipophilic character/water solubility of the three herbicides (log $K_{ow}$=2.65 for diuron; log$K_{ow}$=2.09 for atrazine and log$K_{ow}$=1.08 for MCPA) and also the different $K_{oc}$ values and percentages recovered of each herbicide after the leaching experiments, diuron is not expected to leach a lot in soils as here studied, MCPA is expected to leach easily, atrazine having an intermediate behaviour.

## Acknowledgements

This research was supported by the CICYT Project AMB93-0600

## References

◻    [1] Cheng, H.H. ed. 1990. Pesticides in the Soil Environment: Processes, Impacts, and Modelling. Soil Science Society of America, Inc., publisher: Madison, Wisconsin.

◻    [2] Guenzi, W.D. ed. 1986. Pesticides in soil and water. Soil Science Society of America, Inc., publishers: Madison, Wisconsin.

◻    [3] Hartley, G.S. and Graham-Bryce, I.J. 1988. Physical Principles of Pesticide Behaviour, Vol 2. Academic Press: London.

◻    [4] González Pradas, E.; Villafranca Sánchez, M.; Pérez Cano, V.; Socías Viciana, M. and Valverde García, A. 1992. Soil adsorption of diuron: influence of $NH_4Cl$ and organic matter additions. The Science of the Total Environment, 123/124: 551-560.

◻    [5] Jackson, M.L. 1982. Análisis químico de suelos; Editorial Omega: Barcelona.

◻    [6] Socías Viciana, M. 1990. Adsorption of phosphate ions on soils of the Almeria province. Ph.D. Dissertation, University of Granada, Spain.

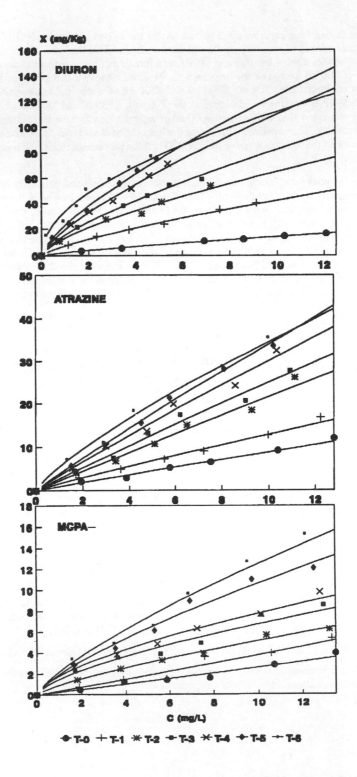

**Figure 1 :** Sorption isotherms of diuron, atrazine and MCPA on the soil samples

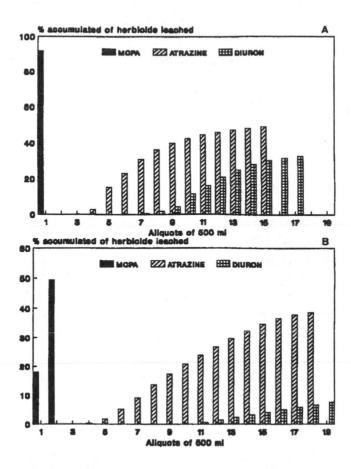

**Figure 2 :** Accumulated percentages of herbicide recovered in the effluents
(A: T-0; B: T-6)

*2.3.8.*        *Comparison of Percolation Studies in Disturbed and Undisturbed Soil*

# E. F. C. Grießbach

UER de Chimie analytique et phytopharmacie
Faculté des Sciences Agronomiques de Gembloux (Belgium)

## Abstract

A comparison of soil percolation studies between disturbed and undisturbed soil is given. The soil columns with disturbed soil were set up according to a slightly modified test protocol from the german federal biological office (BBA, authority for pesticide registration).

The undisturbed soil was taken from an agricultural field soil by driving a steel lysimeter directly into the field soil. The lysimeters were taken to the laboratory and run under the same conditions as the first study.

It was observed that the mobility behaviour of the applied substance (a type of organosilicon compound which is intended to be used as adjuvant in herbicide formulations) differed in these two studies.

These results may have an impact on how to choose the most appropriate test methods to study the mobility behaviour of pesticides and adjuvants used in pesticide formulations and on the interpretation of the test results.

## Introduction

Acetate capped propylheptaethyleneoxide heptamethyltrisiloxane (SE7Ac), has been developed for use as an adjuvant in herbicide formulations to increase efficacy and rainfastness of the active substances. To facilitate the analyses, this work was performed with radiolabelled material ($^{14}$C-SE7Ac). The registration of pesticides or agricultural adjuvants requires information on the mobility of these kinds of substances. This is in order to evaluate the potential for accumulation of these substances in soil, or the pollution of ground water. Percolation studies through soil columns are frequently used to obtain data on the leaching potential of a substance.

Today, percolation studies are done according to authority guidelines (e. g. Biologische Bundesanstalt für Land- und Forstwirtschaft, BBA. Federal Republic of Germany), using standard soils. More and more, however, the trend is towards the use of lysimeter studies using undisturbed field soils.

## Material and Methods

One soil, Gembloux Aba was chosen for this study. (The characteristics of this soil are shown in the following table 1.)

| Soil properties | Disturbed soil Soil 1 and 2 | Undisturbed soil Soil 3 |
|---|---|---|
| particle size | | |
| < 2 μm [g/kg] | 177 | 157 |
| 2 - 50 μm [g/kg] | 754 | 790 |
| > 50 - 2000 μm [g/kg] | 69 | 53 |
| FAO classification | silt loam | silt loam |
| pH water | 6.6 | 7.9 |
| organic matter [g/kg] | 17.2 | 23.6 |

**Table 1** : Soil characteristics

For percolation on disturbed soil, the soil was dried to 3.5% (w/w) and 16% (w/w) initial moisture and sieved to 2 mm particle size (soil 1 and 2 respectively).

Polypropylene columns were used for the study of these two soils. The columns were set up according to the BBA guideline 4-2[1] with a soil core of 40 cm length and 11.5 cm diameter, on top of a 2 cm length layer of sand. The columns were then saturated with CaCl2 $5.10^{-3}$ M solution to 34% water content (the field capacity, v/v). The densities for soils 1 and 2 were 1.3 and 1.1 $g/cm^3$ respectively.

For percolation on undisturbed soil (soil 3), steel lysimeters were used with a 20 cm long soil core of 10 cm diameter. This was extracted from the field by driving the steel lysimeters into the ground and then removing them along with the soil core intact. The lysimeters containing the cores were then placed into polypropylene columns with a 5 cm long sand layer at the base. The density of that soil was 1.5 $g/cm^3$ and the field capacity was 40% water content (v/v). Although the pH of the soil is different, the pH in the leachate of both soils was measured to be between 7.5 and 8.5 during the experiment. 20 mg $^{14}$C-SE7Ac (± 640 000 dpm) were applied to the top 5 cm of soil in two separate layers of 10 mg each. One layer was 3 cm below the surface, and the other layer was 1 cm below the surface. After an adsorption time of about 5h the artificial rainfall was started. It was distributed in six periods of two hours each. Between them there was a two hours' rest. The flow rate for soil 1 and 2 was 9 mm /day, and for soil 3 it was 10 mm CaCl2 $5.10^{-3}$ M/day.

The $^{14}$C-SE7Ac content in the leachate was analysed once per day using liquid scintillation counting as detection method.

Legend for figure 1 and 2 :     q = flow rate [mm/day]
                                 Θ = volumetric moisture content
                                   = field capacity [%v/v] / 100
                                 L = column length [mm]

In applying the factor q/Θ*L to the x-axis units (days), it is possible to compare soil columns with different field capacity, flow rate and length.

The value 1 on the x-axis indicates the time where the pore volume in the columns was replaced once.

## Results and discussion

The results of soils 1, 2 and 3 are presented in figures 1 and 2 respectively.

Soils 1 and 2 show a similar leachate distribution pattern but the beginning of the leaching was delayed by 5-6 days for the initially dry soil (soil 1).

The leachate distribution pattern of $^{14}$C-SE7Ac in disturbed soil (soil 1 and 2) and undisturbed soil (soil 3) is different.

These observations clearly indicate that the leaching behaviour of the substance is strongly influenced by the soil conditions. This may lead to different conclusions being formed of the leaching potential of the substance into ground water (depending on the soil conditions under which the leaching study was carried out).

## *Acknowledgement*

This work was funded by Dow Corning Europe.
My thanks go also to my supervisor, Prof. Copin, for his advice and help.

## *Reference*

□    (1) Biologische Bundesanstalt für Land- und Forstwirtschaft, Bundesrepublik Deutschland: Richtlinien für die amtliche Prüfung von Pflanzenschutzmitteln, (1986), Teil IV, 4-2, Versickerungsverhalten von Pflanzenschutzmitteln.

# LEACHING DISTRIBUTION PATTERN

## SE7Ac

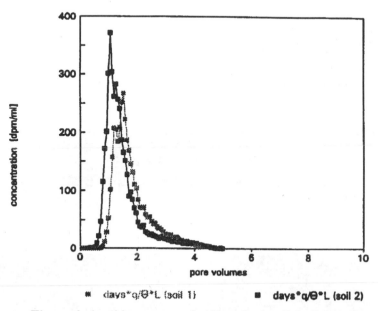

Figure 1: leaching pattern for disturbed soil (soil 1 & 2)

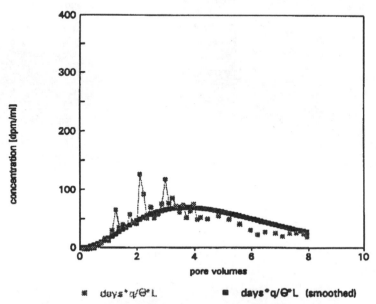

Figure 2: leaching pattern for undisturbed soil (soil 3)

## Legend see previous page

## 2.3.9.   *Ringtesting of the Modified Test Guideline for Adsorption/Desorption of Chemicals in Soils*

## M. Herrmann*, G. Kuhnt**, H. Muntau***

*       German Federal Environmental Agency (UBA), Berlin

**   Geographical Institute, University of Kiel

*** Environmental Institute, Joint Research Centre of the European
      Commission, Ispra

In order to determine soil related properties of chemicals it is of utmost importance to select soil samples which represent most frequent soil types but at the same time also cover a wide range of soil components like particle size pattern, organic carbon content, pH, etc.

For purposes of adsorption/desorption testing a set of five top soils (fig 1) has been identified which fulfil these criteria for the area of the European Union.

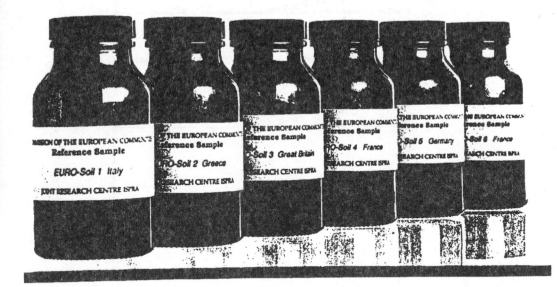

| | | |
|---|---|---|
| 1 = Vertic Cambisol | 3 = Dystric Cambisol | 5 = Orthic Podzol |
| 2 = Rendzina | 4 = Orthic Luvisol | 6 = sub-soil |

**Figure 1 :** Soil samples

## Objective

A laboratory inter-comparison test was conducted during 1989/1990 to prove the feasibility of a revised version of the OECD Test Guideline 106 on adsorption/desorption of chemicals to soil (test sequence is given in fig 2), as well as the suitability of the selected set of test soils. 27 laboratories from nine Member States participated in the exercise in part or as a whole.

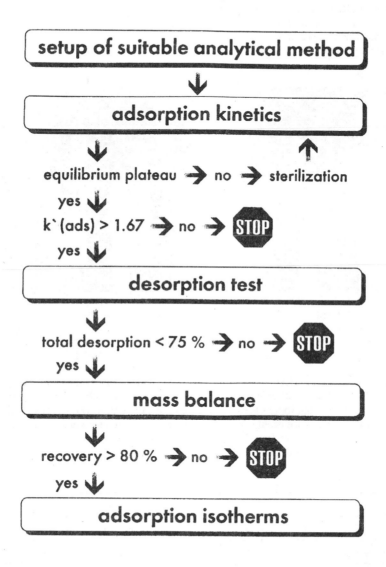

**Figure 2 :** Test sequence of the test programme

Three test substances, Lindane, Atrazine, 2,4-D were chosen by a delegation of European soil experts, covering a wide range of physical-chemical properties relevant for adsorption/desorption, eg. water solubility, partition coefficient n-octanol/water, $pK_a$. Ready-to-use soil samples as well as all test chemicals were provided from one common source, respectively, in order to minimize contributions to standard deviation of the results e.g. caused by different soil preparation techniques.

## Results

Some of the participating laboratories employed radiolabelled test substances, others used non-radiolabelled ones. From the differences between the data of both analytical techniques the contribution to the overall standard deviation can be estimated. The standard deviation of those results gained by non-labelled testing was roughly at least twice as high as for the data provided by participants which employed radiolabelled substances (fig 3). However, it must be stated that, provided a thorough analytical treatment, excellent results can also be achieved with non-radiolabelled material; while on the other hand some laboratories suffered from analytical problems even when employing radiolabelled test substances.

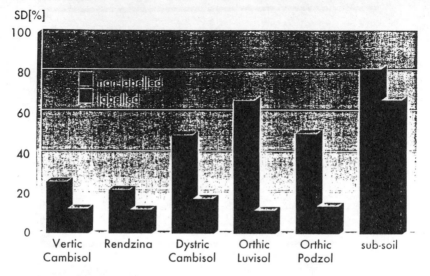

**Figure 3 :** Percentage of standard deviation (SD) from arithmetic mean for adsorption coefficient k'(Atrazine)

Out of the set of six soils the five top soil samples reflected in a satisfying manner the range of sorption relevant properties for the different test compounds, viz organic carbon content, pH, particle size spectrum. In all cases equilibrium between aqueous and solid phase was achieved within 16 hours. The different compositions of pedological parameters among the test soils are reflected by the different interaction with the physico-chemical properties of the test substances. Table 1 shows the arithmetic mean of the adsorption coefficient from Freundlich isotherms.

|                  | LINDANE | ATRAZINE | 2,4-D |
|------------------|---------|----------|-------|
| Vertic Cambisol  | 33.2    | 8.3      | 2.3   |
| Rendzina         | 37.8    | 2.3      | 0.7   |
| Dystric Cambisol | 41.7    | 2.7      | 1.5   |
| Orthic Luvisol   | 10.8    | 0.9      | 0.3   |
| Orthic Podzol    | 230.3   | 40.1     | 45.6  |
| sub-soil         | 0.6     | 0.2      | 0.2   |

**Table 1 :** Adsorption coefficient $K_{Freundlich}$ for soil/substance combinations

A report compiling pedological characteristics of test soils, preparation technique as well as a detailed statistical evaluation of the ringtest will be published by the JRC of the European Commission in Ispra.

## *Conclusions*

■ The statistical evaluation of the ringtest demonstrate that the selected set of five top soils is appropriate to describe the adsorption behaviour of chemicals in the most frequent types of soils throughout the European Union.

■ The sub-soil of extremely low organic carbon content did not reveal any relevant adsorption for all of the three test compounds. The analitical errors normally exceeded the different adsorption behaviour of the test substances. Such kind of soils should not be not used any further.

■ It can be expected that the set of selected soils also represents conditions of other countries of comparable climatic/geographical conditions outside the European Union.

*2.3.10.*    *The adsorption kinetics as a function of initial concentration of pesticide in the soil*

# J. Kozak, O. Vacek & M. Valla

University of Agriculture, Prague, The Czech Republic

## Introduction

Adsorption process shows distinct changes with the time of interaction between sorbate and sorbent. It is very useful to describe the adsorption kinetics and to determine the rate constants of the process. The values of constants could be considered as a very useful input data for many prediction models. The kinetics of organic chemicals adsorption on the soil was subject of many studies (e.g.,Brusseau and Rao,1990, Boesten and Van Der Pas, 1988, Kozak et al.1994 and others). There is always a problem to describe successfully the nonequilibrium adsorption process. One possible approach has been suggested by Moreale and Van Bladel (1979). They exploited for adsorption kinetics description transformation of the relationship between the adsorbed amount and the time into linear equation of hyperbolic tangent:

$$q = \frac{qmax^{Bt}}{1 + Bt} \qquad (1)$$

where: q = adsorbed amount ; qmax = adsorbed amount at equilibrium; t = time; B = constant.

Exponential function was found for the rate of adsorption as a most suitable:

$$\frac{\delta q}{\delta t} = k\,(\,qmax - q\,)^N \qquad (2)$$

where k = rate constant; N = order of reaction.

The aim of this contribution was to present our approach to the description of adsorption kinetics and to show the relationship between pesticide concentration and the rate constant of adsorption on the soil.

## Material and Methods

Adsorption experiments were performed on the model soil, classified as sandy Regosol. The main characteristics of the soil are given in the table 1. The model pesticide used for the adsorption study was chlortoluron [3-(3-chlor-4 methylphenyl) -1,1-di-methylurea]. Air dry soil (10g) was shaken in centrifugation tubes on the action shaker with a solution of herbicide (10 ml) in 0.02M $CaCl_2$. Times of interaction were 0.5, 1, 2, 4, 6 and 24 hours. After centrifugation the herbicide concentration in supernatant were determined by mean of HPLC. The initial concentrations of chlortoluron were: 3, 6, 12, 27, 34 and 54 $\mu g.g^{-1}$.

| ORGANIC CARBON % | pH (H₂O) | CLAY CONT. (< 0.001 mm) % |
|---|---|---|
| 1.95 | 5.4 | 21 |

**Table 1**

## Results and Discussion

In our study the eq.1 was linearised as follows:

At the condition a = qmax and b = B,

$$y = \frac{abx}{1+bx} = \frac{bx.a}{bx\left(\frac{1}{bx}+1\right)} = \frac{a}{\frac{1}{bx}+1} \tag{3}$$

and thus:

$$\frac{1}{y} = \frac{1/bx + 1}{a} = \frac{1}{a}\left(\frac{1}{bx}+1\right) = \frac{1}{abx} + \frac{1}{a} \tag{4}$$

$$\frac{1}{y} \to y; \ \frac{1}{x} \to x; \ \frac{1}{a} = a_0; \ \frac{1}{ab} = a_1 \tag{5}$$

and thus:

$$a = \frac{1}{a_0}; \ b = \frac{a_0}{a_1} \tag{6}$$

A computer programme was prepared for computation of parameters qmax and B, regression coefficient for evaluation of fit of experimental data to the theoretical curves, rate constant and the order of reaction. The programme made it possible to choose any time interval for computation. As a guideline could be used the value of correlation coefficient between experimental data and theoretical curve.

It was possible to describe adsorption kinetics of chlortoluron in two time intervals. At first the interval 0 -120 (min) was chosen, which was consistent with our previous results (Kozak et al. 1994). The typical examples of curves describing the rate of adsorption for concentrations 54 ppm and 24 ppm are shown on Figs 1 and 2 respectively. The rate constants for both time intervals were computed and the results are given in the figures. In our previous study we distinguished the time interval 15 min. which could be considered from the point of view of practical application too short. We tried to determine the relationship between the initial concentration of pesticide and the parameters of the equation (1).

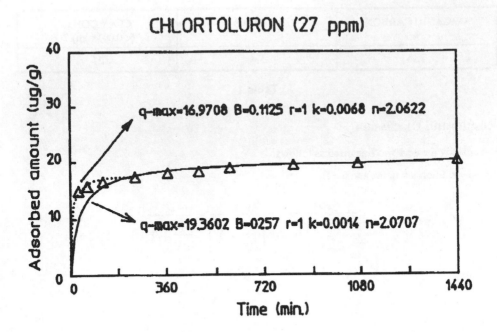

Figure 1

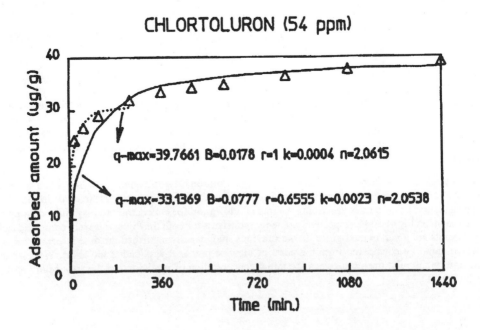

Figure 2

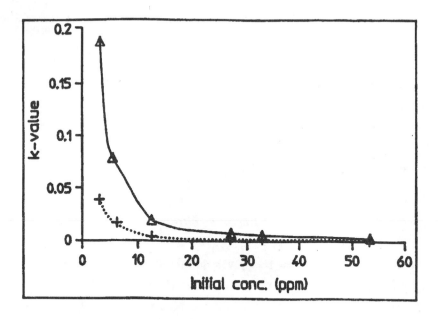

**Figure 3**

On the Fig. 3 and 4 respectively are given the above mentioned parameters. It is apparent that qmax values show a linear relationship which the initial concentration. K-value showed a rapid decrease with the initial concentration. The break-point was usually found at the initial concentration of approximately 15 ppm of the herbicide applied.

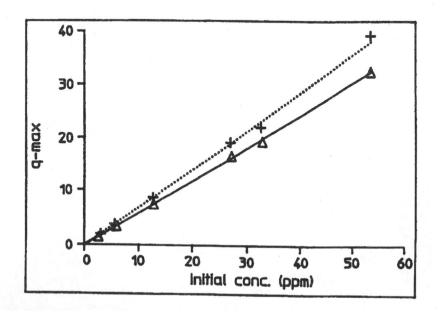

**Figure 4**

It could be concluded that the above mentioned concentration of the studied herbicide is readily adsorbed. This may be related to the amount of active specific adsorption sites for the herbicide under study.This fact must be considered in prediction models.

## Conclusions

The adsorption kinetic study of chlortoluron herbicide showed, that it could be satisfactorily described by the modified equations. It was found that the value of initial concentration showed the distinct influence on the adsorption rate. The hypothesis for the possible explanation was suggested.

## Literature Cited

□    Boesten, J.J.I.T., Van der Pas, L.J.T., 1988: Soil Sci. 146:221-231.

□    Brusseau, M.L., Rao, P.S.C., 1989: Geoderma, 46:169-192.

□    Kozak, J., Vacek, O., Janku,J., 1994: Rostl. Vyr. 40, 1994 (4):333-342.

□    Moreale, A, Van Bladel, R.,1979: Soil Sci.127:1-9.

## 2.3.11.   Behaviour of Chlormequat Adsorption on Montmorillonite in Presence of Copper

# C. Maqueda, E. Morillo and T. Undabeytia

Instituto de Recursos Naturales y Agrobiología.
Apdo. 1052. 41080 Sevilla (Spain).

## Abstract

The competitive adsorption of the cationic pesticide chlormequat (CCC) and the heavy metal Cu on montmorillonite was studied. CCC adsorption on the clay mineral decreased with increasing Cu concentrations, indicating high competition between both cations. Similar fact was observed for Cu adsorption in presence of CCC. In $CaCl_2$ medium, CCC and Cu adsorption on montmorillonite decreases very much, and the isotherms obtained are much closer, being not important the different Cu or CCC concentrations used, since the species that really has a competitive effect towards Cu and CCC is $Ca^{2+}$ from background electrolyte.

## Introduction

The indiscriminate and excessive use of pesticides produces many ecological problems, the contamination of soils being one of them. Also metal contamination of soils is increasing as a result of the use of agrochemicals and the disposal of industrial and domestic wastes to agricultural lands.

The persistence of agrochemicals and heavy metals in soils depends mainly on adsorption-desorption reactions that take place with soil components. However, the presence of metals could influence the adsorbing properties of soils with respect to pesticides, and vice versa, since phenomena of competition, hydrolysis, decomposition, complexation, etc. could take place (Morillo and Maqueda, 1992; Staedler and Schindler, 1993; Undabeytia et al., 1994). Taking into account the surface properties of clay minerals, the study of the mutual influence that metals and pesticides could have on their uptake by clay minerals is necessary to understand their fate in soils.

The purpose of this research is to describe the adsorption phenomena of the pesticide chlormequat on the clay mineral montmorillonite in presence of the heavy metal copper. This study is included within a wide study about the behaviour of different pesticides and various heavy metals which can be present simultaneously.

## Materials and Methods

The clay used was a standard montmorillonite from Arizona (SAZ-1) with CEC of 120 meq/100 g (80% of CEC being satured with $Ca^{2+}$).

The pesticide used was chlorocholine chloride (2- chloroethyltrimethylammonium chloride), commonly known as chlormequat (CCC). This pesticide is a plant-growth regulator very soluble in aqueous medium, and is the active component of two products widely used in agriculture.

The adsorption experiments were done in triplicate in 50-ml polypropylene centrifuge tubes, by mixing 0.1 g of clay mineral with solutions containing various concentrations of Cu (0- 30 ppm) and/or the pesticide (0-1 mM). The samples were shaken for 24 h at 20±1°C. The experiments were carried out also in 0.01 N $CaCl_2$ medium. The adsorbed concentration of Cu and the pesticide was obtained as the difference between their concentrations before and after equilibrium. The concentration of CCC was determined using the spectrophotometric method proposed by Pasarela and Orloski

(1965), based on the formation of a yellow complex with dipicrylamine. The copper concentration and the inorganic cations $Ca^{2+}$, $Mg^{2+}$ and $Na^+$ liberated by montmorillonite during the adsorption of Cu or CCC were determined by atomic absorption spectrometry or flame photometry.

The adsorption isotherms were obtained by plotting the Cu or CCC adsorbed versus their equilibrium concentrations.

## Results and Discussion

The adsorption isotherms for the pesticide chlormequat and the heavy metal copper on montmorillonite are shown in figure 1. The adsorption isotherm of the pesticide (fig. 1a) shows a shape corresponding to the "S" type of Giles et al. (1960). The isotherms obtained by other authors for the same pesticide and clay mineral (Maza et al., 1989) contrast with those of this paper, showing adsorption isotherms of Langmuir type. This discrepancy can be attributed to the longer span of concentrations used by the former which masked the behaviour of the lower part of the isotherms.

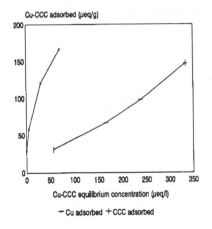

**Figure 1 :** Adsorption isotherms of CCC (a) and Cu (b) on montmorillonite

Figure 1b shows the adsorption isotherm of Cu on montmorillonite. It has been suggested by several authors (Benjamin and Leckie, 1981; García Miragaya et al., 1986; Madrid et al., 1991) that the adsorption of metal ions by layer silicates and other adsorbents may occur at several types of surface sites, with higher energy sites acting at the lower range of surface coverage. Retention of Cu ions by montmorillonite can be expected to occur through two different processes by adsorption on variable charge sites and by exchange of the cations saturating the permanent charge planar sites of montmorillonite.

As CCC is a cationic species, one could expect it to be retained by a cationic exchange that displaces the inorganic cations situated in the silicate interlamellar space. This was tested by determining for every point on the isotherms the concentration of the inorganic cations liberated to the solutions at equilibrium. In general these amounts were between (15-21%) less than adsorbed amount of CCC+Cu, indicating that not all the CCC and Cu were adsorbed through cation exchange, but adsorption onto the external surfaces of the silicate also take place.

Figure 2 shows the adsorption isotherms of CCC from Cu-free solutions in comparison with CCC adsorption in presence of different concentrations of Cu (5, 10, 20 and 30 ppm). For all concentrations of Cu used, the amount of CCC adsorbed decreases (in relation to Cu-free treatment) as the metal concentration increases. So, there is a competitive effect between them for clay interlamellar positions. The same effect is also observed for Cu adsorption on montmorillonite when CCC is present (fig. 3). For similar concentrations of Cu and CCC used (expresed as µeq/l) the percentage of non-adsorbed CCC, due to the presence of Cu, is the same as the percentage of non-adsorbed Cu due to the presence of CCC. It indicates that the competitive effect of one towards the other is similar.

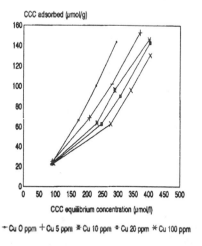

**Figure 2 :** Adsorption isotherms of CCC in presence of Cu in H$_2$O medium

When Cu concentration is the lowest (5 ppm) the amount of Cu adsorbed is nearly the same (fig. 3), independently of pesticide concentration. It is probably due to the adsorption of Cu on edge positions in the clay, sites of high affinity for the metal, and for which the pesticide not compete.

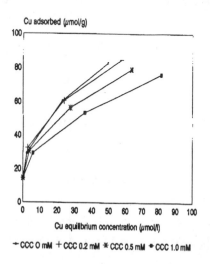

**Figure 3 :** Adsorption isotherms of Cu in presence of CCC in H$_2$O medium

Figure 4 shows the adsorption isotherms of CCC in presence of Cu in $CaCl_2$ medium. The CCC adsorption decreases about 25% when background electrolyte is used. The isotherms carried out with different concentrations of Cu are closer to each other than in aqueous medium. The reason is that the especies that really has a competitive effect towards $Cu^{2+}$ is $Ca^{2+}$ from background electrolite. $Ca^{2+}$ is a hard Lewis acid and the interlayer charge heterogeneous distribution of montmorillonite corresponds to a hard Lewis base, so calcium cations will interact more energetically with montmorillonite than $Cu^{2+}$, which is a soft Lewis acid. The same fact is also observed for Cu adsorption isotherms (fig 5), but they are even closer.

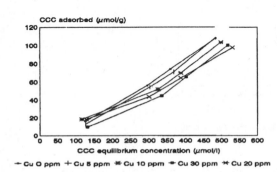

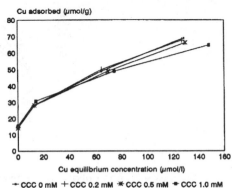

Adsorption isotherms of CCC in presence of Cu in $CaCl_2$ medium

Adsorption of Cu in presence of CCC in $CaCl_2$ medium

# Acknowledgements

We are grateful to M.E.C. (Research Project AMB92-0394) for financial support.

# References

□   Benjamín, M. M. and Leckie, J. O. 1981. J. Colloid Interface Sci. 79: 209-221.

□   García-Miragaya, J., Càrdenas, R. and Page, A. L. 1986. Water, Air and Soil Pollut. 27: 181-190.

□   Giles, G. H., Mac Ewan, T. H., Nekhwa, S. N. and Smith, D. 1960. J. Chem. Soc. 3973-3993.

□   Madrid, L., Diaz-Barrientos, E. and Contreras, M. C. 1991. Aust. J. Soil Res. 29: 239-277.

- Maza, J., Jimenez-Lopez, A. and Bruque, S. 1989. Soil Sci. 147: 11-16.
- Morillo, E. and Maqueda, C. 1992. Sci. Total Environ. 123/124: 133-143.
- Pasarela, N. R. and Orloski, E. J. 1965. Gunter Sweig, Ed, VII: 523-544.
- Stadler, M. and Schindler, W. 1993. Clays and Clay Miner. 41(6): 680-692.
- Undabeytia, T., Morillo, E. and Maqueda, C. 1994. Toxicol. Environ. Chem. 43: 77-84.

## 2.3.12.    *Study of Terbacil Movement in Southern Spain Soils*

# A. Mora, M.C. Hermosín and J. Cornejo

Inst. Recursos Naturales y Agrobiología de Sevilla-C.S.I.C.
Avda. de la Reina Mercedes s/n, Aptdo. 1052
41.080-Sevilla SPAIN

## Introduction

The risk of ground water contamination from agricultural pesticide use is ultimately determined by the relative rates of percolation and degradation within the soil profile and by factors controlling these processes such as climate, soil properties, microbial activity and chemical properties of the pesticides.

Pesticide mobility has been measured in the laboratory by using soil leaching columns, soil thick-layer trays and soil thin-layer chromatography, and so related with the properties which control its movement.

Terbacil is relatively mobile in soils in comparison with other usual herbicides, but there are not many studies about its mobility and noneone on Spanish soils. Therefore the experiments reported here were conducted to improve the knowlege on this particular matter.

## Materials and Methods

### *Herbicide*

Terbacil (3-tert-butyl-5-chloro-6-methyluracil) is widely used for the selective control of many annual and some perennial weed in crops as citrus, apples, peaches, alfalfa and blueberries.

### *Soils*

The soils were selected on the basis of previous work done in our lab which allowed us to select sampling points at the intensive cultivation zone surrounding Do9ana National Park. The samples used in this study corresponding to the upper layer of three soils classified like Xerofluvent (P-1), Typic Rhodoxeralfs (P-2) and Entic Pelloxererts (P-7) whose properties are shown in table I.

### *Adsorption/desorption assays*

Equilibrium adsorption isotherms for all the soils were obtained using the batch procedure. Equilibrium was achieved by shaking duplicate samples of soils (5 gr) with 0.01M $CaCl_2$ solution of terbacil (20 ml). The shaking was carried out for 24 hours. Similar procedure was used to get equilibrium desorption isotherms.

### *Column displacement setup*

Methacrylate columns 30 cm long x 5.5 cm i.d. were made joining segments 5 cm long with silicone sealer, this was done in the form that a silicone ring remained inside the column each 5 cm to prevent movement of water down the walls. Finally a Büchner funnel was glued to the botton of each column. The lower 5 cm were filled with glass wool and coarse quartz sand. The soils were hand-packed uniformly into the columns in 2.5 cm deep increments to a depth of 20 cm.

| Soil | | P-1 | P-2 | P-7 |
|---|---|---|---|---|
| pH (1:1) | | 7.7 | 7.9 | 7.6 |
| o.m. (%) | | 2.50 | 1.24 | 2.05 |
| Sand (%) | | 12.1 | 70.7 | 11.9 |
| | Thick | 2.9 | 53.3 | 2.4 |
| | Thin | 9.2 | 17.4 | 9.5 |
| Silt (%) | | 43.7 | 8.9 | 35.9 |
| Clay (%) | | 44.3 | 20.4 | 52.3 |
| | Kaolin. | 7.1 | 4.5 | 15.7 |
| | Illit. | 29.2 | 12.4 | 18.8 |
| | Smect. | 8.0 | 3.5 | 17.8 |
| C.E.C. (mEq/100gr) | | 19.2 | 9.0 | 28.5 |
| CaCO$_3$ (%) | | 24.16 | 0.64 | 15.76 |

**Table 1** : Selected soil properties

The upper 10 gr of soil from each column were removed and mixed with 1 mgr of terbacil (99%) on approximately 2.5 ml of methanol; laterly, 10 ml of water were added and equilibrate during 24 hours. A separate study showed that no-loss occurs during this process. The soils were then returned to the columns that at the same time were saturated with water (0.01M CaCl$_2$) and allowed to drain free 24 hours. Then the upper part of the column was covered with a 3 cm layer of coarse quartz sand to prevent soil disturbance during the water application and evaporation from the top of the column.

The water (0.01M CaCl$_2$) was added with a rate of 25 ml and the same volume of leachate collected daily during a period of 11 days; after this, 50 ml until the end of the test. A aliquot from each leachate sample was injected on a HPLC system ( 150x3.9 mm C18 column; water:methanol 60:40, 1 ml/min; 282 nm).

At the end of the leaching process the columns were allowed to drain for 36 hours and sectioned into 5 cm deep increments, each soil fraction mixed thoroughly, placed into plastic bags and labeled. Samples not analyzed inmediately were stored at -5°C.

## Results and Discussion

The adsorption/desorption isotherms were fitted to the Freundlich equation, $C_s=K C_e^n$, where K and n are constants and $C_s$ and $C_e$ are adsorbed (mmol/gr soil) and solution (mmol/L) concentrations respectively. The values of the Freundlich adsorption constant K and n are summarized in table II. Although these adsorption coefficients are quite low, the relative differences among the soils are significants indicating that the capability of adsorbing terbacil is P-1 > P-7 >> P-2. The difference in sorption among the three soils can be related to the variation of the soil organic matter and clay content, not only on their quantity but also on their type.

| | $K_F$ | $n_F$ | $K_{F1}$ | $n_{F1}$ | $K_{F2}$ | $n_{F2}$ |
|---|---|---|---|---|---|---|
| P-1 | 0.739 | 0.941 | 0.876 | 0.319 | 0.567 | 0.355 |
| P-2 | 0.233 | 0.542 | 0.201 | 1.018 | 0.187 | 0.309 |
| P-7 | 0.485 | 1.032 | 0.622 | 0.719 | 0.405 | 0.513 |

**Table 2** : Adsorption/desorption Freundlich constants

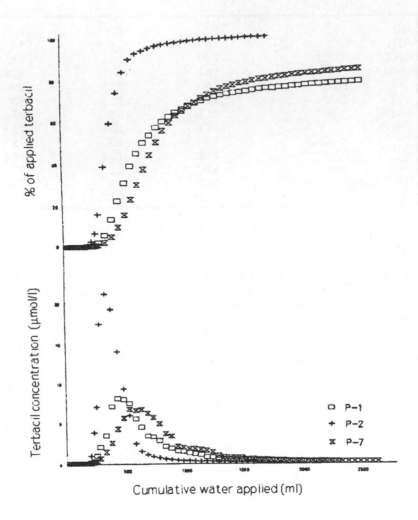

**Figure 1 :** Cumulative curve of terbacil leachate loss (upper) and terbacil leachate concentration curve (lower)

Figure 1 illustrates the terbacil concentration on leachates and the cumulative leachate loss versus added water volume. The great recovery of total terbacil from each soil column suggests that minimun or not degradation ocurred. Basic soil characteristics can be used to explain differences in herbicide movement through the three soils. The soils exhibit differences in several important parameters that can be related to the different solute mobility. Particle size analysis shows that the P-2 soil is composed largely of medium and thick particles whereas P-1 and P-7 soils are dominated by finer particle sizes. Thus, it appears that P-2 soil should have a macropore-dominated structure which explains the bigger slope exhibit by P-2 soil (cumulative curves in figure 1). The situation between P-1 and P-7 soils is not very different in this aspect and we can find two fundamental differences. The first is that P-1 soil exhibits its highest leachate concentration before than P-7 (concentration leachate curves in figure 1) indicating that terbacil is less retained. The second difference is that P-1 curve reachs its plateau before than P-7 one (cumulative curves in figure 1) indicating the adsorption/desorption hysteresis is bigger on P-1, in agreement with adsorption/desorption results.

For cases where the movement of pesticides through soils is under steady-state water flow conditions and the adsorption isotherms obeys the Freundlich equation we can use the retardation term:

$$R = 1 + \frac{\rho K n C^{n-1}}{\theta}$$

where $\rho$ is the soil bulk density and $\theta$ is de pore volume, like a quantitative index of mobility. The retardation term (R) can also be expressed as the number of pore volumes of flow required to achieved breakthrough, assuming that breakthrough of a nonadsorbed compound would occur at one pore volume of flow. In our case we assume that breakthrough occurs when the leaching loss is 50% of applied terbacil. The R values calculated are: 3.63 for P-1, 1.86 for P-2 and 3.82 for P-7. It is interesting to verify that there is not relation between K and R, indicating that for weakly adsorbed pesticides adsorption process is not the most important factor to evaluate their mobility, being the physical processes of great importance.

## Acknowledgment

This work was partially supported by CICYT through proyect AMB 93-81 and by Junta de Andalucia through the Research Group n°4092.

## References

▢   BILKERT, J.N. y RAO, P.S.C. (1985) J. Environ. Sci. Health, B 20(1):1-26.

▢   GARDINER, J.A., RHODES, R.C., ADAMS, J.B. Jr. y SOBOCZENSKI, E.J. (1969) J.Agric. Food Chem., 17:980-986.

▢   SKROCH, W.A., SHEETS, T.J. y SMITH, J.W. (1971) Weed Sci., 19:257-260.

▢   SKROCH, W.A., SHEETS, T.J. y MONACO, T.J. (1975) Weed Sci., 23:53-57.

▢   SWAN, D.G. (1972) Weed Sci., 20:335-337.

▢   VAN GENUTCHEN, M. Th. y CLEARY, R.W. (1979) en Bolt, G.H. (Ed.), "Soil Chemistry B. Physico-Chemical Models", Cap. 10, Elsevier, Amsterdam.

▢   WATERS, W.E. y BURGIS, D.S. (1968) Weed Sci., 16:149-151.

▢   WEBER, J.B., SWAIN, L.R., STREK, H.J. y SARTORI, J.L. (1980) en Camper, N.D.(Ed.) "Research Methods in Weed Science"

▢   WEBER, W.J. Jr., MACGINLEY, P.M. y KATZ, L.E. (1991) Wat. Res., 5:499-528.

*2.3.13.*     *Interactions of Two Triazines with Humic Substances from European Soils along a Gradient Climate*

# A. Piccolo[1], L. Gatta[2], A. Zsolnay[3]

1.   Dipartimento di Scienze Chimico-Agrarie, Università di Napoli "Federico II", Via Università 100, 80055 Portici, Italy.

2.   Dipartimento di Biologia Vegetale, Università di Roma "La Sapienza", Rome, Italy.

3.   Institute of Soil Ecology, GSF, Neuherberg, Germany.

## Abstract

Humic substances (HS) were extracted from three european soils cropped with different monocultures and placed along a climate gradient. Namely, a soil from southern Germany (Munich, M) under hops, a soil form central Italy (Tuscany, T) under vineyards, and a soil from southern Italy (Bari, B) under citrus trees. From each site, HS were extracted from samples of three locations: a. along the crop rows, b. between the rows and c. from an uncropped control soil. The HS extracts were used to study the adsorption interactions with Atrazine and Simazine. Molecular weight distribution of the humic extracts by SDS-PAGE electrophoresis showed that MW decreased in the order: T>B>M. $^{13}$C-NMR spectra showed that the soils climate gradient is reflected in the chemical composition of the corresponding humic samples. The degree of aromaticity increased in the order: M<T<B, showing an enhancing degree of polycondensation (humification) with decreasing latitude. Adsorption isotherms for Atrazine and Simazine by the Freundlich equation showed that adsorption of Atrazine decreased in the order: M>T>B, whereas for Simazine the order was reversed: B>T>M. These results may be explained by the structural characteristics of both HS and the triazines. High aromaticity and high molecular size distribution favour adsorption of Simazine whereas both structural parameters are inversely proportional to Atrazine adsorption. This behaviour is consequential to the stereochemical hindrance of the isopropyl group in Atrazine as compared to the linear ethyl group in the same position in the Simazine molecule. Such results seem to indicate that the charge transfer mechanism is the predominant interaction mode between HS and the two triazine molecules.

## Introduction

Triazines are widely used herbicides worldwide. Their application on European soils is widespread in several types of cultures. The relatively high mobility of triazines down to lower horizons of soils with prevalent loamy texture has created environmental concern in Europe. Little attention has been devoted to the relation of the triazines with the organic components of soils despite the recognition that herbicides mobility in soils is highly related to the soil organic matter content (Khan, 1980) and, thus, to the soil humic substances (Stevenson, 1982). Recent studies (Piccolo et al., 1992) have shown that Atrazine adsorption on humic substances varies with the chemical and macromolecular structure of humic substances.

The aim of this study was to evaluate the interactions of two widely used triazines, Atrazine and Simazine, with humic substances obtained from european soils under different monocultural practices.

# Materials and Methods

Soil samples were selected from the surface horizons (0-40 cm) of agricultural sites under three different monocultures situated in different climatic conditions. A soil near Munich, Germany (M), permanently cultivated with hops. A soil in Tuscany, Italy (T) under permanent vineyards. A soil near Bari, Italy (B) sustaining a permanent citrus fruit production. From each soil, samples were collected from three locations: a. under the crop rows, b. between the crop rows, c. from an uncropped control soil. The soil samples were air dried, passed through a 2 mm, and stored for further analysis.

Humic substances were extracted from soil samples by conventional methods as described by Stevenson (1982). The isolated humic acids (HA) were purified by a double acid-base, precipitation-dissolution procedure, and treated with a HCl-HF 0.5% (v/v) solution, dialyzed against distilled water until chloride-free, and freeze-dried.

Solution-state $^{13}$C-NMR spectra of the purified HA were obtained in a quantitative mode using a Bruker 500 MHz spectrometer with inverse gated decoupling, 45 pulse, acquisition time of 0.1 s, and a relaxation delay (decoupler off) of 1.9 s. Other details on sample preparation for NMR analysis and peaks integration procedures are reported elsewhere (Piccolo et al., 1990). Molecular weight distribution of HA was achieved in a gradient (4 to 20%) of SDS-Polyacrilamide gel electrophoresis as described by Trubetskoj et al. (1991).

Values for adsorption of Atrazine and Simazine on HA were obtained by shaking overnight the purified HA with concentrations of 1, 5, 10, and 25 ppm of Atrazine and of 1, 3, 5, and 10 ppm of Simazine in aqueous solutions of 0.01M CaCl$_2$ (50mg:10ml, HA:solution). The suspension was then centrifuged (4,000 rpm), and 9 ml of the supernatant separated and analyzed for the herbicide concentration. First desorption values were arrived at by adding 10ml of a CaCl$_2$ 0.01M solution to the residue of the adsorption treatment and shaking overnight again. After centrifugation, 9 ml of the supernatant were separated and analyzed for the herbicide concentration. Data relative to the second desorption were obtained by repeating the procedure adopted for the first desorption.

The triazine content in the different solutions were determined by HPLC and a UV-Vis detector of a Spectra-Physics model P2000 and relative software-integrator system. Elution conditions for both Atrazine and Simazine were those described by Piccolo et al. (1992), using a 60:40, acetonitrile:water solution in a isocratic mode, pumped at 2 ml/min in a C$_{18}$ Spheri5 Brownlee 220x4.6 mm column and 30x4.6 mm precolumn. Determination was achieved at 220 nm using an external standard method.

# Results and Discussion

## *Characteristics of Humic Substances*

The $^{13}$C quantitative distribution obtained from NMR spectra of HA from the soil sites of M (hops), T (vineyards), and B, (citrus), are reported in Table 1.

| PPM | Ma | Mb | Mc | Ta | Tb | Tc | Ba | Bb | Bc |
|-----|------|------|------|------|------|------|------|------|------|
| 0-48 | 23.6 | 20.8 | 20.8 | 21.5 | 20.2 | 16.5 | 18.5 | 19.9 | 19.5 |
| 48-105 | 23.3 | 25.3 | 26.9 | 26.9 | 23.9 | 26.2 | 14.6 | 15.4 | 25.2 |
| 105-145 | 27.8 | 26.3 | 25.1 | 27.6 | 29.4 | 27.3 | 38.2 | 38.8 | 27.3 |
| 145-165 | 7.5 | 10.1 | 9.5 | 8.9 | 9.5 | 11.4 | 8.9 | 7.7 | 10.1 |
| 165-190 | 17.8 | 17.6 | 17.6 | 14.9 | 16.8 | 18.5 | 19.8 | 18.1 | 18.1 |
| Aromatic. | 42.9 | 44.1 | 42.1 | 43.0 | 46.8 | 47.6 | 58.7 | 56.8 | 45.5 |

**Table 1** : C-NMR distribution (%) over chemical shift range (PPM) and aromaticity (105-165 PPM/0-165 PPM) of humic extract for Munich (M), Tuscany (T), and Bari (B) samples for the in-rows (a), between-rows (b), and control (c) samples.

The NMR spectra reveal that the chemical composition of the humic samples reflect the climate gradient in the series of samples passing from Munich to Tuscany, and to Bari samples. The degree of aromaticity increases in the order: M<T<B, showing an enhancing content of polycondensed (humified) compounds with decreasing latitude. Conversely, the alkyl carbons (0-48 ppm) and the C-O, C-N carbons (carbohydrates and proteins) (48-105 ppm) are higher in the M samples than in the T and B samples, in the order, thereby reflecting a higher microbial transformation in the latter samples. The content of carboxyl C (165-190 ppm), and index of the acidity reactivity of the humic samples is lightly higher in the B samples than in the other two sites, confirming a somewhat more advanced humification for the most southern samples. As for the locations, the in-rows samples (a) appeared to be higher in aliphatic carbon than the b and c samples except for the B site. The a location was lower in carbohydrates and protein content than the b and c ones in M and B but somewhat higher than the rest in the T site. The a location was also lower than rest of locations in aromaticity in the M and T sites whereas was higher in the B site. In general. it appeared that, where the warmer climate favoured microbial activity, the aromaticity of the a location was higher than the rest of locations whereas the less resistant alkyl carbon decreased and the carbohydrates and proteins were more stable, possibly for a rapid incorporation in the humus core.

The electrophoretic pattern obtained with SDS-PAGE electrophoresis (unshown) reveals that for the 9 HA samples, the molecular dimensions generally decreased in the order: T>B>M. The HA samples of the three locations within each monoculture appeared to have similar molecular size distributions except for the B site whose c sample showed a lower molecular dimension than the a and b samples. The order of molecular dimensions of the HA samples from the three site is consistent with the model attributing a larger polycondensation of humic acids to the soils of warmer climates.

## *Adsorption and desorption*

Data on the adsorption and two desorption of Atrazine and Simazine obtained for the described 9 samples were fitted into the empirical Freundlich equation. The Freundlich adsorption maxima (Kf) of Atrazine (unshown) showed that adsorption was generally highest for the M samples, followed, in the order, by the T and B samples, whereas the intensity of adsorption (1/n) was generally close to the unity for the T and B samples and slightly less for M.

|        | Ma   | Mb   | Mc   | Ta   | Tb   | Tc   | Ba   | Bb   | Bc   |
|--------|------|------|------|------|------|------|------|------|------|
|        |      |      |      |      | Atrazine |    |      |      |      |
| % ADS  | 50.9 | 50.4 | 60.3 | 41.7 | 48.3 | 48.1 | 44.4 | 44.7 | 40.0 |
| % I D  | 34.6 | 36.5 | 16.5 | 31.8 | 21.0 | 20.7 | 36.7 | 27.6 | 30.9 |
| %IID   | 17.3 | 17.5 | 8.3  | 17.9 | 10.9 | 7.1  | 19.9 | 27.4 | 26.7 |
|        |      |      |      |      | Simazine |   |      |      |      |
| % ADS  | 32.0 | 25.2 | 35.4 | 33.4 | 31.9 | 30.5 | 34.3 | 36.4 | 24.7 |
| % ADS  | 31.1 | 30.0 | 26.0 | 26.0 | 25.3 | 24.3 | 22.6 | 17.5 | 27.9 |
| % IID  | 13.5 | 17.4 | 12.7 | 14.5 | 14.3 | 13.3 | 28.3 | 3.4  | 18.1 |

**Table 2 :** Average over the different concentration used, of percent of adsorption and of I and II desorption for Atrazine and Simazine interactions with the HA.

The order of adsorption: M>T>B, was also evident from the mean percent of Atrazine adsorbed on the different HA samples (Tab. 2), whereas the percent of the first and second desorption showed that the least adsorbing site, B, was also the larger desorber followed by T and M. As for the locations within the sites, the order was generally c<b<a for both adsorption and desorption. In general, the adsorption results for Atrazine appeared to be inversely proportional to the aromaticity and molecular dimensions of the samples.

In the case of Simazine, the Freundlich maxima (Kf) calculated from adsorption and desorption data (unshown), revealed that the highest adsorbing materials were the samples from the B site, followed in order, by the T and M samples. These results are also shown in Tab. 2, where it is evident that mean % of adsorption is highest in all B samples (B>T>M) except for the c location for which the order is M>T>B. Conversely, the first desorption was least in mean percent from the B samples except, again, from the c location. Less consistent are the results for the second desorption.

Though the observed mean percent of adsorption of Simazine is somewhat lower than that of Atrazine, the order of adsorption of Simazine (B>T>M) was the reverse of that found for Atrazine (M>T>B). Simazine appeared to be adsorbed to a larger extent by samples of high aromaticity and high molecular size distribution whereas both parameters are inversely proportional to the adsorption of Atrazine. It is interesting the noted reverse behaviour of the Bc sample for the interaction with Simazine. The high aromaticity is accompanied by a reduced molecular size that is the lowest of all samples. For that HA, in fact, the Simazine adsorption is lowest whereas the desorption is relatively large. As for the locations, the a sample seemed to be the highest adsorber of simazine in the B site, followed by b and c samples, whereas their behaviour was similar in the T site, and slightly higher for c in the M site.

The contrasting behaviour of Atrazine and Simazine, may be temptatively explained by the stereochemical hindrance due to the isopropyl group linked to a nitrogen atom in Atrazine as compared to the linear ethyl group in the same position in the simazine molecule. The HA molecular complexity and steric rigidity due to the presence of planar phenyl groups, hinders the approach of Atrazine more than that of Simazine and decrease the formation of the bonds that permit adsorption. This interpretation suggests that a charge transfer mechanism is the predominant interaction mode between humic substances and the two triazine molecules although other modes of interactions (hydrogen bonding, ligand exchange) may not be discarded.

# References

□ Khan, S.U. 1980. Pesticides in the Soil Environment. Elsevier, Amsterdam.

□ Piccolo, A., L. Campanella, B.M. Petronio. 1990. Carbon-13 NMR spectra of soil humic substances extracted by different mechanisms. *Soil Sci. Soc. Am. J.*, 54:750-755.

□ Piccolo, A., G. Celano, C. De Simone. 1992. Interactions of atrazine with humic substances of different origins and their hydrolysed products. *Sci. Total Envir.* 117/118:403-412.

□ Stevenson, F.J. 1982. Humus Chemistry: Genesis, Composition, Reactions. Wiley, New York.

□ Trubetskoj, O.A., O.E. Trubetskaya, T.E. Khomutova. 1991. Isolation, purification and some physical-chemical properties of soil humic substances fractions obtained by polyacrilamide gel electrophoresis. *Soil Biol. Biochem.* 24:893-896.

## 2.3.14. *Influence of Surfactants on the Behaviour of Pesticides in Soil*

# W. Steurbaut

Department Crop Protection Chemistry
University Gent, Belgium

Surfactants are used as formulation additives or spray adjuvants in order to improve the activity and performance of pesticide applications. Surfactants and pesticides can be present in the soil as a result of a soil treatment, of a leaf treatment followed by run off or rain off but also of an atmospheric fall out of worldwide used pesticides.

This presence in the soil can give rise to a whole range of side effects such as adsorption, leaching, degradation, organic matter degradability and influence on microbial life. The influence of surfactants on these side effects caused by pesticides are not well known and only investigated to a limited extend. Therefore, a series of nonionic polyethoxylated (EO) and polypropoxylated (PO) nonylphenol (NP) surfactants with varying nEO and/or mPO-chair length (n = 6, 10, 18, 30 ; m = 0, 20) are studied in combination with a selection of fungicides (propiconazole, diclobutrazol, triadimefon, etaconazole, triadimenol, captan, iprodione, vinclozolin, metalaxyl) in order to evaluate their impact on adsorption and leaching. In a first approach some relatively simple methods were used such as a batch type adsorption method with Freundlich and Langmuir adsorption isotherm determination, and a soil thin layer chromatographic determination of soil mobility (by determination of a frontal Rf value and a mobility factor M.F.). A sandloam soil (60 % sand, 32 % loam, 5,7 % clay, 1,7 % o.m., pH (KCl) 5,1) was used throughout the experiment.

## Soil adsorption of surfactants and fungicides

Adsorption isotherms and the derived constants are presented in table 1 and illustrated for surfactants in figure 1. Both Freundlich and Langmuir approaches are suitable for the determination of surfactant and fungicide adsorption:

| | |
|---|---|
| Freundlich: | $x/m = k\, C^{1/n}$ |
| Langmuir: | $x/m = bKC/(1 + KC)$ |
| $x/m$: | amount pesticide (x) adsorbed by the soil (m) (mg/kg) |
| C: | pesticide concentration in equilibrium solution (mg/l) |
| k, 1/n, K: | constants |
| b: | adsorption maximum (meq/100g or mg/kg) |
| r: | regression coefficient |

The adsorption of surfactants is inversely related to the chain length of the hydrophillic part (n + m) as shown by the following equation:

$$\log b = 0.207 - 0.231\,(n + m) \qquad r^2 = 0.96$$

with b = adsorption maximum (mg/kg) according to Langmuir. Fungicide adsorption is similar for all triazole type fungicides. On the other hand no clear relationship is observed between adsorption and water solubilities or octanol/water partition coefficients.

## Influence of surfactants on fungicide adsorption

Fungicide adsorption can be influenced by the presence of surfactants as shown by the results in table 2. Only the Freundlich approach is applicable for evaluation of the results. Because of the low correlation coefficients no Langmuir constants could be derived. Generally fungicide adsorption decreases in the presence of surfactants. This occurs especially when surfactant concentrations in the solution are exceeding the CMC (critical micelle concentration) as illustrated in figure 2. Probably several factors can interfere (2) the normal solute/adsorbent adsorption process:

- competition between fungicide and surfactant adsorption
- competition between micelles, free fungicide and/or surfactant molecules for adsorption
- decreased fungicide adsorption potential by inclusion of molecules into free micelles.

Especially the last possibility is very specific for surfactants.

## Influence of surfactants on the mobility of fungicides in soil

The leaching behaviour of the selected fungicides, as determined by soil-TLC (table 3), shows a clear relationship with their water solubilities (W.S.) (3):

$$M.F. \cdot 10^4 = 0,6 \, W.S. + 0,98 \cdot 10^4 \qquad\qquad r^2 = 0,98$$

The presence of surfactants have a very diverse influence on the fungicide mobility as illustrated in fig. 3. Short chained surfactants (6 EO) have no or even an adverse effect on fungicide mobility. Increasing chain length give a maximum influence for 18 EO units. mPOnEO-surfactants have a more pronounced influence than their corresponding nEO-analogs. No clear relationship was found between adsorption parameters and mobility when surfactants were present. This can probably been explained by the complex system of interactions when surfactants are involved. Increasing the surfactant concentrations far above the CMC give rise to increased mobility (fig. 4) possibly by the incorporation of fungicide molecules in percolating micelles.

## *Conclusions*

From this study it is clear that surfactants can have a special and very complex influence on fungicide adsorption and mobility in the soil. Due to some specific physicochemical properties (hydrophillic - lipophilic parts, micelle formation) of surfactants special interactions with fungicides can occur which disturb normal adsorption and leaching processes. It is still not clear whether these effects have also a significant impact on the actual behaviour of pesticides under field conditions.

# References

◻ Giles, C.M. (1970)
In: "Sorption and transport processe in soil", S.C.I. Monograph 37, 14-32.

◻ Sharon, M.S., Miles, J.R.W., Harris, C.R., Mc Ewen, F.L. (1980)
Water Res., 14, 1095-1100.

◻ Worthing, C.R., Walker, S.B. (1983)
The pesticide manual, BCPC.

| | Freundlich | | | Langmuir | | | |
|---|---|---|---|---|---|---|---|
| | 1/n | K | r | K (1/ppm) | b (mg/kg) | b (meq/100g) | r |
| NP 6 EO | 0,38 | 464 | 0,98 | 8 | 7143 | 148 | 0,99 |
| NP 10 EO | 0,35 | 454 | 0,90 | 7 | 6135 | 93 | 0,97 |
| NP 18 EO | 0,32 | 463 | 0,97 | 6 | 5181 | 53 | 0,94 |
| NP 30 EO | 0,32 | 413 | 0,99 | 9 | 3937 | 26 | 1,00 |
| NP 20 PO 10 EO | 0,22 | 1059 | 0,99 | 15 | 5492 | 32 | 0,99 |
| NP 20 PO 30 EO | 0,12 | 1520 | 0,94 | 69 | 3404 | 13 | 1,00 |
| propiconazole | 0,79 | 26 | 0,99 | 148 | 263 | 77 | 0,98 |
| diclobutrazol | 0,73 | 15 | 0,97 | 1088 | 270 | 83 | 0,88 |
| etaconazole | 0,87 | 21 | 0,96 | 61 | 417 | 127 | 0,89 |
| triadimefon | 0,87 | 22 | 0,99 | 82 | 385 | 131 | 0,92 |
| triadimenol | 1,12 | 17 | 0,97 | 93 | 385 | 1130 | 0,96 |
| vinclozolin | 1,16 | 36 | 0,92 | 321 | 712 | 249 | 0,89 |
| iprodione | 0,96 | 26 | 1,00 | 181 | 463 | 140 | 0,90 |
| captan | 0,80 | 64 | 0,97 | 289 | 510 | 169 | 0,95 |
| metalaxyl | 1,10 | 14 | 1,00 | 156 | 412 | 148 | 0,92 |

**Table 1** : Adsorption isotherm constants of surfactants and fungicides

| Surfactant | propiconazole | | vinclozolin | | iprodione | | captan | | metalaxyl | |
|---|---|---|---|---|---|---|---|---|---|---|
| | 1/n | K | 1/n | K | 1/n | K | 1/n | K | 1/n | K |
| no surfactant | 0.79 | 25.9 | 1.16 | 36.2 | 0.96 | 26.1 | 0.80 | 63.7 | 1.10 | 14.1 |
| NP 6 EO | 0.48 | 12.8 | 0.94 | 13.9 | 1.09 | 13.7 | 0.91 | 27.2 | 1.16 | 10.3 |
| NP 10 EO | 0.96 | 13.8 | 0.95 | 12.2 | 1.04 | 10.9 | 0.92 | 33.5 | 1.22 | 12.6 |
| NP 18 EO | 0.91 | 15.0 | 0.91 | 14.3 | 0.89 | 17.4 | 0.84 | 38.5 | 1.11 | 12.1 |
| NP 30 EO | 0.97 | 10.8 | 0.87 | 17.0 | 0.91 | 16.1 | 0.99 | 27.1 | 1.22 | 7.9 |
| NP 20 PO 10 EO | 1.01 | 15.4 | 0.94 | 10.4 | 0.74 | 34.0 | 1.09 | 29.9 | 1.15 | 21.5 |
| NP 20 PO 30 EO | 0.83 | 15.3 | 0.96 | 12.2 | 0.83 | 15.6 | 1.08 | 28.4 | 1.52 | 7.7 |

**Table 2 :** Fungicide adsorption isotherms after combined fungicide-surfactant treatment

| Fungicide | Rf | M.F. |
|---|---|---|
| propiconazole | 0.3 a | 1.02 a |
| diclobutrazol | 0.4 a | 1.02 a |
| triadimefon | 0.4 a | 1.01 a |
| etaconazole | 0.4 a | 1.19 ab |
| triadimenol | 0.4 a | 1.33 b |
| captan | 0.3 a | 1.07 a |
| iprodione | 0.2 a | 1.03 a |
| vinclozolin | 0.3 a | 1.00 a |
| metalaxyl | 0.7 b | 5.07 c |

equal letters indicate nonsignificant differences (5 %) according to a Duncan test

**Table 3 :** Mobility of fungicides in soil

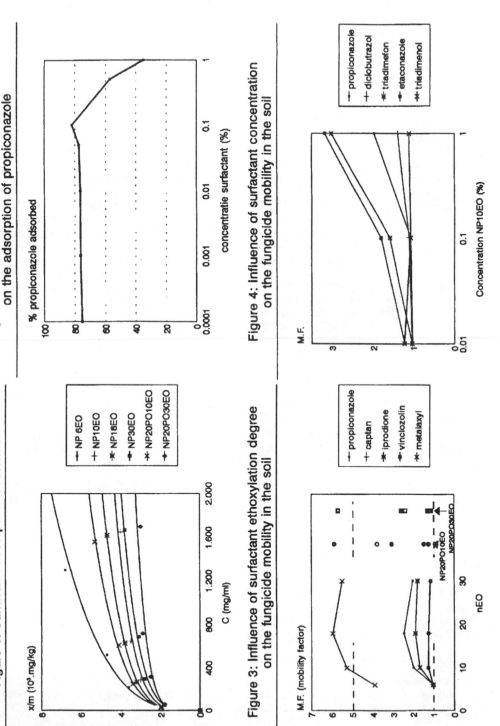

Figure 1: Surfactant adsorption isotherms

Figure 2: Influence of surfactant concentration on the adsorption of propiconazole

Figure 3: Influence of surfactant ethoxylation degree on the fungicide mobility in the soil

Figure 4: Influence of surfactant concentration on the fungicide mobility in the soil

2.3.15.    *Effect of Formulation Additives Atpol and Olbras on Atrazine*
           *Adsorption in a Soil-water System*

# M. K. Swarcewicz

Department of Chemistry, Agricultural University of Szczecin, Poland

## Introduction

Most pesticides are formulated with adjuvants but apart from simple comparisons of perfomance, for example of granules and an emulsifiable concentrate, only sporadic studies have been reported on the effects of an oil adjuvant on the behaviour of soil-applied compounds [Calvet,1980; Hance,1983]. Since surfactants are present in the herbicide formulation they may cause competition between the herbicide and surfactant for adsorption sites, which would affect the movement in soil and bioactivity of the particular herbicide [Bayer and Foy,1982]. Adsorption of herbicide alone in comparison with herbicide mixed with oil adjuvant should provide information both as to the mode of transport, used by the pesticide as well as to less effective hazard assessments as results of oil adjuvant use.

The aim of this study was to determine the influence of oil adjuvants ATPOL and OLBRAS on atrazine adsorption in the soil. The experiments conducted were short-term sorption studies with atrazine in soil suspension. The relation between atrazine concentration at equilibrium and the amount adsorbed in soil allows the constraction of the Freundlich adsorption isotherms.

## Materials

The soil used in the laboratory studies was a top 20 cm layer from a field at the Research Station Ostoja. It was a sandy clay loam with 1.5% organic matter, 40% sand , 26% clay, and pH 6.1 [in 1M KCl], 7.4 me/100g sorption capacity and 25.2% maximum water holding capacity [Bogda et al.,1990]. The herbicide used was a commercial formulation of atrazine [2-chloro-4-ethylamino -6- izopropy-lo-s-triazine] as 50% wettable powder and analytical grade sample. A stock solution of 50 mg/l of atrazine in 0.01 M CaCl$_2$ was prepared by dissolving appropriate amounts of Azoprim 50.

The oil adjuvant ATPOL contains 17% surfactant [Atplus 300F] and 83% paraffin oil 1113, and OLBRAS liquid mixture contains 12% surfactant and 88% esters of natural fatty acids. The both are produced in Poland. The ratio of herbicide concentration to adjuvant concentration was 1:1.

## Methods

Equlibrium adsorption isotherms for all soil-pesticide-adjuvant combinations were measured using the short-term sorption system [a batch experiment]. Equilibrium was achieved by shaking triplicate samples of 4 g soil [60% WHC] with 10 ml of 0-50 mg/l pesticide solution and adjuvant for 2 h. Preliminary experiments had indicated that there was no measurable increase in pesticide adsorption beyond this time. After shaking, the tubes were centrifuged of the supernatant taken for analysis.

### *Analysis of atrazine residues in water*

Atrazine was determined by gas chromatography method with the use of FID, 3% OV-17 glass column a Pye-104, retention time 2,4 min; Capillary Gas Chromatograph Carlo Erba GC 6000 Vega Serie 2,column DB-5, retention time 4.8 min.

## Results and Discussion

The relationship between the calculated concentration in water phase and the amount adsorbed fitted the empirical Freundlich equation :

$$x/m = K \, C^{1/n}$$

where $x/m$ is the amount of atrazine adsorbed [mmol/g], C is equal to the equilibrium solution concentration [mmol/ml], K and $1/n$ are constants. The K value indicates the extent of adsorption while the $1/n$ value shows the degree of non-linearity between adsorption and solution concentration. Taking the logarithm of the above equation one obtains the expression :

$$\log x/m = \log K + 1/n \, \log C$$

The values of log K and $1/n$, which represent intercept and slope of the isotherm, respectively, were obtained from their plots using computer program of multiple regression analysis. Correlation coefficient [r] between log $x/m$ and log C were greater than 0.97 [tab.1]. The values of $1/n$ for all combinations were from 0.97 to 1.03. The K values were similary for formulated atrazine [w.p.] in water and in the presence of oil adjuvant Olbras 2.24 and 2.33, respectively. The K value was increased to 2.72 for Atpol. There were significant differences between slopes with Atpol and water. The slight increase the K value for Atpol shows probably a change of the partition of solutes between surface and solution. On the other hand, comparison of the oil adjuvants that Atpol had contains more surfactants than Olbras.

The linear adsorption isotherms $K_d$

$$x/m = K_d \, C$$

have been generally accepted for low concentration. In this experiment using a soil with low organic matter content and high concentration of atrazine. The $1/n$ values [tab.1] shows in these cases obtained linear adsorption isotherms. The similar results obtained Rao and Davidson,1979, for atrazine.

The results show that, the presence of oil adjuvant increased the sorption coefficient $K_d$ about 19% Atpol, and 17% Olbras, in comparison with the value obtained for water.

Previous studies indicate that in the presence of the oil adjuvant Atpol, the degradation of atrazine in soils was slower [Swarcewicz,1991].

Soil organic carbon content generally correlates well with pesticide adsorption. The soil sorption constant $K_{oc}$ [ Swann et al.,1983; Walker et al.,1988] was described by the equation :

$$K_{oc} = \frac{K_d \, x \, 100}{\% \ organic \ matter}$$

On basis of the $K_{oc}$ values listed in table 1, the extent of atrazine adsorption by soil was in the following order : water < Olbras < Atpol.

This study has shown that oil adjuvant can cause slight change of atrazine sorption in a soil-water system. This type of indirect effect needs much further investigations.

|        | Freundlich isotherm |           |      |      | Linear isotherm |      |          |
|--------|---------------------|-----------|------|------|-----------------|------|----------|
|        | Log K               | SE on logK | 1/n  | K    | $K_d$           | SE   | $K_{oc}$ |
| Water  | 0.350               | 0.034     | 0.97 | 2.24 | 2.11            | 0.13 | 140      |
| Atpol  | 0.435               | 0.031     | 0.97 | 2.72 | 2.51            | 0.18 | 167      |
| Olbras | 0.367               | 0.031     | 1.03 | 2.33 | 2.46            | 0.36 | 160      |

Correlation coefficient r > 0.97
SE - standard error

**Table 1 :** Adsorption isotherm constants for atrazine with oil adjuvants

# References

□    1. Bayer D.E. and C.L.Foy , [1982] Action and fate of adjuvants in soils. In:
     Adjuvants for Herbicides, Weed Sci.Soc.Am.Monog., 84-92

□    2 . Calvet R.,[1980] Adsorption-desorption phenomena.In:Interactions Between Herbicides
     and the Soil.(Ed.) Hance R.J., Academic Press, 21

□    3. Hance R.J.,[1983] Processes in soil which control the availability of pesticides. In:
     Proc.Congr. Plant Protection, Brighton, Vol.2, 537-544

□    4. Bogda A.,T.Chodak and E.Niedzwiecki ,[1990] Some properties and mineralogical com-
     position of Gumieniecka Plain Soil. Soil Sci.Ann (Polish), Vol.26. no.3/4,179

□    5. Rao P.S., and J.M.Davidson, [1979] Adsorption and movement of selected pesticides at
     high concentrations in soils. Weed Res. Vol.13,179

□    6. Swann R,L., D.A.Laskowski ,P.J. McCall , K.Vander Kuy and H.J.Dishburger , [ 1983]
     Res.Rev. Vol.85, 17-26

□    7. Swarcewicz M.[1991] Influence of Atpol on the degradation rates of atrazine and fluazifop
     butyl in soils. Abst.4th Workshop Chem. Biol. Ecotoxic.
     Behaviour Pest. Soil Environment, Rome,Italy

□    8. Walker W.W., C.R. Cripe ,P.H. Pritchard and A.W.Bourguin , [1988]
     Chemosphere Vol.17 no.12, 2255-2270

## 2.3.16.  *Hymexazol-Cation-Clay Complexes*

# V. Vanderheyden[1], Ph. Debongnie[2], L. Pussemier[2], C. Badot[1], and P. Cloos[1]

[1]  unité CATA, Fac. Sc. Agr., UCL, Louvain-la-Neuve

[2]  ISO-IRC, Min. Agr., Tervuren

## Introduction

The adsorption of hymexazol on clays may be enhanced by the presence of some cations. This property, useful for controlled-release formulations, may also affect the fate of the fungicide once it is released into the soil. We have studied the adsorption of hymexazol (structure I) and the closely related compound methylisoxazol (structure II) on the smectite clay hectorite saturated with different cations.

I

Hymexazol

II

Methylisoxazol

## Results

### *Adsorption*

The adsorption enhancement, estimated from the magnitude of the specific IR absorption bands at ca.1530 cm$^{-1}$, followed the series $Ca^{2+} < Al^{3+} < Fe^{3+} < Cu^{2+}$ (table 1). The closely related compound 5-methylisoxazol adsorbed only on Cu-clays, showing the hydroxyl group to be essential for complexation with the other cations.

| Cation saturating the hectorite clay | hymexazol | | methylisoxazol | |
|---|---|---|---|---|
| | CHCl$_3$ | CH$_3$OH | CHCl$_3$ | CH$_3$OH |
| Ca$^{2+}$ | + | traces | | |
| Al$^{3+}$ | ++ | traces | no | no |
| Fe$^{3+}$ | +++ | + | no | no |
| Cu$^{2+}$ | +++ | + | +++ | + |

**Table 1 :** Adsorption on the hectorite films after dipping in CHCl$_3$ or CH$_3$OH solution

## X-ray diffraction

The X-ray diffraction patterns showed that the interlamellar space (12,8 Å), just sufficient for a monolayer of water molecules, was not altered by complexation : the hymexazol molecules in these spaces must therefore lie parallel to the clay layers.

## ESR

The distinctive shapes ($g_{//} \approx 2.35$, $g_{\perp} \approx 2.08$) of the Cu(II)- ESR (Electron Spin Resonance) spectra of hectorite-Cu films, placed either perpendicular or parallel to the magnetic field, indicate that each copper ion is bound to six atoms, with two longer bonds perpendicular to the clay layer, and four shorter bonds parallel to it. Presumably, the two longer bonds involve oxygen atoms from the clay structure, while the four shorter bonds involve water and/or organic molecules. The shift observed upon adsorption of hymexazol ($\Delta$ $g_{//} \approx 2.35$ -0.06) shows that coordination bonds are formed between the copper ion and the nitrogen atom of hymexazol. The same shift is observed with methylisoxazol, but only on dried films.

These results, together with the X-ray diffraction patterns, suggest that the structure of the hectorite-Cu-hymexazol complexes is as in structure III.

**III**

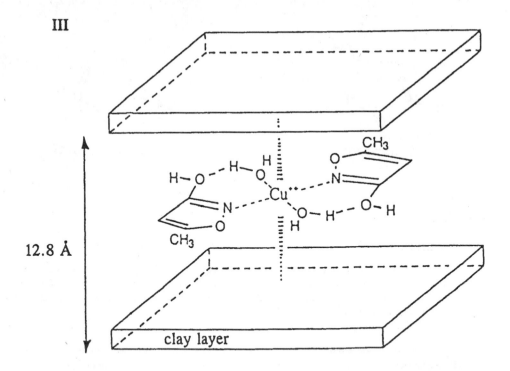

12.8 Å

clay layer

## IR

Adsorption of hymexazol onto hectorite-Cu, -Fe or -Al films induced both a large shift (between -23 and -27 cm$^{-1}$) of its C=N vibration band at ca. 1530 cm$^{-1}$ (table 2), and an increase in the intensity of the asymmetrical deformation band of the methyl C-H bonds at ca. 1440 cm$^{-1}$. This indicates a diminution of the double bond character, probably due to the formation of the N-cation and/or O-cation bond(s). However, this interpretation may have to be reconsidered, because with methylisoxazol, drying the films induced the expected increase of the methyl C-H band, but the shift of the C=N bond ($+10$ cm$^{-1}$) was in the direction opposite to that expected.

| | $\nu_{C=N}$ | $\nu_{C=C}$ | $\delta_{CH3}$ |
|---|---|---|---|
| Hymexazol | 1529 | 1635 | 1440 |
| Hect-Cu-hymexazol | 1502 | 1630 | 1433 |
| Hect-Fe-hymexazol | 1506 | 1624 | 1417 |
| Hect-Al-hymexazol | 1506 | 1630 | 1433 |
| Methylisoxazol | 1475 | 1601 | 1444 |
| Hect-Cu-methylisoxazol | 1485 | 1601 | 1444 |
| Hect-Cu-methylisoxazol (dried) | 1485 | 1597 | 1423 |

**Table 2 :** IR absorption wavelengths (cm$^{-1}$) of hymexazol and methylisoxazol, in KBr pellets and adsorbed onto hectorite films

## Release Experiments

The first trials indicated that the release of hymexazol from the hectorite-Cu films into water was very slow (> 25 days)

## *Conclusions*

- The physico-chemical process described here may influence the fate of hymexazol and other pesticides in the environment, Fe(III) and Al(III) being among the most common cations in soils. Cu(II) is rarer, but may be important in special cases such as copper-treated vineyards.

- The clay-Cu complexes have a potential as controlled-release formulations. This is being studied further at our Institute.

- The counter-example of methylisoxazole shows that the IR spectra must be interpreted with caution.

## 2.4.        Short Communications

### 2.4.1.        *Influence of solid/liquid ratio on experimental error of sorption coefficients in relation to OECD guideline 106*

### J.J.T.I. Boesten

DLO Winand Staring Centre for Integrated Land Soil and Water Research,
P.O. Box 125, 6700 AC Wageningen, Netherlands

The OECD guideline 106 (OECD, 1981) implies that sorption is derived from the measured decrease in concentration in the liquid phase of a soil suspension. Boesten (1990) analyzed the experimental error for the sorption coefficient for such a method theoretically. The result was that both the systematic and random error in the sorption coefficient are controlled by the product, P, of the solid/liquid ratio and the sorption coefficient (P is dimensionless). Both types of error increase with decreasing P and tend to infinity if P tends to zero. Assuming realistic lower limits of the systematic and random errors and accepting 10% uncertainty in the sorption coefficient, leads to the requirement that P should be larger than 0.3. If a solid/liquid ratio of 0.2 $dm^3 kg^{-1}$ is used (as suggested by the OECD guideline), P > 0.3 implies that sorption coefficients below 1.5 $dm^3 kg^{-1}$ cannot be measured with acceptable accuracy.

Boesten (1990) reviewed about 30 measurements on the influence of the solid/liquid ratio on pesticide adsorption isotherms. In some 85% of the cases the ratio had no influence on the measured isotherms, whereas in about 15% of the cases an increase in the ratio resulted in sorption decreasing considerably. This literature study suggests that the solid/liquid ratio has usually no influence on sorption isotherms of pesticides as measured in soil suspensions. For the cases where an influence was found, the measurements made at the highest solid/liquid ratios are the most relevant, because these highest ratios are closest to those occurring in the field. Combination of this with the results of the error analysis leads to the recommendation to use a standard solid/liquid ratio of 1 kg $dm^{-3}$ instead of 0.2 kg $dm^{-3}$.

## *References*

□       Boesten, J.J.T.I., 1990. Influence of solid/liquid ratio on the experimental error of sorption coefficients in pesticide/soil systems. *Pesticide Science* 30: 31-41.

□       Organisation for Economic Co-operation and Development (OECD), 1981. Adsorption/desorption, OECD guideline for testing of chemicals, no 106. OECD, Paris, 23 pp.

### 2.4.2. *Investigation on S - Triazine Herbicide Interactions with some Metal Ions and Humic Acid Employing Electrochemical Detection Modes*

# I. Grabec[1]; B. Ogorevc[1]; V. Hudnik[1]; I. Turk[2]; F. Lobnik[2]

[1]    National Institute of Chemistry, POB 30, SLO- 61115 Ljubljana (Slovenia)

[2]    Biotehnical Faculty, Department of Agronomy, University of Ljubljana, Ljubljana, Slovenia

In soil, regarded as a complex bio-physico-chemical system, numerous endo- and exo-geneous components are involved into susceptile chains of kinetically and thermodynamically interdependent processes and equilibria. In this context, the knowledge and understanding of bioinorganic coordination reactions are obviously essential. In general, it is well known that such reactions can influence the chemical (activity), physical (transport) and biological (bioavailability, toxicity) nature of all soil components participating in an interaction process. Among various endo- and exo-geneous substances of concern, triazine herbicides, metal ions, e.g. Cu (II), Zn (II), Cd (II) and Pb (II) and humic substances, are surely of high interest. It is also widely accepted that the electrochemical analytical techinques (e.g. voltammetry, polarography) can serve as a good tool for interaction and speciation studies due to their unique inherent ability to discern among different oxidation states and physico-chemical forms of metal ions and ligands.

The objectives of this work are the possible interactions between selected s-triazine herbicides (e.g. atrazine), including their natural decomposition products, and above listed metal ions in an aqueous solution. Different conditions have been considered e.g. pH, solution composition, concentration levels as well as the presence of certain soil endo-geneous substances (humic acid). Polarographic and voltammetric techniques have been utilized to carry out an investigation on coordination behaviour and complex formation parameters of the studied species by using mercury drop and carbon paste electrodes.

*2.4.3.        Role of Soil Clay Fraction in Pesticide Adsorption: Defining a Kclay*

# M.C. Hermosin and J. Cornejo

Instituto de Recursos Naturales y Agrobiología de Sevilla.
Apartado 1052. Sevilla 41080. Spain.

The clay fraction of soils has been repeatedly emphasized as important in pesticide soil adsorption
(1-6). However, a definition of a clay adsorption coefficient, Kclay, similar to that for organic carbon,
Koc, has not been generally attempted. Blume (7) in his "Handbuch des Bodenschutzes", in the chapter
devoted to soil contamination, reported Kclay values for many pesticides but no explanation was
given on how and why this coefficient was calculated.

The subject of this communication is to discuss on the suitability of defining a K clay, according to
the expression:

$$K \text{ clay} = Kf \text{ or } Kd \; / \; \% \text{ clay} \times 100$$

Applying this expression to results found by the authors (1-4) and in the literature (5, 6), a great decrese
of the variation coefficient is shown when Kclay is considered instead Kf or Kd. Some examples are
shown in Figure 1.

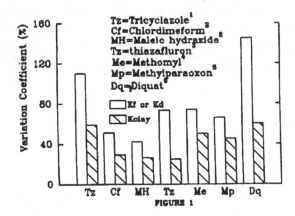

FIGURE 1

The pesticide characteristics favoring Kclay definition are: cationic, polar and both polar and
protonable character, regardless to the water solubility.

The soil characteristics determining the suitability of the Kclay definition are: a similar clay
mineralogy (oxides or layer silicates) and similar organic carbon content. However, the first condition
is more important than the second one.

There must be kept in mind that soil clay fraction is constituted by organic and inorganic components.
As the clay minerals are of different nature of the silt and sand minerals, the organic matter associated
to the clay fraction is different to that in silt and sand fractions (8,9).

# References

□ 1. ROLDAN, I., HERMOSIN, M.C. & CORNEJO, J. Sci. Total Environ. 132: 217 (1993).

□ 2. HERMOSIN, M.C. & CORNEJO, J. Toxicol. Environ. Chem. 25: 45 (1989).

□ 3. COX, L., HERMOSIN, M. C. & CORNEJO, J. Fresenius Environ. Bull. 3: 129 (1994).

□ 4. COX, L., HERMOSIN, M.C. & CORNEJO, J. Chemosphere 27: 837 (1993).

□ 5. SANCHEZ-MARTIN, M. J. & SANCHEZ-CAMAZANO, M. Soil Sci. 152: 283 (1991).

□ 6. KOOKANA, R. S. & AYLMORE, L. A. G. Aust. J. Soil Res. 31: 97 (1993).

□ 7. BLUME, H.P. Handbuch des Bodenschutzes, Ecomed-Verlag, Germany, pp.:324-340 (1992).

□ 8. CHRISTENSEN, B. T. Adv. Soil Sci. 20: 1-90 (1992).

□ 9. PRESTON, C. M., NEWMAN, R. H. & ROTHER, P. Soil Sci. 157: 26 (1994).

## 2.4.4.    Correlation between Molecular Structure and Soil Retention of Five Phenylpyridylureas

# A. Hocquet[1], J. Fournier[2]

[1]    ENS Cachan, Département de chimie, LPPM, F94235 Cachan Cedex

[2]    CREPA, 8, rue Becquerel, F49070 Beaucouzé

We focus our study on five phenylpyridylureas (cytokinin-like plant growth regulators) with varying substituents and atom positions.

Our work is divided in two parts :

- Analysis of soil extracts of those molecules in order to know the retention of these products by a laboratory soil, in such conditions that adsorption is favored.

- Structural study of these molecules using varoious techniques (X-ray diffraction, Infra-red and UV-visible spectroscopies, $^{15}N$, $^{13}C$ and $^1H$ NMR, TLC, molecular mechanics and semi-empirical calculations) in different media (solid state, solution in hydrogen bond accepting and hydrogen bond donating solvents, isolated molecules) in order to elucidate :

  — the electronic, configurational and conformational structures of these molecules compared with each others and with other ureas,

  — the intermolecular forces that those molecules present in the studied media.

The goal is to correlate these two studies in order to make hypotheses upon the intermolecular forces involved between those molecules and the argilo-humic complex.

In order to undestand soil-pesticide interactions, structural analysis should not be limited to polarity. We thus present another way of stuctural study which could prove itself complementary of the traditional sorption studies and of the quantitative structure-activity relationships.

# SECTION III

# TRANSFORMATION

Chairman :        F. ANDREUX
                  (Univ. de Bourgogne, France)

# TRANSFORMATION

## 3.1.  Introductory Presentation

*3.1.1.*    *Kinetics of Pesticide Biodegradation in Soils : Principles and Applications*

# J. P.E. Anderson[1]

BAYER AG, Crop Protection, Institute for Environmental Biology,
51368 Leverkusen, Germany

## Abstract

The rate of pesticide biodegradation in any layer of any biochemically intact soil is determined by the structure of the chemical, the quantities and distribution of bioreactive sites that can degrade it, the availability of the pesticide to these sites, and if the pesticide is being degraded in intact, living cells, the activity level of these cells. Interacting with these variables to speed, slow or stop biodegradation are the quantity of pesticide applied to the soil, soil temperature, soil moisture, composition of the soil atmosphere, soil pH, the availability of cofactors, soil nutrients and electron acceptors, and soil microhabitat geometry.

## Principles

The biodegradation of pesticides in soils is a catabolic process in which biologically formed entities, specifically enzymes or intact cells, convert the molecules of insecticides, herbicides, fungicides, nematicides or acaricides into less complex intermediates or end products. When pesticides degrade in soils, there are a number of points of ecological interest. These include the speed of degradation, the intermediates and end products formed, the biochemical reactions involved, and the pathways followed. Because of space limitations, this paper will concentrate on the variables and mechanisms that influence *the rates of pesticide biodegradation* in soils.

The discussion can be started with the contention: The variables that determine the rates of degradation of any biodegradable pesticide in *any* layer of *any* soil are:

1. the *structure* of the pesticide

2. the *availability* of the pesticide to enzymes or cells that can degrade it

---

1    Author's Note : This paper was presented as a lecture at the 5th International COST Workshop on Environmental Behavior of Pesticides and Regulatory Aspects held in Brussels, April 1994. The written version has been held in the form of a lecture, and no attempt has been made to review the literature or aknowledge the contributions of colleagues in this or related fields. The data presented here come exclusively from our laboratories and are cited as published work.

3.      the *quantities* of enzymes or cells (mostly microorganisms) that can degrade it, and

4.      the *"activity level"* of these enzymes or cells.

## *Influence of pesticide structure on biodegradation rates*

The structure of a pesticide determines its degree of biodegradability, its intrinsic toxicity, its nutrient value and its physical behavior in soil. *All* of these variables interact to determine the kinetics of biodegradation.

The structure of a pesticide determines if it is readily or poorly biodegradable. With readily biodegradable pesticides, the enzymes necessary for catabolism are found in many species and many genera of common soil microorganisms. The microbial genera that can degrade such pesticides are found in most soils, regardless of their latitude or longitude of origin. When added to soils, the catabolism of readily biodegradable pesticides starts immediately and proceeds at measurable rates under all but the most extreme environmental conditions.

With poorly biodegradable pesticides, the enzymes necessary for degradation are not widely distributed among species or genera of common soil microorganisms. However, microorganisms with the capacity to degrade these compounds can be found in most soils. For degradation of poorly biodegradable pesticides to proceed at measurable rates, "special conditions", such as soil flooding or the co-metabolism of high-energy nutrients are necessary.

The structure of a pesticide determines its intrinsic toxicity, or the degree of activity of the pesticide *against* the enzymes or cells that can attack and degrade it. If the pesticide is extremely toxic, it denatures or kills many of the enzymes or cells that are attacking it. Accordingly, this can slow or partially inhibit biodegradation.

As with all biologically active substances, pesticide toxicity is not only dependent on structure, it is dependent on concentration. With pesticides that are intrinsically toxic to microbial cells, but are biodegradable, the speed of degradation is often determined by the concentration of chemical at the microhabitat level. If concentrations are too high, some of the attacking enzymes or cells are killed, and the chemical is initially degraded at a slow rate. However, once concentrations decrease to tolerable levels, i.e. by dilution caused by diffusion out of the microhabitat or by inactivation by adsorption, the rates of degradation increase.

The structure of a pesticide determines its nutrient value. When pesticides that can serve as C and energy sources are applied to soils and degraded by the microflora, populations of cells responsible for degradation can increase in size. This causes a local increase in the rate of degradation. Counteracting and masking the effects of local rate increases are rate decreases caused by the progressive reduction of the quantity of pesticide available for degradation. In most pesticide degradation experiments, chemicals are only applied once to soils. Samples of the soil are taken at different times and extracted and analyzed for degradation products. In such single application experiments, the effects of cell population growth on the rate of degradation of the pesticide usually go unnoticed. Under some conditions, especially in monocultures, pesticide are often applied repeatedly to the same soil. Biodegradable pesticides that can provide nutrients and/or energy for microbial growth can induce the build-up of unnaturally large "specialized" populations, and these degrade each new application of the chemical at a higher rate. In the past, this has resulted in product weakening or product failure. In the literature, this phenomena is known as accelerated or enhanced pesticide degradation.

Structure determines energy that must be invested to break the bonds of a chemical. To break bonds of many poorly biodegradable pesticides, cells attacking the molecules must invest energy. With pesticides of this sort, microorganisms that can degrade the compound occur in most soils. However, to initiate and sustain degradation, a second, energy-rich material must be cometabolized. Examples of such pesticides are DDT and dieldrin.

Structure determines the physical behavior of a pesticide in soil. It determines the amount of chemical lost to biodegradation through volatilization from the soil, and more important, it determines the degree of adsorption of the compound to soil solids, and desorption into the soil solution. For most non-ionic pesticides, the degree of adsorption is related to the amount of organic matter in the soil. In soils with high organic matter content, adsorption is usually high. Conversely, in soils with low organic matter content, the degree of pesticide adsorption is usually low. Adsorption of ionized pesticides usually depends on the quantities and qualities of clay minerals in the soil.

### Influence of pesticide bioavailability on degradation in soils

In general, if a pesticide is tightly adsorbed to solid surfaces of a soil, it is not available for biodegradation. Thus, the amount of pesticide in the soil solution determines its availability for biodegradation.

When pesticides are dissolved in the soil solution, degradation rates are determined by their mobility. High mobility increases, and low mobility decreases the probability of contact between pesticide molecules and the enzymes or cells that can degrade them. The amount of available water in the soil determines degradation rates. Water acts as a solvent for all pesticides, regardless of their water solubility. Increasing the water content of soil increases the *total amount* of pesticide in solution, which increases the probability of contact between pesticide molecules and the enzymes or cells that can degrade them.

### Influence of enzymes or cell quantities on pesticide degradation rates in soils

In any soil, the rate of biodegradation of a pesticide depends on the quantities and distribution of enzymes or cells that can degrade it. Enzymes that can attack a pesticide can be exuded into the soil solution, or retained inside of cells. Cells that can attack a pesticide can be mobile or sessile, and can be aerobic, microaerophilic, or strictly anaerobic.

### Influence of activity levels of enzymes or cells attacking the pesticide on degradation rates

The activity levels of the enzymes or cells that can degrade a pesticide determines its rate of biodegradation. In general, the higher the level of biological activity in a soil, the faster the rate of degradation. In soils, biological activity levels are determined by temperature, available moisture, composition of the soil atmosphere, quantities of available nutrients and soil pH.

## Proof of Statements

To "prove" the above statements, data from a series of studies with two biodegradable, $^{14}$C-labelled herbicides, Diallate and Triallate, will be presented. In laboratory tests, both herbicides were added to samples of soil and incubated in an aerobic, all-glass system which allowed full recovery of the applied radioactivity. The incubation system, extraction procedures, and analytical procedures are described elsewhere (1). Except where stated differently, a sandy brown podzol with 1.25 % C, 0.12 % N and a pH of 5.4 was used. In all but 1 experiment, the soil was treated with 1 mg a.i./ kg (dry wt) soil. In the podzol (and in all other soils tested), the major biodegradation products of Diallate and Triallate were $^{14}$CO$_2$ and $^{14}$C-labelled "bound" residue.

### Influence of pesticide structure on biodegradation rates

Diallate and Triallate differ in structure by a single chlorine atom. In the podzol, the influence of this slight structural difference resulted in 50 % of the applied Diallate being degraded in 3 weeks, and 50 % of the applied Triallate being degraded in 17 weeks. Thus, with these herbicides, as with most other non-polymerized organic chemicals, slight changes in structure can cause meaningful or even drastic changes in biodegradation rates.

## Structure and intrinsic toxicity of pesticides

To test the toxicities of Diallate and Triallate to the enzymes and cells that can degrade them, samples of the podzol were treated with concentrations of the herbicides which ranged from 1 to 2000 mg/kg. The amounts of each herbicide degraded to [14]C-labelled products were determined after 7 days. For both herbicides, there were direct relationships between the quantities of herbicide added to the soil, and the amounts degraded (2). In this test, there was no indication of concentration dependent decreases in the quantities of herbicide degraded. If a downward break in either curve had been observed, it would allow the toxic concentration of herbicide to be estimated. Since this did not occur, there was no indication of intoxication of the degradation process by either herbicide. Thus, differences in toxicity were not responsible for the differences in the biodegradation rates of Diallate and Triallate.

## Influence of availability on the rates of pesticide biodegradation

In the above experiments, the absolute quantities of Diallate or Triallate degraded increased with increases in the absolute quantities added to the soil. This suggests that there is a simple relationship between the quantities of pesticide in a soil and the rates of degradation.

If this relationship is true, why didn't identical concentrations of Diallate and Triallate degrade at identical rates? As an approach to answering this question, the distribution of the herbicides between the solid and liquid phases of the soil were examined. For this purpose, "adsorption isotherms" of both herbicides were prepared. To mimic the situation in the toxicity tests, a fixed ratio of soil and water (10 g dry wt soil and 20 ml water) were treated with

different concentrations of Diallate or Triallate. The resulting soil slurries were shaken at 20°C until partitioning of the herbicides between soil solids and soil water was complete. In these tests, as in the toxicity tests above, the major variable was the quantity of Diallate or Triallate in the soil-water system.

Fig. 1 shows the distribution of Diallate or Triallate between the solid (adsorbed herbicide) and liquid phases (herbicide in solution) of the soil-water slurries. At all concentrations, there was more Diallate *in solution* than Triallate. Coupling this information to the results from the concentration experiment, it can be hypothesized that one reason for the differences in the rates of degradation of Diallate and Triallate was that at any chosen concentration, more Diallate is in solution and available for degradation than Triallate.

Can the above concepts be used to explain why under practical conditions, soil moisture has such a profound influence on the rates of pesticide degradation? To answer this question, results from an experiment with soil water content and Diallate degradation is examined (Fig. 2). For these tests, the podzol was treated with [14]C-Diallate and samples were incubated at 20°C and ca. 2.5, 9.8, 12.2, 16.5 or 20 % water content, in the dark (3). The data in Fig. 2, which summarize the results after 4 weeks, show that increasing the amount of water in the soil increased the rate of degradation. In separate experiments, the same trends were found with [14]C-Triallate, however, with Triallate, the quantities degraded within 4 weeks were less (3).

To visualize the mechanism by which water influences the rates degradation of [14]C-Diallate, an isotherm was prepared in which the amount of soil and the amount of pesticide were held constant, and the quantity of water was varied. For this purpose, a large quantity of soil was treated with 1 mg [14]C-Diallate/ kg (dry wt) soil. Ten g dry wt portions of this treated soil were amended with different quantities of water, and the samples were shaken until equilibration had occurred. The distribution of the herbicide between the soil solids and soil water was then determined. As is shown in Fig. 3, the quantity of water in the system directly influenced the *total* amount of herbicide in the soil solution. The higher the quantity of water in the system, the higher the absolute quantity of Diallate in solution. Coupling this back to the degradation data in Fig. 2, it can be concluded that the amount of herbicide in solution determines its bioavailability, which in turn, determines its rate of degradation (3). This principle holds for all pesticides.

## Influence of quantities of enzymes or cells on rates of pesticide degradation

In any soil, Diallate and Triallate are immediately attacked by soil enzymes or soil microorganisms. Currently, it is quite difficult to determine whether exo-enzymes, intact cells, or both are responsible for *in situ* degradation. However, demonstration of the relationship between enzyme or cell quantities and the rates of degradation of these herbicides is possible. For this purpose, two types of experiment can be conducted. In the first, the quantities of enzymes and cells in the soil are changed by amending the soils with nutrients. In the second, the quantity of microbial biomass in the soil can be decreased by starvation.

In one experiment, the podzol was treated with $^{14}$C-Diallate and subsamples of this soil were left unamended, or received amendments of glucose or a carbohydrate mixture (4). The samples were incubated at 20°C and - 1.8 bar water potential (ca. 12.2 % water content), in the dark for 70 days. In soils supplemented with nutrients, the quantities of metabolically active cells, measured as microbial biomass C, first increased and then decreased (carbohydrate mixture), or increased and remained relatively stable (glucose). In contrast to this, the biomass in the unamended soil decreased throughout the experiment. A comparison of biomass data to degradation rates showed that in the 2 nutrient amended soils where microbial biomasses increased, Diallate was degraded at higher rates than in the unamended soil. A positive relationship between biomass increases and degradation rates is evident (4).

Unfortunately, the nutrient experiment had 2 major flaws. First, the addition of solid but biodegradable organic matter to soil not only changed the quantities of enzymes and cells in the soil, it also changes the adsorptive properties of the soil. This interferes with the availability of the herbicide to the microflora, and this influences biodegradation rates. Secondly, the addition of nutrients to the soils changed the activity levels of the organisms in the soils.

To avoid the difficulties cause by adding nutrients, the biomass in samples of the podzol was manipulated by starvation. We noted above that the microbial biomass in unamended laboratory soil decreased with time. In experiments not shown here (4, 5), it was found that this was due to carbon starvation. Using the principle of starvation, samples of the podzol, which started with about 650 mg microbial C/kg, were held at 45, 33 and 20°C and ca. 12.2 % water content for 90 days. After 90 days, 80, 130 and 330 mg metabolically active microbial C/kg soil remained in the samples, respectively. These podzol samples, and a fresh sample which contained 655 mg microbial C/kg, were treated with $^{14}$C-Diallate or $^{14}$C-Triallate and further incubated at 20°C and 12.2 % water content for 70 days. At given intervals, the degradation, volatilization and binding of radioactive residues in the soils was measured.

The results from the 10th week of both experiments are summarized in Fig 4. The data show, that there were straight line correlations between the initial quantities of microbial biomass in the soil samples and the rates of degradation of both herbicides. It is concluded from these data, that the rates of degradation of both herbicides are determined by the quantities of enzymes or cells in the microbial biomass that can degrade them (4).

## Influence of the "activity levels" of enzymes or cells on the rates of pesticide biodegradation

How do the activity levels of enzymes or cells in soil influence the rates of pesticide degradation? This question can be answered by adding energy to the soil in the form of heat. In one such test, samples of the podzol were treated with $^{14}$C-Diallate and $^{14}$C-Triallate held at different constant temperatures for 70 days. The distribution of radioactivity over time was followed (1, 6).

Data from the 10th week of the test with Diallate are summarized in Fig. 5. The data show, that above 27°C, volatilization became a major route of loss of Diallate from the soil. This removed the compound from biodegradation. Below 27°C, and down to 4°C, decreasing the temperature decreased the rates of degradation of $^{14}$C-Diallate to $^{14}$CO$_2$ and $^{14}$C-bound residue. The reason for these rate differences are found in Fig. 6.

Referring to the Fig. 6, which concentrates on the data between 4 and 27°C, it can be seen that the biomass in the soil held at 27°C was much less than that in the soil held at 4°C. However, the rates of degradation were highest at 27°C and lowest at 4°C. These results seem to contradict previous data, which showed the positive influence of enzyme or cell quantities on the rates of pesticide degradation. However, close examination of Fig. 6 gives an explanation for this apparent contradiction. Counteracting and vastly over-compensating for the effects of different biomass sizes were the effects of temperature on the activity levels of the different biomasses. The activity levels, measured at the end of the experiment using the temperatures at which the soils were incubated during the degradation experiments, showed that there were direct correlations between the activity levels of the different biomasses and the amounts of Diallate or Triallate degraded (4). Thus, a further and very important variable that determines the rates of pesticide degradation in soils is the activity level of the enzymes or cells attacking the chemical.

## Modelling Pesticide Biodegradation in Soils

To be useful for modelling the environmental fate of pesticides, the principles given above should apply to any biodegradable pesticide in any microbiologically active soil. To test the applicability to other soils, 11 fields containing different types of soil were sampled. The soils were treated with 1 mg $^{14}$C-Diallate/kg (dry wt) soil and held at 20°C and 1.5 to 1.8 bar water potential in the dark for up to 70 days. At appropriate intervals, samples were taken to determine the distribution of parent compound and its metabolites in the soil and in the air above the soil. In addition to degradation tests, the initial quantities of metabolically active biomass in the soils were measured and 22°C adsorption isotherms for $^{14}$C-Diallate were prepared (6).

Fig. 7 shows the Diallate recovered from the soil, *plus the Diallate recovered from the air above the soil (i.e. volatilized Diallate)*. Analyzed as the total Diallate in the soil-air system, it was found that the greatest amounts were recovered when the herbicide was applied to soils with low biomass and either low availability (high adsorption constant: Diallate retained in the soil) or high availability (low adsorption constant: Diallate volatilized into the air). In respect to biodegradation, the greatest amounts of Diallate were degraded to $^{14}CO_2$ and $^{14}$C-bound residue in the soils with high availability and high quantities of microbial biomass (data not shown here).

For modelling pesticide biodegradation, the variables influencing rates can be classified as:

1.    Key variables:
      a.    the structure of the pesticide
      b.    the bioavailability of the pesticide
      c.    the quantities of enzymes or cells that can degrade the pesticide, and
      d.    the activity levels of the enzymes or cells.

2.    Driving variables:
      a.    temperature
      b.    available moisture
      c.    available nutrients
      d.    available electron acceptors (not discussed here), and
      e.    pH (not discussed here).

# Future Needs

To completely understand the biodegradation of pesticides in natural habitats, a number of areas need further research. The microbial biomass measurements and the adsorption constants used in this paper at best give crude estimates of the potential of the soil to degrade a pesticide and the availability of a pesticide to organisms that can degrade it. Intensified work is needed here. In addition, further and innovative work is needed on: definition of the structural features of pesticides that determine the biochemical lability of pesticides; description and analysis of soils microhabitats with emphasis on the microbiology of anaerobic sites in otherwise aerobic soil layers; measurement of the bioavailability of pesticides as related to the geometry of micro- and macrohabitats and; the in situ dynamics of microbial populations during pesticide degradation, including the build-up and survival of groups that have developed in response to repeated pesticide applications.

# *References*

□   (1) Anderson, J.P.E. (1975). Einfluß von Temperatur und Feuchte auf Verdampfung, Abbau un Festlegung von Diallat im Boden. *Pflanzenkrankheit Pflanzenschutz*, Sonderheft VI: 141-146.

□   (2) Anderson, J.P.E. (1980). Relationship between herbicide concentration and the rates of enzymatic degradation of $^{14}$C-Diallate and $^{14}$C-Triallate in soil. *Archives of Environ. Contam. Toxicol.* 9: 259-268.

□   (3) Anderson, J.P.E. (1981). Soil moisture and the rates of biodegradation of Diallate and Triallate. *Soil Biol. Biochem.* 13: 155-166.

□   (4) Anderson, J.P.E. (1984). Herbicide degradation in soil: influence of microbial biomass. *Soil Biol. Biochem.* 16: 483-489.

□   (5) Anderson, J.P.E. (1987). Handling and storage of soils for pesticide experiments. In *"Pesticide Effects on Soil Microflora"*, Taylor & Francis, pp. 45-60 (L. Somerville, M.P. Greaves, Eds)

□   (6) Frehse, H. and Anderson, J.P.E. (1983). Pesticide residues in soil - problems between concept and concern. *IUPAC Pest. Chem., Human Welfare and Environ.* 4: 23-32, (R. Greenhalgh, N. Drescher, Eds).

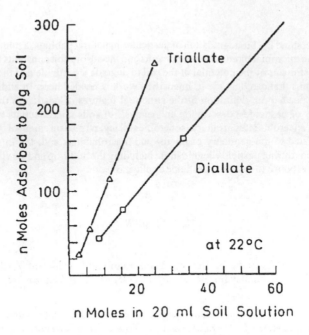

**Figure 1 :** Adsorption Isotherms of Diallate and Triallate in a sandy brown
podzol with 1.25 % C and a pH of 5.4

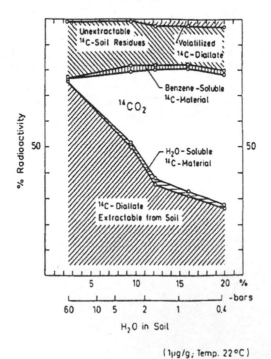

(1µg/g; Temp. 22°C)

**Figure 2 :** Influence of moisture on the dissipation of Diallate from soil

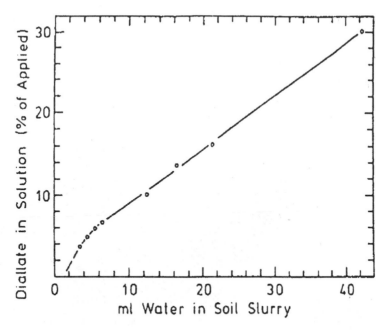

**Figure 3 :** Relationship between soil water content and the quantity of Diallate in solution

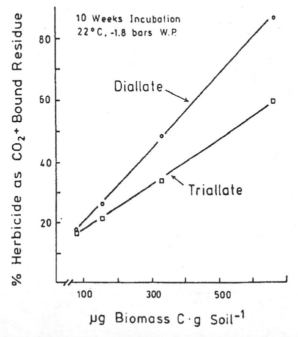

**Figure 4 :** Relationship between the quantity of microbial biomass in soil and the rates of degradation of $^{14}$C-Diallate or $^{14}$C-Triallate

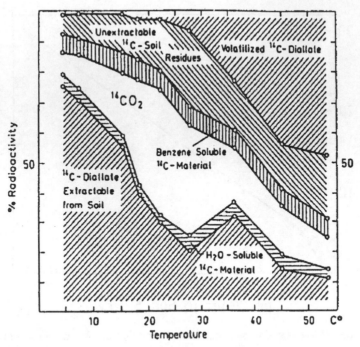

**Figure 5 :** Influence of temperature on the rates of degradation of $^{14}$C-Diallate in soil

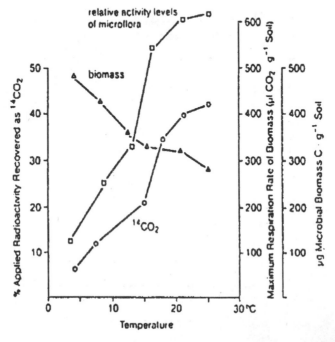

**Figure 6 :** Influence of temperature on microbial activity levels, microbial biomass, and the rates of degradation of $^{14}$C-Diallate in soil

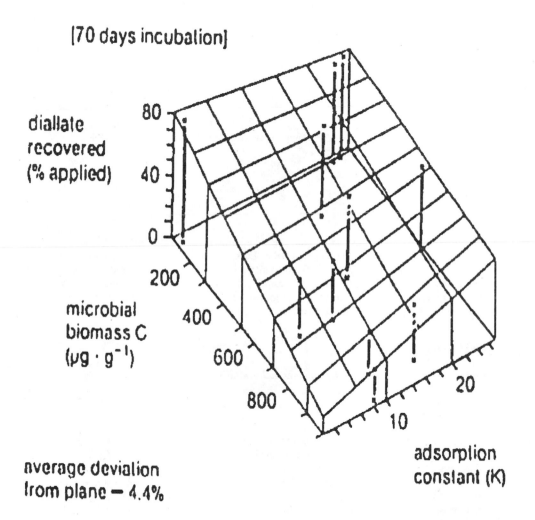

[70 days incubation]

diallate
recovered
(% applied)

microbial
biomass C
(µg · g⁻¹)

average deviation
from plane — 4.4%

adsorption
constant (K)

**Figure 7** : Relationship between the quantity of microbial biomass,
the degree of Diallate adsorption and the total amount of Diallate
recovered from soils and the atmosphere above the soils

# 3.2.     Platform Presentations

*3.2.1.     Addition Products of Chloroanilines and Dealkylated Atrazines on Humic Like Polymers : a Model of Pesticide Bound Residues in Soil and Water Media*

## W. Völkel[1,2], A. Lieurade[1], J.-M. Portal[1], M. Mansour[2] & F. Andreux[1,3]

[1]    Centre de Pédologie Biologique, CNRS, BP5, F-54501 Vandoeuvre-lès-Nancy cedex, France

[2]    Institut für Oekologische Chemie, GSF, Schulstrasse 10, D-85356 Freising, Germany

[3]    Université de Bourgogne, Centre des Sciences de la Terre, 6, Boulevard Gabriel, F-21000 Dijon, France

## Summary

A model reaction is presented to explain the formation of pesticide bound residues from 3,4-dichloroaniline (DCA) and de-isopropylatrazine (DIA). These pesticide metabolites were reacted through oxidative polycondensation with catechol and methylcatechol. Infrared and proton-NMR spectra revealed the complex structure of the resulting humic-like polymers, and the covalent binding of the incorporated pesticide molecules. A large part of them were bound to quinonic structures of the same type as in monomeric derivatives.

## Key Words

*Bound residues, humic substances, pesticides, chloroanilines, atrazine, catechol.*

## Introduction

According to the IUPAC [1] "Non-extractable residues (sometimes referred to as 'bound' or 'non-extracted' residues) in plants and soils are defined as chemical species originating from pesticides used according to good agricultural practices, that are unextracted by methods which do not significantly change the chemical nature of these residues". Consequently if one is interested in this nature of bound residues (BR), one is confronted with the problem of characterizing a species that cannot be isolated. In a first approach the following questions can be asked: (i) what is the bound species, and to which species is it bound, and (ii) what are the kinds of bindings that occur in BR?

The answer to the first question was partly given by the above definition but the metabolites are sometimes more reactive than the original pesticide molecules. As far as the soil is concerned, it has been shown that the amounts of BR are often correlated with the soil organic matter (SOM) content [2]. This SOM mainly consists of the classical fractions fulvic acids (FAs) and humic acids (HAs), and humin, each of them being regarded as a mixture of large random polymers [3]. A pesticide or its metabolite can react with either the humic polymers or their precursors. Polyphenols are the most important group of precursors, among which catechol and 4-methylcatechol were chosen for the present study.

Concerning the second question, polyphenols can be easily oxidized into quinones, to which nucleophilic partners like nitrogen-, oxygen-, sulfur- or phosphorus-compounds can be added. Primary amines are bound by nucleophilic addition, according to a Michael analogue reaction. There are some pesticides with free amino-groups, such as metribuzine, but more frequently, pesticide metabolites, such as chlorinated anilines and dealkylated triazines, are concerned. In the present study, 3,4- dichloroaniline (DCA) and de-isopropylatrazine (DIA) were reacted with the above diphenols. The addition products which are formed in a first step, are called here "monomers". They are ortho-quinones substituted with 3,4-dichlorophenylamino and [(2-chloro-6-isopropylamino-)-1,3,5-triazin-4- yl]-amino groups, respectively [4,5]. If the oxidation reaction proceeds, polymeric products are formed. The best known and controlled polymerization reaction results in a humic-like polymer which includes pesticide BR. In other cases, the pesticide can be reacted with the already formed polymer [6, 7].

## Material and Methods

The conditions to produce the monomers in aqueous phosphate buffer solution at a pH between 6 and 7 and the isolation of the products by preparative HPLC on RP-18 were formerly described [5]. In the case of polymers, the aqueous solutions of 0.5 M $NaH_2PO_4$ and $Na_2HPO_4$ (produced by Merck) were mixed to obtain a 250-ml solution with the desired pH, which varied between 6 and 8. The pesticide compound, either DCA (produced by Aldrich) in amounts from 10 to 240 mg, or DIA (produced by Cluzeau Info Labo), in amounts from 32 to 43 mg, was dissolved in 1 ml of ethanol, and added to the buffer solution. After complete dissolution of occasional precipitates, 330 mg ( 3 mmole) of catechol (produced by Riedel de Haen), recristallyzed in n-hexane for the reactions with DCA or 1 650 mg for those with DIA or 1 860 m g of 4- methylcatechol (produced by EGA) for both DCA and DIA, were added. The reaction was run at room temperature in the dark during 7 days under a constant stream of oxygen. The solutions turned to an intense red colour, which changed to dark brown after several hours.

After reaction, the solution of DCA-polymer was dialysed for one week against distilled water in a Servapor 10-14 kD tubing, then extracted with $CH_2Cl_2$, to remove unreacted monomers, and finally freeze-dried. In the case of DIA the solutions were acidified to pH 1.5, to precipitate the HAs, then centrifuged at 27000 g during 20 min. The solid was separated from the solution, washed with 3 N HCl and centrifuged again. The joined solutions were dialyzed at 2 kD and the solid fractions were dispersed in distilled water and dialyzed at 4 kD. Each dialysed fraction was freeze- dried and was considered as synthetic FAs and HAs, respectively.

Infrared spectra were obtained on a Beckman IR 4250 spectrophotometer. The pellets contained 1 mg of polymer and 200 mg KBr. The elemental analysis (C, H, N) was carried out on a CHN-1106 Carlo Erba analyser. The proton-NMR spectra were obtained on a Bruker AC 400 MHz in deuterated methanol ($CD_3OD$).

## Results and Discussion

The polymerization reactions were based on mechanisms that take place in different natural processes, such as humification [8], formation of pigments [9], plants defense [10] or formation of the exoskeleton of insects [11]. This universal reaction, which can be carried out in the laboratory, is controled by pH, and results in random polymers. The formation of monomers which precedes the polycondensation can be carried out with abiotic or enzymatic catalysis [12]. The optimum conditions for the production of monomers were a pH between 6 and 7, with pesticide and phenols in about equimolar amounts and a reaction time of several hours. The formation of a polymer with an incorporated pesticide molecule required a pH between 6.5 and 8. A large excess of phenol and a reaction time of about one week was necessary to complete the polymerization.

## Comparison of synthetic polymers with natural humic acids

Infrared spectroscopy and NMR spectroscopy always showed broad signals for these kinds of polymers, due to the overlaping of many signals for similar individual bindings. Therefore, only general structural characteristics could be identified. The IR-spectra of natural HAs showed more vibrations for aliphatic and for amide groups than the catechol polymers. The methylcatechol polymer showed vibrations of aliphatic C-H of the methyl group. A very close model for HAs would therefore be a mixture of catechol and methylcatechol with a bound amino acid. The simple model compound allowed to show the incorporation of either DCA or DIA into HAs.

The $^1$H-NMR spectra of the polymers supplied information about the bindings of a pesticide molecule, when compared with the spectra of small basic units. They also revealed numerous informations about the polymer itself. The spectra of the catechol polymer were dominated by a large signal from 6 to 8 ppm with a maximum at 6.9 ppm. There were additional groups of very narrow peaks, which could be attributed to catechol either in terminal position or strongly adsorbed on the polymer surface, and redissolved in methanol. There were also signals at 6.85 and 6.84 ppm and smaller ones at 6.46 and 6.20 ppm, which belonged to substituted quinones, and smaller signals at 0.9 and 1.3 ppm in the aliphatic region. Generally the resonances suggested that the skeleton of the polymer was formed of aromatic and quinonic structures with few aliphatic groups resulting from the opening of the catechol-rings [13].

The polymers of methylcatechol showed a greater part of aliphatic protons, most probably deriving from the methyl groups. The relative values of the aliphatic region (1-2.5 ppm) and the aromatic region (6.5-7.5) showed clear differences between these two polymers. Concerning the methylcatechol polymer, if it is assumed that methylcatechol was substituted in the 3- and 4-positions, there should be a ratio of 3 aliphatic protons to 2 aromatic protons. As values higher than 1.5 were found, it could be supposed that the methylcatechol was also substituted in the 6-position. Another possibility would be a large number of opened aromatic rings. The maximum for the aromatic protons was shifted to 6.66 ppm in comparison to 6.9 ppm for the catechol polymers, indicating a higher proportion of quinones in the methylcatechol polymer.

## Incorporation of DCA in the catechol-polymer

The most important parameters for the incorporation of DCA during the polymerisation of catechol were the concentration of DCA and the pH of the solution. At a constant pH of 7.25, 10 mg (0.06 mmole) to 240 mg (1.44 mmole) of DCA were added to 330 mg (3 mmole) of catechol in the buffer solution. The amount of DCA incorporated increased with increasing amounts of DCA but the yield of polymer decreased (Table 1), and the monomeric amino-substituted quinone was the main product.

| mmole DCA | 0 | 0.06 | 0.08 | 0.12 | 0.15 | 0.37 | 0.45 | 0.50 | 0.62 | 1.51 |
|-----------|---|------|------|------|------|------|------|------|------|------|
| C/N ratio | - | 40.1 | 41.8 | 35.8 | 33.2 | 30.5 | 32.1 | 21.4 | 16.7 | 15.3 |

**Table 1** : Influence of the amount of 3,4-dichloroaniline (DCA) reacted with
3 mmole of catechol on the C/N ratio of the polymers,
as calculated from elemental analysis.

This tendency was illustrated by the [1]H-NMR spectra: in the polymer prepared with less DCA, the resonances at fields lower than 7 ppm might come from monomer analogous structures (Figure 1a). When compared with the respective shift values of the monomer in deuterated methanol (in parentheses), the shift values were at 7.02 (7.07 dd), 7.17 (7.28 d) and 7.45 (7.49 d) ppm for the aniline ring-protons, and at 5.83 (6.05 s) ppm for the quinonic protons. In the polymer prepared with twice more DCA, the variety of aromatic and quinonic protons increased, with sharper and better resolved peaks (Figure 1b). This suggested a progressive change with higher amounts of DCA, from substituted quinones (monomer-like structures) to other environments of binding. However, all signals indicated that DCA was strongly bound to the polymer, and that free DCA was never observed (Table 2).

| DCA (ppm) | DCA with quinone (ppm) | DCA with polymer (ppm) |
|-----------|------------------------|------------------------|
| 6.56      | 7.07                   | 6.45, 7.02, 7.1-7.2*   |
| 6.78      | 7.28                   | 6.64, 7.17, 7.4-7.5*   |
| 7.12      | 7.49                   | 6.89, 7.45, 7.6-7.8*   |

* in 3- to 4-fold repetition; always according to the pattern of the splitting double duplet, duplet and duplet

**Table 2 :** Proton-NMR data (in CD3OD) of the aniline ring of 3,4-dichloroaniline (DCA), free and bound to ortho-quinone and to catechol polymer (0.5 : 3 mmole).

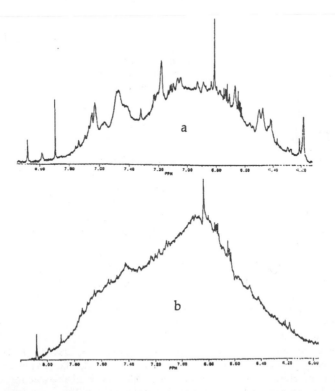

**Figure 1 :** Proton-NMR spectra of catechol polymers synthesized from 3,4-dichloro-aniline (0.25 mmole) and catechol (3 mmole) at pH 7.0 (a) and pH 8.5 (b).

The pH values influenced the structure of the polymer as well as the incorporation of DCA. Between pH 6.5 and 7, like in Figure 1a, DCA might be covalently bound to quinones. At pH higher than 8.5, a broad resonance from 6 to 8 ppm with fewer structural details was observed. The elemental analysis of the polymers (Table 3) confirmed that at higher pH less DCA was incorporated. The fast auto-reaction of catechol became prevalent over the addition of aniline to the catechol rings. The incorporation of monomers in the polymers was also carried out successfully, confirming that the addition of anilines to quinones was a possible first step of this incorporation. Finally it could be concluded that besides high DCA concentrations, pH values between 6.5 and 7 favoured the incorporation of DCA in the polymers, in the form of monomer-like structural units.

| Reaction pH | 6.8 | 7.2 | 8.2 | 8.7 |
|---|---|---|---|---|
| C/N ratio | 19.7 | 25.9 | 29.9 | 36.0 |

**Table 3 :** Influence of pH on the amount of 3,4-dichloroaniline incorporated in the catechol polymers, as expressed by C/N ratio

| DIA in $CD_3OD$ (ppm) | DIA with quinone in $CD_3CN$ (ppm) | DIA with polymer in $CD_3OD$ (ppm) |
|---|---|---|
| 1.15 | 1.32 | 1.28, 1.3-1.4 |
| 3.34 | 3.47 | 3.63, 3.3-4.3 |

**Table 4 :** Proton-NMR data of the alkyl chain for de-isopropylatrazine (DIA), free and bound to methyl-orthoquinone and to the catechol-polymer (0.08 mmole : 6 mmole)

## *Addition of DIA to humic-like polymers*

As for DCA, the dependance of DIA on pH and concentration could be shown. In this case the comparison of HAs and FAs was considered. The proton signals from the triazine methylene group were located around 3.5 ppm, where no proton signal of the polymer was observed. This had the advantage to be always detected, eventhough DIA was incorporated in relatively low amounts. Unfortunately the pattern of these alkyl chain protons was complex, already in the original molecule, because of conformational isomers [14], and made difficult the identification of the binding of DIA in the polymer.

Some signals in the $^1$H-NMR spectra suggested that DIA was bound to a quinone, but many further occuring resonances showed other kinds of bindings, excluding free DIA. In the spectrum of the FA the signals for the $-CH_2-$ protons were sharper than in that of the HA, were more intense than any other resonance, and showed greater variations (3.5-4.3 ppm), as shown in Table 4. Obviously DIA was preferably bound to FAs, as also suggested by Bertin et al. [7]. It could also be seen that HAs contained more aromatic protons and relatively less aliphatic protons than FAs, in accordance to the results obtained for natural HAs. Less DIA (C/N: 50-160) than DCA was incorporated, as revealed by elementary analysis. This fact can be explained by the delocalisation of electrons in the triazine ring, which lowered the electron density of the $NH_2$-group. This supported the observation that in real soils anilines were bound faster and in higher amounts to SOM than atrazine and its derivatives [12].

# Conclusion

The polymerization reaction of catechol represents an adequate model to simulate oxidative humification processes. The resulting products were similar to natural humic like polymers, but a differenciation between humic and fulvic acids could be made. The incorporation of DCA and DIA followed similar mechanisms, which corresponded to the formation of covalent bound residues. Proton-NMR spectroscopy showed that these residues were bound to quinone-like structural units, although other kinds of bindings were suspected. Further comparisons with other small synthetic addition products should be made.

# Acknowledgement

The authors wish to thank Dr. N. Hertkorn and E. Schindlbeck from the IÖC-GSF Institute in Munich for the realization of the NMR-spectra.

# References

☐ [1] T.R. Roberts, Non-extractable pesticide residues in soils and plants. *Pure and Applied Chem.*, 56 (7), 1984, 945-956.

☐ [2] A. Moreale, A. Gallez, R. Van Bladel, Influence de la matière organique et de la teneur en eau des sols sur les phénomènes de sorption de l'aniline, *Weed Res.* 17, 1977, 349-356.

☐ [3] W. Ziechmann, in: F.H. Frimmel, R.F. Christman, *Humic substances and their role in the environment*, John Wiley & Sons Ltd., Chichester, S. Dahlem Konferenzen 1988, p. 113.

☐ [4] W. Völkel, Th. Choné, J.-M. Portal, B. Gérard, M. Mansour, F. Andreux, Reaction of the pesticide metabolite 3,4-dichloroaniline with humic acid monomers, *Fresenius Envir. Bull. 2*, 1993, 262-267.

☐ [5] W. Völkel, A. Lieurade, F. Andreux, Réactions d'addition d'herbicides et de métabolites aminés, sur un précurseur des substances humiques, le méthylcatéchol, *C. R. Ac. Sci., Paris*, 1994 (to be published)

☐ [6] N. Senesi, C. Testini, Theoretical aspects and experimental evidence of the capacity of humic substances to bind herbicides by charge-tranfer mechanisms (electron-donor-acceptor processes), *Chemosphere*, 13, 1984, 461-468.

☐ [7] G. Bertin, M. Schiavon, J.-M. Portal, F. Andreux, *Contribution to the study of non-extractable pesticide residues in soils. Incorporation of atrazine in model humic acids prepared from catechol*, in: J. Berthelin, ed., Diversity of environmental biogeochemistry, Elsevier, Amsterdam, 10991, 105-110.

☐ [8] F.J. Stevenson, *Humus Chemistry. Genesis. Composition. Reactions*, John Wiley & Sons, New-York, 1982.

☐ [9] R.A. Nicolaus, *Melanins*, Herman, Paris, Actualités Scientifiques et Industrielles nbr 1336, 1968.

☐ [10] F.J. Marner, Chemische Kriegslisten zur Abwehr von Schadinsekten, *Chemie in unserer Zeit 2*, 1993, 88-95.

☐    [11] M.G. Peter, Die molekulare Architektur des Exoskeletts von Insekten, *Chemie in unserer Zeit 4*, 1993, 189-197.

☐    [12] Ph. Adrian, E.S. Lahaniatis, F. Andreux, M. Mansour, I. Scheunert, F. Korte, Reaction of the soil pollutant 4-chloroaniline with the humic acid monomer catechol, *Chemosphere* 18, 1989, 1599-1609.

☐    [13] F. Andreux, Utilisation de molécules modèles de synthèse dans l'étude des processus d'insolubilisation et de biodégradation des polycondensats humiques, *Science du Sol*, 4, 1981, 271-292.

☐    [14] G.J. Welhouse, W.F. Bleam, NMR Spectroscopic Investigation of Hydrogen Bonding in Atrazine, *Environ. Sci. Technol.* 26, 1992, 959-964.

☐    [15] R. Barta, Fate of herbicide-derived chloroabilines in soil, *J. Agric. Food Chem.* 19, 2, 1971, 385-387.

## 3.2.2.      Assessment of Isoproturon Degradation by Soil Fungi

# R. Steiman[1], F. Seigle-Murandi[1], J-L. Benoit-Guyod[2], G. Merlin[3] and M. Kadri[1,2]

Groupe pour l'Etude du Devenir des Xénobiotiques dans l'Environnement (GEDEXE)

[1]     Laboratoire de Botanique, Cryptogamie, Biologie Cellulaire et Génétique

[2]     Laboratoire de Toxicologie et Ecotoxicologie
      UFR de Pharmacie de Grenoble, Université J. Fourier, BP 138, 38243 Meylan Cedex

[3]     Laboratoire de Biologie et Biochimie Appliquée, Ecole Supérieure d'Ingénieurs
      en Génie de l'Environnement et Construction (ESIGEC), Université de Savoie,
      BP 1104, 73011 Chambéry Cedex

## Abstract

As a part of a study conducted on the fate of xenobiotics in the environment, a selection of 74 strains of Micromycetes mostly isolated from soil and belonging to various taxonomic groups have been incubated with isoproturon. Toxicity assessment was obtained by cultivation on malt extract agar or potato dextrose agar media containing isoproturon at concentrations ranging from 100 to 1000 mg $L^{-1}$. Growth and morphological changes were recorded after 15 days of incubation at 24°C. Partial or total inhibition was observed for 36 % of the strains even at the lowest concentration used. 82 % were inhibited at 500 mg $L^{-1}$ and 16 % were not inhibited by 1 g $L^{-1}$ of isoproturon.

125 strains have also been cultivated in liquid synthetic medium with isoproturon (100 mg $L^{-1}$) for 5 days. Evaluation of isoproturon in the culture media was made by HPLC. On the whole, only 8 % of the strains depleted 70 % or more of isoproturon. The best results were obtained with the following genera : *Rhizoctonia, Cunninghamella* and *Phoma* that depleted more than 80 % of the substrate after only 24 h for some of them. The very fast depletion of isoproturon from the culture media was only due to biotic phenomena as no adsorption occurred on the fungal biomass. So, when isoproturon disappeared, it was most probably degraded.

## Introduction

3-(4-isopropylphenyl)-1, 1-dimethyurea or isoproturon (iPr) is a selective herbicide belonging to the family of substituted ureas. It acts principally after root absorption. It is used pre or post emergence for the control of annual grasses and broad leaved weeds in barley, wheat corn, rice, soya and rye at 1.0 - 2.0 Kg/ha.

IPr is stable under natural light as a consequence of its low quantum yield of photolysis (Kulsrestha and Mukerjee, 1986) and its UV absorption spectrum (upper limit at 290 nm). Thus, it is not surprising that studies on the fate of iPr in sterile and non sterile soils reveal that biodegradation is the main process for its disappearance (Fournier et al., 1975; Davies and Marsh, 1980). But these global studies do not allow to precise which are the efficient microorganisms, bacteria or fungi or both. The aim of this study was first to evaluate the toxicity of iPr towards micromycetes at concentrations far exceeding agricultural concentrations, then to assess its depletion from the culture media.

## Material and Methods

### Microorganisms

The fungal strains were taken from the mycotheque of our Laboratory (CMPG : Collection Mycology Pharmacy Grenoble). Table 1 give their taxonomic distribution. Most of them were *Fungi imperfecti* isolated from various substrates : mainly from soil, decayed wood, food, etc... (De Hoog et al., 1985; Gams et al., 1990; Seigle-Murandi et al. 1980, 1981). Stock cultures were maintained at 4°C on solid malt extract medium (1.5 %) (MEA) or Potato Dextrose Agar medium (PDA).

### Assessment of toxicity of isoproturon on solid media

A selection of 74 micromycetes belonging to various taxonomic groups was grown on MEA or PDA media containing iPr at final concentrations 0, 0.1, 0.5 and 1.0 g $L^{-1}$. IPr (5 g $L^{-1}$) was dissolved in DMSO/$C_2H_5OH$ (50/50, v/v) sterelized though a millipore syringe (pore size 0.22 mm) and added aseptically to the media before inoculation. Petri dishes were incubated for 15 days at 24°C. The percentage of inhibition was calculated by comparing the growth diameter of the reference and treated strains. Morphological changes were also recorded. Each experiment was made in triplicate.

### Depletion of isoproturon in liquid medium

125 strains were cultivated in the synthetic medium of Galzy and Slonimsky (GS), pH 4.5 with glucose (5 g $L^{-1}$) in 125 ml Erlenmeyer flasks containing 25 ml of medium. Cultures were incubated with shaking for 2 days (180 rpm, orbital shaker) in order to allow the biomasses to grow and reach 4 g $L^{-1}$ ± 10 % (dry weight). At this stage, no glucose remained in the medium. IPr was added to the 2-day-old cultures to a final concentration of 100 mg $L^{-1}$. The depletion of iPr was evaluated after 7 days of cultivation (5 with iPr). Each series of experiment included flasks extracted at time = 0 and abiotic references. Temperature was 24°C, light was 1220 lux with a photoperiod of 12 h per day. Each experiment was made in triplicate. In the same conditions, 19 strains were cultivated in malt extract (ME) medium.

After filtration of the media on Millipore membranes (0.45 μm), depletion of iPr was assessed by HPLC analysis performed with a liquid chromatograph Shimadzu equipped with a pump LC 6A, an automatic injector Shimadzu SIL-9A and a UV detector SPD 6A. The separation column was 4.0 mm inside diameter x 150 mm long filled with μ-Bondapak $C_{18}$. The mobile phase was methanol : water (75:25 v/v). The flow rate was 1 ml $min^{-1}$ and analyses were performed at 240 nm. Each sample was injected at least 3 times and the mean taken. Results were within a ± 10 % range.

For adsorption determinations, the biomass collected on the filter was extracted with methanol and analyzed by HPLC as previously.

## Results and Discussion

The aim of this work was first to look at the toxicity of iPr against micromycetes. Toxicity assay on solid media showed that this compound was relatively toxic for the fungi. In Figure 1, results have been expressed according to the taxonomic groups. The most sensitive groups were Ascomycetes and Stilbellales, the less sensitive Zygomycetes. Other groups showed intermediary sensitivity. Inside the groups, some species were particularly resistant and were not inhibited even at concentration 1 g $L^{-1}$ : *Aureobasidium pullulans, Scytalidium lignicola, Colletotrichum musœ, Coniophora putanea, Trichoderma pseudokoningii, Aspergillus parasiticus, Fusarium solani, Fusarium cœruleum* and among Zygomycetes, *Cunninghamella bainieri, C. blakesleeana, C. elegans.* Moreover, the growth of *Coniophora putanea* was stimulated by iPr when used at 0.1 and 0.5 g $L^{-1}$. Morphological changes occurred when strains were sensitive to isoproturon. Conidiogenesis was the first inhibited generally around 0.5 g L$^{-1}$ then growth was inhibited at a higher concentration. For strains whose growth was not affected by isoproturon, conidiogenesis was not affected even at the higher concentration.

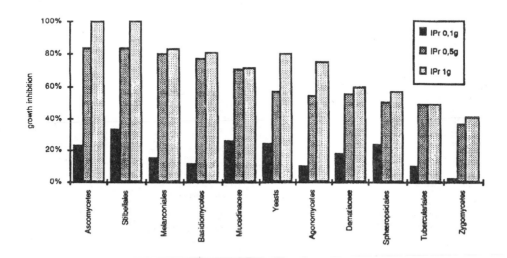

**Figure 1 :** Toxicity of iPr on fungi cultivated on MEA medium.
Results are given by taxonomic groups

As a function of the results obtained, cultivation in GS liquid medium with iPr (100 mg L$^{-1}$) was made with more strains, especially Zygomycetes. After 5 days of cultivation, abiotic degradation, mainly photodegradation, was negligible and always lower than 5 %. Results given in Table 1 are expressed as the percentage of depletion of iPr in the media. Methanol extraction of the drained biomass allowed to check that no adsorption occurred. So depletion of iPr corresponded to biodegradation or biotransformation into another metabolite(s).

Inside the taxonomic groups, the various genera and species depleted very unequally. The most efficient strains (i.e. > 70 % of disappearance of isoproturon) were : *Rhizoctonia solani* (83 %), *Ceriporiopsis subvermispora* (75 %), *Coniophora puteana* (70 %), *Pholiota abstrusa* (78 %), *Cicinobolus cesatii* (77 %), *Phoma herbarum* (85 %) and in Zygomycetes, *Cunninghamella blakesleeanus* (96 %), *C. echinulata* (93 %) and 2 strains of *C. elegans* (98 and 91 %).

**AGONOMYCETES : 31 %**

| | |
|---|---|
| Rhizoctonia solani | 83 % |
| Rhizoctonia sp. | 13 % |
| Rhizoctonia sp. | 5 % |
| Sclerotium sp. | 23 % |

**ASCOMYCETES : 21 %**

| | |
|---|---|
| Chœtomium globosum | 3 % |
| Cryphonectria parasitica | 60 % |
| Dichotomomyces cejpii | 30 % |
| Emericella variecolor var. variecolor | 10 % |
| Eurotium repens | 0 % |
| Neurospora crassa | 23 % |
| Neurospora sitophila | 9 % |
| Pyronema omphalodes | 35 % |
| Sclerotinia sclerotiorum | 23 % |
| Sporormiella australis | 14 % |
| Sporormiella minimoides | 22 % |

**BASIDIOMYCETES : 35 %**

| | |
|---|---|
| Bjerkandera adusta | 57 % |
| Ceriporiopsis subvermispora | 75 % |
| Ceriporiopsis subvermispora | 26 % |
| Chondrostereum purpureum | 1 % |
| Coniophora puteana | 70 % |
| Dichomitus squalens | 26 % |
| Fibroporia vaillantii haploide | 33 % |
| Lœtisaria arvalis | 21 % |
| Phanerochœte chrysosporium | 2 % |
| Phanerochœte chrysosporium | 0 % |
| Pholiota abstrusa | 78 % |
| Tyromyces fissilis | 35 % |

**DEMATIACEÆ : 21 %**

| | |
|---|---|
| Acremonium murorum | 10 % |
| Alternaria alternata | 8 % |
| Alternaria raphani | 28 % |
| Aspergillus niger | 26 % |
| Aureobasidium pullulans | 19 % |
| Cladosporium cladosporioides | 29 % |
| Cladosporium herbarum | 41 % |
| Drechslera spicifera | 29 % |
| Embellisia annulata | 28 % |
| Oidiodendron echinulatum | 4 % |
| Oidiodendron tenuissimum | 22 % |
| Scytalidium lignicola | 13 % |

**MELANCONIALES : 31 %**

| | |
|---|---|
| Cicinobolus cesatti | 77 % |
| Colletotrichum dematium | 55 % |
| Colletotrichum gloeosporioides | 18 % |
| Colletotrichum musœ | 26 % |
| Colletotrichum musœ | 15 % |
| Pestalotia palmarum | 32 % |
| Pestalotia truncata | 18 % |
| Pestalotiopsis sp. | 40 % |
| Pestalotiopsis versicolor | 12 % |
| Truncatella sp. | 14 % |

**MUCEDINACEÆ : 19 %**

| | |
|---|---|
| Acremonium chrysogenum | 0 % |
| Acremonium roseum | 31 % |
| Aspergillus parasiticus | 27 % |
| Aspergillus penicilloides | 16 % |
| Aspergillus terreus | 9 % |
| Aspergillus ustus | 20 % |
| Aspergillus versicolor | 37 % |
| Beauveria alba | 29 % |
| Calcarisporium arbuscula var. luteola | 22 % |
| Chrysosporium pannorum | 17 % |
| Cladobotryum verticillatum | 30 % |
| Pœcilomyces variotii | 17 % |
| Penicillium atramentosum | 5 % |

| | |
|---|---|
| Penicillium aurantiogriseum | 7 % |
| Penicillium chrysogenum | 16 % |
| Penicillium cyaneum | 32 % |
| Penicillium italicum | 8 % |
| Penicillium purpurrescens | 10 % |
| Sporothrix cyanescens | 0 % |
| Trichoderma harzianum | 0 % |
| Trichoderma pseudokoningii | 4 % |
| Trichophyton tonsurans | 1 % |
| Verticillium lecanii | 57 % |
| Verticillium leptobactrum | 49 % |

**SPHÆROPSIDALES : 30 %**

| | |
|---|---|
| Ascochyta imperfecta | 27 % |
| Coniothyrium sp. | 33 % |
| Coniothyrium sp. | 25 % |
| Coniothyrium sporulosum | 9 % |
| Macrodiplodia zeœ var. macrospora | 12 % |
| Phoma glomerata | 47 % |
| Phoma herbarum | 85 % |
| Trematophoma sp. | 0 % |

**STILBELLALES : 5 %**

| | |
|---|---|
| Dicyma ampullifera | 13 % |
| Doratomyces stemonitis | 2 % |
| Doratomyces microsporus | 0 % |
| Trichurus spiralis | 4 % |

**TUBERCULARIALES : 8 %**

| | |
|---|---|
| Cylindrocarpon destructans | 11 % |
| Cylindrocarpon ianthothele var. majus | 2 % |
| Cylindrocarpon macrosporum | 8 % |
| Epicoccum nigrum | 26 % |
| Fusarium coeruleum | 0 % |
| Fusarium fusarioides | 3 % |
| Fusarium moniliforme var. subglutinans | 7 % |
| Fusarium oxysporum | 6 % |
| Fusarium solani | 22 % |
| Fusarium sulphureum | 2 % |
| Myrothecium gramineum | 3 % |

**YEASTS : 16 %**

| | |
|---|---|
| Cryptococcus albidus | 26 % |
| Cryptococcus magnus | 26 % |
| Rhodotorula aurantiaca | 25 % |
| Rhodotorula glutinis | 0 % |
| Rhodotorula rubra | 18 % |
| Saccharomyces cerevisiœ (CCY 28-3) | 0 % |
| Sporobolomyces roseus | 24 % |
| Sporobolomyces roseus | 9 % |

**ZYGOMYCETES : 31 %**

| | |
|---|---|
| Absidia glauca | 0 % |
| Absidia spinosa | 49 % |
| Cunninghamella bainieri | 54 % |
| Cunninghamella blakesleeana | 96 % |
| Cunninghamella echinulata | 93 % |
| Cunninghamella elegans | 98 % |
| Cunninghamella elegans | 91 % |
| Helicostylum piriforme | 1 % |
| Mortierella bainieri | 4 % |
| Mortierella hyalina | 12 % |
| Mortierella isabellina | 24 % |
| Mortierella ramanniana | 4 % |
| Mucor genevensis | 8 % |
| Mucor mucedo | 9 % |
| Mucor racemosus | 2 % |
| Phycomyces blakesleeanus | 8 % |
| Phycomyces blakesleeanus (-) | 15 % |
| Rhizopus arrhizus | 41 % |
| Rhizopus stolonifer | 8 % |
| Syncephalastrum racemosum | 25 % |
| Sygorhynchus moelleri | 3 % |

Comparison of the results obtained after toxicity assay on MEA medium and depletion assay after cultivation in liquid medium showed that there was no correlation between these two series of results; i.e. strains insensitive to iPr could deplete iPr (*Coniophora puteana*, the genus *Cunninghamella*) as well as not to deplete it or very meanly (*A. pullulans, S. lignicola, C. musæ, T. koningii, A. parasiticus, F. cœruleum, F. solani*). We previously obtained similar results when several strains of micromycetes were completely inhibited when cultivated on solid media and nevertheless some of them depleted very rapidly and efficiently pentachlorophenol (PCP) in liquid media (Seigle-Murandi et al., 1993). A rapid assay on solid media was developed to screen fungi for resistance to toxic chemicals and thus select strains able to treat contaminated soils or wastewater (Alleman et al., 1993). As a function of the results obtained with PCP and with iPr, we can assume that toxicity observed for chemicals on solid media does not allow any prevision on the degradation behaviour of the chemical in liquid media i.e. the possibility of treatments of polluted soils or wastewaters.

A last assay compared the depletion of iPr by 19 fungal strains chosen in different taxonomic groups when cultivated in 2 different culture media : GS and ME media. All strains maintained their depletion properties in GS medium but a few were optimized in ME medium (Table 2) which shows the importance of the culture media for degradation studies.

| | Disappearance of isoproturon | |
| --- | --- | --- |
| | GS | ME |
| *Bjerkandera adusta* | 57 % | 80 % |
| *Colletotrichum musæ* | 15 % | 66 % |
| *Coniothyrium sp.* | 33 % | 72 % |
| *Cryphonectria parasitica* | 60 % | 85 % |
| *Cunninghamella blakesleeana* | 96 % | 94 % |
| *Cunninghamella echinulata* | 93 % | 98 % |
| *Cunninghamella elegans* | 91 % | 95 % |
| *Cunninghamella elegans* | 98 % | 95 % |
| *Pestalotia palmarum* | 32 % | 94 % |
| *Pestalotia truncata* | 18 % | 62 % |
| *Pestalotiopsis sp.* | 40 % | 67 % |
| *Pholiota abstrusa* | 78 % | 98 % |
| *Phoma herbarum* | 85 % | 66 % |
| *Pyronema omphalodes* | 35 % | 74 % |
| *Rhizoctonia solani* | 83 % | 100 % |
| *Rhizopus arrhizus* | 41 % | 80 % |
| *Sclerotium sp.* | 23 % | 75 % |
| *Syncephalastrum racemosum* | 25 % | 72 % |
| *Tyromyces fissilis* | 35 % | 80 % |
| **mean** | **55 %** | **82 %** |

**Table 2 :** Depletion of iPr by 19 fungal strains in GS medium and in malt extract (ME)

Finally, it is clearly established that iPr is by far less toxic for fungi than for plants. That is in accordance with previous works on the persistence and metabolism of iPr in soil (Mudd et al., 1983) which had shown that at agricultural field levels, iPr had no harmful effects on microorganisms of soil or fertility of soil (Mudd et al., 1985). The susceptibility of non-target biota like cyanobacteria to herbicide has also been investigated (Nirmal Kumar, 1991). Toxicity of iPr for plants is explained by the inhibition of photosystem II (PS II) at concentrations as low as 10 µM (0,3 mg/L) (Ponte-Freitas et al., 1991,

1993; Haddad et al., 1992). The lack of PSII in fungi explain that they are much less sensitive to iPr than plants. However cellular target explaining fungal toxicity at the 100 mg/L level remain to be found.

Most fungal degradation studies have focused on white-rot fungi and more specially on one species, *Phanerochœte chrysosporium* because of the production of extracellular and non specific enzymes (Barr and Aust, 1994). The results obtained here and previous work of our group (Seigle-Murandi et al., 1992, Steiman et al., 1994; Benoit-Guyod et al., 1994), suggest that a large number of fungi can potentially be used to treat contaminated soils and wastewaters.

# References

▫   Alleman, B., Logan, B. and Gilbertson, R.L. 1993. A rapid method to screen fungi resistance to toxic chemicals. *Biodegradation*, 4, 125-129.

▫   Barr, D.P. and Aust, S.D. 1994. Mechanisms white rot fungi use to degrade pollutants. *Environ. Sci. Technol.*, 28, 78A-87A.

▫   Benoit-Guyod, J.-L., Seigle-Murandi, F., Steiman, R. Sage, L. and Toé, A. 1994. Biodegradation of pentachlorophenol by micromycetes. III Deuteromycetes. *Env. Tox. Water Qual.*, 9, 33-44.

▫   Davies, H.A. and Marsh, J.A.P. 1980. Effects of Chlorpropham, Chlortoluron and Isoproturon on respiration and transformation of nitrogen in two soils. *Bull. Environ. Contam. Toxicol.*, 25, 706-712.

▫   De Hoog, G.S., Seigle-Murandi, F., Steiman, R. and Eriksson, K.-E.L. 1985. A new species of *Embellisia* from the North Sea. *Anton. Leeuwenhoek Int. J. Gen. M.*, 51, 409-413.

▫   Fournier, J.C., Soulas, G. et Catroux, G. 1975. Dégradation microbienne de l'isoproturon dans des modèles de laboratoire. *Chemosphere*, 4, 207-214.

▫   Galzy, P. et Slonimsky, P. 1957. Variations physiologiques de la levure au cours de la croissance sur l'acide lactique comme seule source de carbone. *CR Acad. Sc.*, 245D, 2423-2426.

▫   Gams, W., Steiman, R. and Seigle-Murandi, F. 1990. The Hyphomycete genus Goidanichiella. *Mycotaxon*, 38, 149-159.

▫   Haddad, G., Ponte-Freitas, A., Ravanel, P. and Tissut, M. 1992. Leaf penetration and foliar transfer of isoproturon, from a stirred medium to the D1-protein target. *Plant Physiol. Biochem.*, 30, 173-180.

▫   Kulsrestha, G. and Mukerjee, S.K. 1986. The photochemical decomposition of the herbicide isoproturon. *Pestic. Sci.*, 17, 489-494.

▫   Mudd, P.J., Greaves M.P. and Wright S.J.L. 1985. Effects of isoproturon in the rhizosphere of wheat. *Weed Res.*, 25, 423-432.

▫   Mudd, P.J., Hance, R.J. and Wright, S.J.L. 1983. The persistence and metabolism of isoproturon in soil. *Weed Res.*, 23, 239-246.

▫   Nirmal Kumar, J.L., 1991. Response of *Anabaena* sp. 310 to isoproturon. *J. Indian Bot. Soc.*, 70, 277-280.

▫   Ponte-Freitas, A., Haddad, G., Ravanel, P., and Tissut M. 1993. Penetration of isoproturon and inhibition of photosynthesis after droplet deposition on leaf fragments. *Pest. Biochem. Physiol.*, 45, 54-61.

□   Ponte-Freitas, A., Haddad, G., Tissut, M., and Ravanel, P. 1991. Distribution of isoproturon, a photosystem II inhibitor, inside wheat leaf fragments. *Plant Physiol. Biochem.*, 29, 67-74.

□   Rognon, J., Thizy, A., Poignant, P. et Pillon, D. 1972. Essais de desherbage des céréales avec la N-(isopropyl-4-phényl)-N'-N'-diméthylurée.*Medelingen Fakulteit Landbouwwetenschappen Gent 32*, n° 2, 663-669.

□   Seigle-Murandi, F., Nicot, J., Sorin, L. and Genest, L.C. 1980. Association mycologique dans la salle de la Verna et le tunnel de l'EDF du réseau de la Pierre Saint Martin. *Rev. Ecol. Biol. Sol*, 17, 149-157.

□   Seigle-Murandi, F., Nicot, J., Sorin, L. and Lacharme, J. 1981. Mycoflore des cerneaux de noix destinés à l'alimentation. *Cryptogamie, Mycologie*, 2, 217-237.

□   Seigle-Murandi, F., Steiman, R., Benoit-Guyod, J.-L. and Guiraud, P. 1993. Fungal degradation of pentachlorophenol by micromycetes. *J. Biotechnol.*, 30, 27-35.

□   Seigle-Murandi, F., Steiman, R., Benoit-Guyod, J.-L., Guiraud, P. 1992. Biodegradation of pentachlorophenol by micromycetes. I. Zygomycetes. *Env. Tox. Water Qual.*, 9, 33-44.

□   Steiman, R., Benoit-Guyod, J.-L., Seigle-Murandi, F., Sage, L. and Toé A. 1994. Biodegradation of pentachlorophenol by micromycetes. II. Ascomycetes, basidiomycetes ans yeasts. *Env. Tox. Water Qual.*, 9, 33-44.

□   Thizy, A., Pillon, D., Poignant, P. et Rognon, J. 1972. Activité herbicide sélective de la N-(isopropyl-4-phényl)-N', N'-diméthylurée. *C.R. Acad. Sc.*, 274 série D, 2053-2056.

3.2.3.    *"Microcosms" - a test system to determine the fate of pesticides in undisturbed soil columns*

# R. Schroll[1], G. Cao[2], A. Mora[3], T. Langenbach[4], I. Scheunert[1]

[1]    GSF-Institut für Bodenökologie, Ingolstädter Landstr.1,
       D-85758 Oberschleißheim/Neuherberg, Germany

[2]    Chinese Academy of Agricultural Sciences, Beijing, P.R.China

[3]    Instituto de Recursos Naturales y Agrobiologia de Sevilla, Spain

[4]    Instituto de Microbiologia, UFRJ, Rio de Janeiro, Brazil

## Abstract

Leaching, volatilization and biodegradation processes, as well as metabolism of [14]C-labeled chemicals, can be monitored in undisturbed soil columns with a new test-system called "microcosms". Up to now the new system has been established in laboratories in Germany, Brazil, China, Israel and Spain to study the behaviour of pesticides in soils of different climate zones. The first results obtained with the model compound terbuthylazine will be presented.

## Introduction

Chemization of agriculture can be observed all over the globe. But the behaviour of pesticides in different kinds of soils is not well understood. The application of results obtained in temperate soils to tropical areas has been proved unsatisfactory. Subtropical and tropical soils differ from temperate soils mainly by the quality of clay minerals, the amount and quality of iron oxides and organic matter. Moreover, one can expect that the types of microbial communities are different from those in temperate soils. All these factors, combined with different soil moisture and temperature regimes, may influence the rates of volatilization, degradation, accumulation and transport of pesticides.

Therefore, it became necessary to study the fate of [14]C-labeled pesticides on a large scale in different soils in countries from different climate zones under identical conditions, in order to guarantee a comparability of the results and therefore to determine the influence of different types of soils on the fate of pesticides.

As it is known from previous experiments (Schroll et al., 1992a) the transformation of results obtained with existing laboratory test systems to the field situation is very difficult. Therefore a new system for testing [14]C-labeled pesticides had to be designed which should combine experiences from laboratory model systems (Schroll and Scheunert, 1992) with the advantages of lysimeter experiments (Schroll et al., 1992b).

Natural soils possess a defined structur with a porous system. But this structure disappears when the soils are homogenized. However, many tests, such as sorption- and desorption tests for pesticides, are carried out in disturbed soils. In these cases, the determined maximal adsorption capacity is not always active in a natural soil. In contrast, soil homogenization activates the soil microorganisms, possibly leading to a faster degradation of pesticides than in a soil with the native structure. In the new test-system the soils are be tested with their natural structur undisturbed.

Another demand on the new system is its economical feasibility as well as a fast and uncomplicated establishment in laboratories in different countries. Precipitation, fluctuation of soil humidity as well as an intensive gas exchange above the soil surface and a simulation of a groundwater level should be included in the system to simulate natural conditions.

In contrast to the systems of Nash (1983) and Branham (1985), no plants should be included in the test system. In the microagro-ecosystems of both authors it is impossible to distinguish between the volatilization of pesticides from plants and soil. Also it is not possible to prevent the uptake of $^{14}CO_2$ - formed by the total degradation of the $^{14}C$-labeled pesticide in soil - by plants; this effect would lead to an underestimation of the biodegradation of the pesticides in soils. On the other hand, plants themselves could probably influence the fate of pesticides in soils and thus obscure the influence of the soil.

## Material and Methods

### *Test System "Microcosms"*

The new test-system (Figure 1) itself consists of a round metal cylinder with a height of 30cm and a diameter of 15cm. At the bottom of the cylinder a filter paper and a metal sieve are placed to keep the soil in the cylinder. A defined underpressure caused by a membrane pump - necessary to subdue the capillary forces - at the bottom of the sieve guarantees the leaching of the perculate out of the soil. The leachate is collected in 1 l glass bottles.

The metal cylinder is closed with a special cover which allows precipitation as well as an intensive gas exchange. 14 small specially formed openings in the bottom of the microcosms cover guarantee a drop application of the precipitation water. A ventilator placed in the middle of the microcosm cover causes a wind speed of about 1m/sec over the soil surface. A membrane pump draws the air through the cover, resulting in an air exchange of 23 l/h leading to an air exchange rate of more than 60 times per hour in the microcosm atmosphere. Both the wind speed and the air exchange rate were selected according to the German BBA - guideline for volatilization tests.

Between the microcosm cover and the membrane pump several traps are placed to adsorb volatile $^{14}C$-labeled organic compounds (parent compound and metabolites) and $^{14}CO_2$ resulting from total degradation of the pesticides separately. A new special equipment on the top of the traps guarantees that volatile organic solvents such as Ethyleneglycol-monomethylether, which is added to the first two traps to fix the volatile organic compounds, do not evaporate from the traps. To fix $^{14}CO_2$ the traps Nr. 3 and 4 are filled with a mixture of Diethyleneglycolmonobuthylether and Ethanolamine.

In front of the inlet of the microcosm cover an Erlenmeyer flask filled with a saturated salt solution is placed to achieve a defined air humidity for the air drawn into the microcosms. To guarantee constant temperature the microcosms can be placed in refrigerators.

### *Testprocedure*

The 30cm long metall cylinder is pressed carefully into the soil surface to a depth of 28cm, resulting in a free space of 2cm above the soil in the column (=atmosphere of the micro-cosm), and then removed carefully. Before the application of the $^{14}C$-labeled pesticide (the pesticide was dissolved in 10ml water), precipitation, according to a uniformly distributed average rainfall of 600mm per year, is simulated for 2 weeks (3 times a week) in order to condition the soil column, followed by a week of no precipitation in order to let the soil surface become dry again

according to a real field situation. After the application of the $^{14}C$-labeled terbuthylazine (10 uCi per test) in an amount of 980 g/ha and another week without precipitation, rainfall was simulated again three times a week.

Aliquots of the cocktail for the volatilization and the cocktail for the $^{14}CO_2$ were taken about 3 times a week, mixed with the scintillation cocktails (Ultima Gold XR and Permablend) and measured in a scintillation counter. The main amount of the volatilization cocktail was stored for every week and then washed with water and dichlormethane. The organic phase was concentrated to dryness, resolved in a mixture of methanol and water (1:1; v:v) and injected to the HPLC to identify the volatilized $^{14}C$-labeled organic compounds.

Radioactivity in the leachate was determined every week. At the end of the test period, the soil was pressed out of the metal cylinder with special equipment and cut into several soil layers. The radioactivity in the different soil layers was first determined by combustion of soil aliquots and then the soil layers were extracted separately to determine the amount of extractable $^{14}C$-labeled parent compound and metabolites.

## Results

The types of soils used in the first step of the international network up to now are shown in table 1: With the exception of the Brazilian soil the soils used were mainly sandy soils. First results for biodegradation, transport and volatilization are shown from figure 2 to 4. In all soils typical patterns of mineralization could be observed. As expected leaching of terbuthylazine was very poor in all soils used. The main amount of residues could be detected in the first soil layer at the end of the test periods. According to the application method (Lembrich, 1994) the highest volatilization rates could not be observed on the day of application but three days after application. As the volatilization of pesticides from wet sandy soils is higher than from dry soils (Dörfler, 1991) the fluctuation of the volatilization rates can be explained by the change of the water content in the soils after simulated precipitation. But in general the volatilization decreases in an exponential form in all soils.

|                | German soil | Brazilian soil | Chinese soil | Spanish soil |
|----------------|-------------|----------------|--------------|--------------|
| sand %         | 85          | 58             | 63           | 88           |
| silt %         | 10          | 9              | 21           | 9            |
| clay %         | 5           | 33             | 16           | 3            |
| pH-value       | 4.9         | 4.3            | 8.5          | 5.6          |
| % org. matter  | 1.8         | 2.9            | 1.7          | 1.5          |

**Table 1 :** Characterization of the soils used

## *Conclusion*

The microcosm test system for $^{14}C$-labeled pesticides can give a large amount of information about the fate of pesticides in undisturbed soil columns. But more tests have to be done on different types of soils with different types of pesticides and also under the climatic conditions of the native environment to create a sufficient data base for predicting the behaviour of pesticides in soils of different climate zones.

# Literature

□ R. Schroll, T. Langenbach, G. Cao, U. Dörfler, P. Schneider, I. Scheunert: "Fate of $^{14}$C-terbuthylazine in soil-plant systems", Sci. Total Environ. 123/124, (1992a), pp 377-389 R. Schroll, I. Scheunert: "A laboratory system to determine separately the uptake of organic chemicals from soil by plant roots and by plant leaves", *Chemosphere* 24/1, (1992), pp 97-108

□ R. Schroll, U. Dörfler, P. Schneider, I. Scheunert: "A lysimeter system to determine leaching and volatilization of $^{14}$C-labeled compounds", *BCPC Mono. 53, Lysimeter studies of pesticides in the soil* (1992b)

□ R.G. Nash: "Determining environmental fate of pesticides with microagroecosystems" *Res. Rev.* 85, (1985), pp 199-215

□ B.E. Branham, D.J. Wehner, W.A. Torello, A.J. Turgeon: "A microecosystem for fertilizer and pesticide fate research", *Agron. J.* 77, (1985), pp 176-180

□ D. Lembrich: unpublished results (1994)

□ U. Dörfler, R. Adler-Köhler, P. Schneider, I. Scheunert, F. Korte: "A laboratory model system for determining the volatility of pesticides from soil and plant surfaces", *Chemosphere* 23/4, (1991), pp 485-496

# Microcosm

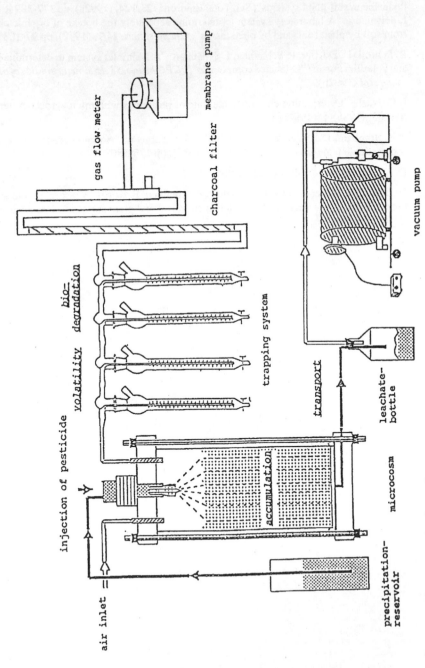

Fig. 1: Microcosm test system

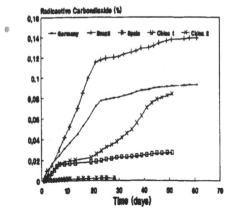

Fig. 2: Radioactive Carbondioxide in
% of the applied amount

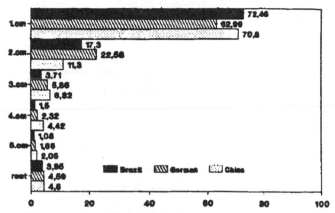

Fig.3: Distribution of radioactivity in the
different soil layers

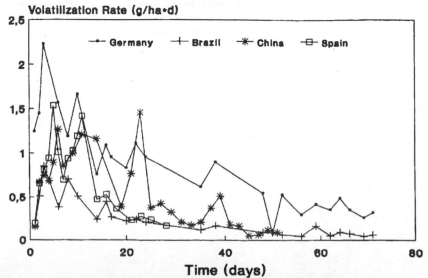

Time (days)
Fig.4: Volatilization rates for terbuthylazine

## 3.3.        Poster Presentations

### 3.3.1.        *Transformation mechanism of a sulfonylurea herbicide :*
*the thifensulfuron methyl*

## J. Bastide, J-P. Cambon and D. Vega

GERAP URA CNRS 461
52 Av de Villeneuve 66860 Perpignan France

The understanding of the processes governing the behaviour of agrochemicals in soils requires the knowledge of the degradation mechanisms. The relative role of microorganisms and chemical hydrolysis in the degradation of a new sulfonylurea (thifensulfuron methyl) was estimated. The rates of transformation thifensulfuron methyl were measured (Table 1) in sterilized and no sterilized soils and in solutions.

| Half-lives (hours) | | | | | |
|---|---|---|---|---|---|
| Soil pH | No sterile soil | Sterilized soil | Solution pH | Solution a | Solution b |
| 6.3 | 38.5 | 680 | 6 | 1950 | |
| 7.8 | 39.6 | 495 | 8 | | 739 |

**Table 1** : Half-lives of thifensulfuron methyl in soils and solutions

The thifensulfuron methyl stability in aqueous solution was markedly influenced by this pH . In no sterilized soil the first degradation step was rapid. In sterilized soil the stability of thifensulfuron methyl was higher. Since the degradation rate in sterilized soil and in solution was very low , a biological factor was certainly responsible for soil degradation.

Different experiments were carried out to isolate microorganisms from the soil : enrichment procedures or soil percolation were unsuccessfull.

The degradation rates of thifensulfuron methyl was depending on thifensulfuron methyl concentration and soil moisture. In soil (at 28°C and 25% of humidity) and solution the first step of degradation always produced thifensulfuron by hydrolysis of ester function. The biological factor giving the hydrolysis of ester function may be an extracellular esterase.

In solution the chemical hydrolysis gave the same pathway than the other sulfonylurea herbicides : cleavage of sulfonylurea bridge to give product (II) and hydrolysis of methoxy group of triazine ring giving product (III). This product gives product (IV) by opening of the triazine ring.

The different hydrolysis products were searched on soils treated with thifensulfuron methyl. Different extraction methods were developped to extract all transformation products. The first method methanol- water-acetic acid allows the extraction of product I (74%) ; II (78%) ; III (7%).

The second method allows the extraction of II (90 %) ; III (95 %) . But during the extraction the product I was transformed into product III (yield 100).

Using the two extraction methods the amount of different products in soil may be measured : Products I ; II and III were found in soil after 48 days of incubation with thifensulfuron methyl. The thifensulfuron pathway was fair related to a chemical degradation.

3.3.2.        *Mobility of Thiazafluron as Related to Sorption and Porosity*
              *Properties of Soils*

# L. Cox, R. Celis, M.C. Hermosín and J. Cornejo

Instituto de Recursos Naturales y Agrobiología de Sevilla,
C.S.I.C. Apartado 1052, 41080 Sevilla, Spain.

The interrelationships between the leaching patterns of the herbicide thiazafluron (1,3-dimethyl-1-(5-trifluoromethyl-1,3,4-thiadiazol-2-yl)urea) in three soils of Southern Spain and the sorption parameters of this herbicide and pore size distributions of these soils has been investigated. Adsorption data were fitted to Freundlich equation. Mobility has been studied by leaching in handpacked soil columns under saturated-unsaturated flow conditions and the retardation factor Rf calculated as the pore volumes required for leaching 50 % of the applied herbicide. Rf values and Freundlich constants Kf for thiazafluron and the three soils studied correlated directly. The different pore size distribution of the soils, studied by mercury porosimetry, contributed to the different leaching pattern and to the extent of adsorption processes.

## Introduction

The problem of pesticide contamination is of great importance due to the increasing number of chemicals used in agriculture. The study of pesticide adsorption-desorption processes and movement throughout the soil profile contributes to the knowledge of the fate of these organic compounds in soil and aquatic environments (1). These processes, in general, have been shown to be inversely related (2).

Thiazafluron is an urea derivative herbicide and adsorbs on soils in moderate amounts. Previous studies indicated that soil clay content is a very important soil property affecting its adsorption (3,4). The aim of this study was to compare the leaching patterns of this herbicide in different soils, paying special attention to the influence of the different adsorption capacity and pore size distribution of the soils.

## Materials and Methods

The herbicide thiazafluron used in this study was the high purity compound supplied by Ciba-Geigy. This herbicide is a crystaline solid of melting point 136-137°C, v.p. 267 $\mu$Pa at 20°C and water solubility at 20°C 2.1 g/Kg (5). The soils used in this study were sampled, air dried, sieved to pass a 2 mm mesh and stored in a refrigerator. Physico- chemical properties are given in Table 1.

The distribution of pore radii of soil samples from $4 \ 10^4$ to 3.7 nm was determined using a Carlo Erba 2000 mercury depression and intrusion porosimeter. 0.25 g of each soil sample were previously heated at 90°C during 24 h and then outgassed at room temperature for 30 min. A value for the surface tension of mercury of 0.48 $Nm^{-1}$ and a contact angle on soils of 141.3° was used with the Laplace equation assuming cylindrical pores in the calculations.

| Soil property | Soil 1 | Soil 2 | Soil 3 |
|---|---|---|---|
| % Organic Matter | 1.24 | 2.50 | 2.05 |
| pH | 7.9 | 7.7 | 7.6 |
| % Sand | 70.7 | 12.1 | 11.9 |
| % Silt | 8.9 | 43.6 | 35.8 |
| % Clay | 20.4 | 44.3 | 52.3 |

**Table 1** : Physicochemical properties of the soils.

Adsorption studies were performed using duplicate samples of 2.5 g of each soil sample treated with 10 ml of thiazafluron initial solution concentrations (Ci) 0.05, 0.1, 0.3, 0.5, 0.8, 1, 1.5 and 1.8 mM in 0.01 M $CaCl_2$. The suspensions were shaken at $20 \pm 2°$ C for 24 hours, centrifuged and equilibrium concentration (Ce) determined in the supernatant by UV spectroscopy (absorption maximum, 266 nm). Differences between Ci and Ce were assumed to be adsorbed. Desorption was measured after adsorption using the 1 mM initial solution concentration by three successive dilutions with 0.01M $CaCl_2$.

Soil columns were made up with six 5 cm long pieces of 5 cm inner diameter sealed with silicon. The first ring was filled with sea sand and the last ring was filled with sea sand and glass wool in order to avoid the drainage of soil particles into the leachates. Columns were handpacked with 120 g of soil per ring, in the case of the heavier soils (soil 1 and 3) and with 130 g/ring (soil 2). Leaching experiments were run in triplicates, and a blank column (without herbicide) was used for each soil. Columns were conditioned with 250 ml of 0.01 M $CaCl_2$, allowed to drain for 24 hours, and the amount of thiazafluron corresponding to the maximum application rate in soils (12 Kg/Ha) was applied to the top of the columns. The total amount of $CaCl_2$ applied was 2,5 L, 25 ml daily during the first 10 days and 50 ml daily until the end of the leaching experiment. Leachates were collected daily, filtered through 0.2 μm diameter Dynagard filters and analyzed by HPLC under the following conditions: Nova-Pack column of 150 mm length x 3.9 mm i.d.; column packing, C18; flow rate, 1 ml/min; eluent system, 60: 40 water- methanol mixture.

## Results and Discussion

Table 2 gives the adsorption coefficients for thiazafluron in the soils studied obtained by fitting adsorption isotherms to Freundlich equation. Desorption percentages after three successive dilutions are also given in this Table. As shown in Table 2 and confirmed in previous studies on the adsorption-desorption of thiazafluron by 20 soil samples including soils 1, 2 and 3 (4), $K_f$ values increased when the clay content of the soils increased. This is mainly due to the high water solubility of thiazafluron, since high polar solutes are more likely to adsorb on mineral sorbents than on organic matter. Desorption percentages shown in Table 2 indicate that only a small amount of the herbicide adsorbed is recovered after three desorption cycles. This has been attributed mainly through irreversible binding of thiazafluron molecules to soil clay (4).

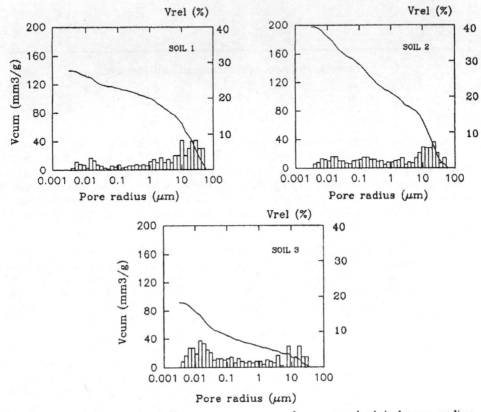

Figure 1. Cumulative and relative mercury pore volume vs calculated pore radius

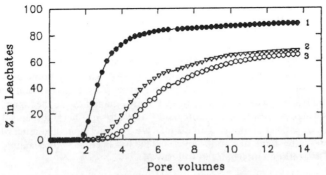

Figure 2. Thiazafluron cumulative BTCs

| Soil | $K_f$ | $n_f$ | r | % D | Rf |
|------|-------|-------|------|------|-----|
| 1 | 0.41 | 0.73 | 0.97 | 13.7 | 4.7 |
| 2 | 1.00 | 0.76 | 1.00 | 14.7 | 6.3 |
| 3 | 1.33 | 0.80 | 1.00 | 14.2 | 7.9 |

**Table 2 :** Thiazafluron adsorption coefficients, Kf (mmol/Kg) and nf, calculated from Freundlich equation. Desorption percentages (% D) calculated at 1 mM thiazafluron initial solution concentration. Retention factors Rf, calculated as the pore volumes required to leach 50 % of the applied herbicide.

From Figure 1, it can be seen that the pore size distributions for the three soils studied were significantly different. The volume of greater pores ($r > 1$ µm) for soils 1 and 2 are nearly the same but soil 1 has a lower volume of smaller pores ($r < 1$ µm). Pore size distribution of soil 3 is characterized by the presence of pores only in the lower limit of pore radius studied (about 300 Å). Thus, the mean pore radius, $r_m$, for the three soil samples decreases in the order $r_m(3) < r_m(2) < r_m(1)$.

Thiazafluron BTCs for the three soils studied are given in Figure 2. As shown in this figure, thiazafluron moves more rapidly in soil 1 than in the other soils. The retention factors Rf, calculated as the pore volumes required to leach 50 % of the applied herbicide (6), are given in Table 2 and correlated directly with adsorption coefficients Kf, indicating that adsorption and mobility of thiazafluron in soils are inversely related. The smaller $r_m$ values observed in soils 2 and 3 would give rise to slower leaching of the herbicide in these soils, which would also contribute to greater extent of adsorption. This slower leaching could also result in an extended degradation of the herbicide, which could also explain the lower recovery percentages found for these soils.

## Acknowledgements

This project has been supported by CICYT through research Project AM93-97 and by Junta de Andalucía through Research Group No. 4092.

## References

□    1. Beck, A. J.; Johnston, A.E.J. and Jones, K.C. (1993). Movement of nonionic organic chemicals in agricultural soils. *Critical Reviews in Environmental Science and Technology* 23(3): 219-248.

□    2. Bilkert, J.N. and Rao, P.S.C. (1985). Sorption and leaching of three non-fumigant nematicides in soils. *J. Environ. Sci. Health* B20 (1): 1-26.

□    3. Cox, L.; Hermosín, M.C. and Cornejo, J. (1994). Retention of thiazafluron by surface horizons of some Spanish soils. *Fresenius Environ. Bull.* 3: 129-134.

□    4. Cox, L.; Hermosín, M.C. and J. Cornejo (1994). Adsorption and desortion of thiazafluron by soils. *International Journal of Environmental Analytical Chemistry* (in press).

□    5. Worthing and Hance, 1991. The Pesticide Manual. BCPC, Surrey, U.K..

□    6. Brusseau, M.L. and Rao, P.S.C. (1989). Sorption nonideality during organic contaminant transport in porous media. *Critical Reviews in Environmental Control* 19(1): 33-99.

### 3.3.3.     Phototransformation of Metamitron in the Presence of Soils

# L. Cox[*], M.C. Hermosín[*], J. Cornejo[*] and M. Mansour[**]

[*]     Instituto de Recursos Naturales y Agrobiología de Sevilla,
        C.S.I.C. Aptdo. 1052, 41080 Sevilla, Spain.

[**]    G.S.F.-Forschungszentrum für Umwelt und Gesundheit. Institut für
        Ökologische Chemie. Schulstraße 10, D-85356 Freising-Attaching, Germany.

The herbicide metamitron (4-amino-3-methyl-6-phenyl-1,2,4- triazin-5(4H)-one) photodegradates rapidly in aqueous solution leading to the formation of a main degradation product, the deaminated compound desaminometamitron. The influence of soils in this process has been investigated. Photodegradation of metamitron was slower in the presence of soils, although some differences were observed between the different soil samples studied which indicate that iron oxides of the soils may play an important role in the photolysis of metamitron in soil-water suspensions.

## Introduction

Photodegradation processes are involved in dissipation of pesticides in water, soils and plants. Therefore, the study of these abiotic transformations makes an appreciable contribution in determining the final fate of these xenobiotics, specially in aquatic environments. The pesticide metamitron is an herbicide of pronnounced selectivity due to specific detoxification reactions at the 4-amino group that occur in tolerant plants (SCHMIDT and FEDTKE, 1977). This herbicide is degraded in soils primarily through the activities of soil microorganisms (WALKER and BROWN, 1981), and this degradation gives the deaminated and less bioactive compound desaminometamitron. Metamitron and other 4-amino-1,2,4-triazin-5-one herbicides have been shown to photodegradate in aqueous solution, leading to the formation of the respective deaminated compounds as main degradation products (ROSEN and SIEWERSKI,1971; PAPE and ZABIK, 1972).

The relevance of photodegradation processes in the fate of this soil applied herbicide is due to the high water solubility and moderate persistence of metamitron (WORTHING and HANCE, 1991), which indicate that it might be found as an environmental contaminant in agriculture runoff waters, were photolysis processes play an important role. Thus, the aim of this study was to assess the influence of soils on the phototransformation of the herbicide metamitron in aqueous medium.

## Materials and Methods

Metamitron is a crystalline solid of m.p. 166.6°C, v.p. 13 mPa up to 70°C and solubility at 20°C 1.8 g/L water. The metamitron used in this study was the high purity compound obtained from BAYER AG.

The soils selected for this study correspond to an intensively cultivated area located in Southern Spain. Soils were sampled, air dried, sieved to pass a 2 mm mesh and stored in a refrigerator. Their physicochemical properties were determined by the usual methodology and the clay mineralogy by X-Ray diffraction on oriented specimen (HERMOSIN et al., 1987) and are given in Table 1.

Photodegradation experiments were performed in a SUNTEST photoreactor from Heraeus equipped with a xenon lamp and a permanent filter selecting wavelenght > 290 nm in order to simulate the emission spectrum of sunlight. 250 ml of 50 ppm metamitron aqueous solution and 50 g of each soil were irradiated in quartz flasks for 3.5 hours. In every case, dark control suspensions of each soil were used.

The decrease in metamitron concentration during photolysis experiments and the evolution of the main photoproduct was determined by HPLC under the following conditions: ODS Hypersil 5 μm (200 x 4.6 mm) column; eluent system, 80/20 water- acetonitrile mixture at flow rate 1 ml/min; UV detection at 306 nm. Mass Spectrometry has been used to iddentify the main degradation product.

| Soil | pH | %O.M. | %Clay | %Illite | %Smectite | %Kaolinite | %Fe₂O₃ | C.E.C |
|------|-----|-------|-------|---------|-----------|------------|--------|-------|
| 1 | 7.7 | 2.24 | 44.3 | 29.2 | 8.0 | 7.1 | 1.2 | 19.2 |
| 2 | 7.9 | 0.99 | 20.4 | 12.4 | 3.5 | 4.5 | 1.4 | 9.0 |
| 3 | 7.6 | 2.54 | 52.3 | 18.8 | 17.8 | 15.7 | 0.7 | 28.5 |
| 4 | 5.9 | 0.49 | 20.0 | 17.0 | 1.0 | 2.0 | 8.5 | 2.8 |

**Table 1** : Physicochemical properties of the soils

## Results and Discussion

Figure 1 shows the evolution of 50 ppm metamitron aqueous solution with time under irradiation with the xenon lamp of the photoreactor, the evolution of metamitron in the dark control under the same conditions and the evolution of a main degradation product, which has been iddentified in the literature (ROSEN and SIEWERSKI,1971; PAPE and ZABIK, 1972) as desaminometamitron (DAM) and confirmed by us by mass spectrometry. As it can be seen, metamitron is rapidly degraded in the presence of light and the decrease in the solution concentration gives rise to an increase in the evolution of DAM. It should be noticed as well that DAM is much more stable than the parent compound. The UV absorption maximum of metamitron (306 nm) falls between the range of the emission spectra of sunlight, indicating that direct photolysis of the compound is occurring.

Figure 2 shows the evolution of 50 ppm metamitron in the presence of the different soils studied when irradiated under the same conditions, together with the evolution of the 50 ppm aqueous solution. Curves followed first order kinetics and the decrease in metamitron concentration also gave rise to an increase in DAM. As shown in this figure, the presence of soils protects Metamitron from photodegradation, due to the "screen" or light attenuation effect of soil particles (MILLER and ZEPP, 1979). Eventhough, some differences in metamitron photolysis rate were found between the different soils. Dark controls did not show significant differences between the soils, indicating that these differences are mainly due to differences in the photolysis rate and not to differences in the extent of adsorption and/or chemical degradation processes. These differences could be atributed to indirect photolysis, due to absorption of light energy by soil components and transmission to metamitron molecules and/or to the formation of reactive species that enter into a chemical reaction with the compound. The slowest photolysis rate was found in the presence of soil 2, of low organic matter and clay content. The evolution of metamitron was similar in the presence of soils 1 and 3, of similar organic matter and clay content but different clay mineralogy. The different photolysis rate in the presence of soils 1, 2 and 3 suggests that clay minerals and organic matter may act as photosensitizers in the photolysis of metamitron. The higher photolysis rate in the presence of soil 4, of high iron oxide content, seems to indicate that iron oxides may play an important role in this process. Iron oxides may act as photosensitizers of the photodegradation of metamitron, and this effect seems to be higher than the corresponding to clay minerals and organic matter. This indirect photolysis could be caused by active oxygen species which are supposed to be formed from iron oxides when exposed to ultraviolet light (CUNNINGHAM et al., 1988; SUN and PIGNATELLO, 1993).

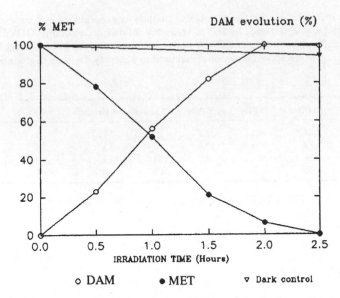

**Figure 1 :** Evolution of metamitron (MET) and desaminometamitron
(DAM) after 2.5 hours of irradiation time.
Evolution of MET in the dark control under the same conditions.

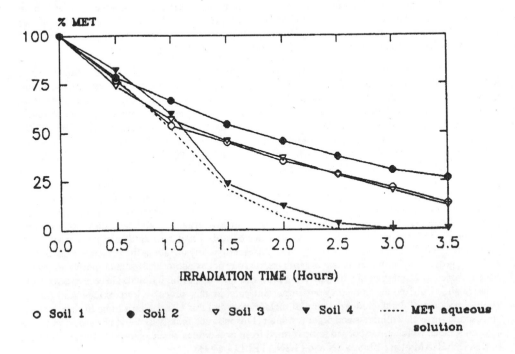

**Figure 2 :** Evolution of metamitron in soil suspensions as a function of irradiation time.

# Acknowledgements

This project has been supported by CICYT through Project AM93-81 and through "Acción Integrada" HA93-97.

# References

□ 1. Cunningham, K.M.; Goldberg, M.C. and Weiner, E.R. 1988. Environ. Sci. Technol. 22: 1090.

□ 2. Hermosín, M.C.; Cornejo, J. and Pérez-Rodríguez, J.L. 1987. Soil Sci. 144: 250.

□ 3. Miller, G.C. and Zepp, R.G. 1979. Water Res. 13: 453.

□ 4. Pape, B.E. and Zabik, M.J. 1972. J. Agric. Food Chem. 20: 72.

□ 5. Rosen, J.D. and Siewerski, M. 1971. Bulletin of Environmental Contamination and Toxicology 6(5): 406.

□ 6. Schmidt, R.R. and Fedtke, C. 1977. Pesticide Science 8: 611.

□ 7. Sun, Y. and Pignatello, J.J. 1993. Environ. Sci. Technol. 27: 304.

□ 8. Walker, A. and Brown, P.A. 1981. Proc. EWRS Symp. Theory and Practice of the Use of Soil Applied Herbicides, 1981.

□ 9. Worthing, C.R. and Hance, R.J., (1991). The Pesticide Manual. BCPC, Surrey, U.K..

3.3.4.     *Studies on the Degradation of [$^{14}$C]isoproturon in a Subsurface Soil under Different Conditions*

# H. Ellßel, R. Kubiak & K.W. Eichhorn

Staatliche Lehr- und Forschungsanstalt, FB Phytomedizin,
Breitenweg 71, 67435 Neustadt/Wstr., Germany

## Summary

A laboratory experiment was conducted to study the influence of climatic conditions, $O_2$ and $CO_2$-concentration, and A.I. concentration on the degradation of [$^{14}$C]isoproturon in a subsoil sample.

Degradation rate was enhanced when:

a)     temperature was enhanced

b)     $CO_2$-concentration was enhanced

c)     A.I. concentration was reduced

## Introduction

In recent years, interest in the determination of the degradation of pesticides in the subsurface environment has increased. Several experiments have shown the possibility of degradation in the unsaturated zone. In most cases, degradation is much smaller in subsurface soil than in surface soil. The question arises as to which are the important factors influencing degradation processes in the deeper soil layers. A laboratory experiment was conducted to investigate the degradation of [$^{14}$C]iso-proturon in subsoil material, taken from a depth of 75 - 85 cm, under subsoil conditions. These subsoil conditions were investigated over a long period measuring temperature, moisture, and content of $O_2$ and $CO_2$ in the soil air (Fig. 1a, b, c). Mean values of those results (a "summer scenario" and a "winter scenario") were used for the laboratory experiment reported here.

## Material and Methods

The soil used was a loamy sand (72.1 % sand, 19.9 % silt and 8.0 % clay) having less than 0.1 % organic carbon. [phenylring-U-$^{14}$C]isoproturon (Fig. 2) with a specific radioactivity of 2.7 MBq/mg was used, formulated as Arelon$^R$. Two concentrations of isoproturon (2.00 and 0.02 mg A.I./kg dry soil) were mixed into the subsoil material. For incubation, 300 ml Erlenmeyer flasks and the adsorption system described by ANDERSON (1975) were used containing 100 g of soil.

In order to investigate the effect of $CO_2$ on the degradation rate, 0 % and 2 % were maintained in the incubation flasks during the 100 day-experiment period (Fig. 3a, b). The variants listed in Table 1 were used in two replicates.

The formation of $^{14}CO_2$ was investigated after 2, 4, 8, 16, 32, 64, and 100 days using the method described by ANDERSON (1975).

The radioactivity in the soil was investigated after 100 days. For this purpose, the soil was extracted three times with methanol/water (80/20, vol/vol), the solutions were combined and the methanol was removed by evaporation. The remaining aqueous phase was extracted three times with 100 ml dichloromethane, both phases were separated and the organic phase was concentrated to approx. 1 ml. These extracts were used for TLC (silica gel F254, chloroform/ethyl acetate (1/2, vol/vol)) and chromatograms were recorded using a Linear Analyzer (Berthold). Determination was focused on the A.I. and the main metabolite N-methyl-N'-(4-isopropylphenyl)-urea (HOE 064145). Determination limit was 1 µg/kg. Radioactivity in liquid phases was determined by liquid scintillation counting (LKB) using suitable scintillators. Radioactivity in the soil was determined the same way after combustion (Packard Oxidizer).

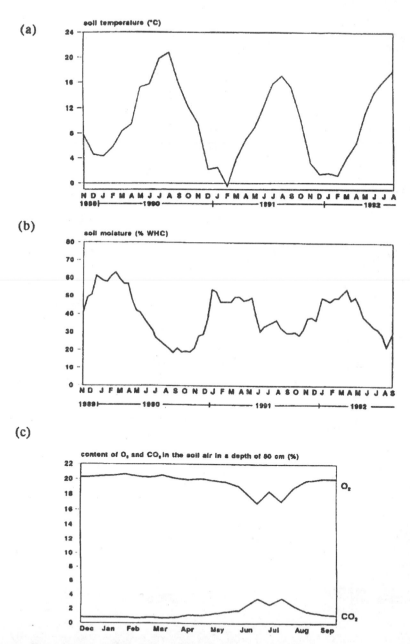

**Figure 1 :** Temperature (a), moisture (related to the water holding capacity of 24.9 %) (b), and content of $O_2$ and $CO_2$ in the soil air (c) in a depth of 80 cm

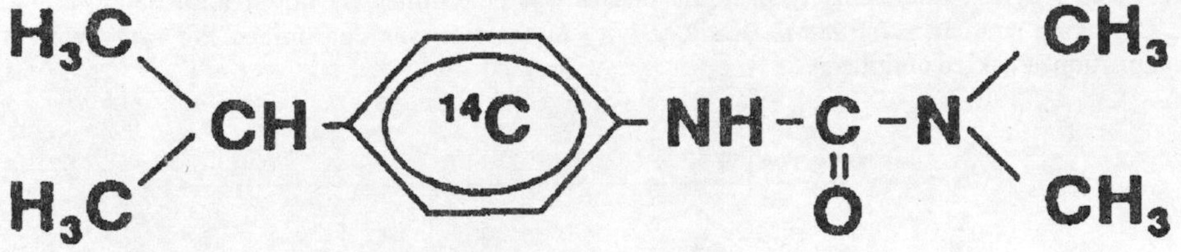

**Figure 2 :** Structural formula of isoproturon and labelling position ($^{14}$C)

| temp./moisture | "summer" 15°C/33 % WHC | | | | "winter" 5°C/44 % WHC | | | |
|---|---|---|---|---|---|---|---|---|
| CO$_2$ content of the incubation air | 0 % | | 2 % | | 0 % | | 2 % | |
| amount of A.I. added to the soil[1] | A | B | A | B | A | B | A | B |

[1]    A = 2.00 mg a.i./kg
      B = 0.02 mg a.i./kg

**Table 1 :** Variants of the experiment

(a)                                    (b)

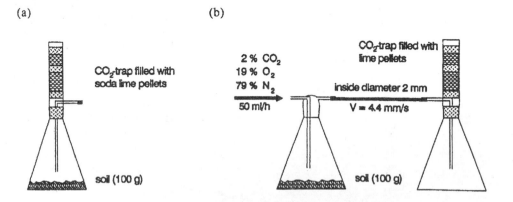

**Figure 3 :** Incubation system for 0 % CO$_2$ (a) and 2 % CO$_2$ (b)

## Results and Discussion

The mineralization of the [phenylring-U-$^{14}$C] labelling position was enhanced by a higher temperature (15°C, "summer scenario") compared with the "winter scenario" (5°C) for both isoproturon concentrations. Since it is well known that isoproturon is degraded by soil microorganisms (SOULAS & REUDET, 1977), this result was due to enhanced microbial activity with rising temperature (Fig. 4).

These findings were confirmed by the soil investigation at the end of the experiment. As expected, isoproturon degraded faster under summer than under winter conditions. However, in spite of the low temperature, less than 55 % of the isoproturon applied remained at the end of the experiment. An interesting result was that, at the higher concentration (2.00 mg/kg), isoproturon degradation in soil was slower as compared with the lower concentration of 0.02 mg/kg. Unexpectedly, degradation of isoproturon was accelerated by additional $CO_2$ in the atmosphere of the vessel (Fig. 5).

Since it is known that the atmosphere in such an experimental system will become free of $CO_2$ a few hours after closing it with the glass tube filled with lime pellets (Fig. 3a), the degradation capacity may be limited by this experimental factor. The results of isoproturon degradation, as influenced by temperature, isoproturon concentration, or $CO_2$-content, indicated that, depending on the experimental conditions a wide range of results are possible. Since it is known that the $CO_2$ content in deeper soil layers may be 2 % or more under field conditions (Fig. 1c), this laboratory variant reflected more the real outdoor conditions than the 0 %-variant. The results obtained here should be taken into account when results of pesticide degradation in deeper soil layers are necessary for the mathematical simulation of leaching processes in soils.

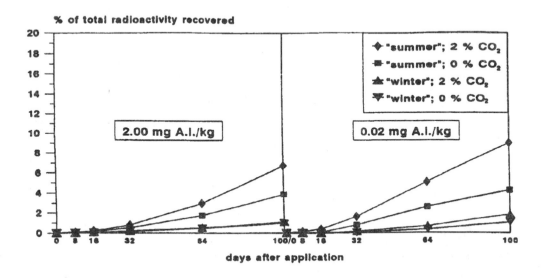

**Figure 4 :** $^{14}CO_2$ production after application of [$^{14}$C]isoproturon to the subsoil

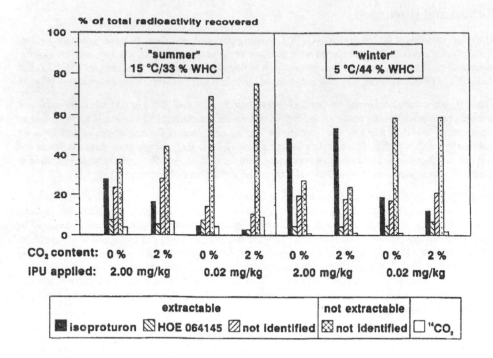

**Figure 5 :** Radioactive balance 100 days after application of
[$^{14}$C]isoproturon to the subsoil

## Acknowledgements

The authors wish to thank the AgrEvo Co. for providing $^{14}$C-labelled isoproturon.

## Literature

□    ANDERSON, J.P.E., Can. Microbiol. 21: 314, 1975

□    SOULAS, G & M.A. REUDET, Z Pflanzenkrankh. Pflanzensch. 8: 227, 1977

*3.3.5.*   *Collaborative Study on the Effect of Ethylene Oxide Sterilization on Soil Components*

# M.Gennari[1], M. Nègre[1], C. Gessa[2], A. Pusino[3], C. Crecchio[4] & P. Ruggiero[4]

[1]   DI.VA.P.R.A. Chimica Agraria, Università di Torino, Via P. Giuria 15, 10126 Torino, Italy

[2]   Istituto di Chimica Agraria, Università di Bologna, Via Berti Pichat 10, 40126 Bologna, Italy

[3]   Dipartimento di Agrochimica e Agrobiologia, Università di Reggio Calabria, Piazza S. Francesco di Sales 4, 89061 Gallina (RC), Italy

[4]   Istituto di Chimica Agraria, Università di Bari, Via Amendola 165/a, 70126 Bari, Italy

## Introduction

The tests for biodegradability of pesticides are usually carried out by comparing dissipation curves in sterile and non sterile soils. None of the methods currently in use for soil sterilization (thermal, electromagnetic or chemical treatments) is completly free from drawbacks and a good sterilization is usually accompanied by modifications of the physico-chemical soil properties. The sterilization of soil with epoxides such as ethylene and propylene oxides has been considered by several authors (Clark, 1950; Allison, 1951; Skipper and Westerman, 1973; Lopes and Wollum, 1976; Dao et al., 1982; Wolf et al., 1989). The general observation is that the main variations of the soil properties induced by sterilization are an increase of the pH and of the organic matter content of the soil.

In this paper, we present a synthesis of a study we carried out to obtain more information on the effect of ethylene oxide sterilization on soils. In particular, we studied the interaction of ethylene oxide with montmorillonite and humic acids. In order to verify if chemical modifications of soil constituents induced by sterilization had effects on adsorption capacity of the soil, the adsorption of a herbicide was tested. Acifluorfen was considered appropriate for this study because previous reports ( Ruggiero et al., 1992; Pusino et al., 1992; Gennari and Nègre, 1990) have pointed out its high capacity of adsorption on soil organic matter.

## Materials and Methods

The clay was a Montmorillonite 25 (bentonite) from Upton (Wyoming) supplied by Ward's Natural Science Establishment, Rochester New York. Humic acids were extracted from a Andosol and from a brown soil. The soils were two Entisol , an Istosol and an Andosol . Sterilization of self-supporting clay films, humic acids and soils were performed by 8 hours treatment with 12 % ethylene oxide in freon , at $4.10^5$ Pa, 40°C and 65 % relative moisture.

The procedures used for the preparation of clays and humic acids, the microbiological investigation, the adsorption experiments and the analytical methods have been described in details in our previous papers (Gennari et al.,1987; Pusino et al., 1990; Nègre et al., 1994).

# Results and Discussion

## *Efficiency of the ethylene oxide sterilization*

The efficiency of the sterilization has been checked in soil by determination of the following groups of microorganisms: fungi, actinomycetes, aerobic and anaerobic nitrogen fixers, ammonifiers, nitrifying, denitrifying, proteolytic, amylolytic, pectinolytic, aerobic and anaerobic cellulolytic. The absence of these microorganisms in the sterilized soil indicated that ethylene oxide had a good biocidal effect.

## *Effect of sterilization on montmorillonite*

The IR spectra of the sterilized Cu-, Al-, and Fe- saturated montmorillonites showed some new absorption bands characteristic of the IR spectrum of polyethylene glycol indicating that ethylene oxide polymerized on these clay surfaces. Such effect was not observed on the Na-, K- and Ca-saturated montmorillonites.

X-Ray diffraction data confirmed these results. The sterilization induced an increase of the basal spacing of the Cu-, Al-, and Fe-saturated clays, whereas no effect was observed on the sterilized Na-, K-, and Ca-clays.

A decrease of the cation-exchange capacity values of the Cu-, Al-, and Fe- saturated montmorillonites was observed after sterilization in the order Cu< Al< Fe while the cation exchange capacity values of the Na-, K- and Ca-saturated montmorillonites was unchanged. This result, which was unexpected because of the high solubility of polyethylene glycol in water, suggests that the polymer, at least in part, remained in the clay and behaved as a hydrophobic screen towards the hydrated cations, part of which were resistant towards the exchange.

## *Effect of sterilization on humic acids*

FT-IR and CP-MAS $^{13}$C-NMR analyses of sterilized and non sterile humic acids have evidentiated that the ethylene oxide treatment introduced the hydroxyethyl group in the humic acids molecules. This has been attributed to the reaction of ethylene oxide with the hydrogen atoms of the OH and COOH groups to give:

$$humic\text{-}O\text{-}CH_2\text{-}CH_2OH$$
$$acid \quad \text{-}C\text{-}O\text{-}CH_2\text{-}CH_2OH$$
$$O$$

The reaction of ethylene oxide with labile hydrogen atoms of humic acid is likely to be responsible for the observed increase of humic acids pH by about 0.5 units caused by sterilization.

## *Effect of sterilization on soil properties*

Soils presenting a large range of organic matter content were submitted to ethylene oxide sterilization. The properties of a soil having a low organic carbon content (0.6 %) and a basic pH (7.9) were not affected by sterilization. The effects of the sterilization on soils having an acidic pH ( 4.5 to 5.6) was an increase of their pH and organic carbon content. The increase of the organic carbon content caused by the reaction of ethylene oxide with humic acids was higher, the lower the pH.

## *Effect of sterilization on the adsorption of acifluorfen*

Isotherms of adsorption of acifluorfen on humic acids showed that the sterilization caused a decrease of the amount of herbicide adsorbed. This result suggests that the adsorption sites of the organic surfaces are partially inactivated by sterilization. Another explanation is that the limited adsorption on sterilized humic acids could be in relation with the increase of their pH.

Indeed, the amount of adsorbed acifluorfen on humic acids has been seen to decrease when pH increases and to be higher at pH values near its pKa (3.5) ( Ruggiero et al., 1990).

Adsorption of acifluorfen on soils was not affected by the sterilization for soils with a low or medium organic carbon content (< 2 %) independently on the increase of the soil pH or organic matter content. In contrast, sterilization of soils with a high organic carbon content (about 10 %) caused a decrease of the adsorption of acifluorfen indicating that these soils had a behaviour similar to that of pure humic acids.

## Conclusion

Ethylene oxide sterilization of soil has be seen to have a good biocidal effect.

Ethylene oxide has been seen to polymerize to polyethyleneglycol in the interlayer of homoionic clays saturated with $Al^{3+}$, $Fe^{3+}$ and $Cu^{2+}$. This reaction was not observed on the montmorillonites saturated with $Na^+$, $K^+$ and $Ca^{2+}$.

Ethylene oxide does not polymerize on humic acids but reacts with their acidic hydrogen atoms forming hydroxyalkylated compounds.

From a pratical point of view, the polymerisation of ethylene oxide on the inorganic surfaces would not be significant since it does not occur on the Na-, K-, and Ca-clays which are the most widely present in agricultural soils.

Reaction of ethylene oxide with humic acids has to be taken into account when comparing sterilized and unsterile soils to separate the abiotic and biotic degradation of pesticides. In particular, ethylene oxide sterilization has to be used carefully for soils with a high organic matter content and for acidic soils and when the transformation of herbicides is highly pH-dependent. On the other hand, all sterilization methods now available (autoclaving, gamma irradiation, potassium azide treatment, etc) alter the soil properties and little is known on the mechanisms of these modifications (Aldrich and Martin, 1952; Lopes and Wollum, 1976; Skipper and Westerman, 1973; Wolf et al., 1989). The use of ethylene oxide, whose mechanisms of interaction with soil are known seems to be an appropriate sterilization procedure.

## Acknowledgments

This study was partly financed by the Ministero dell'Università e della Ricerca Scientifica (60%) of Italy.

## References

□    Aldrich, D.G. and J.P. Martin. 1952: Effect of fumigation on some chemical properties of soils. *Soil Sci.* 73: 149-159.

□    Allison, L.E. 1951. Vapor sterilization of soil with ethylene oxide. *Soil Sci.* 72: 341-351.

□    Clark, F.E. 1950. Changes induced in soil by ethylene oxide sterilization. *Soil Sci.* 70: 345-349.

□    Dao, T.H., D.B. Marx, T.L. Lavy, and J. Dragun. 1982. Effect and statistical evaluation of soil sterilization on aniline and diuron adsorptive isotherms. *Soil Sci. Soc. Am. J.*, 46: 963-969.

□    Gennari, M., M. NXgre and R. Ambrosoli. 1987. Effect of ethylene oxide on soil microbial content and some soil characteristics. *Plant and Soil.* 102: 197-200.

□    Gennari. M. and M. NXgre. Acifluorfen persistence in soil. Proc. of the 3rd workshop: *study and prediction of pesticides behaviour in soil, plants and aquatic systems*, Munchen, FRG, 30 May- 1st June 1990. 221-236.

□    Lopes, A. S. and A. G. Wollum. 1976. Comparative effects of methyl bromide, propylene oxide and autoclave sterilization on specific soil chemical characteristics. *Turrialba* 26 (4): 351-355.

□    Nègre, M., M. Gennari, C. Crecchio and P. Ruggiero. Effect of ethylene oxide sterilization on soil organic matter. To be published.

□    Pusino, A., M. Gennari, A. Premoli and C. Gessa. 1993. Formation of polyethylen glycol on montmorillonite by sterilization with ethylen oxide. *Clays and Clay Minerals.* 38 (2): 213-215.

□    Pusino, A., W. Liu, Z. Fang and C. Gessa. 1993. Effect of metal-binding ability on the adsorption of acifluorfen on soil. *J. Agric. Food Chem.*, 41: 502-505.

□    Ruggiero, P., C. Crecchio, R. Mininni and M.D.R. Pizzigallo. 1992. Adsorption of the herbicide acifluorfen on humic acids. *Sci. Total Environ.* 123/124: 93-100.

□    Skipper, H. D. and D. T. Westerman. 1973. Comparative effects of propylene oxide, sodium azide and autoclaving on selected soil properties. *Soil Biol. Biochem.* 5: 409-414.

□    Wolf, D.C., T. H. Dao, H. D. Scott and T.L. Lavy. 1989. Influence of sterilization methods on selected soil microbiological, physical and chemical properties. *J. Environ. Qual.* 18: 39-44.

*3.3.6.*     *Potential for Pesticide Degradation in the Unsaturated Zone*
             *of a Sandy Soil*

# K. J. Lewis and J. S. Dyson

Zeneca Agrochemicals, Jealott's Hill Research Station, Bracknell, Berkshire RG12 6EY, U.K.

## Introduction

Biological activity in the unsaturated zone has the potential to reduce considerably the impact of pesticides in soil and groundwater. However, current use of mathematical models for predicting the fate of chemicals in soil often ignores pesticide degradation in subsurface soil.

The aim of this study was to examine the potential for pesticide degradation in the unsaturated zone of a sandy soil and the influence of environmental conditions on degradation processes. Soil sampled from the unsaturated zone was therefore studied, under laboratory conditions, to examine (i) the soil's physicochemical and microbial properties and (ii) the mineralisation of 2,4-dichlorophenoxyacetic acid (2,4-D) and atrazine.

## Soil Sampling

The soil was sampled down to 150 cm from a field site at Frensham, Surrey, UK. This site has been under grass for 17 years, and comprises a sandy loam overlaying a shallow aquifer. Soil samples (25 cm x 5 cm) were obtained by coring horizontally into the face of a 150 cm deep soil pit, in 25 cm increments down the soil profile.

## Physicochemical Properties

The physicochemical properties of all soil samples were determined including organic matter content, pH, cation exchange capacity, moisture holding capacity (mhc) and textural classification.

Most significantly, surface soil (0-25 cm) contained 1.9% organic matter, whilst subsoil samples (25-150 cm) had much lower levels of organic matter (0.2-0.8%). The subsoil can therefore be considered to constitute an oligotrophic environment.

Surface soil (0-25 cm) also differs markedly from subsoil in both moisture content and moisture holding capacity, both properties being markedly lower in the subsurface.

Therefore, as a microbial habitat this subsoil has a low nutrient status and lower moisture availability.

## Microbial Properties

Soil microbial biomass was determined by substrate induced respiration (Anderson and Domsch, 1978). Levels of microbial biomass decline dramatically from surface soil (320 mg-C/kg soil) to subsoil (32 mg-C/kg soil, 125-150 cm), showing an approximately exponential decline.

This decline of microbial biomass down the soil profile occurs in a manner similar to that for organic matter and probably results from the nutrient status of the soil.

## Pesticide Degradation

Degradation of two pesticides was determined, in all soil samples, by monitoring the mineralisation of atrazine and 2,4-D to carbon dioxide in a closed system (Anderson, 1990), under standard laboratory conditions (40% mhc at zero suction and 20°C in the dark). Pesticides were applied as [$^{14}$C-ring-UL]-labelled 2,4-D (0.5-1.0 mg/kg dry weight soil) and [$^{14}$C-ring-UL]-labelled atrazine (0.7 mg/kg dry weight soil).

Both pesticides were mineralised throughout the soil profile indicating the potential for pesticide degradation in the unsaturated zone.

The quantity of degradation varied, with 2,4-D giving consistently higher levels of mineralisation (55-90 % at 56 days) than atrazine (1-5 % at 106 days). This is presumably due to each compound being subject to different degradation processes. Levels of atrazine mineralisation were found to decline down the soil profile in a similar pattern to the levels of microbial biomass. By contrast, 2,4-D mineralisation declined down the profile to a depth of 125 cm, but below this depth the rate of mineralisation increased rapidly over days 7 and 14, leading to high levels of mineralisation in the 125-150 cm soil layer at the end of the incubation period. For this compound, therefore, final levels of mineralisation did not correlate with levels of microbial biomass, throughout the incubation period.

Addition of glucose (0.5 g/kg dry weight soil) reduced 2,4-D mineralisation in subsoil by 10-20%, with little effect (a reduction of 2%) in surface soil, indicating that nutrient status may have a critical role in the degradation process.

## Influence of Environmental Conditions on Pesticide Degradation

The effects of environmental conditions on subsoil degradation processes are now being determined using the systems described and for environmental conditions measured in the field. These conditions include actual subsoil temperatures, moisture levels and oxygen content (Takagi et al. 1993).

Initial studies have examined 2,4-D mineralisation at differing moisture levels (15, 20, 25, 30, 40% mhc at zero suction). In soil layers from 0-100 cm, 2,4-D mineralisation declines with decreasing moisture levels. Below this depth there is a reduction only at the lowest moisture levels (15 and 20 % mhc).

At higher moisture levels (25-40% mhc) patterns of 2,4-D mineralisation were as described previously, with higher levels of mineralisation in the deeper subsoil layers. However, this pattern is not repeated at the lowest water content (15 % mhc), with no increased rates of mineralisation in the deepest soil layer (125-150 cm).

This preliminary investigation highlights the type of controls which may influence microbial processes in subsoil. By continuing these experiments to include variations in the other soil conditions, a more complete understanding of degradation processes in subsoil will be made. This should result in more accurate predictions about the fate and behaviour of chemicals in soil, particularly potential movement to groundwater.

## *Conclusions*

Soil from the unsaturated zone (subsoil) constitutes an oligotrophic environment with less than 1% organic matter.

Subsoil contains significant microbial populations, levels declining with increasing soil depth.

Pesticide degradation occurs in subsoil under laboratory conditions, indicating the potential for degradation in the field.

The quantity and pattern of pesticide degradation in subsoil is compound specific.

Both the pattern and quantity of degradation are influenced by environmental conditions such as soil moisture.

Work on other environmental conditions will allow a more accurate prediction as to the fate of pesticides in subsoil.

## References

□ Anderson, J.P.E and Domsch, K. (1978) Soil Biology and Biochemistry, 10: 215-221

□ Anderson, J.P.E. (1990) Advances in Applied Technology, 4: 129-145

□ Takagi, K., Anderson, J.P.E., Lewis, K.J., Lewis, F.J., Dictor M-C., and Soulas, G. (1992) International Symposium on Environmental Aspects of Pesticide Microbiology, Sigtuna Sweden, pp270-277

## Acknowledgements

The authors wish to thank the Commission of the European Communities for funding this project. Bayer AG and INRA (Dijon) collaborators in the project and Zeneca Agrochemicals for both financial support and encouragement.

*3.3.7.*     *Factors Involved in Biodegradation of PCNB by Selected Micromycetes*

# D. Lièvremont[1], F. Seigle-Murandi[1]*, J.-L. Benoit-Guyod[2] & R. Steiman[1]

Groupe pour l'Etude du Devenir des Xénobiotiques dans l'Environnement (GEDEXE)

[1] Laboratoire de Botanique, Cryptogamie, Biologie Cellulaire et Génétique

[2] Laboratoire de Toxicologie et Ecotoxicologie

UFR de Pharmacie de Grenoble, Université J. Fourier, BP 138, 38243 Meylan Cédex, France

## Abstract

We monitored rates of degradation of PCNB in liquid cultures amended with 100 mg of PCNB per liter and inoculated with either *Sporothrix cyanescens* (Deuteromycete), *Mucor racemosus* or *Rhizopus arrhizus* (Zygomycetes). Concentrations of PCNB were determined by High Performance Liquid Chromatography analysis over 20 days. Fast removal of PCNB (68 - 81 %) was observed in cultures of *S. cyanescens* and *M. racemosus* thus outlining the absence of lag phase. We also monitored rates of PCNB adsorption on fresh mycelia of *S. cyanescens* and *M. racemosus* and on lyophilized mycelia of *S. cyanescens* and *Rhizopus arrhizus*. Except for living *M. racemosus* which mean adsorption rate was 70.53 µmol of PCNB per g of dried biomass (corresponding to 20.81 µg/mg), low mean concentrations of PCNB in the sorbed phase were recorded (3.31-4.92 µmol/g corresponding to 0.98-1.45 µg/mg) thus demonstrating that PCNB adsorption was a minor process.

## Introduction

In the last few decades, inadvertent release of harmful chemicals in the environment leading to worldwide contamination of ecosystems has become the focus of considerable attention because of the risk facing public health.

Bioremediation, though the use of appropriate and active microorganisms to transform these chemicals, have shown real promises. Some fungi have been shown to possess the ability to mineralize recalcitrant compounds. Recently, these problems were addressed in the laboratory. As a consequence, we have conducted an extensive study on the biodegradation of PCNB (Seigle-Murandi et al 1992; Steiman et al. 1992) on more than a thousand strains of micromycetes belonging to different taxonomic groups. PCNB is a commercial fungicide used against soilborne plant pathogens such as *Rhizoctonia solani* or *Sclerotium rolfsii* (Nakanishi & Oku 1969; Katan & Lockwood 1970; Ingham 1985; Carling et al. 1990).

The purposes of this study were (i) to demonstrate the ability of *S. cyanescens, M. racemosus* and *R. arrhizus* to transform PCNB and (ii) to determine PCNB adsorption rates on fresh mycelia of *S. cyanescens* and *M. racemosus* and lyophilized mycelia of *S. cyanescens* and *R. arrhizus*.

## Materials and methods

### *Chemicals*

PCNB was purchased from Janssen (Beerse, Belgium). Its purity was assessed by HPLC before use (> 99 %). Other products were from Prolabo (Paris, France). All chemicals were of the highest purity available.

### *Culture Conditions*

Fungal strains evaluated in this study were obtained from our laboratory collection (Collection Mycology Pharmacy Grenoble). They were maintained on Malt Extract Agar (MEA) or Potato Dextrose Agar (PDA) medium at 4°C. *S. cyanescens* was reactivated on solid PDA medium for eight days at 24°C while *M. racemosus* was on MEA medium. Cultures were set up in 250 ml flasks filed with 70 ml of GS medium. Each of them contained scraped inocula consisted of mycelium and spores to which was added aseptically after two days a PCNB solution to a final concentration of 100 mg/l. Cultures were cultivated during 22 days with shaking. Control cultures consisted of non-inoculated flasks. Each series of experiments was made in triplicate or more.

In adsorption experiments, 50 mg of fungal lyophilized powder were added aseptically to 25 ml of GS medium in 100 ml Erlenmeyer flasks, at ambient temperature (24°C). The flasks were agitated over 5 days on a rotary shaker (180 rpm, shaking diameter 25 mm). Control cultures consisted of cell-free flasks. PCNB solution was aseptically added to obtain a final concentration of 100 mg/L. Each series of experiments was carried out in triplicate.

PCNB was extracted from the media with bidistilled ethyl acetate (3 x 15 ml). Extraction efficiencies in controls were higher than 95 %. The extracts were evaporated to dryness at 40°C under a gentle stream of nitrogen. The residue was redissolved in methanol (5 ml). PCNB was desorbed from mycelia with one volume (10 ml) of methanol on a rotating shaker for 1 h.

PCNB in methanol was quantified by HPLC. Generally, quantitation was performed immediately. The column was a C-18 (4.6 mm D.I x 300 mm filled with 10μ-Bondasorb). The mobile phase was pure methanol. Flow rate was 1 ml/min. PCNB was detected at 303 nm with a retention time of approximately 2.2 min.

## Results and Discussion

### *Biodegradation of PCNB*

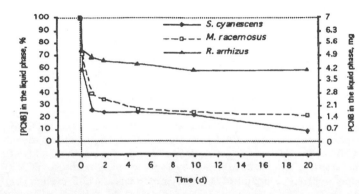

Fig. 1. Profiles of PCNB concentration in GS medium by *S. cyanescens*, *M. racemosus* and *R. arrhizus* as a function of time. The concentration of PCNB was 100 mg.l⁻¹. Each point represents a mean of three or four replicates.

Concentrations of PCNB were determined at various times over a 20 day period. Experimental data are depicted in figure 1. Two patterns are observed. First one concerned both *S. cyanescens* and *M. racemosus*. They demonstrated that the concentration of PCNB extractable from the liquid media was dramatically decreased. 68 to 81 % of the PCNB initially present was removed after 20 days compared to uninoculated controls. The time course showed that most of the total fungal transformation of PCNB occurred during the first day (51-64 %). This absence of lag phase proved that the non acclimated fungal strains *S. cyanescens* and *M. racemosus* were very efficient degraders of PCNB. Apparently, the biological system was able to adapt to the high input concentration of the chemical. Rates of degradation were generally comparable. Nevertheless *S. cyanescens* removed PCNB to a slightly greater extent. *R. arrhizus* inscribed in a second pattern in which we also reported an absence of lag phase along with a major removal which was performed equally in the first 24 hours. Nevertheless, efficiency of the removal was far lower compared with the two other species (31 % vs 68 to 81 %). As a result of fungal activity, we could say that an unknown portion of PCNB was converted to transformation products although these latter were not identified. Thus, cultures might merely alter PCNB structure just sufficiently to prevent its detection, rather than perform a more complete mineralization. The percentage of PCNB lost through abiotic degradation such volatilization process was slight. Abiotic conversion of PCNB was always less than 10 % during the 20 d period. In any event, the amount of PCNB removed from inoculated cultures was higher than the amount removed in the non-inoculated control cultures.

## *Adsorption of PCNB*

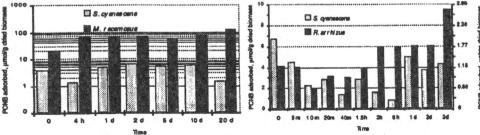

Fig. 2. Evolution of PCNB adsorption in GS medium on fresh biomass of *S. cyanescens* and *M. racemosus*. Each point represents a mean of three replicates.

Fig. 3. Evolution of PCNB adsorption in GS medium on lyophilized biomass of *S. cyanescens* and *R. arrhizus*. Each point represents a mean of three replicates.

As illustrated in Figure 2, behaviour of PCNB adsorption was rather different between *S. cyanescens* and *M. racemosus*. Quantities of sorbed PCNB recovered in organic extracts from *S. cyanescens* were low. Nevertheless, at extraction time 0, PCNB was already sorbed (4.23 µmol/g or 1.25 µg/mg). The observed process was rapid. Effects of longer contact times were provided by extraction time 1d to 20d. We could define that equilibrium was soon reached by *S. cyanescens* contrary to the pattern observed by *M. racemosus*. Apparent adsorption of PCNB was far greater on *M. racemosus* (mean value of 70.53 µmol/g or 20.81 µg/mg). With this strain, quantities of sorbed PCNB rose as the time was getting longer. Apparently, the reversibility of sorption process was not effective. It might be explained by a very tenacious binding of PCNB to the live cells. In contrast, apparent desorption of PCNB seemed to appear with *S. cyanescens*.

Uptake of PCNB by lyophilized biomass was represented on figure 3. The apparent sorption by *S. cyanescens* live biomass was nearly equal than the respective uptake by the same lyophilized dead cells. We could think that the absence of metabolic protection would promote a greater accumulation by dead cells. Our results could indicate no change in surface adsorptive properties between live and dead *S. cyanescens* cells. *R. arrhizus* is an extensively used fungus in biosorption experiments. It has been shown to exhibit biosorption of a wide range of compounds (Tsezos & Seto 1986; Bell & Tsezos 1987; Fourest & Roux 1992). Thus, it was of interest to conduct adsorption of PCNB on it and precisely on dead lyophilized biomass. It must be pointed out that uptake increased with time as noted with *M. racemosus* live biomass. It was not shown to be reversible. Overall, adsorption of PCNB on dead biomass of *R. arrhizus* remained low (mean value of 4.93 $\mu$mol/g or 1.45 $\mu$g/mg). The strain did not prove its high efficiency.

In summary, none of the strains tested possessed a significant sorptive uptake capacity towards PCNB. We considered that biosorption was responsible for only an insignificant portion of the observed pollutant removal.

## Conclusions

Experimental results revealed that PCNB could be readily removed from liquid media by both *S. cyanescens* and *M. racemosus* and with similar patterns. Removal was achieved to a much lower extent by *R. arrhizus*. Abiotic conversion was slight. Regardless of differences induced by species or physiological state, the extent of biosorption by fungal strains remained low.

In summary, the present study thus concludes that the most effective mean for removing of PCNB from liquid media was the biodegradation process. Biological removal of this chemical by adsorption was much less effective and could not remove significant amounts of the pollutant.

## References

□   Bell JP & Tsezos M (1987) Removal of hazardous organic pollutants by biomass adsorption. *J. Wat. Pollut. Control Fed.* 59 : 191-198

□   Carling DE, Helm DJ & Leiner RH (1990) In vitro sensitivity of *Rhizoctonia solani* and other multinucleate and binucleate *Rhizoctonia* to selected fungicides. *Plant Disease* 74 : 860-863

□   Fourest E & Roux JC (1992) Heavy metal biosorption by fungal mycelial by-products : mechanisms and influence of pH. *Appl. Microb. Biotechnol.* 37 : 399-403

□   Ingham ER (1985) Review of the effects of 12 selected biocides on target and non-target soil organisms. *Crop Prot.* 4 : 3-32

□   Katan J & Lockwood JL (1970) Effect of pentachloronitrobenzene on colonization on alfalfa residues by fungi and *Streptomyces* in soil. *Phytopathology* 60 : 1578-1582

□   Nakanishi T & Oku T (1969) Metabolism and accumulation of PCNB by phytopathogenic fungi in relation to selective toxicity. *Phytopathology* 59 : 1761-1762

□   Seigle-Murandi F, Steiman R, Benoit-Guyod JL, Muntalif B & Sage L (1992) Relationship between the biodegradative capability of soil micromycetes for pentachlorophenol and for pentachloronitrobenzene. *Sci. Total Environ.* 123/124 : 291-298

□   Steiman R, Benoit-Guyod JL, Seigle-Murandi F & Muntalif B (1992) Degradation of PCNB by micromycetes isolated from soil. *Sci. Total Environ.* 123/124 : 299-308

▫    Tsezos M & Seto W (1986) The adsorption of chloroethanes by microbial biomass. *Water Res.* 20 : 851-858

3.3.8.      *Accelerated Degradation of Carbofuran in Soils with Prior Carbamate Insecticides Exposure*

# C. Morel-Chevillet, D. Pautrel, M.P. Charnay, C. Catroux and J.C. Fournier

INRA, Laboratoire de microbiologie des sols, 17 rue Sully, 21034 Dijon, France

## Introduction

Repeated soil application of the same pesticide can determine the growth of microorganisms able to use it and occasionally chemically similar compounds as carbon and energy sources. An increase in the size of the degrading microflora and consequent accelerated degradation of the pesticide can reduce the period of effective target control.

In France, failure of the insecticide carbofuran (2,3-dihydro-2,2-dimethyl-7-benzofuranyl-methylcarbamate) to control various insect pests was observed in corn monoculture and other agricultural crops. Previous studies in the Dijon Laboratory have shown that carbofuran degradation in most French soils is mainly due to microbial action. Microorganisms which degraded the pesticide were isolated and characterized.

Our experiments had two objectives :

■      to estimate the risk of carbamates and especially carbofuran degradation after cross treatments with various carbamate compounds,

■      to study the diversity and ecology of the microorganisms involved in the degradation of different active ingredients.

## Material and Methods

### Soils

—      Montardon is a silty-sand soil, pH 6.2 from the Southwest of France collected from a plot untreated with carbamate pesticides for eight years,

—      Dijon is a silty-clay soil, pH 7.4 from the Burgundy region which has never been treated with carbamates.

### Pesticides

Sixteen different pesticides including 14 methylcarbamates, aldicarb and chlorpropham were studied (all compounds were 98-99 % pure). Carbonyl-1-($^{14}$C)carbofuran (specific activity, 26.3 MBq/mmol; 99 % pure) was used for radiometric measurements.

### Experimental Procedures

Fresh soil samples of 50 g were put in bottles and treated with 0.68 mmoles (equivalent to 150 μg of carbofuran) of each unlabelled pesticide dissolved in 50 μl of methanol. Control samples were treated only with methanol. Samples were left for one hour in a ventilated area to remove the solvent before adjusting the soil moisture to 90% of water holding capacity. Samples were then placed in closed 2 liters jars and incubated at 20°C, in the dark for 29 days. Air in these jars was periodically renewed. After this incubation period, carbofuran-degrading microorganisms were enumerated in pretreated samples using the MPN method with $^{14}$C-carbofuran as the sole source of carbone and energy (50 mg and 10 MBq/liter) in a mineral, basal medium (pH 6.5). Residual activity in the medium was measured with a liquid scintillation counter after 21 days of incubation at 28° C. Thirty days after the pretreatment, soil samples were retreated with labelled carbofuran (0.68 mmoles and 1.18 MBq per sample). Released $^{14}CO_2$ was trapped in 0.2M NaOH and radioactivity was measured as previously shown.

## Results and Discussion

Only preliminary results are presented.

In the Montardon soil, the half-life of labelled carbofuran was 17-22 days in control samples. The mineralization rates were enhanced by pretreatment with most methylcarbamate compounds (table 1). The rates of carbofuran degradation was most enhanced (half-life 1-3 days) in samples which had been pretreated with isoprocarb. Carbofuran, furathiocarb, benfuracarb, and bendiocarb pretreatments also enhanced the rate of carbofuran degradation (half-life 3-6 days) while carbaryl, ethiofencarb, methiocarb and promecarb had a lesser effect (half-life 6-9 days). The methylcarbamates pirimicarb and formetanate as well as aldicarb and chlorpropham had no effect on subsequent carbofuran degradation in this soil.

As shown on table 2, the highest population of carbofuran degrading microorganisms was found in samples pretreated with isoprocarb (3000 fold that of untreated controls). With most of the other methylcarbamate pretreatments the carbofuran-degrading population was about a 100 fold that of control samples.

| Soil | Half-life | $^{14}C$-carbofuran mineralization (%/day) | |
|---|---|---|---|
| pretreatment | (days) | Initial rate | Maximal rate |
| control | 17-22 | 0.97 | 5.64 |
| bendiocarb | 3-6 | 10.59 | 15.07 |
| benfuracarb | 3-6 | 11.23 | 14.66 |
| carbaryl | 6-9 | 2.53 | 9.87 |
| carbofuran | 3-6 | 10.67 | 14.33 |
| carbosulfan | 3-6 | 6.21 | 13.15 |
| dioxacarb | 3-6 | 7.11 | 12.53 |
| ethiofencarb | 6-9 | 3.82 | 9.45 |
| fenobucarb | 3-6 | 6.78 | 14.63 |
| formetanate, HCL | 17-22 | 0.93 | 5.69 |
| furathiocarb | 3-6 | 10.80 | 15.60 |
| isoprocarb | 1-3 | 13.06 | 19.18 |
| methiocarb | 6-9 | 3.23 | 9.19 |
| pirimicarb | 17-22 | 0.95 | 6.00 |
| promecarb | 6-9 | 3.90 | 12.89 |

**Table 1 :** Carbofuran degradation in Montardon soil after pretreatment with various methylcarbamates

| Soil pretreatment | Carbofuran-degrading microorganisms /g soil | |
|---|---|---|
| control | 1.5 | $10^3$ |
| bendiocarb | 9.5 | $10^4$ |
| benfuracarb | 2.5 | $10^5$ |
| carbaryl | 4.5 | $10^5$ |
| carbofuran | 2.0 | $10^5$ |
| carbosulfan | 4.5 | $10^4$ |
| furathiocarb | 2.0 | $10^5$ |
| isoprocarb | 4.5 | $10^6$ |
| methiocarb | 1.5 | $10^5$ |

**Table 2 :** Number of carbofuran-degrading microorganisms in Montardon soil samples after pretreatment with various methylcarbamates

In Dijon soil accelerated degradation of carbofuran was not seen after pretreatment with carbofuran itself or with other carbamates (table 3). Microorganisms able to use carbofuran as the sole carbon and energy source were not detected in this soil.

| Soil | Half-life | $^{14}$C-carbofuran mineralization (%/day) | |
|---|---|---|---|
| pretreatment | (days) | Initial rate | Maximal rate |
| control | 20-27 | 2.20 | 2.45 |
| aldicarb | 20-27 | 2.22 | 2.23 |
| benfuracarb | 20-27 | 2.38 | 2.93 |
| carbaryl | 20-27 | 2.65 | 2.92 |
| carbofuran | 20-27 | 2.20 | 2.63 |
| chlorpropham | 20-27 | 2.49 | 2.74 |

**Table 3 :** Carbofuran degradation in Dijon soil after pretreatment with various carbamate pesticides

Preliminary results of these experiments indicate that a similar microbial population may be involved in the accelerated hydrolysis of most methylcarbamates pesticides. However, the availability of these compounds in the soil as the microbial toxicity of their phenol residues could influence the microbial diversity. Therefore, new strains have to be isolated and characterized after each specific methylcarbamate treatment.

## References

□   [1] Naïbo, B. 1988. Biodégradation accélérée des pesticides dans le sol. *Phytoma* 401 : 23-25.

□   [2] Charnay, M.P., B. Naïbo, C. Catroux and J.C. Fournier. 1992. Study of accelerated carbofuran degradation in some French soils. p. 193-197. *In* J.P.E. Anderson, D.J. Arnold, F.

Lewis, and L. Tortensson (ed.), Proc. Int. Symp. Environ. Asp. Pest. Microbiol. Sigtuna, Sweden.

□       [3] Charnay, M.P. and J.C. Fournier, 1994. Study of the relation between carbofuran degradation and microbial or physicochemical characteristics of some French soils.

□       [4] Charnay, M.P. 1993. La dégradation accélérée du carbofuran dans les sols : aspects microbiologiques et possibilités de contrôle du phénomène. *Ph.D. Thesis*, Université Claude Bernard-Lyon 1, France.

□       [5] Parekh, N.R., A. Hartmann, M.P. Charnay and J.C. Fournier. 1994. Genotypic and phenotypic diversity among carbofuran-degrading soil bacteria (submitted).

3.3.9.      *Plasmid-Profiling of Carbofuran-Degrading Soil Bacteria*

# N. R. Parekh, A. Hartmann and J-C. Fournier

Institut National de la Recherche Agronomique, Laboratoire de Microbiologie des Sols,
17 rue Sully, BP 1540, 21034 Dijon-Cedex. France.

## Introduction

Field and laboratory experiments have shown that repeated application of several N-methylcarbamate compounds readily induces rapid biological degradation of these chemicals in subsequent treatments and thus reduces their efficacy (Chapman & Harris 1990). Carbofuran (2,3-dihydro-2,2,dimethyl-7-benzofuranyl methylcarbamate) is a soil-applied insecticide and nematicide which is used extensively to control insect pests of corn, rice and other agricultural crops. In the case of carbofuran a single application of the recommended dose is sufficient to induce accelerated microbial degradation (Suett & Jukes 1993). Carbofuran can be degraded by microorganisms in at least three different ways but hydrolysis of the methylcarbamate linkage is the primary mechanism of inactivation in previously treated soils (Chaudhry & Ali 1988).

There have been many reports of the isolation and characterisation of carbofuran-degrading microorganisms and different enzymes which hydrolyse N-methyl carbamate compounds have been described (reviewed by Chapalamadugu & Chaudhry (1992). Only the gene which codes for carbofuran hydrolase (*mcd* gene) in *Achromobacter* sp. WM111 has been cloned and this is located on a 100 kb plasmid designated pPDL11 (Tomasek & Karns 1989). Plasmid profiling combined with Southern hybridisation can provide information on the distribution of plasmids among environmental isolates and indicate the presence and location of specific plasmid-encoded genes. We have used these methods to determine the presence and distribution of plasmids and of the carbofuran-hydrolase (*mcd*) gene among seventy-nine bacteria from English and French soils which initially hydrolysed carbofuran to carbofuran phenol.

Part of this work has been submitted to Applied and Environmental Biology (Parekh et al. 1994b) for publication.

## Methods and Materials

Eighty-one bacteria which initially hydrolysed carbofuran to carbofuran phenol were studied; seventy-nine were isolated from English and French soils by enrichment with carbofuran as the sole source of carbon and nitrogen (Parekh et al. 1994a, Charnay 1993) and two were reference carbofuran-degrading strains, *Achromobacter* sp. strain WM111 (Karns et al. 1986) and methylotrophic bacterium strain ER2 (Topp et al. 1993). Following repeated sub-culture, all isolates were tested for their ability to degrade carbofuran and its breakdown product methylamine as sole sources of carbon and nitrogen using radiorespiratory methods. Plasmid profiling of carbofuran-degrading bacteria was done using a horizontal gel electrophoresis method (Wheatcroft et al. 1990). The presence and location of the carbofuran hydrolase gene among the isolates was determined by Southern blotting plasmidic DNA and hybridising it with a DNA probe specific for the carbofuran hydrolase (*mcd*) gene.

## Results and Discussion

Carbofuran-degrading isolates in this study were mainly Gram-negative bacteria and have been shown to be phenotypically diverse (Parekh et al. 1994b). After sub-culture, fifty two isolates, including both reference strains, degraded both carbofuran and methylamine, five isolates had lost their ability to degrade carbofuran but degraded methylamine and twenty four isolates did not degrade carbofuran or methylamine.

Eight distinct plasmid profiles with 2-4 plasmids ranging in size from 76 to about 1340 kb were visualised in 52 of the 57 isolates which degraded carbofuran and/or methylamine. Our results showed that phenotypically different carbofuran-degrading bacteria had the same plasmid profiles. Twenty four of these isolates (including both reference strains) contained an approximately 100 kb plasmid which hybridised with the *mcd* gene probe; the *mcd* gene encoding plasmid was always associated with an approximately 580 kb plasmid. The presence of an approximately 100 kb plasmid which encodes the carbofuran hydrolase (*mcd* ) gene in phenotypically diverse soil isolates from geographically separate countries suggests that this plasmid is transmissible or can be mobilised between a wide range of Gram-negative bacteria. Transfer of such catabolic plasmids may lead to the rapid evolution of pesticide-degrading capacity within the soil microbial population and thus may be involved in the phenomenon of enhanced degradation (Karns 1990). Twenty eight isolates did not hybridise with the *mcd* gene probe and thus hydrolyse carbofuran using an enzyme or enzymes encoded by genes with no homology to the carbofuran hydrolase (*mcd*) gene. It may be interesting to determine the substrate specificity of the carbofuran hydrolase activity in these isolates and compare them to previously described enzymes.

One or two plasmids ranging in size from 40 to about 1300 kb were visualised in only eight of the 24 isolates which had lost their ability to degrade carbofuran and did not degrade methylamine. None of these isolates hybridised with the *mcd* gene probe or had plasmid profiles which were similar to those seen in bacteria which degraded carbofuran and/or methylamine.

## Acknowledgements

This work was supported by the Agriculture Food Research Council (AFRC), U.K. and Institut National de la Recherche Agronomique (INRA), France. N. R. Parekh was funded by an AFRC/INRA fellowship scheme during the period of this work.

## References

◻    Chapalamadugu, S., and G. R. Chaudhry, 1992. Microbiological and biotechnological aspects of metabolism of carbamates and organophosphates. *Critical Reviews in Biotechnology* 12. 357-389.

◻    Chapman, R. A., and C. R. Harris, 1990. Enhanced degradation of insecticides in soil: Factors influencing the development and effects of enhanced microbial activity, p. 82- 97. In K. D. Racke & J. R. Coats (ed.)*Enhanced Biodegradation of Pesticides in the Environment*. American Chemical Society Symp. Ser. 426.

◻    Charnay, M-P, 1993. La dégradation accélérée du carbofuran dans les sols : aspects microbiologique et possibilités de contréle du phénoméne. *Ph.D Thesis*, Université Claude Bernard - Lyon 1, France.

◻    Karns, J. S., W. W. Mulbry, J. O. Nelson, and P. C. Kearney, 1986. Metabolism of carbofuran by a pure bacterial culture. *Pesticide Biochemistry and Physiology* 25. 211- 217.

◻    Karns, J. S, 1990. Molecular genetics of pesticide degradation by soil bacteria. p 141- 152. In K. D. Racke & J. R. Coats (ed.)*Enhanced Biodegradation of Pesticides in the Environment*. American Chemical Society Symp. Ser. 426.

□    Parekh, N. R., D. L. Suett, S. J. Roberts, T. Mckeown, E. D. Shaw, and A. A. Jukes, 1994a. Carbofuran-degrading bacteria from previously treated field soils. *Journal of Applied Bacteriology* 76. in press.

□    Parekh, N. R., A. Hartmann, M-P, Charnay, and J- C. Fournier, 1994b. Phenotypic and genotypic diversity among carbofuran-degrading soil bacteria. Submitted to *Applied and Environmental Microbiology*.

□    Suett, D. L., and A. A. Jukes, 1993. Accelerated degradation of soil insecticides : comparison of field performance and laboratory behaviour, p. 31-41. In M. Mansour (ed.), *Fate and Prediction of Environmental Chemicals in Soils, Plants and Aquatic Systems*. Lewis Publishers, Boca Raton.

□    Tomasek, P. H. and J. S. Karns, 1989. Cloning of a carbofuran hydrolase gene from *Achromobacter* strain WM111 and its expression in Gram-negative bacteria. *Journal of Bacteriology* 171. 4038-4044.

□    Topp, E., R. S. Hanson, D. B. Ringelberg, D. C. White, and R. Wheatcroft, 1993. Isolation and characterization of an N-methylcarbamate insecticide-degrading methylotrophic bacterium. *Applied and Environmental Microbioliology* 59. 3339- 3349.

□    Wheatcroft, R., D. G. McRae, and R. W. Miller, 1990. Changes in the *Rhizobium meliloti* genome and the ability to detect supercoiled plasmids during bacteroid development. *Molecular Plant-Microbe Interactions* 3. 9-17.

*3.3.10.*     *Chemical and Photochemical Biocides Decomposition in Aqueous Environments : Determination of the Products Formed*

# R. Perraud, M. Papazian & P. Foster

Groupe de Recherche sur l'Environnement et la Chimie Appliquée
Université Joseph Fourier - 1, rue FranYois Raoult - 38000 GRENOBLE FRANCE

## Introduction

Our laboratory is interested in the fate of novel synthetic biocides, particularly in methylene bis thiocyanate (MTC), in thiocyanomethylthiobenzothiazol (TCMTB), and in bromonitropropanediol (BNPD) which show fungicidal and bactericidal properties. We have undertaken a study on the decomposition of these compounds which, in special cases, are good substitutes for polychlorinated molecules [1].

MTC: NC-S-CH2-S-CN

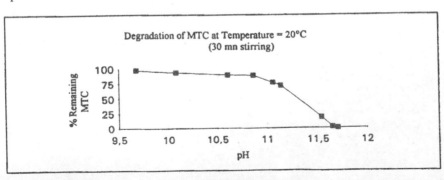

TCMTB :

BNPD :

While syntheses of these molecules have been published [1], [2], [3], [4], studies of the fate of these molecules in the environment have just begun. In the present work we investigate the conditions for the hydrolytic and photochemical decomposition of these biocides in aqueous media and in the atmosphere, determining the main decomposition products formed.

## Hydrolytic Decompositions of MTC, TCMTB and BNPD

### *pH and temperature dependent hydrolytic Decompositions*

#### 1 - Evolution of the decompositions

While MTC, TCMTB and BNPD stay intact in acid environment, it undergoes more or less considerable decompositions in basic media, depending on pH and temperature. We show that at ambient temperature (20°C) and determined pH, these compounds are completely decomposed as we can see in the following figures :

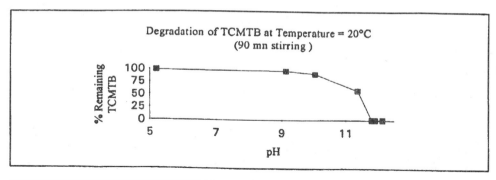

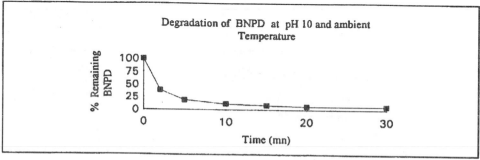

## 2 - Decomposition products of MTC, TCMTB and BNPD

— Optimum decomposition conditions for MTC are at very basic pH (pH ≈ 12).
Using different analytical means (HPLC or GLC-MS-coupling) formation of its decomposition products could be shown. While these compounds are formed directly in the decomposition of the biocide, we also find N,N-dimethylthioformamide as a transformation product of dimethylformamide, the latter having been present in the formulation without any doubt.
CS$_2$ and 1,2,4-trithiolane have been trapped in cartridges filled with adsorbant (Hayesep Q) and then transferred to a gas chromatograph by means of thermodesorption.

— Like MTC, TCMTB is stable in acid environment. We have found out that in a particular range of pH (between 11 and 12) and of temperature, a variation of 0.1 of the pH value can cause more than a 50% rise of decomposition of the active molecule. As mercaptobenzothiazole (MBT) is appearing, TCMTB is disappearing.

— Concerning BNPD, its decomposition is complete since pH = 10 at ambient temperature. We found the formation of a main volatil organic compound : the formaldehyde.

| | Products in condensed phase | Products in atmospheric phase |
|---|---|---|
| **MTC** | ions SCN⁻  1,2,4-Trithiolane | CS₂  1,2,4- trithiolane |
| **TCMTB** | ions SCN⁻  MBT : | 1,2,4- trithiolane |
| **BNPD** | | Formaldehyde : HCHO |

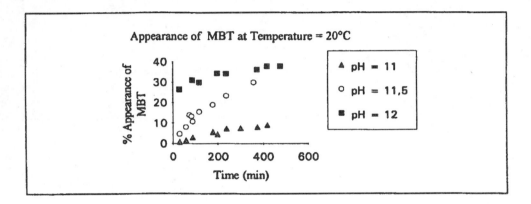

## *Photochemical Decompositions in Aqueous Media*

### 1 - Solar light irradiation of MTC and TCMTB

We have tried to simulate the natural conditions by irradiating solutions of MTC and TCMTB with neon glow lamps having wave lengths between 350 and 700 nm.

Working conditions : The solutions of the biocides which have been previously adjusted to a concentration of 90 ppm and to pH 5 to 6 and ambient temperature are installed in a black chamber of $0.35 \, m^3$ for photochemical simulation purposes. For each of the biocides we prepared a blank (just of water and the compound) and a solution containing traces of a photosensitizer ($H_2O_2$ at a concentration of 100 µm. The four solutions are irradied during 260 hours, samples being taken every 15 minutes for analytical purposes.

Under these conditions MTC stays intact. On the other hand, we notice interesting decomposition kinetics for TCMTB where the role of the sunlight and of hydrogen peroxide must be stressed.

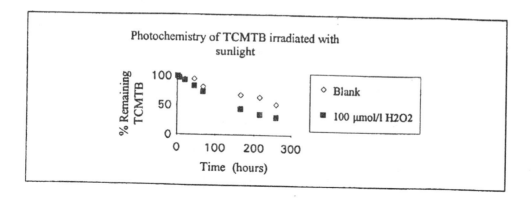

The following table resumes the values of the rate constants of the decomposition of TCMTB exposed to artificial sunlight, with and without H2O2, and the corresponding life times :

$$\tau = [TCMTB]/(-d[TCMTB]/dt) = 1/k$$

| BLANK (without H₂O₂) | $k_B = (6{,}50 \pm 1{,}30).10^{-7} \ s^{-1}$ | $\tau = (430 \pm 70)$ hours |
|---|---|---|
| +100µm of H₂O₂ | $k_H = (1{,}30 \pm 0{,}40).10^{-6} \ s^{-1}$ | $\tau = (220 \pm 6)$ hours |

As a result, TCMTB decomposes twice as fast in the presence of traces of H₂O₂, nevertheless, decomposition time is very long (220 hours).

## 2 - UV light irradiation of TCMTB, MTC and BNPD

As it was expected, the TCMTB is decomposed very efficiently. But as the MTC doesn't react under solar light irradiation, we have used a stronger irradiation.Under those new conditions : UV light at 254 nm, we found an interesting decomposition of the MTC. We have also tested the BNPD, and we found that it also decomposes in this case.

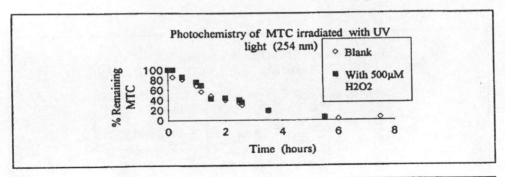

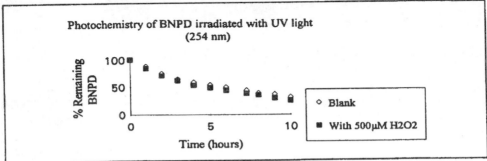

# Conclusion

MTC, TCMTB, and BNPD have been under investigation in terms of their decomposition behaviour under different conditions, in aqueous media. We have confirmed the presence of decomposition products in the condensed phase and in the gazeous phase close to this aqueous phase. The photo-chemical decomposition of these compounds proceeds slugishly in short terms, but points out the efficiency of traces of photosensitizers in aqueous media. The latter should be used more in order to advance the decomposition of toxic products into simple compounds.

# References

□    [1] Kennedy,M.J.High Performance Liquid Chromatography Analysis of Preservative Treated Timber for 2-(Thiocyanomethylthio)benzothiazole and Methylenebisthiocyanate. *Analyst*, Vol.111,701-705, (1986).

□    [2] Hurbin,V. Optimisation de Synthèse de Molécules Organiques Thiocyanées Biocides. Etude en Réacteur Pilote Automatisé. Dégradation des Actifs. Application au TCMTB et au MTC. *Thèse à l' UJF*, Grenoble I, Génie des Procédés, (1991).

□    [3] Perraud,R and Chraibi,N. Rapport sur la Synthèse du Thiocyanomethylthiobenzothiazole. Laboratoire du GRECA pour l'entreprise JOUD, (1988 ).

□    [4] Perraud,R and Lemche,M. Rapport sur la Synthèse du Méthylènebisthiocyanate. Labora-toire du GRECA pour l'entreprise JOUD, (1989).

### 3.3.11.    Photodegradation study of the herbicides Terbutylazine, Cyanazine, and Terbutryne

# Y. Sanlaville, P. Schmitt[*], M. Mansour[*], P. Meallier

Université Claude Bernard Lyon I, Laboratoire de photochimie industrielle
43, boulevard du 11 novembre 1918, F-69622 Lyon, France

[*] GSF - Institut für ökologische Chemie
Schulstraße 10, D-85356 Freising-Attaching, Germany

## Key Words

Photodegradation, Humic Substances, Hydrogene Peroxyde, S-triazines.

## Introduction

Triazines are among the most used pesticides worldwide. Because they can be found in many environmental compartments, their fate in ecosystems and the caracterisation of their degradation pathways are to be determined. It is also important to remove these products from drinking water, in order to reach the required quality normes. (1, 2)

| | R1 | R2 | R3 |
|---|---|---|---|
| Cyanazine | Cl | $C_2H_5$ | $CCN(CH_3)_2$ |
| Terbutylazine | Cl | $C_2H_5$ | $C_4H_9$ |
| Terbutryne | $SCH_3$ | $C_2H_5$ | $C_4H_9$ |

After their application, these chemicals can be removed from the environment in different ways. A possible degradation pathway for S-triazines herbicides in water is suggested to be the photooxydation through hydroxy radicals. Oxydative N-dealkylation was shown to be an important primary degradation pathway (3). The photooxydation with hydrogene peroxyde as a photosensitiser, which generates oxygenated species, mainly hydroxy radicals, is a possible way to eliminates contaminants of water and has got to be tested on these compounds family. The knowledge of the reaction kinetics, the determination of degradation products and the quantum yields of photodegradation are important parameters for the prediction of the persistance of active compounds in the environment and their elimination and will be therefore determined.

## Analytical Methods

The triazines were quantified with an HPLC equipped with a variable wavelength UV detector.
Detection : cyanazine : 220 nm, Terbutylazine : 223 nm, Terbutryn : 225 nm
Eluant : Cyanazine and Terbutylazine : ACN/$H_2O$ 50/50, Terbutryn : ACN/$H_2O$ 50/50 with $5.10^{-3}$
mol. of ammonium acetate, pH 7,2.
Irradiation system : Suntest Heraeus (Hanau, Germany), >290 nm.

## Photodegradation in presence of Humic Substances

One of the environment naturally occuring important compounds are the humic substances. Humic
substances with an accelerating effect on the degradation can photosensibilise the disparition of
S-triazines through the formation of Oxygenated Species, and particularly hydroxy-radicals, which
conduct to oxydation of the lateral chains of the triazines.

The fulvic acid we used was extracted from a soil of the Munich region according to the IHSS method.

The figure 1 shows the effect of the fulvic acid on the terbutylazine :

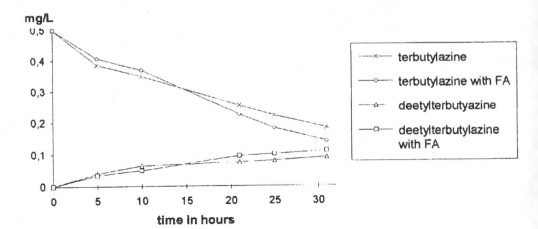

**Figure 1 :** photodegradation of terbutylazine and formation of deetylterbutylazine
in the presence of fulvic acid (FA) 5 ppm

The deisopropylterbutylazine and the hydroxyterbutylazine were also determined, but not quantified. The humic substances favorised the dealkylation processes against the hydroxylation pathways as the first step of degradation.

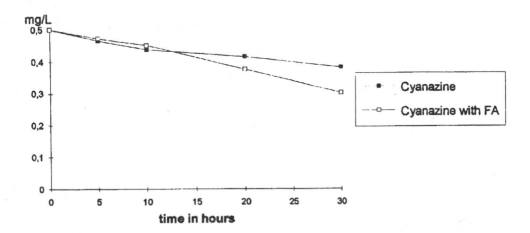

These humic substances showed also an accelerating effect on the photodegradation of Cyanazine (fig. 2).

**Figure 2 :** photodegradation of Cyanazine 0,5 ppm with fulvic acid (FA) 5 ppm

## Photooxydation with Hydrogene Peroxyde

The quantum yield of photodegradation of the triazines were determined at 254 nm with an optical bench (Amko, Germany) equipped with a 450 W Xenon lamp. The UV irradiation of $H_2O_2$ is known to produce oxygenated species ($OH°$, $H_2O°$, $O°^-$, $O_3°^-$) and is intended to be used in decontamination processes. (4)

The table 1 shows the increasing of the quantum yield of photodegradation through hydrogene peroxyde :

| Product | without $H_2O_2$ | with $H_2O_2$ | Ratio $H_2O_2$/product |
|---------|------------------|---------------|------------------------|
| Terbutylazine | 0,092 | 0,262 | 103,2 |
| Cyanazine | 0,048 | 0,209 | 108,1 |
| Terbutryn | 0,032 | 0,205 | 108,4 |

**Table 1** : quantum yield of photodegradation of Cyanazine, Terbutylazine and Terbutryn each 5 ppm, with 2,245 mmol.$L^{-1}$ $H_2O_2$ at 254 nm

## Conclusion

The humic substances we used showed an accelarating effect on the photolysis of the herbicides Cyanazine and Terbutylazine. For the terbutylazine, it has been found that the presence of humic substances enhanced the formation of dealkylated products, particularly of deetylterbutylazine.

The irradiation at 254 nm in presence of $H_2O_2$ showed an increasing of the quantum yield of photodegradation that reach values between 0,205 and 0,262 for the three pesticides.

## References

□      [1] Meallier, P.; Mamouni, A. and M. Mansour *Chemosphere* 20, 267273 (1990).

□      [2] Pellizetti et all.; *Chemosphere* 20, 891-910 (1992)

□      [3] Mansour M (ed.) : *Fate and prediction of environmental chemicals in soils, plants and aquatic systems.* Lewis Publishers, Boca Raton, Ann Arbor (USA), London, Tokyo (1993).

□      [4] F.J. Beltran, G. Ovejero and B. Acedo; *Wat. res.*, Vol. 27, N° 6, pp 1013.1021, 1993.

### 3.3.12.    Photocatalytic Degradation of Terbuthylazine in Soil and on $TiO_2$

# L. Scrano[*], S.A. Bufo[**] and M. Mansour[***]

[*]     IstitutoTecnico Agrario 70018,Alberobello, Italy.

[**]    Istituto di Chimica Agraria, Università degli Studi,via Amendola 165/A,
        70126 Bari, Italy.

[***]   G.S.F. - Forschungszentrum für Umwelt und Gesundheit GmbH.
        Institut für Ökologische Chemie,Schulstrasse 10, 85356 Freising, Germany.

## Introduction

Photochemical investigations in heterogeneous systems constitute a considerable interest for the mineralization, trasformation and elimination of pesticides in the different environmental compartments (1,2,3,4,5). Because some herbicides of the Triazine family are currently in use and they persist in the soil and water and even reach the ground water, phototransformation concerning the fate of them may assume importance as a method of envinronmental decontamination (6).

The study of degradation of persistent pesticides presents some difficulties because they decompose very slowly in the soil/water environment. The use of light irradiation in the presence of $TiO_2$ permits the increase of degradation rate (6) of these substances and the rapid examination of their photoproducts.

In this work photocatalitic oxidation of terbuthylazine in aqueous environment in the presence of selected soils and with the addition of $TiO_2$ was investigated under aerated conditions. The formation of intermediates products was also examined.

## Materials and Methods

Four soils characterized by different clay and organic matter contents (Table 1) were selected. 10g/L duplicate suspensions of each soil were prepared in 5mg/L ($2.17.10^{-5}$ M) aqueous solutions of test substances. 0.5g/L of $TiO_2$ were added to one of the duplicate suspensions. The other one was used as control sample.

The photocatalytic oxidation of the herbicides was performed in a solar simulator (Suntest), at 40°C, and in a pyrex reactor furnished of UV-Hg Lamp (290nm) at 20°C. The concentrations of terbuthylazine (TBA), and hydroxyterbuthylazine (HTBA) and deethylterbuthylazine (DTBA), as the most probable intermediate products (7), were determined at various illumination time by using a liquid chromatograph equipped with UV-VIS detector (215nm) and a C-18 column (5μm), 25cm long, 4.6mm i.d. A gradient mobile phase from 50 to 70% acetonitrile in water with a flow rate of 1 ml/min was programmed. Control samples were kept in the dark by covering the reactors with aluminum foil to confirm that a given product was only the result of the photochemical reactions.

| Sample | Soil Type | pH | O.M. (%) | Clay (%) | I* (%) | M* (%) | K* (%) | Fe₂O₃ (%) | C.E.C. (cmol.Kg⁻¹) |
|--------|-----------|-----|------|------|------|------|------|-------|---------|
| P1 | Xerofluvent | 7.7 | 2.24 | 44.3 | 29.2 | 8.0 | 7.1 | 1.17 | 19.2 |
| P2 | Typic Rhodoxeralfs | 7.9 | 0.99 | 20.4 | 12.4 | 3.5 | 4.5 | 1.40 | 9.0 |
| P7 | Entic Pelloxererts | 7.6 | 2.54 | 52.3 | 18.8 | 17.8 | 15.7 | 0.71 | 28.5 |
| P10 | Salorthidic fluvaquent | 7.5 | 0.83 | 67.7 | 37.2 | 11.5 | 16.9 | 1.07 | 24.1 |

\* I = Illite; M = Montmorillonite; K = Kaolinite

**Table 1 :** Characteristics of investigated soils

## Results

Tables 2 and 3 show the experimental results obtained using Suntest and UV-test. The degradation is more effective by using UV-Lamp, but the photochemical reactions under solar simulator and UV irradiations seem to be coherent.

| t (h) | C.10⁵ (mol.L⁻¹) | | | | | | | | | | | |
|---|---|---|---|---|---|---|---|---|---|---|---|---|
| | P1 | | | P2 | | | P7 | | | P10 | | |
| | TBA | HTBA | DTBA | TBA | HTBA | DTBA | TBA | HTBA | DTBA | TBA | HTBA | DTBA |
| 0 | 2.17 | ud | ud | 2.17 | ud | ud | 2.17 | ud | ud | 2.17 | ud | ud |
| 0,5 | 1.51 | 0.43 | ud | 1.09 | ud | 0.38 | 1.04 | ud | 0.28 | 1.22 | ud | 0.52 |
| 1 | 1.30 | 0.48 | 0.29 | 0.87 | 0.53 | 0.63 | 0.82 | ud | 0.62 | 0.56 | 0.43 | 0.62 |
| 2 | 0.95 | 0.37 | 0.44 | 0.50 | 0.61 | 0.71 | 0.74 | 0.41 | 0.71 | 0.13 | 0.46 | 0.73 |
| 4 | 0.63 | 0.32 | 0.53 | 0.17 | 0.39 | 0.54 | 0.35 | 0.43 | 0.54 | ud | 0.39 | 0.29 |
| 8 | 0.41 | 0.27 | 0.61 | ud | 0.19 | 0.28 | 0.13 | 0.40 | 0.50 | ud | 0.18 | 0.10 |
| 24 | 0.22 | 0.21 | 0.55 | ud | ud | ud | ud | 0.19 | 0.38 | ud | ud | ud |
| 32 | 0.19 | ud | ud | ud | ud | ud | ud | ud | ud | ud | ud | ud |

ud = undetectable

**Table 2 :** Illumination times (t) and concentrations (C) of TBA, HTBA and DTBA in aqueous solutions in the presence of soil and TiO₂ - SUNTEST

Table 4 shows the variation of herbicide concentrations in the aqueous solutions in the presence of the soils but without light irradiation. In this case HTBA and DTBA were always undetectable, so it may be that the generalized decrease of the TBA concentrations in the samples kept in the dark is only due to soil sorption phenomena. Naturally, sorption also occurs in the irradiated samples but the decomposition process continuosly removes the herbicide from the soil/water repartition equilibrium. Owing to the photoreaction, the concentrations of TBA decrease with the illumination time until to undetectable values (Tables 2 and 3). The organic radicals which are formed on the surface of the catalyst give rise not only to HTBA and DTBA but also to further intermediates progressively degraded, until stable compound are formed (7). This phenomenon can explain the HTBA and DTBA depletion during the photoreaction.

| t | C.$10^5$ (mol.L$^{-1}$) | | | | | | | | | | | |
|---|---|---|---|---|---|---|---|---|---|---|---|---|
| (h) | P1 | | | P2 | | | P7 | | | P10 | | |
|  | TBA | HTBA | DTBA | TBA | HTBA | DTBA | TBA | HTBA | DTBA | TBA | HTBA | DTBA |
| 0 | 2.17 | ud | ud | 2.17 | ud | ud | 2.17 | ud | ud | 2.17 | ud | ud |
| 0,5 | 1.22 | 0.42 | 0.81 | 0.52 | 0.33 | 0.22 | 0.35 | 0.44 | 0.46 | 0.61 | 0.35 | 0.51 |
| 1 | 0.85 | 0.35 | 0.48 | 0.22 | 0.31 | 0.26 | 0.26 | 0.38 | 0.45 | 0.24 | 0.41 | 0.47 |
| 2 | 0.53 | 0.31 | 0.47 | 0.17 | 0.28 | 0.32 | 0.20 | 0.35 | 0.41 | ud | 0.28 | 0.45 |
| 5 | 0.25 | 0.28 | 0.43 | ud | 0.17 | ud | 0.13 | 0.31 | 0.29 | ud | 0.19 | 0.38 |
| 7 | 0.18 | 0.23 | 0.28 | ud | ud | ud | ud | 0.27 | 0.25 | ud | ud | 0.15 |
| 9 | 0.15 | ud | ud | ud | ud | ud | ud | ud | ud | ud | ud | ud |

ud = undetectable

**Table 3 :** Illumination times (t) and concentrations (C) of TBA, HTBA and DTBA in aqueous solutions in the presence of soil and TiO$_2$ - UV-Hg Lamp

To understand the trends of photochemical reactions it is necessary to study the degradation kinetics. The general rate equation can be written:

$$-dC/dt = K.C^n \qquad (1)$$

where "C" is the concentration of the herbicide in the solution at time "t"; "K" is the kinetic constant.

The knowledge of the reaction order is essential for finding the correct integrated rate equation. By trying to fit data to various integrated rate equations it is possible to verify the reaction order. From the values of determination coefficients ($R^2$) it results that the degradation process for the investigated samples can be better described by the second order reaction kinetics (Table 5).

Equation (1) can also be written:

$$dd/dt = K.(C_{max}-d)^n \qquad (2)$$

where "d" is the amount of disappeared herbicide, and "$C_{max}$" is the maximun degradable quantity of the herbicide, i.e. the fraction which could be photodegraded if the reaction would be carried to completion.

| t | C.$10^5$ (mol.L$^{-1}$) | | | | | | | | | | | |
|---|---|---|---|---|---|---|---|---|---|---|---|---|
| (h) | P1 | | | P2 | | | P7 | | | P10 | | |
|  | TBA | HTBA | DTBA | TBA | HTBA | DTBA | TBA | HTBA | DTBA | TBA | HTBA | DTBA |
| 0 | 2.17 | ud | ud | 2.17 | ud | ud | 2.17 | ud | ud | 2.17 | ud | ud |
| 0,5 | 1.34 | ud | ud | 1.35 | ud | ud | 1.15 | ud | ud | 0.91 | ud | ud |
| 1 | 1.26 | ud | ud | 1.26 | ud | ud | 1.13 | ud | ud | 0.87 | ud | ud |
| 2 | 1.22 | ud | ud | 1.25 | ud | ud | 1.09 | ud | ud | 0.74 | ud | ud |
| 5 | 1.17 | ud | ud | 1.22 | ud | ud | 1.04 | ud | ud | 0.65 | ud | ud |
| 7 | 1.13 | ud | ud | 1.22 | ud | ud | 1.00 | ud | ud | 0.65 | ud | ud |
| 9 | 1.09 | ud | ud | 1.22 | ud | ud | 1.00 | ud | ud | 0.65 | ud | ud |

ud = undetectable

**Table 4 :** Reaction times (t) and concentrations (C) of TBA, HTBA and DTBA in aqueous solutions in the presence of soil without illumination

Integrating equation (2) for n = 2 and solving for "d" yields:

$$d = C_{max}.t/(t+t_{0.5}) \tag{3}$$

where:

$$t_{0.5} = halftime = 1/C_{max}.K \tag{4}$$

In aqueous solutions "$C_{max}$" has the same value as the concentration of the organic substance before the beginning of photodegradation process (C0). Because of the sorption phenomena, in the presence of soil "$C_{max}$" could not have the same value of the herbicide concentration in the solution used for preparing the soil suspension ($C_0$). Equation (3) allows the evaluation of the maximum degradable quantity, half-time and, indirectly, the value of the kinetic constant, by rearranging equation (4).

From the values of determination coefficients ($R^2$) in Table 5 it results that the sorption reactions can also be described by the second order kinetic equation. In this case "$C_{max}$" assumes the significance of maximum sorbable quantity (Figure 1).

Comparing the results of Table 5 with data in Table 1 it is possible to see that the organic matter content is the most effective characteristic of soil in the degradation process. Soils with higher content of organic matter (P1 and P7) seem to protect mostly the herbicide from the photochemical reaction. Probably, this evidence is due to competitive reactions of free radicals with the soil organic substances.

| Sample | n | $R^2$ | Suntest $C_{max}$ | $t_{0.5}$ | K | n | $R^2$ | UV-Hg Lamp $C_{max}$ | $t_{0.5}$ | K | n | $R^2$ | in the dark $C_{max}$ | $t_{0.5}$ | K |
|---|---|---|---|---|---|---|---|---|---|---|---|---|---|---|---|
| P1 | 2 | 0.999 | 2.06 | 1.30 | 0.37 | 2 | 0.999 | 2.10 | 0.65 | 0.71 | 2 | 0.999 | 1.10 | 0.28 | 3.26 |
| P2 | 2 | 0.999 | 2.21 | 0.50 | 0.90 | 2 | 0.999 | 2.22 | 0.16 | 2.76 | 2 | 0.999 | 0.96 | 0.05 | 19.30 |
| P7 | 2 | 0.999 | 2.23 | 0.76 | 0.59 | 2 | 0.999 | 2.20 | 0.19 | 2.42 | 2 | 0.999 | 1.18 | 0.12 | 6.86 |
| P10 | 2 | 0.999 | 2.19 | 0.27 | 1.67 | 2 | 0.999 | 2.21 | 0.13 | 3.50 | 2 | 0.999 | 1.53 | 0.10 | 6.52 |

**Table 5 :** Parameters of reaction kinetic: reaction order (n), determination coefficient ($R^2$), halftime ($t_{0.5}$), maximum degradable quantity ($C_{max}$), kinetic constant (K)

Instead, the higher surface extent of soils which are richest of clay minerals and show the higher value of exchange capacity (P7 in the P1, P7 group; P10 in the P2, P10 group), increases the degradation rate. In this case the presence of Montmorillonite and Kaolinite is more effective than Illite.

Finally, it seems that also the presence of $Fe_2O_3$ can exert some protection effect on the herbicide decreasing its degradation rate.

The values of maximum sorbable quantity in Table 5 agree with the percent of clay content in soils but, the sorption reaction cannot protect the herbicide from the degradation process. In fact, the calculated maximum degradable quantities have almost the same value of the concentrations ($C0_2$) of the herbicide in the aqueous solutions (Figures 2 and 3).

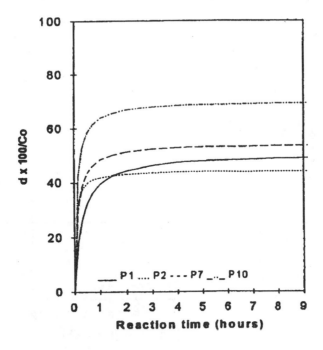

**Figure 1 :** Trend of sorption reactions (equ. (3)) - in the dark

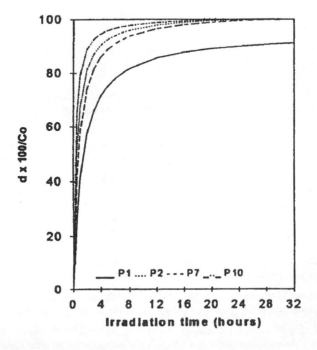

**Figure 2 :** Trends of degradation process (equ. (3)) - SUNTEST

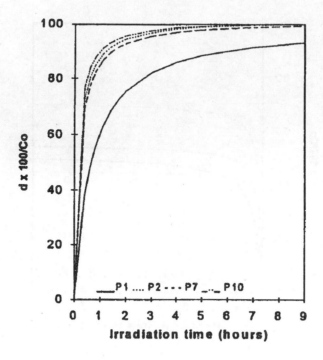

**Figure 3 :** Trends of degradation process (equ. (3)) - UV-Hg Lamp

## References

□   1. Méallier, P., Mamouni, A. and M. Mansour. Photodegradation des molecules phytosanitaires. VII: Photodégradation du Carbétamide seul et en présence d'adjuvants de formulation.
*Chemosphere* 20, 267-273 (1990).

□   2. Méallier, P., Mamouni, A. and M. Mansour. Photodegradation of pesticides. VII: Photodegradation of carbetamide: Photoproducts. *Chemosphere* 26, 1917-1923 (1993).

□   3. Moza, P. N., Sukul, P., Hustert, K. and A. Ketrupp. Photooxidation of metalaxil in aqueous
solution in the presence of hydrogen peroxide and Titanium Dioxide. *Chemosphere* (1993),
in press.

□   4. Pusino, A. and C. Gessa. Photolysis of acifluorfen in aqueous solution. *Pest. Sci.*, 32, 1-5
(1991).

□   5. Kawaguchi, H.. Photooxidation of 2-propanil in aqueous solution in the presence of
hydrogen peroxide. *Chemosphere*, 27, 577-584 (1993).

□   6. Mamoumi, A., Mansour, M., Méallier, P., and Ph. Schmitt. Abiotique degradation pathways
of selected pesticides in the presence of oxygen species in aqueous solutions. (1993), in press.

□   7. Pelizzetti, E., Minero, C., Carlin, V., Vincenti, M. and E. Pramauro. Identification of
photocatalytic degradation pathways of 2-Cl-s-triazine herbicides and detection of their
decomposition intermediates. *Chemosphere* 24, 891-910 (1992).

# 3.4.     Short Communications

*3.4.1.      Bound Residues : Can Photodegradation Assist or Substitute Biodegradation?*

## F. Andreux

Université de Bourgogne, Centre des Sciences de la Terre
6, Boulevard Gabriel 21000 Dijon, France

In soil and water media, pesticides can undergo either biological transformations or abiotic transformations, or both. Biological transformations are mainly governed by microorganisms, or their exoenzymes bond to soil components. They can result either in the partial degradation, or in the total mineralization and desappearance of the pesticide. Biological transformations can also occur in plants, following the absorption of pesticides by active roots. In a first step, enzymic oxidation, reduction, hydrolysis or acylation can occur. In a second step, the pesticide can form *conjugated residues* with simple biomolecules, as well as with structural polymers such as lignin (Khan, 1980).

There are several common points between the formation of conjugates in plants and bound or unextractable residues in soils. The main of them are *their low or unforeseeable reversibility* and *their poorly identified structure*. As mentioned earlier (Andreux et al., 1993 ; Völkel et al., 1994), important progress has been made in our knowledge about bound residues formed with chloroanilines. These molecules are constitutive (*"metabolites"*) of about twenty pesticides, and present a generally recognized toxicity. One reason for this toxicity is their high reactivity with oxidizing systems. In soils, chloroanilines easily form substituted imino-quinones (called *"monomers"*), through nucleophilic addition on oxidized phenolic rings. The products of such reaction are humic-like polymers that act as a trap for the chloroaniline molecules, resulting in their detoxification.

The problem is the permanence of such detoxification. Are we sure that the incorporation of toxic molecules in polymers is irreversible? The stability of the monomers was studied in the case of 3,4-dichloroaniline (DCA) added to catechol, to form a substitued dichloro-phenylamino-cyclohexadiene-dione (DCPC). Both DCA and DCPC were labelled with $^{14}C$, and incubated in a crop soil for more than one year. The kinetics of evolved $^{14}CO_2$ were monitored, as shown on Fig. 1. Their was no evidence of depleting effect on total soil respiration, at least with DCPC. The decomposition of the molecules proceeded very slowly: after 119 days at 20°C, 2.1% of DCA, and less than 0.03% of DCPC, were decomposed. This reaction would suggest that polymerization contributed to an efficient elimination of the soluble toxic DCA. However, there was still a proportion which remained accessible to microbes, and could be released in the soil solution. It is therefore necessary to eliminate the free material, for instance using photochemical degradation. This is probably easier in water media than in soils, but it is now well known that the residues are preferentially bound to the fine clay-size fraction (Völkel et al., 1994), and therefore can be dispersed in soil gravitationnal and run-off water.

Limited information is available about such possibilities. The degradation of 4-chloroaniline and 2,4,6-trichloroaniline bound to different soil fractions was studied earlier by Ter Meer Beck (1986). Table 1 shows that the photodegradability of 2,4,6-trichloroaniline was very easy when it was free, but was slower with soil-bound material. Eventhough humin was a strongly bound fraction, it was as biodegradable as photodegradable. This suggests that photolytic methods applied to adsorbed substrates could be combined with biological methods, to prevent the release of bound residues from the soil humin fraction. Possibly, other abiotic reactions, such as catalytic degradation in the presence of oxygen could supplement the effect of both biological and photochemical reactions.

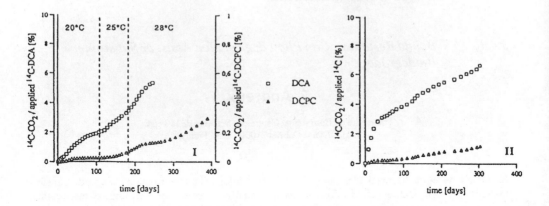

**Fig. 1 :** Cumulative curve of $^{14}$C-CO$_2$ evolved from DCA and DCPC during incubation at stepwise increasing temperature, from 20 to 28°C (I), and at 28°C (II).

| Soil fraction | Conc. of bound $^{14}$C μg/g[a] | Aerobic mineralization 28 days %$^{14}$CO$_2$ | Photodegradation λ = >290 nm 17 hr %$^{14}$CO$_2$ |
|---|---|---|---|
| Inorganic fraction | 0.019 | 1.08 | 0.77 |
| Organic fractions | | | |
| Humic acids | 1.03 | 0.39 | 0.09 |
| Humin | 0.24 | 1.39 | 1.11 |
| Fulvic acids | 0.22 | 5.05 | 1.48 |
| Total soil | 0.05 | 1.91 | 1.44 |
| Free | | | |
| 2.4.6-trichloroaniline | - | 0.614 | 29.88 |

Source : Ter Meer-Bekk.48

[a] μg (Equivalent to the parent compound)/g dry soil fraction.

**Table 1 :** Concentrations of soil-bound residues of $^{14}$C-labelled 2,4,6-trichloro-aniline in soil fractions, and their mineralization and photodegradation (after Ter Meer Bekk, 1986)

# References

◻    F. Andreux, I. Scheunert, Ph. Adrian, M. Schiavon, *The binding of pesticide residues to natural organic matter, their movement, and their bioavailability*, in: Fate and prediction of environmental chemicals in soils, plants, and aquatic systems, M. Mansour (Ed.), Lewis Publishers, Boca Raton, 1993, 13, 133-148.

◻    S. U. Khan, Plant uptake of unextractable (bound) residues from an organic soil treated with prometryn, *J. Agric. Food Chem.*, 28, 1980, 1096-1098.

□    C. Ter Meer-Bekk, *Bildung, Charakterizierung und Bedeutung sogenannter "Gebundener Rückstände" in Boden*, Doctor Thesis, Technische Universität München, Germany, 1986.

□    W. Völkel, Th. Choné, F. Andreux, M. Mansour, F. Korte, Influence of temperature on the degradation and formation of bound residues of 3,4-dichloroaniline in soil, *Soil Biol. Biochem.* *(submitted for publication)*.

### 3.4.2.     *Degradation study of mecoprop in subsoil*

# I. S. Fomsgaard

Danish Institute of Plant and Soil Science, Flakkebjerg, DK-4200 Slagelse

In the present study degradation of mecoprop in subsoil and in surface soil was investigated in a danish soil profile. The experiments were part of an EEC joint programme. The purpose of the experiments was to elucidate the correlation between degradation rate of a herbicide (mecoprop) and soil microbial activity shown by degradation of $^{14}$C-labelled Na-acetate.

Soil samples were taken at three depths, 15, 45 and 75 cm. All samples were taken in metal tubes. Samples from 45 and 75 cm were kept undisturbed where samples from 15 cm were sieved to remove plant remains. $^{14}$C-labelled mecoprop (0.05 mg kg$^{-1}$) was added to the subsoil samples by injection into the soil core and to the topsoil samples by mixing in an erlenmeyer flask. All samples were incubated at 10°C (four replicates), evolved $^{14}CO_2$ was measured as shown in fig. 1. At 153 days the order of reaction for the evolution of $^{14}CO_2$ changed. Analysis of variance (ANOVA) of the summarized areas below the curves for $^{14}CO_2$-evolution from mecoprop at 150 days showed a significant difference between the degradation rate for mecoprop at the three depths with the highest degradation in the plough layer and decreasing degradation rates at increasing soil depths.

At each depth further four replicates of soil samples were taken for determination of microbial activity by degradation of $^{14}$C-Na-acetate. Na-acetate is easily degraded by most microorganisms since it is a natural substance in their metabolism. Degradation of Na-acetate in the plough layer (15 cm) was significantly more rapid after 2 hours than in the deeper layers (45 and 75 cm) (fig. 2). After 10-20 hours the evolution of $^{14}CO_2$ from $^{14}$C-Na-acetate was more rapid in soil from 45 and 75 cm than in soil from 15 cm. This could be due to differences in utilization of Na-acetate for the microorganism at differ- ent depths possibly caused by the fact that subsurface organisms are generally oligotrophic. The initial capacity (after 2 hours) of degrading $^{14}$C-Na-acetate (5 mg kg$^{-1}$) to $^{14}CO_2$ was considered the best expression for microbial activity.

A plot of summarized areas of %$^{14}CO_2$ from mecoprop after 150 days versus degradation rate of Na-acetate at 2 hours (fig. 3) show a curve that does not seem to be rectilinear. More values would be necessary to be able to describe the correlation.

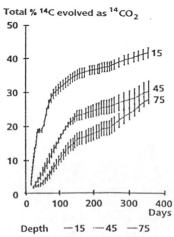

Depth     —15  —45  —75

Degradation of $^{14}$C-mecoprop
0.05 mg/kg. Site 1

Fig. 1

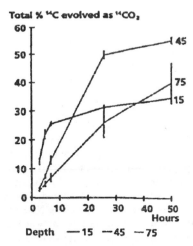

**Total % ¹⁴C evolved as ¹⁴CO₂**

Depth —— 15  —— 45  —— 75

**Degradation of ¹⁴C-Na-acetate
5 mg/kg. Site 1**

Fig. 2

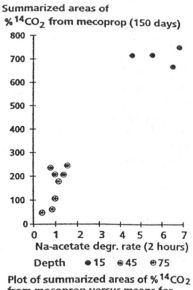

Summarized areas of
% $^{14}CO_2$ from mecoprop (150 days)

Depth    ● 15    ◉ 45    ⊕ 75

Plot of summarized areas of % $^{14}CO_2$
from mecoprop versus means for
Na-acetate degradation rate. Site 1.

Fig. 3

### 3.4.3.     Microbial breakdown of pesticides affected by altered oxygen conditions

# J.P.M. Vink

Ministry of Transport, Public Works & Water Management
Rijkswaterstaat, Directorate Flevoland, Research dept.
P.O. Box 600, 8200 AP Lelystad, The Netherlands

Transformation of pesticides, as induced by microbial metabolism, will occur for most organic pesticides applied to soil surfaces. These transformation kinetics can be described and simulated using non-linear models based on microbial growth, competition within the population, substrate availability and temperature inhibition. Transformation plays an important role in transport behaviour of pesticides and emission from soil layers. However, when leached into environments with lower oxigen concentrations (e.g. subsoils, surface waters and saturated sediments) transformation pathways of these compounds may change drastically as a result of altered, mostly unfavourable conditions for micro-organisms. In sediments of large water bodies in The Netherlands, pesticides and pesticide residues were detected, in spite of vast dillution factors for water and suspended solids. Current characteristics, such as $DT_{50}$, can no longer be applied to evaluate ecological risks in these systems. Research is needed on persistency of pesticides in low-oxygenous environments and aquous systems.

Transformation of pesticides in saturated soils, ground water and surface waters may have a substantial effect on their occurrence in aquous ecosystems. Concentrations of pesticides in extracted ground water, used for drinking water, were reported on numerous occasions. Sediments and deep waters can become anaerobic during the summer months, causing microbial transformation to proceed along different pathways. Only very limited information is available on the transformation of pesticides in water saturated sediments or surface waters, and a reasonable certainty about the formation of harmless metabolites does not exist. Therefore, the influence of transformation in the subsoil on concentrations in low-oxygenous environments has not been evaluated.

To study transformation behaviour of pesticides in soil, water saturated sediments and surface water of varying composition, an air-tight box, 100 l capacity, is constructed to contain disturbed (prepared) and undisturbed (unprepared) samples. Inert nitrogen gas is used to expel oxygen from the system, a valve avoiding excessive air pressures. Samples are subjected to very low oxygen levels between zero and 1%. Photodecomposition is limited by using black paint and dark stained glass for a window. Oxygen concentrations are measured with a polarographic sensor, enabeling reliable monitoring of oxygen levels within this concentration range. A small fan provides sufficient gas turbulence in the system to warrant accurate, continuous oxygen measurement. Transformation behaviour of three groups of pesticides were tested: aldicarb (carbamate), MCPP (phenoxy-acid) and simazine (s-triazine). Incubation tests were carried out for oxic and anoxic undisturbed soil and anoxic lake bottom sediment. Results show a remarkable decline in transformation rates as oxygen concentrations decrease. Under low oxygen concentrations, only marginal concentrations of the first and second metabolites of aldicarb were formed. Half-life time of simazine and MCPP increased from 15 to 50 days and from 4 to 40 days respectivily.

An attempt has to be made to obtain, primarily, some general insights on the relationship between transformation kinetics of pesticides in low-oxygenous environments and the characteristics of the transport medium. Subsoils, aquifers and surface waters have to be categorized in terms of pH, redox and suspended solids dependency.

# SECTION IV

# GROUND AND SURFACE
# WATER MONITORING

Chairman :  G. PUCHWEIN
(Bundes Anstalt für Agrobiologie - Austria)

# GROUND AND SURFACE WATER MONITORING

## 4.1.    Introductory Presentation

*4.1.1.    Monitoring Ground and Surface Waters : Sampling Strategy and Implementation in Legislation*

### A.M.A. van der Linden

National Institute of Public Health and Environmental Protection,
Laboratory of Soil and Groundwater Research, P.O.Box 1, NL-3720 BA Bilthoven.

## Abstract

The number of pesticides that have been shown to occur in ground and surface water is growing steadily. Monitoring programmes, that have been conducted recently, indicate that contamination is not incidental and that transport of pesticides and their metabolites deserves more attention; both scientifically and politically.

From the scientific point of view more attention should be paid to: sampling strategies, sampling methods, analytical methods and interpretation and explanation of observed results. Due to the very high costs involved in sampling and chemical analysis, sampling strategies are very important. Interpretation and explanation of the results should at least verify the agricultural use of the pesticide as the source of the contamination.

Results of monitoring programmes do not have an explicit role in pesticide legislation in most countries. Up till now, observations probably influenced expert judgement and in this way may have influenced registration. In the Netherlands, an explicit role is given to monitoring results in the decision tree 'Leaching to Shallow Groundwater'; criteria for the use of monitoring results are, however, not yet finalized. Also, in the (concept) Uniform Principles a pronounced role is given to monitoring results (especially negative results). Here again criteria are lacking.

In this paper some suggestions for sampling strategies are given as well as criteria that can be used to qualify monitoring results. The suggested role of monitoring results in decision making is highlighted and implications for (re)registration of pesticides are indicated.

## Introduction

Groundwater and surface water is often referred to as a source for drinking water, and for this function it should be free of (toxic concentrations of) hazardous chemicals. Not only for the function 'source of drinking water', but also for other functions the quality of the water should be beyond discussion. Important functions in this respect are:

■    its role as process water in industry;

■    its role as irrigation water in agriculture;

■        its function as a habitat for organisms.

For other functions the quality of the water is somewhat less important.

In national and international legislation maximal permissible concentrations for individual or total concentrations of pesticides (including metabolites, degradation and reaction products) are set drinking water in order to safeguard people from possible harmful effects. For instance in the USA drinking water should comply with WHO-standards and in countries of the European Union the maximum permissible concentration is 0.1 $\mu$g/L for individual compounds or 0.5 $\mu$g/L for the sum of compounds. Whereas the drinking water itself has received broad attention, its sources - mainly groundwater and surface water - are mostly not protected to this extent yet. Van Haasteren (1993) argues that groundwater should at least comply with the values set for drinking water and that groundwater should be suitable for drinking water without special purification. For several pesticides, especially polar pesticides, such a purification would technically be very difficult and quite expensive.

From monitoring programmes in several countries (see, amongst others, Friesel, (1986) and Fielding (1992)) it has become clear that both surface water and groundwater may sometimes be contaminated with pesticides and that concentrations in many cases may exceed maximal permissible concentrations. In Table 1 a compilation of monitoring results of several pesticides in Dutch groundwaters is given. For the data reported in Table 1 it is checked that the pesticides were used according to label instructions; for results of other monitoring programmes this is not necessarily the case.

In the near future results of monitoring programmes will become more important in the process of (re-)registration of pesticides. In the Netherlands monitoring will be implemented in legislation in January 1995. The exact role of monitoring data in the evaluation of possible leaching to groundwater is indicated in Figure 1. Positive findings trigger further research in case of relatively low concentrations or a denial of (re-)registration in case of relatively high concentrations. For a more detailed discussion of Figure 1 the reader is referred to Brouwer, et al. (1993). For surface water the role of monitoring data in legislation is not that clear yet. In practice, positive findings in surface water trigger a stop in the intake of surface water by drinking water companies. In the latest draft of the Uniform Principles it is stated that monitoring data will be taken into account in the ultimate decision on authorization of plant protection products. According to this draft adequate monitoring should be performed to show that concentrations stay below the threshold values stated above.

These recent developments urge for some international standardization regarding sampling strategies, sample taking, analyses and interpretation of the results. This contribution is intended to contribute to the standardization discussion.

| pesticide | number of samples[a] | number of positives[b] | concentration (μg/l) | |
|---|---|---|---|---|
| | | | minimum[c] | maximum |
| 1,3-dichloropropene | 207 | 9 | < 0.1 | 6.8 |
| 1,2-dichloropropane | 183 | 110 | < 0.1 | 200 |
| aldicarb | 67 | 4 | < 0.05 | 0.4 |
| atrazine | 365 | 217 | < 0.02 | 0.59 |
| bentazone | 328 | 140 | < 0.02 | 98 |
| dichlobenil | 22 | 20 | < 0.02 | 6 |
| ETU | 193 | 83 | < 0.1 | 42 |
| MCPA | 27 | 0 | < 0.2 | < 0.5 |
| mecoprop | 48 | 1 | < 0.2 | 2 |
| metolachlor | 186 | 41 | < 0.03 | 0.5 |

[a]   Number of samples included in the RIVM groundwater monitoring programme in
      the period 1985 -1991.
[b]   Number of samples with a concentration equal to or above the limit of
      determination.
[c]   < indicates a concentration below the limit of determination. The value reported is
      the lowest limit of determination achieved in the monitoring programme.

**Table 1 :** Residues of selected pesticides in Dutch groundwaters. Data refer to
concentrations (μg/L) found in shallow groundwater (maximal depth c. 3 m
below soil surface). Reference: Cornelese and Van Maaren (1993).


# Monitoring

## *Strategy*

The monitoring strategy for groundwater and surface water may be somewhat different from
each other. The objective(s) of the monitoring programme determine(s) to a large extent the final
set-up of the programme. The possible differences in strategies are mainly due to differences in
the objectives and to differences in 'response time' to an application of pesticides. For surface
water the most relevant response time is in the order of a few minutes to several days; for shallow
groundwater the characteristic response time usually is several weeks to two years and for deeper
groundwater several years to several decades. In case of deep groundwater tables (more than 5
m) one may choose to sample soil moisture (in a zone definitely below the root zone) and estimate
from these data the load of the groundwater assuming that further transformation is negligible.
An overview of several aspects of the different strategies is given in Table 2.

| surface water | groundwater |
|---|---|
| - define purpose | - define purpose |
| - select compounds | - select compounds |
| - select water body | - select area / region |
| - start interviewing farmers | - start interviewing farmers |
| - select sampling points | - select fields |
| - select period (application period) | - estimate arrival times |
| - sample flow-proportional | - select sampling periods |
| - sample frequently during period | - sample area ad random |
|  | - prepare mixed samples |

**Table 2 :** Strategy aspects of monitoring groundwater and surface water

## *Sampling methods*

In many cases the objective of a monitoring programme is to evaluate possible threats to one or several functions of the water body. According to current evaluation procedures (comparison of environmental concentrations (PEC) with no effect (or no concern) concentrations (NEC)) almost always a time averaged, a space averaged or a time and space averaged concentration is necessary. Time and space may be dependent on the objective of the monitoring programme; in the draft Uniform Principles it is suggested that the most relevant concentration for groundwater would be: an average concentration over a period of one year. The most relevant space in this case would be: the field of application.

Equipment for sampling of surface water - with options for time and flow proportional sampling - is commercially available. One only needs to check whether the apparatus is inert to the compounds of interest. In most cases an apparatus with glass or stainless steel tubing and collection vessels will be suitable. Cooling of collected samples down to c. 2°C might in some cases be required. Further mixing of samples after collection should be avoided.

Even within one field the spatial variability of soil characteristics might be substantial and, correspondingly, a great variability in concentrations might be expected. In order to obtain a reliable estimate of the space averaged concentration, a vast number of groundwater samples should be taken and the sample spot should be chosen ad random in the field. In order to reduce analysis costs, samples may be mixed to composed samples. At least four composed samples should be analyzed. If one wants to have also a time averaged concentration, the sampling should be repeated. In the following (in italics) a method is described that can be used to obtain space averaged concentrations. Of course sampling equipment should be thoroughly cleaned between two samplings to avoid cross contamination, and control samples should be taken using the same equipment. After collection sampled should be analyzed (or at least extracted) as soon as possible. Transport and, if necessary, storage should be done under cold conditions.

## *Method to obtain a space averaged concentration*

*Using ad random functions on a computer or data from ad random tables, four parallel transects are drawn over a field. These transects should not be parallel to either sides of the field. On each of the transects five ad random spots are indicated and in this way a total of twenty sample spots are identified. On these spots, using a 10 cm hand auger, a hole is drilled to a depth of about 50 cm and a 10 cm outer diameter tube is installed in order to prevent topsoil from falling down. Using a hand auger of 6 cm the hole is drilled further until a depth of 75 cm below the groundwater table (in some soils a suction auger is necessary for the zone below the groundwater table). Then a stainless steel filter of 50 cm length equipped with a stainless steel capillary (a so-called sampling lance) is installed to cover the saturated zone between 25 and 75 cm below the groundwater table. The capillary is firstly attached to a vacuum in order to rinse the filter and secondly to a collection bottle, also put under vacuum. In this way, depending on the size of the collection bottle and the depth of the filter (maximum depth c. 5 m), quite easily several hundred millilitres of water can be collected. After collection, the 20 samples are brought to the laboratory and four composed samples (of five samples each) are prepared using equal volumes of collected groundwater. The composed samples then are transferred to the analysing laboratory. Upon receival of the results, average values and confidence intervals can be calculated.*

In case of groundwater at greater depths (more than 5 m) in practise it is not possible to take as many as 20 (or more) groundwater samples (for routine monitoring). In this case one must either rely on results of fewer samples (still a minimum of four samples is required if one is interested in average concentrations) or soil moisture samples at a certain depth below the root zone should be collected. This may be achieved by collection of (wet) soil samples and extraction soil water in either of the three following ways:

—    centrifuging soil moisture out of the wet soil samples;

—    extraction of soil moisture by vacuum or over-pressure;

—    replacement of soil moisture by heavy fluids and collection of the replaced moisture.

The soil moisture obtained by either of these methods may be handled in the same way as described above.

## *Analytical methods*

Analyses should be performed in a laboratory that complies with GLP requirements. The analytical method applied should have a lower limit of determination that is not higher than 50 % of the maximal permissible level and, preferably the limit of determination is not higher than 10% of this level. Positive samples should be analyzed by two fully independent methods; one of the two with a mass specific detection method (MS). In the total monitoring programme for a single compound sufficient control samples (c. 10% of all analyses) should be incorporated. Control samples should include blanks (for instance distilled water), standard series, reference samples, samples spiked at the level of interest and blanks from a non-contaminated source. An advantage would be that the analysing laboratory is involved in ring tests on the compound of interest. In case of doubt, contra expertise might be invoked from an independent analysing laboratory.

## *Interpretation of results*

Results of the monitoring programmes must not be reported as results of the analyses only. If the results are to be used in the evaluation of the (re-)registration of agricultural pesticides, much more information on the monitoring programme is necessary. A report on the monitoring programme should at least contain information on:

—    a review of other monitoring data on the same compound(s);

—    the agricultural use of the pesticide (application rates, application period, method of application, etc.);

—    a description of the area (or field) in which the samples are taken (including: soil types, crops, slopes, drains, etc.);

—    climatic conditions in the period between application and sample taking

—    the history of the area (or field);

—    a comparison of obtained results with results expected on the basis of knowledge on pesticide mobility and persistence in the medium of interest

—    the severity of the results (a comparison of observed concentrations with threshold values or no environmental concern concentrations (NEC).

Furthermore it would be helpfull to indicate in the report the representativity of the monitoring results (the domain for which the results are thought to be representative), both in time and space.

## Conclusions

Groundwater and surface water have many functions in our society and for most of these functions the quality of the water should be beyond doubt. In this respect not only the current use of the water, but also any possible future use of the water should be regarded. Therefore, monitoring should assist in the prevention of water from being contaminated and not only function as a check on the quality just before use.

Because of the increasing role of monitoring data in the (re-)registration of pesticides, (international) harmonization of monitoring of surface water and groundwater should receive much attention in the coming few years. This harmonization should not only include sampling strategies and analytical methods, but also methods for interpretation (especially statistical methods).

Political bodies have to define more clearly in which way the maximal permissible concentrations should be used (or interpreted). For instance they should indicate whether confidence limits should be included in the analyses or not.

## References

□   Brouwer, WWM, Van Vliet, PJM, Linders, JBHJ, Van den Berg, R. 1993. Environmental risk assessment for plant protection products: the Dutch approach. *Environmental Outlook* 4: 30-35.

□   Cornelese, AA, Van Maaren, HLJ. 1993. Veldonderzoek bestrijdingsmiddelen; resultaten 1991. RIVM report no. 725803006. (in Dutch).

□   EEC 1980. Commission of the European Communities. Directive on the quality of drinking water for human consumption. No. L 299/11-29.

□   Fielding, M. (ed). 1992. Pesticides in ground and drinking water. Water pollution research report 27. Commission of the European Communities. ISBN 2-87263-068-6.

□   Friesel, P. 1986. Grundwasserqualitätsbeeinträchtigungen durch die Anwendung van Pflanzenschutz- und Schädlingsbekämpfungsmitteln (PSM). *Bundesgesundheitsblatt,* 29: 424-427. (In German)

□   Van Haasteren, JA. 1993. Pesticides and ground water. Council of Europe Press, ISBN 92-871-2384-5

## Legend

Figure 1 : Dutch decision tree 'Risk of leaching to shallow ground water.

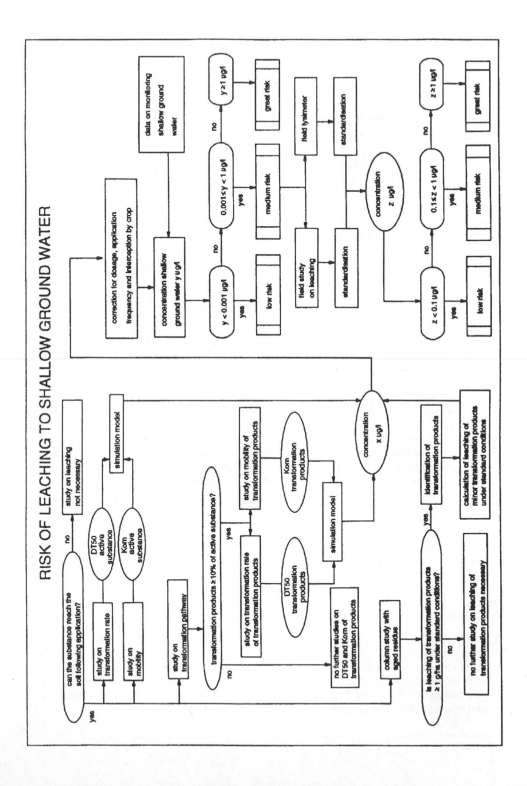

RISK OF LEACHING TO SHALLOW GROUND WATER

# 4.2.    Platform Presentations

*4.2.1.      Monitoring Pesticide Runoff and Leaching in a Surface-Dominated Catchment: Implications for Aquatic Risk Assessment[1]*

# P. Matthiessen[a], D. Brooke[b], R. Clare[c], M. Mills[d] and R. Williams[e]

[a]    Ministry of Agriculture, Fisheries & Food, Fisheries Laboratory, Remembrance Avenue, Burnham-on-Crouch, Essex CM0 8HA, UK.
[b]    Building Research Establishment, Bucknall's Lane, Garston, Watford WD2 7JR, UK.
[c]    ADAS Rosemaund, Preston Wynne, Hereford HR1 3PG, UK.
[d]    National Rivers Authority, Welsh Region, St. Mellons Business Park, Cardiff CF3 0LT, UK.
[e]    Institute of Hydrology, Wallingford, Oxfordshire OX10 8BB, UK.

## Abstract

This paper describes an intensive project to measure the translocation of pesticides from agricultural fields to a stream in a surface water-dominated catchment at ADAS Rosemaund in Herefordshire, UK. The purpose of the research, which started in 1987, was to gather data on pesticide concentrations in soils, soil water, drainage water and a stream from a catchment where inputs were well-defined, in order to assist the development of computerised translocation models. In the 6 years that the project has been in existence, approximately 20 pesticides have been studied in detail, with attention focusing on the hour-by-hour changes in concentration which occur in the stream during and after rainfall events. Most data relate to the water-soluble herbicides, but experiments have also been conducted with organochlorines, organophosphates, carbamates, pyrethroids and certain fungicides. All pesticide applications have been made using Good Agricultural Practice.

In summary, the data show that almost all the pesticides tested appear transiently in the stream after rain, generally at maximum concentrations in excess of 0.5 $\mu$g l$^{-1}$. Peak concentrations for some pesticides have reached 50 $\mu$g l$^{-1}$ and above, but concentrations generally return to baseline values (< 0.1 $\mu$g l$^{-1}$) within 24 hours of the cessation of rain. Studies of soil water have shown that a proportion of the incident rainfall is able to by-pass the main soil blocks and rapidly gain field-drain depth by means of preferential flow paths. The majority of the translocated pesticide is carried by this route to the field drains. Soil water regimes of this type are not uncommon, particularly on clay soils which are prone to cracking, but this phenomenon is not generally taken into account when pesticides are being assessed for the risk they pose to the aquatic environment.

The concentrations of some pesticides seen in the Rosemaund stream are more than an academic curiosity because the use of crustacean *in-situ* bioassays has demonstrated lethal effects caused by some of the more acutely toxic insecticides. It is therefore important that findings such as these are recognised in pesticide risk assessment strategies. One approach has been to develop improved models of pesticide translocation, and the Rosemaund data have been used successfully to validate a fugacity model which predicts mean pesticide concentrations during rainfall events.

# Introduction

When assessing the potential risks which new pesticides could pose to aquatic organisms, it is clearly essential to be able to predict the exposure which the latter may experience. Although the exposure which results from accidental oversprays of streams and ponds, or from spray drift, may be rather simply quantified, that derived from residues which leach down the soil profile into field drains and thence to ditches and streams is much harder to predict. Indeed, many pesticide manufacturers and regulatory authorities do not attempt such a calculation with new compounds, but merely make a semi-qualitative assessment of leaching potential. Perhaps the most well-known of these estimates of leaching is the Groundwater Ubiquity Score (GUS) (Gustafson, 1989) which is a simple synthesis of the soil organic carbon adsorption coefficient ($K_{oc}$) and the degradation half life in soil. There are a variety of more complex models, many of them computerised, which take into account additional properties of the chemical and soil in question (OECD, 1989), but most are too data-intensive for use in the early stages of risk evaluation. Furthermore, they tend to assume that the soil behaves as a homogeneous block, rather like a chromatographic column, whereas it is known that many soils do not display these characteristics. For example, in clay soils which are prone to cracking, rainwater falling in the autumn is able partially to by-pass the surface soil layers and rapidly reach the level of the field drains. Pesticides which are dissolved by this water may therefore not reach equilibrium with the bulk soil, but would be preferentially transported to both the field drains and the groundwater.

The research outlined in this paper was primarily undertaken to measure pesticide translocation under the conditions described above at ADAS Rosemaund in the west of England, in order to provide reliable data which could be used to validate simple predictive models of 'worst case' pesticide exposure from leaching in headwater streams. Such models would be of use in the early stages of risk assessment before field data are available. Some of the Rosemaund data have already been published elsewhere (Brooke & Matthiessen, 1991; Williams et al., 1991 a & b; Matthiessen et al., 1992; Di Guardo et al., 1994; Matthiessen et al., 1994; Williams et al., 1994), but this paper presents a brief overview of the results obtained and conclusions reached to date.

# Materials and Methods

The methods used have been fully described by Matthiessen et al. (1992) and in subsequent papers (see above), and only a very brief summary will be given here. The 180 ha catchment used for the experiments almost exactly encompasses a mixed farm run by ADAS Rosemaund, situated near Hereford in the west of England. It is gently sloping (76-115 m altitude range), and a stream arising on the site drains into the River Lugg after accepting water from field drains, surface runoff and base flow. Most fields contain a network of drains at 1 m depth with permeable backfill to within 0.3-0.5 m of the surface. The dominant Bromyard soil is a silty clay loam with low organic matter (2-3% at the surface, little below 0.35 m). It is prone to cracking when dry and is underlain at depths of 0.6-1.5 m by generally impermeable siltstone. Most leaching water therefore reaches the stream via the field drains. Annual average rainfall is 664 mm, and the stream flows almost continuously. The farm supports a wide range of arable crops (typically winter cereals, grass, oilseed rape, beans and hops) and pesticides are applied according to Good Agricultural Practice guidelines.

The pesticide translocation project has focussed on a wide range of substances, many of which are prone to leaching (e.g. phenoxy herbicides) but others which might appear less likely to migrate through soil (e.g. trifluralin). The pesticides discussed in this paper and their relevant physicochemical and degradation properties are listed in Table 1.

| Pesticide name (and group) | Water solubility (mg/l) | Vapour pressure (Pa) | Soil half-life (days) | Soil organic carbon adsorption coefficient $K_{oc}$ $(1 \ kg^{-1})^a$ | Groundwater Ubiquity Score[b] |
|---|---|---|---|---|---|
| Aldicarb (carbamate) | 6000 | $4 \cdot 10^{-3}$ | 30 | 30 (5.5) | 3.73 'probable leacher' |
| Atrazine (triazine) | 30 | $4 \cdot 10^{-5}$ | 60 | 122 (89.8) | 3.40 'probable leacher' |
| Carbofuran (carbamate) | 351 | $1.5 \cdot 10^{-3}$ | 50 | 22 (16) | 4.51 'probable leacher' |
| 2,4-D (phenoxy acid) | 890 | $1.1 \cdot 10^{-3}$ | 10 | 20 (265) | 2.70 'transition' |
| Dicamba (benzoic acid) | 5600 | $4.5 \cdot 10^{-3}$ | 14 | 4 (66) | 3.89 'probable leacher' |
| Dichlorprop (phenoxy acid) | 350 | $<1 \cdot 10^{-5}$ | 10 | 20 | 2.70 'transition' |
| Dimethoate (organophosphate) | 25000 | $1 \cdot 10^{-3}$ | 7 | 20 (3.9) | 2.28 'transition' |
| Isoproturon (urea) | 55 | $3.3 \cdot 10^{-6}$ | 20 | 129 (73) | 2.46 'transition' |
| Lindane (organo chlorine) | 6.5 | $3 \cdot 10^{-3}$ | 266 | 1100 (2587) | 2.32 'transition' |
| MCPA (phenoxy acid) | 825 | $2 \cdot 10^{-4}$ | 7 | 20 (0.041;204. pH dependent) | 2.28 'transition' |
| Mecoprop (phenoxy acid) | 620 | $3.1 \cdot 10^{-4}$ | 7 | 20;79.4 (0.52;81.8. pH dependent) | 2.28; 1.77 'transition' |
| Simazine (triazine) | 5 | $8.1 \cdot 10^{-7}$ | 60 | 130 (37.4) | 3.35 'probable leacher' |
| Triclopyr (pyridine) | 440 | $1.7 \cdot 10^{-4}$ | 46 | 27 | 4.27 'probable leacher' |
| Trifluralin (dinitroaniline) | 0.5 | $6 \cdot 10^{-3}$ | 60 | 8000 (48380) | 0.17 'improbable leacher' |

[a]   $K_{oc}$ values in parentheses estimated from Karickhoff's equation; $K_{oc}=K_{ow} \cdot 0.41$

[b]   Gustafson (1989)

[c]   Data obtained from : Wauchope *et al.* (1992); Howard (1991); Di Guardo *et al.* (1994)

**Table 1 :** Pesticides used in the Rosemaund experiments[c]

The pesticides were applied by tractor-mounted hydraulic sprayer (liquids) or broadcast (granules) during autumn or spring to a variety of crops within the catchment. Application rates ranged from 0.195 kg ha$^{-1}$ (triclopyr) to 3.0 kg ha$^{-1}$ (carbofuran). Chemical monitoring was concentrated on 2 field drains near the upper end of the catchment, and on 2 points in the stream (just below the monitored drains, and at the point where the stream leaves the farm). Samples were also taken of soil and soil water. Drain and stream water samples were taken automatically at hourly intervals by suction sampler, generally when rainfall exceeded approximately 10 mm in 24 h. Pesticide concentrations in soil and water were measured by a variety of methods including gas chromatography, high performance liquid chromatography and gas chromatography/mass spectrometry. Rainfall and drain/stream flowrates were measured automatically by an on-site weather station and gauging weirs, respectively. On some occasions (dichlorprop and carbofuran experiments), the toxicity of stream water was measured *in situ* with a crustacean (*Gammarus pulex*) feeding rate bioassay.

## Results and Discussion

Space precludes presentation of soil data, but these show that pesticide concentrations in the top metre generally reflected application rates and known degradation characteristics. Of more relevance are the data on soil water (also not presented here) which revealed a tendency for most pesticide residues to reach field drain depth at µg l$^{-1}$ concentrations within days to weeks of application. Investigations of soil structure (Carter & Beard, 1992) and hydraulic conductivity showed that the permeability of the upper soil layers was poor during the early winter months, but that the soil at this time was extensively traversed by macropores (cracks, roots and worm holes) which were responsible for by-pass flow to the field drains. Soils in which by-pass flow is an important component of the hydrological regime are common (approximately 28% of UK soils - J. Hollis, pers. comm. 1993).

The highest mean concentrations for each pesticide observed during all rainfall events are summarised in Table 2. Typically, a rainfall event produced a corresponding pesticide peak in drain and stream water within a few hours. Peak concentrations generally occurred either just before or simultaneously with the peak in water flowrate, but declined rapidly again and reached pre-event levels within 12-24h.

| Pesticide[a] | Amount applied (kg) | Area (ha) | Rainfall (mm) | Lag from spraying (days) | Drain (D) or stream (S) | Max. conc. ($\mu$g l$^{-1}$) | Flow-weighted mean conc. ($\mu$g l$^{-1}$) | Pesticide mass translocated during event (g) |
|---|---|---|---|---|---|---|---|---|
| Aldicarb (2;2) | 1.10 | 2.00 | 17.5 | 47 | D | 3.1 | 1.0 | 0.002 |
|  | 1.10 | 2.00 | 16.5 | 2 | S | 2.8 | 0.9 | 0.44 |
| Atrazine (7;2) | 5.60 | 2.00 | 4.0 | 39 | D | 56.5 | 35.7 | 0.085 |
|  | 14.56 | 5.20 | 72.5 | 42 | S | 5.7 | 2.0 | 9.9 |
| Carbofuran (5;1) | 3.00 | 1.00 | 9.0 | 53 | D | 58.4 | 37.2 | 0.09 |
|  | 9.00 | 3.00 | 72.5 | 36 | S | 26.8 | 10.4 | 52.6 |
| 2,4-D (3;2) | 5.18 | 5.18 | 9.0 | 26 | S | 5.1 | 2.6 | 3.2 |
| Dicamba (2;2) | 2.21 | 5.20 | 9.0 | 26 | S | 2.2 | 1.2 | — |
| Dichlorprop (1;1) | 13.52 | 5.20 | 12.0 | 60 | S | 1.0 | 0.3 | 0.021 |
| Dimethoate (4;1) | 1.56 | 5.20 | 15.0 | 41 | D | 0.6 | 0.2[b] | — |
|  | 4.80 | 16.0 | 10.5 | 27 | S | 3.0 | 1.2 | 0.1 |
| Isoproturon (11;12) | 2.14 | 2.14 | 10.5 | 8 | D | 13.7 | 6.7 | 0.15 |
|  | 9.88 | 5.20 | 10.5 | 27 | S | 17.2 | 10.6 | 0.9 |
| Lindane (6;1) | 1.20 | 2.14 | 28.5 | 7 | D | 4.5 | 1.2 | 0.007 |
|  | 2.91 | 5.20 | 40.0 | 44 | S | 0.7 | 0.2[b] | — |
| MCPA (4;1) | 8.75 | 5.21 | 11.0 | 19 | D | 46.8 | 14.6 | 0.70 |
|  | 26.88 | 16.0 | 15.5 | 4 | S | 12.4 | 1.9 | 0.096 |
| Mecoprop (4;2) | 11.00 | 5.50 | 25.0 | 2 | S | 11.7 | 4.2 | 7.5 |
| Simazine (8;7) | 6.39 | 3.76 | 13.5 | 7 | S | 68.0 | 22.4 | 60.6 |
| Triclopyr (1;1) | 1.01 | 5.18 | 13.5 | 78 | S | <0.01 | <0.01 | — |
| Trifluralin (3;1) | 2.20 | 2.00 | 10.5 | 3 | D | 14.1 | 3.5 | 0.032 |

[a]    Values in parentheses refer to the number of rainfall events for which data exist,
       and the number of pesticide applications, respectively
[b]    Simple mean - no flow data available

**Table 2 :** Pesticide concentrations in drain and stream water during rainfall
events when the highest mean concentrations were measured.

There are several alternative explanations for these observations, but they all involve the need to invoke by-pass flow. The important point is that the highest mean concentrations in the stream during rainfall events generally exceeded 1 $\mu$g l$^{-1}$, and peak values frequently exceeded 10 $\mu$g l$^{-1}$, although the total mass translocated during an event never exceeded 1% of the applied amount (mean = 0.10%). Furthermore, these concentrations cannot be fully explained in terms of the Groundwater Ubiquity Score. For example, trifluralin whose GUS of 0.17 suggests that it will not leach was in fact found at a peak level of 14.1 $\mu$g l$^{-1}$ in drain water, putting it on a par with several predicted (and actual) leachers. Subsequent unpublished data from ADAS Rosemaund for other 'non-leachers' such as deltamethrin confirm that by-pass flow can produce transient pesticide peaks in the stream at $\mu$g l$^{-1}$ levels, and it is very likely that a proportion of this translocation is via residues adsorbed on soil particles which can be carried down macropores and into field drains.

A proportion of such adsorbed residues may not be bioavailable, but there is no doubt that some pesticide concentrations seen in the stream (particularly of the insecticides) can cause significant biological effects. The *Gammarus pulex* bioassay has only been deployed during 3 experiments (dichlorprop, carbofuran, and chlorpyriphos - the latter unpublished; D. Sheahan pers. comm. 1994), but although it did not respond to the herbicide, the two insecticides both depressed feeding activity and caused mass mortality (Matthiessen et al., 1994). *G. pulex* is an ecologically important detritivore, and headwater streams like the one at Rosemaund may often be of conservation significance, so it seems at least possible that aquatic risk assessments of some pesticides have, in the past, been insufficiently rigorous. This is unsurprising given that regulatory authorities are only now attempting to calculate the aquatic exposure which may result from leaching, having previously merely made qualitative estimates. Furthermore, there has been a tendency to dismiss the biological significance of transient pesticide peaks such as those seen at Rosemaund and to focus on the generally lower concentrations seen in large rivers and in raw potable water supplies.

The most practical way to integrate the available data and make reliable exposure predictions is to use computerised models (OECD, 1989), but it is essential that such models should have been validated with field data like those in Table 2. It is also important to be able to make use of the limited physicochemical and degradation data which are available before a pesticide has been used in the field. Early attempts to develop a simple fugacity-based model using Rosemaund data (Brooke & Matthiessen, 1991) were only partially successful, but a more recent fugacity model called SoilFug shows considerable promise (Di Guardo *et al.*, 1994). SoilFug calculates the mean pesticide concentration in water flowing from a treated catchment during a rainfall event, and comparisons with data from Rosemaund and from two Italian catchments (also prone to by-pass flow) show that it is acceptably accurate for most pesticides tested. Exceptions are the phenoxy acid herbicides which dissociate at environmental pH and may adsorb to soils more strongly than predicted by $K_{oc}$, but predictions for the others listed in Table 2 generally lie within an order of magnitude of observed values. It is not yet clear why an equilibrium model such as SoilFug should satisfactorily predict pesticide concentrations that result from processes which are manifestly not at equilibrium. It nevertheless appears to be a tool which, if used with caution, can assist the initial prediction of risks which new pesticides may pose to aquatic life.

In summary, the observations at ADAS Rosemaund have helped to change our understanding of how pesticides translocate into streams at sometimes biologically significant concentrations, and have illuminated both the need for more accurate models that can be used for risk assessment, and the fact that such models are a practical possibility.

## Acknowledgments

This project would not have been possible without the dedicated efforts of a large number of people. Key participants have included Colin Allchin, Graham Beard, John Bell, Lal Bhardwaj, Steve Bird, Andree Carter, Michael Conyers, Mike Crookes, Antonio Di Guardo, Peter Glendinning, Caroline Hack, Graham Harris, Roy Harrison, John Hollis, John Kilpatrick, Ilga Nielsen, Vikram Paul, John Rea, Dick Rycroft, David Sheahan, Christopher Smith, Lisa Stewart, Alan Turnbull, Conrad Völkner, Brad Willis and all the field staff of ADAS Rosemaund. Important sources of funding have included the UK Pesticides Safety Directorate, the European Science Foundation, the UK Ministry of the Environment, the UK Ministry of Agriculture, Fisheries & Food, and the UK National Rivers Authority.

# References

□     Brooke, D. & Matthiessen, P. (1991). Development and validation of a modified fugacity model of pesticide leaching from farmland. *Pesticide Science* 31, 349-361.

□     Carter, A. D. & Beard, G. R. (1992). Interim report on the soil water sampling and soil characterisation programme within a small catchment at Rosemaund EHF (1990-1991). Soil Survey and Land Research Centre, Silsoe, Beds, UK (unpublished).

□     Di Guardo, A., Williams, R. J., Matthiessen, P., Brooke, D. N., & Calamari, D. (1994). Simulation of pesticide runoff at Rosemaund Farm (UK) using the SoilFug model. Submitted to *Environmental Science and Pollution Research*.

□     Gustafson, D. (1989). Ground water ubiquity score: a simple method for assessing pesticide leachability. *Journal of Environmental Toxicology and Chemistry* 8, 339-357.

□     Howard, P. H. (ed.) (1991). *Handbook of Fate and Exposure Data for Organic Chemicals. Vol. III Pesticides*. Lewis Publishers, Michigan, 684 pp.

□     Matthiessen, P., Allchin, C., Williams, R. J., Bird, S. C., Brooke, D. N. & Glendinning, P. J. (1992). The translocation of some herbicides between soil and water in a small catchment. *Journal of the Institution of Water and Environmental Management* 6, 496-504.

□     Matthiessen, P., Sheahan, D., Harrison, R., Kirby, M., Rycroft, R., Turnbull, A., Völkner, C. & Williams, R. (1994). Use of a *Gammarus pulex* bioassay to measure the effects of transient carbofuran runoff from farmland. *Ecotoxicology and Environmental Safety* (in press).

□     OECD (1989). Compendium of environmental exposure assessment methods for chemicals. *OECD Environment Monograph* no. 27, Organisation for Economic Cooperation and Development, Paris, 350 pp.

□     Wauchope, R. D., Buttler, T. M., Hornsby, A. G., Augustijn Beckers, P. W. M. & Burt, J. P. (1992). The SCS/ARS/CES pesticide properties database for environmental decision-making. *Reviews of Environmental Contamination and Toxicology* 123, 1-164.

□     Williams, R. J., Brooke, D. N., Glendinning, P. J., Matthiessen, P., Mills, M. J. & Turnbull, A. (1991a). Measurement and modelling of pesticide residues at Rosemaund Farm. In: *Proceedings of the 1991 Brighton Crop Protection Conference* 2, 507-514, British Crop Protection Council.

□     Williams, R. J., Bird, S. C. & Clare, R. W. (1991b). Simazine concentrations in a stream draining an agricultural catchment. *Journal of the Institution of Water and Environmental Management* 5, 80-84.

□     Williams, R. J., Brooke, D. N., Matthiessen, P., Mills, M., Turnbull, A. & Harrison, R. M. (1994). Pesticide transport to surface waters within an agricultural catchment. Submitted to the *Journal of the Institution of Water and Environmental Management*.

## 4.2.2.        *Atrazine Transfer from a Rural Catchment to Water*

# C. Cann

CEMAGREF - 17 Avenue de Cucillé - 35044 RENNES CEDEX - FRANCE

Analysis has shown last years that the level of pesticides concentration, and mainly atrazine concentration, is high in water resources. The atrazine concentration is often above the level of 100 nanogrammes per litre which is the highest concentration allowed for drinking water and, sometimes, above the level of 2 microgrammes, highest acceptable concentration for drinking water according to the W.H.O.

Those high levels of atrazine concentrations has been observed more often in the west part of France, during several studies (GILLET, Ministère de la Santé). The schist and granite bedrock give very weak groundwater resources and, thus, water supply is made from surface water at 80 %. The whole resources are quite used for water supply and thus must be protected against water pollution.

The intensification of agriculture has been very strong in this part of France for forty years with production of vegetables and mainly poultry, pigs and milk. To feed all those animals, maize cultivation increased up to use 25 % of land now. For this purpose grasslands has been ploughed and wetlands has been drained. Atrazine is the most used herbicide with maize cultivation. In 1990, 2 500 tons were used in France of which 480 tons used in Brittany (6 % of France surface). It can explain why atrazine concentration in water is often higher than in other parts of France.

As people are worrying about that, the CEMAGREF decided to carry on a research about atrazine transfer from fields to water resources on a rural catchment. We got the financial support from the department of Morbihan, from the association of water supply of the department and from the french departments of agriculture and of research for that.

The COET DAN catchment where is carried this research is located in NAIZIN in central Brittany. It is 12 square kilometers large and lie on a schist bedrock. The soil is mainly loam with some clay from place to place. The steam run between smooth hills from 131 m to 67 m at the outlet. Fields are small and sometimes still separated by hedges, especially in the lower part. Maize cultivation has increased a lot for last twenty year. 37 % of the agricultural land is used for this cultivation now.

The CEMAGREF is studying water transfer on this catchment since 1971. For this purpose various equipment as raingauges, gauging stations, limnimeters, piezometers, a meteo station has been settled with data recorders and data loggers. Use of land is also recorded every year. Transfer of nitrogen and phosphorus are studied since 1976 thanks to recording of farmers practices, the analysis of water in the stream, in wells from manual samples at the beginning and from automatic samplers now. Study of pesticides transfer started in 1991 on this catchment.

In order to pinpoint the use of pesticides, an enquiry has been hold at the beginning of the research towards all the farmers working on the fields of the catchment. We got accurate data from 42 of them out of 47. It covers 84 % of surface of the basin. It shows that all of them use atrazine for the maize and only for the ma5ze and that they use it in the same way as it is advised by technicians from cooperatives. They use fairly 1.5 kilogrammes per ha.

In this basin, 368 kg of atrazine were used in 1991 and, in 1992, it rised to 405 kg resulting of the increase of maize cultivation.

To know the load of atrazine in soil, samples of soil taken in the very top layer and 25 cm deep in 6 fields, three times pen year, has been analysed.

The soil content of atrazine varies from 0 to 484 µg/kg of soil. The results appear in the following graphic.

It shows that atrazine has remained in soils until the end of 1992 although none of the six fields were cultivated in maize during this year.

It shows also that atrazine content in the top layer is almost always higher that the content in the lower layer. The difference is very high in fields cultivated in maize. It is more difficult to observe a difference after ploughing of course. It means that atrazine is very slowly leached down in soils.

Computing with 5 000 tons/ha of arable soil from which 2 500 tons of top layer and 2 500 tons of low layer, the stock of atrazine in soil is about :

- 750 g/ha in june, after spreading in concerned fields (half of the spreaded quantity)
- 75 g/ha in october in such fields ($1/20^{th}$ of the spreaded quantity)
- 38 g/ha in february ($1/40^{th}$ of the spreaded quantity)
- 5 g/ha in june the following year without new spreading ($1/300^{th}$ of the spreaded quantity)

Atrazine has been searched also in sediments taken in the stream's bed at the same time. From 32 samples, atrazine were found in 10 samples only and the higher content was 32 µg/kg of wet sediment. It has been searched also in six samples of suspended solids taken during a flood. Atrazine was found in three of them, the highest content was 22 µg/kg of dry matter.

Those concentrations are quite lower than concentrations in soils. From these results, we can observe that atrazine is almost not transported under adsorbed form suspended solids and sediments in the stream. The main part of atrazine in soils is thus desorbed during erosion and transfer of suspended solids by water.

## Atrazine in Water

Samples taken twice in five wells from 6 to 16 meters deep show that pollution of groundwater by atrazine is very weak : atrazine was not found in four wells and was found at 125 and 350 ng/l in the last one. It could perhaps be explained by preferential ways of advance in the soil around this well.

The water quality has been observed at the outlet of the catchment where is settled a gauging station with a weir, a limnimeter, a data logger and automatical samplers.

Results of water analysis show that we must make the difference between two kinds of situation : during floods and out of floods.

Out of floods, 18 samples were taken between august 1991 and july 1992 and 10 samples between september 1992 and june 1993. Atrazine was not found at the detection level of 50 ng/l in one of them only. In the 27 other samples, the concentration was between 70 and 380 ng/l. We cannot observe any tendency along the year for these concentrations which variations seems not to be put in connection with the stock in fields.

Concentrations are largely higher during floods. In order to survey the increase and the decrease of concentrations, automatic samplers has been settled which programme starting depends of a water level probe. When the water level rise to the level of the probe, during floods, the sampler start to take individual samples according to the settled programme. Various programmes have been used with different timesteps.

The variations of atrazine concentrations have been followed in such a way during 14 floods. During every flood the concentration increase. We can never make sure that we measured the total increase because concentrations vary very quickly as we can see but we got sometimes a fairly continuous slope of the increase or decrease of concentration in such a way that we can think that we measured the global tendency of variations.

During some floods, the increase is not very strong. In some cases, the maximal observed concentration is no more than twice the concentration before the flood, but in most cases the increase is higher.

That means that quick flows of water by overland runoff or subsurface flow bring more atrazine to the stream than springs and groundwater. That is coherent with the observations of slowness of leaching of atrazine and its absence in well's water.

During some other floods, the increase has been quite higher, the 28/10/91, the 12/11/91 and even more the 25/05/93 and mainly the 11/06/93. This last day, the concentration rised up to at least 65 070 ng/l which mean more than 200 times the concentration before the flood. In the previous flood, the concentration rised up to 9220 ng/l at least although it was only 290 ng/l, half an hour before.

Those two floods occured at the end of spring, just after the spreading of atrazine on maize cultivation. This can explain why the increase was very higher during these two floods than during floods of autumn or winter. The increase of atrazine concentration occured also early in the flood while the discharge was increasing as well.

The discharge before the flood was very weak in the two cases. The baseflow provided by groundwater give thus only a very small part of the total discharge during the flood. At the beginning of the flood, the main part of the water flow come to the stream by overland runoff. The very large increase of atrazine come thus to the stream by this way.

It means that this overland flow bring to water atrazine carried from the very top layer of soil. Although there is very few atrazine in suspended solids, because atrazine is not strongly adsorbed on soil, those big flows of atrazine can thus come to water with the same kind of flow as suspended solids from erosion.

During autumn and winter, atrazine concentration did not rise so much. the highest measured concentration in this season has been 4 485 ng/l, the 12 november 1991. This time the concentration rised from 290 ng/l to 4 460 ng/l in two hours. The increase is distinctly less sharp than during spring floods. It was even less sharp during other autumn and winter floods.

It can also be noticed that, in autumn and winter, the concentration of atrazine does not increase at the beginning of the flood but at the end of the increase of discharge and even often after the peak of discharge. The peak of concentration occurs at the same time or after the maximal discharge. It means that, in this case, the water the most loaded with atrazine does not come to the stream by the quickest flow which is usually overland runoff. Subsurface flow may bring most of the atrazine from fields to streams.

We can notice also that the decrease of atrazine concentration after the peak is very slow in autumn and winter floods. It takes often several days for the concentration for coming back to the same level as it was before the flood. This kind of water, coming to the stream only several hours and until several days after the rain is typical of subsurface flow.

In spring floods, if the decrease of concentration is quicker and earlier in the flood as it still happens for elements brought by overland runoff, the concentration level off to a very high rate during several day before decreasing to the initial level. It means that after the overland runoff has brought the most loaded water at the beginning of the flood, the subsurface flow goes on bringing loaded water during some times too.

The subsurface flow occurs when water infiltrated in the soil encounter a less permeable layer and, thus, an important part of water remains above this level and move laterally to the stream. It happens sometimes when the soil is made of several layers of different kinds of soils but it happens mainly in ploughed fields where the ploughing make a large macroporosity in the top layer of soil allowing the water to infiltrate easily until the ploughing sill. There the firmed layer of soil does not allow the water to infiltrate down so easily and it begin to move laterally in the top layer of soil. The movement of water is slower in the soil than on the soil and such flow get the stream later than the overland flow of course.

## Interpretation

All these results are coherent and could be explained in this way:
When atrazine is spreaded on fields in may and june, most of it stay in the top layer of soil. When rain occurs then, a large quantity can be taken away from soil by water as it is not strongly adsorbed and bring to the stream by overland runoff. An other part is leached down slowly by water infliltrating in the soil from which a part will join the stream by subsurface flow and another part will remain in the soil. This last part will be degradated slowly but can also be taken away by water of following rainfalls along all the year and go to the stream by subsurface runoff in autumn and winter or leach deeper. The speed of degradation is such versus the speed of leaching that only a very small part can reach the groundwater level.

Thanks to the monitoring of discharge and water concentrations at the same place in the stream, it is possible to quantify the flow of atrazine with the following results :

|                                | 1991-1992 | 1992-1993 |
|--------------------------------|-----------|-----------|
| Flow at the outlet out of floods | 183 g     | 370 g     |
| Flow at the outlet during floods | 189 g     | 2 025 g   |
| Global flow                    | 372 g     | 2 395 g   |
| leakage coefficient            | 0,10 %    | 0,59 %    |

In both cases the flow in the stream is very small by comparison with the spreaded quantity and there is however hightly too hight concentration levels in water. As a consequence, it will not be efficient to try to improve water quality by the way of reducing the quantity of atrazine widespread per hectare.

The leakage of atrazine from fields to water is quite different in the two hydrological years. It is logical as the year 1991-1992 was a very dry year, 608 mm of rainfalls instead of 713 mm as an average and the year 1992-1993 was close to the average, for the rainfalls (751 mm). The importance of floods for carrying atrazine to streams explain this difference.

## *Conclusion*

As the concentration of atrazine in the water of streams and rivers can change very quickly, especially with floods and according to the kind of flows which make the discharge, we can say that :

1)  For the survey of water quality, it is necessary to note for every sample, the precise time, with hour and minute, of sampling, the discharge and to make the relation between the concentration and the variation of discharge. Without that, the result of analysis has no sense.

2)  Analysis must be done on samples taken at a precise time in streams and rivers and never on a mixture of several samples.

3) As the cost of analysis is very high and therefore it is impossible to analyse samples taken every quarter of an hour in every river, it is important to improve research to set up equations in order to rely variations of concentrations with meteorological, hydrological, and agronomical variables.

4) Such equations could be used to build models in order to make forecastings at short term, in order to help water resources management and to make simulations at long term in order to improve agricultural practices for a water protection purpose.

## *References*

☐  CANN C. - 1993 - *Suivi de la qualité de l'eau - Etude menée sur le BVRE du COET DAN -* CEMAGREF - Déchets Solides - RENNES - 162 p + annexes

☐  CANN C. - VILLEBONNET C. - 1994 - *Suivi de la qualité de l'eau, 2ème année - Etude menée sur le BVRE du COET DAN -* CEMAGREF - Déchets Solides - RENNES - 120 p.

☐  GILLET H. - 1991 - *Contamination des eaux par les pesticides en Bretagne - Etude d'évaluation des contaminants par bassins versants -* SRPV - DRAF Bretagne - 7 p + annexes

☐  JONVEL S. - 1992 - *Transferts de pesticides vers les Eaux de Surface en Milieu Rural. (Etude menée sur le Bassin Versant Expérimental du Coët-Dan dans le Morbihan) -* ENITRTS - CEMAGREF - DECHETS SOLIDES - RENNES - 92p. + annexes.

4.2.3.          *Occurence of Pesticides in Surface and Groundwater in Finland*

# J.-P. Hirvi

Water and Environment Research Institute, Helsinki - Finland

## (Full text not received)

*A national pesticide project was carried out in Finland in 1991 and 1992, Rain water, river water and groundwater samples were collected for the analyses of polar herbicides, semi-polar and non-polar pesticides. The project was finanzieted by the Ministry of the Environment and promoted by the Nordic Council of Ministers.*

*The results of the rain water studies showed that the polar herbicides were detected in the highest concentrations (2 - 190 ng/l) in the samples collected during the first half of June. The found herbicides could be ranged in order of decreasing concentration as follows; dichlorprop > MCPA >mecoprop > 2,4-D. Same kind of concentration "top" but occurring in the shift of June and July could be measured for some semi-polar and non-polar pesticides, especially for atrazine, lindane and simazine (10 - 20 ng/l). The concentration of atrazine was highest in the snow sample (100 ng/l). However, atrazine was not measured in the river waters after the snow melting.*

*In the river waters the polar herbicides as dichlorprop, MCPA and mecoprop were found in the concentrations between 200 - 900 ng/l. The concentration top was observed in the shift of June and July. The herbicides could be measured over the detection limits only in those rivers having mean discharge lower than 40 m³/s. Any semi- and non-polar pesticides cound not be detected in the river water samples.*

*Totally 75 groundwater wells were sampled and studied. Only in one of the wells the pesticide contamination was evident. The wells were chosen to be located in the sandy soils and in the cultivated areas of the southern Finland. The wells were also selected by using comprehensive water quality data, eg. those wells affected by surface water runoff could be eliminated. The contaminated well was sampled in August 1991, August and November 1992, and in every sample a metabolite of dichlobenil (2,6-diclorobenzamide = BAM) was detected in the concentration of 600 ng/l. The contamination seemed to be problem of only this well, not of the whole groundwater resource representing by the water of the well.*

*The results of the pesticide studies succested that low concentration levels, mainly polar herbicides were found in the surface waters showing a maximum concentration during June. Furthermore, the groundwaters could be said to be clean of the herbicides and pesticides commonly used in the studied cultivation areas.*

4.2.4.     *Monitoring and Modelling Pesticide Occurrences in a Rural Source*
           *of Drinking Water*

# M. Fielding, D.B. Oakes, J. Weddepohl,
# C. Cable and K. Moore

WRc, Henley Road, Medmenham, Marlow, SL7 2HD, UK

*Around four hundred pesticides are approved for use in the UK and some of these have the potential
to reach drinking water sources. In some rivers pesticide concentrations have been detected above
the EC MAC of 0.1 µg/l. A programme of research was funded by the UK DoE between 1987 and
1993 to investigate the occurrence and fate of pesticides in the rural catchment of the River Leam in
central England. Many pesticides, from both agricultural and non-agricultural usage, were detected
at concentrations above the MAC in the main river and its tributaries.*

*The factors that govern the occurrence of pesticides in water sources, such as the catchment
hydrology, pesticide usage, seasonal effects and pesticide properties were incorporated in a mathematical model to simulate pesticide transport and fate through the catchment into the water. A GIS
based Digital Terrain Model was used to derive flow paths along which pesticide transport was
simulated. The model was able to accurately simulate the concentrations of many of the pesticides
found in the river, and hence to provide a correlation between usage and occurrence. It was shown
that over 95% of the river load of one of the most problematical pesticides, isoproturon, resulted from
applications within 20m of the river, thereby demonstrating the feasibility of catchment control.*

## Introduction

Approximately four hundred pesticides are approved for use in the UK and some of these have the
potential to reach water sources used for drinking water supplies. In some rivers pesticides concentrations above the EC MAC of 0.1 µg/l have been detected. The subject of pesticides in water sources
is one of the most important problems relating to UK water supplies, and the Department of the
Environment commissioned a research programme at WRc to investigate the occurrence and fate of
pesticides in rivers. The River Leam in central England was selected for the study because assistance
and supporting data were readily available from the water utility, Severn Trent Water. The work on
monitoring and modelling has produced valuable information which will be of use in related studies,
and forms the basis for this paper. The work programme also included studies on other chemicals
used in pesticide formulations and pesticide transformation products, but these aspects will not be
discussed here. For further information the reader is referred to the Cable et. al. (1994).

## The River Leam Catchment

The River Leam flows into the Warwickshire Avon between the towns of Leamington and Warwick
and water, for treatment at Campion Hills Drinking Water Treatment Works, is abstracted just
upstream of Leamington. The catchment covers an area of 36 547 hectares and is predominantly rural,
with cereal cultivation and animal husbandry being the main farming activities. The soils are
principally of the Denchworth, Evesham and Whimple associations, heavy clays with permeabilities
declining rapidly with depth which result in the generation of surface runoff in the top metre or less.
The land use areas are given for broad categories in Table 1.

| Category | Area (ha) |
|---|---|
| Crops, grass, fallow & set-aside | 29564 |
| Woods & other land | 815 |
| Total agricultural area | 30379 |
| Non-agricultural land | 6168 |

**Table 1 :** Land Use in the River Leam catchment

Pesticide applications were obtained from specialist surveys conducted annually in England and Wales by WRc and Produce Studies Ltd. Data for the River Leam catchment are given in Tables 2 and 3.

| Pesticide | Use (kg) | Pesticide | Use (kg) | Pesticide | Use (kg) |
|---|---|---|---|---|---|
| Atrazine | 106 | Fenpropimorph | 2290 | Prochloraz | 1057 |
| Chlormequat | 9204 | Glyphosate | 3399 | Propiconazole | 975 |
| Chlorothalonil | 3476 | Isoproturon | 15312 | Simazine | 710 |
| Chlorortoluron | 1322 | Mancozeb | 1267 | Thiabendazole | 1593 |
| Chlorpyriphos | 981 | Mecoprop | 2771 | Tri-allate | 1807 |
| Diquat | 1002 | Pendimethalin | 1470 | Trifluralin | 1311 |

**Table 2 :** Major agricultural pesticide applications (Active Ingredient)
in the River Leam catchment in 1992.

Reliable data on herbicide usage in the non-agricultural sector is difficult to obtain. The survey of potential users, including leisure and manufacturing industries, the various utilities and road and rail authorities was used to compile what was considered to be a reasonably valid data set.

## Pesticide sampling and usage

In order to collect data for the study an intensive sampling program was carried out in 1992. Grab samples of river water were taken from various points along the main river and tributaries using twin Winchester bottles with the bottle necks a few inches below the water surface. During sampling, the water temperature was measured and additional samples were taken for determination of total organic carbon, nitrate and suspended solids. The pH of the samples was measured before storing them at 5 C while awaiting three multi-residue pesticide determinations. Results for the major contaminant pesticides are given in Table 4.

| Pesticide | Use (kg) | Pesticide | Use (kg) |
|-----------|----------|-----------|----------|
| Atrazine | 1.5 | Iprodione | 45.0 |
| Carbaryl | 26.3 | Maleic hydrazide | 84.0 |
| 2,4-D | 96.5 | Mecoprop | 51.5 |
| Dichlobenil | 48.9 | Thiabendazole | 35.6 |
| Diuron | 68.9 | Thiophanate-methyl | 38.0 |
| Glyphosate | 97.6 | Triclopyr | 73.0 |

**Table 3** : Major non-agricultural pesticide applications (Active Ingredient) in the River Leam catchment in 1992

| Pesticide | Max. conc. (ng/l) | Pesticide | Max. conc. (ng/l) |
|-----------|-------------------|-----------|-------------------|
| Atrazine | 580 | Isoproturon | 7450 [*] (1520) |
| Carbendazim | 170 | Linuron | 1940 |
| Chlorotoluron | 410 | Mecoprop | 110000 [*] (930) |
| 2,4-D | 250 | MCPA | 86400 [*] (1160) |
| Diuron | 470 | Simazine | 400 |
| Flutriafol | 220 | Trietazine | 170 |

[*]  transient peaks in the River Itchen tributary. The next highest concentration observed is given in brackets

**Table 4** : Occurrence of pesticides in the River Leam catchment in 1991-1992

Most of the river water samples contained a range of pesticides at significantly above the limits of detection. Neutral and basic pesticides detected at above 0.1 µg/l were atrazine, simazine, trietazine, and flutriafol. Uron and carbamate pesticides detected at above 0.1 µg/l were carbendazim, chlorotoluron, isoproturon , diuron and linuron. Acid herbicides detected at above 0.1 µg/l were mecoprop, MCPA and 2,4-D. Of these pesticides, atrazine, isoproturon, linuron, mecoprop and MCPA were detected at above 0.5 µg/l, whilst the latter four were also detected at above 1.0 µg/l. Levels considerably above 1.0 µg/l were detected in a single sample taken during January 1992 in one of the tributaries well upstream from the water treatment intake works, as shown in Table 4. These relatively high concentrations are unlikely to be associated with normal applications of pesticides, and are thought to result from accidental spills or deliberate misuse.

Some data were also provided by Severn Trent Water for concentrations in the river at the point of outflow from the catchment where water is abstracted for public supply, and these data were used in the modelling study which is described in the next section. It should be noted that an activated carbon plant at the treatment works ensures that the water going into supply complies fully with the EC MAC.

# Predictive modelling

It is difficult and costly to monitor for pesticides in water sources, and data on occurrences are necessarily limited. In order to be able to understand the spatial and temporal variations in pesticides in rivers and to allow interpretation in terms of cause and effect some form of mathematical modelling is required. Mathematical models allow the factors that govern the occurrence of pesticides in water, such as usage, seasonal effects, meteorology, and pesticide physico-chemical properties to be investigated. In addition, mathematical models may be used to aid monitoring strategies, to predict future trends and to assess various options for catchment control where this is thought to be appropriate. Models may also be used to fill in information gaps caused by the financial limitations imposed upon sampling frequency.

It should be borne in mind that modelling cannot act as a surrogate for monitoring as all models require measured data for calibration and validation. In any case most pesticide models are still in the developmental stage and are not yet reliable enough to be widely used without careful control. Water companies too, are required by law to monitor water supplies to ensure compliance with current legislation.

# Development of the predictive model

For the reasons given previously, the development of a reliable predictive tool was seen as a priority for the study described. The model would need to simulate both the hydrological processes governing directions and rates of water flow across the catchment, as well the physical processes of pesticide fate and transport to an acceptable level of accuracy. The runoff flow paths through the catchment will depend on the topography, and to allow the model to simulate the runoff process a GIS Digital Terrain Model (DTM) was used to generate topographic heights on a 250m grid covering the catchment. The DTM also produced the river network and the watershed boundary to define the catchment area. A flow route to the river system is thereby provided for each 250m grid square in the catchment, a necessary requirement for a pesticide transport model. It became apparent however that a 250m grid was inadequate for detailed spatial modelling and a finer spatial resolution was required. The 250m model for the Leam catchment required 5900 mesh points which is a practical size for modelling purposes; a 2m grid would require over 90 million mesh points which is impractical for the DTM and the pesticide transport model. It was therefore decided to model pesticide transport along a typical flowpath of width 1 m. The DTM generates flowpaths and was used to estimate the average flowpath length, of 1350m, in the catchment area. A key assumption underlying this approach is that pesticide concentrations in the runoff discharge to a river are relatively independent of the flow path length, and this was clearly validated by comparison of results from different runs of the model using a variety of such lengths. The assumptions underlying the model may be summarised:

- water and pesticides move through the upper zone of the soil under hydraulic gradients resulting from catchment topography and rainfall recharge;

- the catchment may be represented by a typical flowpath of width 1m and length 1350m, starting on high ground and terminating at the river;

- during movement through the soil the processes of pesticide adsorption onto organic carbon, and chemical and biochemical degradation occur;

- at the river boundary, water and dissolved pesticides flowing out of the soil are multiplied by the total length of the river and equated to the flow and mass of the pesticide in the river.

The flowpath model was based on the method described by Haith (1980) and adapted by subdividing the flowpath into 1 $m^2$ cells with a Haith-type calculation applied to each cell. The runoff (with dissolved pesticide) from one cell becomes an input to the next downgradient and the concentration in the river is equated to that in the runoff from the final cell in the flowpath. The model used a 5-day time step. Runoff was generated from 5-day total of effective rainfall (rainfall-evaporation±soil moisture deficit) derived from a 50-year synthetic rainfall sequence. This sequence was based on the statistical features of actual rainfall in central England and used to allow the effect of climatic variations on pesticide concentrations to be considered. The mean annual effective rainfall for the River Leam catchment was 180 mm/yr.

Spatial details of pesticide applications were not available, but agriculture is well distributed over the catchment and so pesticide usage could be assumed to be relatively uniform over the entire area. The temporal distribution of pesticide usage was obtained from the specialist surveys described previously.

Other model parameters included soil depth, soil bulk density, soil organic carbon, soil water content, pesticide Koc (partition coefficient between water and organic carbon) and pesticide half life. Soil properties for the catchment were obtained from published data, and were not subject to change as part of model calibration. Values for pesticide Koc and half life were taken from Moore (1992). Results for the major pesticides are given in Table 5 and compared with the Severn Trent Water data at the discharge point from the catchment..

| Pesticide | Measured | Simulated | Pesticide | Measured | Simulated |
|-----------|----------|-----------|-----------|----------|-----------|
| Isoproturon | 2400 | 2630 | Diuron | 650 | 660 |
| Chlorotoluron | 300 | 250 | Flutriafol | 220 | 20 |
| Mecoprop | 680 | 440 | Carbendazim | 70 | 40 |

**Table 5 :** Measured and simulated maximum concentrations (ng/l) in the River Leam catchment at Campion Hill water treatment works intake

In most cases the model predictions were good, particularly in view of the fact that the available catchment and pesticide data used in the model were taken directly from published literature and not adjusted by fitting the model to observations. For flutriafol the physico-chemical properties database is very limited and must be considered suspect. Although not shown here, the plots of measured against simulated concentrations showed good agreement. Variations in predicted annual peak concentrations were inferred to result from the annual variations in rainfall recharge, particularly the incidence of runoff immediately after pesticide application. Low peak concentrations result when there is little runoff at the time of application and biochemical decay reduces the loading on the catchment prior to transport into the river.

An important future use of models such as the one described here will be to assess the effect of catchment control options such as restrictions on pesticide usage or the imposition of protection zones in which no pesticide applications would be permitted. The Campion Hills Water Treatment Works includes an activated carbon unit for the removal of pesticides and other organic compounds. Many other river intakes for drinking water supplies do not have this treatment facility, however, and catchment control is an option which must be considered in these cases. Additionally, it would be prudent to limit pesticides in rivers for ecological reasons. Runs of the model for isoproturon showed that control at source could be a feasible option in the River Leam catchment. It was predicted that 80% of the pesticide entering the river results from applications within 10m of the river, and that 16% results from those applications in the zone between 10m and 20m from the river. The rest of the catchment contributes only 4% of the total reaching the river. This illustrates the importance of river margins for attenuating pesticides in runoff, and shows that in this instance pesticide contamination of surface water sources could be significantly reduced by the imposition of relatively small protection zones.

In order to make the results of modelling more widely useful, the model was used to assess the overall relationships between applications, pesticide properties and risk of river water contamination. A

Surface Water Risk Index diagram, similar to the GUS index for groundwaters, was constructed from successive model runs and is shown in Figure 1. This index can be used as a means of assessing risks of surface water contamination due to both existing, and new pesticides.

In Figure 1, the intersection of selected values of Koc and half life gives the maximum catchment averaged pesticide application rate which can be applied without causing river concentrations to exceed 0.1 µg/l. An alternative interpretation is that combinations of Koc and half life which fall below or to the left of any of the specified application rate curves are acceptable; combinations of properties falling above or to the right of the curves will result in exceedances of the EC MAC.

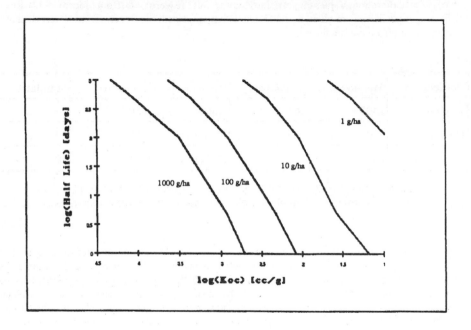

**Figure 1 :** Surface Water Risk Index graph for the Leam catchment

# Conclusions

A wide range of pesticides was detected in river samples from the River Leam catchment. Most were herbicides, although a few fungicides were also encountered. Many occurred at levels in excess of 0.1 µg/l and several exceeded 1.0 µg/l during part of the year. The pesticides detected and their concentrations were related to usage, catchment characteristics and rainfall recharge through a mathematical model describing the transport and fate processes.

The model utilised measured catchment data and was based on a relatively simple flow path calculation. The model was able to predict quite accurately the temporal variations and absolute concentrations of many pesticides. The timing of rainfall after pesticide application, rather than the quantity, was identified as a major factor controlling the severity of pesticide contamination to surface waters.

It was demonstrated with the model that 96% of the isoproturon load in the river resulted from applications within 20m of the river; the remainder of the catchment contributed only 4%.

A Surface Water Risk Index, analogous to the GUS index for groundwaters was produced to allow calculations of maximum pesticide application rates which would ensure compliance of the MAC for pesticides in the River Leam.

# References

□ Cable, C.J., Fielding, M., Gibby, S., Hegarty, B.F., Moore, K., Oakes, D.B. and Watts, C.D.,1994. Pesticides in Drinking Water Sources. WRc Report 3376.

□ Haith, D.A., 1980. A mathematical model for estimating pesticide losses in runoff. *Journal of Environmental Quality*, 9, 428-433.

□ Moore, K.,1992. Availability of multi-residue methods for analysis of pesticides in water sources and supplies. WRc Report FR0312.

## 4.3.      Poster Presentations

*4.3.1.*      *Use of Immunoaffinity Techniques as Separation and Concentration Tools of Pesticides in Environmental Samples*

## Y. Dehon, D. Portetelle and A. Copin

UER Microbiologie
UER Chimie Analytique et Phytopharmacie
Faculté des Sciences Agronomiques de Gembloux
Avenue Maréchal Juin, 6  -  B-5030 Gembloux  -  BELGIUM

Antibodies are commonly used for the detection and quantification of a large variety of molecules ranging from high molecular weigth proteins to small hormones or pesticides molecules.

Unfortunately, most commercial tests for pesticides are not sensitive enough to detect concentrations down to 0.1 µg/L in waters.

Therefore, we develop efficient and easy-to-use techniques to concentrate pesticides in samples by means of specific antibodies.

Commonly, monoclonal antibodies are fixed on a solid matrix packed in columns. Water samples are slowly passed through, columns are washed and the analyte retained by the antibodies is eluted by means of pH or ionic strength variations.

Analytes can be concentrated up to a hundred times while other organic compounds of the sample are not.

In another system; we fix antibodies on magnetic iron oxide particles. This magnetic separation technique is particularly useful when the samples include low water solubility materials that clog columns or foul filters and is independent to flow rate, pressure, ...

Coupling of magnetic particles system with enzyme labelled pesticides or antibodies, allows to increase sensitivity of ELISA tests by a factor 20.

Those techniques are developed in our laboratory for the detection and quantification of isoproturon in water.

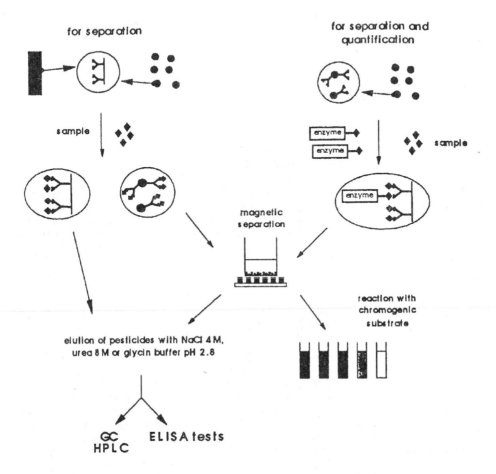

for separation

for separation and quantification

sample

enzyme
enzyme

sample

magnetic separation

enzyme

reaction with chromogenic substrate

elution of pesticides with NaCl 4 M, urea 8 M or glycin buffer pH 2.8

GC
HPLC          ELISA tests

advantages

selective concentration
high concentration factor (x100)
no organic solvents required

advantages

increased sensitivity of ELISA test
separation and quantification in one step
fast and easy to perform technique

disadvantages

risks of unspecific adsorption
of pesticides on the particles

disadvantages

risks of biological degradation of antibodies or enzyme
risks of unspecific adsorption of pesticides on the particles

Current applications in our laboratory :

extraction and quantification of isoproturon
in water samples

increasing sensitivity of ELISA test by a factor 20

 chromatographic support     pesticide molecules     specific antibody     magnetic particles

*4.3.2.*      *Organochlorine Compounds in Surface and Ground Waters in the Zagreb Area*

# S. Fingler, V. Drevenkar and Z. Fröbe

Institute for Medical Research and Occupational Health,
2 Ksaverska St., 41000 Zagreb, CROATIA

## Abstract

The monitoring of selected organochlorine pesticides, chlorophenols and polychlorinated biphenyls (PCBs) in the Sava river and small streams and lakes in the Zagreb city area showed that all waters contained traces of PCBs (3-25 ng/L) and organochlorine pesticides (<1-2 ng/L). The chlorophenol concentrations in small streams (<1-2868 ng/L) were higher than those in the river (<1-18 ng/L) and lake (<1-31 ng/L) waters. The highest concentrations were measured in streams adjacent to industrial facilities which were loaded with industrial waste waters. Those streams pose a constant threat to the ground water purity because of polluted water infiltration into the subsurface and its inefficient purification by natural filtration.

## Introduction

Zagreb, the capital and major industrial centre of Croatia, is a continental city. It is situated between the slopes of a 1035 m high mountain in the north and the river Sava in the south, expanding beyond the river bank. The city area is crossed by a number of small streams springing in the surrounding hills and flowing into the Sava river. Some of those streams flow through densely populated area as well as through industrial zones receiving untreated industrial waste waters. The Sava river is also heavily polluted mainly because of discharge of municipal and industrial liquid waste upstream and in the city area (1). The Zagreb drinking water comes from underground resources, supplemented by springs in the hills with adjoining reservoirs. The purity of the ground waters is likely to be jeopardized by infiltration of the polluted waters from the Sava river and local streams, as well as by improper industrial and domestic waste discharge and occasional chemical spills (2). The principal purification stage of ground waters in the city area is natural filtration through the subsurface sediment layers. This stage is not always efficient enough as indicated by the presence of chlorophenols in most of the river, ground and drinking water samples collected in the city area during the past years (3,4). To trace the origin of selected organochlorine compounds in ground waters the monitoring of chlorophenols, organochlorine pesticides and polychlorinated biphenyls (PCBs) in the Sava river and in small streams and lakes in the city area was initiated in 1992. It included the chlorophenols 4-chlorophenol (4-CP), 2,4-dichlorophenol (2,4,-DCP), 2,4,5- and 2,4,6-trichlorophenols (2,4,5- and 2,4,6-TCP), 2,3,4,6-te-trachlorophenol (2,3,4,6-TeCP) and pentachlorophenol (PCP) and the organochlorine pesticides hexachlorobenzene (HCB), the hexachlorocyclohexane (HCH) group of isomers and compounds of DDT-type.

## Experimental

Chlorophenols, organochlorine pesticides and PCBs were analysed in 13 samples of each stream and lake waters. Four stream water samples were taken from the Sava river and nine from six small streams.

Chlorophenols were measured also in six ground water samples collected from wells positioned at a distance of 0, 6 and 120 m from the bed of a small stream receiving untreated waste water from a pharmaceutical industry and a baker's yeast production plant. The ground water was sampled at a depth of 10.5 m where the maximum velocity of flow was measured. At the same location three samples of soil (depth 0-30 cm) and six samples of sediment (depth 5-17 m) were collected in order to test the sorption capability of those natural sorbents for 2,4,6-TCP, 2,3,4,6-TeCP and PCP.

Chlorophenols, organochlorine pesticides and PCBs were accumulated from water samples and analysed by gas chromatography according to the procedures described earlier (4,5). PCBs were quantitated against a standard mixture consisting of Aroclor 1242 and Aroclor 1260 in a 2.5:1 ratio.

For sorption experiments a solution of 2,4,6-TCP, 2,3,4,6-TeCP and PCP (concentrations of single compounds 250-280 μg/L) in deionized water was prepared and mixed with sorbents in the water to solid phase ratio 10:1. After 24 hours of shaking at room temperature the equilibrium concentration of chlorophenols in the aqueous phase, separated by centrifugation, was determined, and chlorophenol distribution coefficients ($K_d$) were calculated.

## Results and Discussion

The results of PCB and chlorophenol analyses in surface waters in the Zagreb city area, excepting the streams adjacent to industrial plants, are summarized in Table 1. Traces of PCBs were detected in all samples and the highest concentrations were measured in the Sava river. All samples contained also organochlorine pesticides but the concentrations of single compounds were not higher than 2 ng/L. Gamma-HCH appeared most frequently as a consequence of the use of a commercial Lindane formulation. It was followed by alpha-HCH, HCB and 4,4'-DDE. The incidence of chlorophenols was higher in the streams than in the lakes. The chlorophenol concentrations in small streams (3-119 ng/L) were higher than those in the river (<4-18 ng/L) or lake (<1-6 ng/L) waters. However, as shown in Table 2, the highest chlorophenol concentrations were measured in small streams adjacent to industrial plants, which pose a constant threat to the environment and especially to the ground water quality.

The chlorophenol concentrations in ground water wells in the city area measured earlier were highest for 2,3,4,6-TeCP (up to 270 ng/L) and PCP (up to 411 ng/L) (3). In ground water samples collected in the present study from the wells positioned at increasing distances from an industrially polluted small stream the highest concentration was measured for PCP (70 ng/L) in the well adjacent to the stream bed.

| CONCENTRATION, ng/L | | | | |
|---|---|---|---|---|
| COMPOUND | River/streams (N=10) | | Lakes (N=13) | |
|  | Range (n)[*] | Median[**] | Range (n)[*] | Median[**] |
| PCBs | 7 - 25 (10) | 11 | 3 - 14 (13) | 9 |
| 4-CP | 40 -119 ( 2) | 0 | 6 - 31 ( 5) | 0 |
| 2,4-DCP | 4 - 32 ( 6) | 4 | 2 - 6 ( 4) | 0 |
| 2,4,6-TCP | 1 - 18 (7) | 2 | 1 - 5 (3) | 0 |
| 2,4,5-TCP | <2 - 4 (3) | 0 | <2 - 5 (3) | 0 |
| 2,3,4,6-TeCP | <1 - 6 ( 8) | 3 | <1 - 2 ( 8) | <1 |
| PCP | 2 - 23 ( 6) | 2 | <1 - 5 (7) | <1 |

N = number of samples

[*] Ranges apply to positive samples and n is the number of positive samples

[**] 0 stands for values below detection limits

**Table 1** : Chlorophenol and PCBs concentrations in river/stream and lake waters in the Zagreb city area

| COMPOUND | CONCENTRATION RANGE, ng/L |
|----------|---------------------------|
| 4-CP | 5 - 1328 |
| 2,4-DCP | 20 - 32 |
| 2,4,6-TCP | 16 - 178 |
| 2,4,5-TCP | 16 - 31 |
| 2,3,4,6-TeCP | 8 - 115 |
| PCP | 27 - 2868 |

**Table 2** : Chlorophenol concentrations in small streams
(3 water samples) adjacent to industrial facilities in the Zagreb city area

The same sample contained also 2,4-DCP (13 ng/L) and 2,3,4,6-TeCP (4 ng/L). The chlorophenol concentrations in waters from more distant wells were lower: 4 ng/L for and 2,4-DCP and PCP and 8 ng/L for 4-CP. The appearance and levels of chlorophenols in analysed ground waters could be connected with the stream water pollution and depended on the stream water dilution i.e. on the stream water flow rate.

In sorption experiments the pH value of aqueous chlorophenol solutions after equilibrating with soil and sediment samples ranged between 6.5 and 8.0 favouring the phenolate form of 2,4,6-TCP (pKa=6.15), 2,3,4,6-TeCP (pKa=5.40) and PCP (pK=5.25). Although more hydrophilic than parent chlorophenols phenolate anions are sorbed in natural sorbents and the sorption intensity strongly depends on the sorbent organic matter content (6,7). A measurable chlorophenol sorption was noticed only in one sample of surface soil containing 12.2% of organic matter. $K_d$ values of investigated chlorophenols increased with the compound hydrophobicity and were 10.9 for 2,4,6-TCP, 16.7 for 2,3,4,6-TeCP and 36.7 for PCP corresponding to the sorption of 53, 63 and 77% of single chlorophenol amounts initially present in the aqueous phase. However, the sorption in two other surface soil samples containing 2.8 and 2.1% of organic matter, and in samples of deeper sediment layers, consisting of 92-98% of clay and silt and containing 0.5-8.4% of organic matter, was negligible. Consequently, if the stream water infiltrates to the ground water in deeper sediment layers the purification of water by natural filtration is inefficient resulting in greater mobility of pollutants in the aquifer material and their enhanced concentrations in ground water.

# References

□    1. M. Ahel, W. Giger, *Kem. Ind.* 34(1985)295-309.

□    2. M. Ahel, *Bull. Environ. Contam. Toxicol.* 47(1991)586-593.

□    3. S. Fingler, V. Drevenkar, *Toxicol. Environ. Chem.* 17(1988)319-328.

□    4. S. Fingler, V. Drevenkar, B. Tkalcevic, Z. Šmit, *Bull. Environ. Contam. Toxicol.* 49(1992)805-812.

□    5. S. Fingler, V. Drevenkar, Z. Vasilic, *Mikrochim. Acta II*(1987)163-175.

□    6. R.P. Schwarzenbach, Sorption behaviour of neutral and ionizable hydrophobic organic compounds, in: A. Bjorseth and G. Angeletti (Eds.), *Organic Micropollutants in the Aquatic Environment*, Proccedings of the Fourth European Symposium, Vienna 1985, D. Reidel Publishing Company, Dordrecht 1986, pp.168-177.

□    7. Z. Fröbe, S. Fingler, V. Drevenkar, *Sci. Total Environ.*, in print.

*4.3.3.*        *Strategies for Monitoring Pesticides in Surface Water*

# J. Kreuger

Swedish University of Agricultural Sciences, Dept. of Soil Sciences,
S-750 07 Uppsala, Sweden

During the late 1980's various monitoring programs were set up to improve knowledge of pesticide residues in surface waters within the Nordic countries. Pesticides have been found in surface waters in all four Nordic countries. During the last six years, 35 different pesticides have been detected, including 25 herbicides, 7 insecticides and 3 fungicides.

For adequate exposure assessment, however, as a part of risk evaluation, we need good quality data on pesticide exposure patterns and characteristics. The ecological effects of pesticides on flora and fauna in surface waters are dependent on both peak concentrations and the time of exposure. The objectives of the majority of the above mentioned investigations have been to find whether concentrations of pesticides could be detected in surface waters on single occasions. This is not enough to assess the ecological risks posed by pesticides in surface waters. To constitute a basis for exposure assessment, monitoring in the future should improve sampling strategies to increase evaluation possibilities.

Also when regulatory pollution control measurements are determined, transport calculations are useful in evaluating possible changes. Minimum background data for adequate evaluation should include the following: watershed size, land use pattern, soil type, precipitation, water flow rate, amount and type of pesticides used and spraying season. For better recommendations to the users on how to minimise losses of pesticides to the water bodies, there is also a strong need to increase our knowledge of the different transport pathways within a watershed, including all possible processes (spills, runoff, leaching, wind drift, etc.).

In 1990, a watershed in the southernmost part of Sweden was selected for intensive monitoring of pesticide exposure patterns. The watershed has an area of 9 km$^2$ consisting of 95% arable land. Information on pesticide usage (type of pesticides, amounts and spraying occasions) and handling within this area is collected annually from the farmers. About 40 different substances are used and 80-90% (by weight) of these are included in the analyses. At the outlet of the watershed an automatic water sampler collects, on a daily or weekly basis, integrated water samples from May to September. Also, at different sites within the watershed, samples have been collected to assess point sources. Twenty-five different pesticides have been detected throughout the sampling season, with peak concentrations at the time of spraying and elevated concentrations also during storm flow events.

*4.3.4.        A surface and ground water monitoring strategy for the Walloon
              Region of Belgium designed to determine the relative contribution
              of agricultural and non-agricultural practices on water
              contamination by pesticides*

# P. Nadin

Comité régional Phyto, Faculté des Sciences agronomiques, Université catholique de Louvain, Belgium

## Pesticides-water: general context in the Walloon Region

The basic quality standards for surface water and drinking water sources are laid down by national legislation[1-2]. These are for surface water in general : organochlorinated (OC) compounds < 0.01 ppb per compound and < 0.03 ppb total and acetylcholinesterase (AChE) inhibitors < 0.5 ppb total ; and for surface water to be used as drinking water source : OC compounds < 1, 2.5, 5 ppb for the total of parathion, HCH and dieldrin (depending on purification process). When they exist, ecotoxicological values, such as LC50 or NOEC for the most sensitive aquatic organism, can serve as references in surface water. Where surface water is used as a drinking water supply, the drinking water quality standard (MAC 0.1 ppb) can also be used[3].

Standards for the quality of ground water are in preparation in Wallonia[4]. Since more than 80% of the drinking water is supplied by ground water, the drinking water quality standard (MAC), defined by a regional legislation[3], is commonly used as a reference in ground water.

## Actual water monitoring networks in the Walloon Region

Water monitoring is not centralised in Belgium since 'Environment' falls within regional scope. However, in the context of national and international agreements, pesticide residues in surface and ground water have been surveyed on a national basis in '90 and '91.

### *Surface water*

### National survey '91[5]

5 series of samples taken from 20 sampling points (8 boundary points) in Wallonia were analysed from May to December '91. Samples were analysed for the presence of 13 substances (atrazine, azinphos-methyl, dichlorvos, endosulfan, fenitrothion, fenthion, HCH, malathion, parathion, pentachlorophenol, simazine, trifluraline and triphenyltin compounds).

### Regional monitoring[6]

Since '91, the Division in charge of industrial pollution control at the Ministry of the Walloon Region of Belgium has analysed surface water for the presence of pesticides in samples taken from the same 20 sampling points. Each year, 6 series of samples were analysed for the presence of groups of compounds such as AChE inhibitors, OC compounds and s-triazines.

### Water agency monitoring

The River Meuse basin is regularly monitored by a national supplier of drinking water[7] (*Compagnie intercommunale bruxelloise des Eaux*), essentially for s-triazines (atrazine, simazine) and ureas (diuron, isopoturon) herbicides. International Rhine & Meuse Water Agency Association (*RIWA*) also monitors regularly the presence of some 50-60 pesticide residues in the River Meuse[8].

### Ground water

**National survey '90** [9]

In '90, a national survey of ground water has been carried out by the main instances involved in water protection and drinking water production. 1 to 6 samples (following catchment vulnerability) have been analysed in 10 catchments in Wallonia. Groups of compounds have been monitored by multiresidue analyses: AChE inhibitors, OC compounds, s-triazines, ureas, chlorophenoxy herbicides... giving a total of some 25 pesticides.

**Water agency monitoring** [10]

Some 180 catchments are regularly monitored by the main Walloon drinking water Agency (*Société wallonne des Distributions d'Eau*). The frequency of pesticide residue analyses depends on the catchment vulnerability and varies from 1 to 6 times per year. Samples are systematically analysed for traces of atrazine and simazine.

## Inventory of pollutions

Detailed results of water surveys for pesticide residues have been published elsewhere [5 to 10].

### In surface water

Residues of pesticides are regularly found, generally as cocktails in traces, sometimes at higher concentrations. In the context of drinking water production, the presence of herbicide residues (s-triazines and ureas) is the main problem. An average concentration of atrazine in rivers is generally between 0.1 and 1 ppb, simazine exceeds MAC (0.1 ppb) during several months per year in the Meuse River. For this reason, drinking water production from Meuse water implies a specific purification treatment based on active charcoal adsorption.

Residues of OC compounds (lindane, endosulfan, pentachlorophenol) exceed the national basic quality standard for surface water (0.01 ppb) in almost all rivers. In some cases, essentially in the most industrialised rivers, they have been found at high concentrations, sometimes exceeding LC50 for aquatic organisms (max. lindane 2.2 ppb, max. pentachlorophenol 3 ppb).

Organophosphorus compounds are rarely found in concentrations exceeding trace level. Dichlorvos (max. 1.5 ppb) and parathion occasionally exceed MAC and LC50 for aquatic organisms (both 0.1 ppb).

Triphenyltin compounds are present during several months in almost all rivers. In some cases, their concentration exceeds MAC (0.1 ppb) and LC50 for aquatic organism (1 ppb).

### In ground water

Traces of herbicides, principally s-triazines, are regularly found in ground water. According to drink-water distributors [10], some 20% catchments in Wallonia have exceeded MAC (0.1 ppb) and nearly 50% in volume of the walloon production is threatened with triazine contamination. Meanwhile, the national survey has shown that wallonian catchments are, in general, well shielded even though MAC has locally been exceeded (mecoprop, atrazine). Traces of lindane are regularly found.

## Limits of the walloon water monitoring network

Although monitoring networks give satisfactory results of the quality of the water in regard to pesticide contamination, they do not determine the precise origin and importance of contaminations. Acute pollutions could arise from industrial wastes or agricultural 'accidents' (sprayer clearing...); many pesticides are used for agricultural or non agricultural purposes; contamination can be general or localised...

Regional water monitoring networks could be adapted to identify, localise and evaluate sources of pollution in order to adopt proper protection measures for surface and ground water resources.

As seen in point 1, samples are routinely analysed only for 'obligatory' compounds in surface water in Wallonia but a ground water monitoring strategy is not defined.

The *Comité régional Phyto* has published a list of priority compounds to be monitored in surface and ground water [11]. Among some 300 commercialised pesticides, some 20 compounds (or groups of compounds) have been selected according to their importance, utilisation and environmental behaviour.

## Lists of priority compounds

In surface water :

■   AChE inhibitors : multiresidue enzymatic analyse with GC-MS confirmation for parathion, malathion, azinphos-methyl, aldicarbe, carbofuran.
■   OC insecticides : lindane, endosulfan.
■   Phenyltin fungicides.
■   s-Triazines herbicides : atrazine, simazine.
■   Ureas herbicides: chlortoluron, diuron, isoproturon.
■   Chlorophenoxy herbicides : mecoprop, MCPA, 2,4 D.

In ground water

■   AChE inhibitors : multiresidue analyse with confirmation following uses.
■   s-Triazines herbicides : atrazine, simazine.
■   Ureas herbicides : diuron, isoproturon.
■   ETU fungicides.
■   Bentazon.
■   Chlorophenoxy herbicides : mecoprop, MCPA, 2,4 D.

These lists could be regionally adapted according to agricultural and non-agricultural practices, natural resource vulnerability and weather conditions. Analytical methods, could integrate quick multiresidue analyses (enzymatic or immunoenzymatic). GLP and a general method for laboratory measurements could be developed so that results could be compared. In some pilot zones (basin sections), sampling points could be added near potential contamination sources and sampling periods could be adapted to human activities and weather conditions.

## *Bibliography*

□   1 Arrêté royal du 4 novembre 1987, fixant les normes de qualité de base pour les eaux du réseau hydrographique public. *Moniteur belge* 21/11/87 et 09/01/88.

□   2 Arrêté royal du 25 septembre 1984, fixant les normes générales définissant les objectifs de qualité des eaux douces de surface destinées à la production d'eau alimentaire. *Moniteur belge* 03/09/84.

□ 3 Arrêté de l'Executif regional wallon du 20 juillet 1989, relatif à la qualité de l'eau distribuée par reseau. *Moniteur belge* 17/02/90.

□ 4 Projet d'arrêté du gouvernement wallon relatif aux mesures de protection dans les zones de prise d'eau, de prévention et de surveillance de certaines prises d'eau souterraine.

□ 5 A. DEMEYERE & C. PLASMAN, Rapport sur les recherches de pesticides dans les eaux de surface en Belgique. Ministère de la Santé publique et de l'Environnement, 1993, 55 p.

□ 6 Relevé hydrologique. Rapports intermédiaires N6, novembre-decembre. Ministère de la Région wallonne, Direction générale des Ressources naturelles et de l'Environnement, Division des Pollutions industrielles, Namur, 1991-1992.

□ 7 C. BERTINCHAMPS & R. SAVOIR. Exemples de micropollution de la Meuse en amont de Namur, *Actes des Journées de l'ANSEAU*, Bruxelles, 10-11/06/1993.

□ 8 RIWA. Samenwerkende Rijn- en Maaswaterleidingbedrijven. Jaarslag 1992. Deel B : de Maas. Amsterdam, 1993, 121 p.

□ 9 A. DEMEYERE. Rapport sur les recherches de pesticides dans les eaux de captage souterraines destinées à la production d'eau alimentaire. Ministère de la Santé publique et de l'Environnement, Bruxelles, 1990, 64 p.

□ 10 M. ROGER. Contamination des nappes souterraines par les micropolluants, *Actes des journées de l'ANSEAU*, Bruxelles, 10-11/06/93.

□ 11 Comité régional Phyto. *Produits phytosanitaires & qualité des eaux*. Ed. CEBEDOC, Liège, 1993, 96 p.

# 4.4.       Short Communications

## 4.4.1.      *Survey in Upper Austria: Soil and Groundwater Situation*

### G. Puchwein

Federal Institute of Agrobiology, Linz, Austria

In recent years special attention has been given in Austria to the protection of water resources and the conservation of soil quality. While the subject of ground, surface and drinking water is regulated by federal laws and decrees (1 - 3), soil protection measures fall under the responsibility of provincial governments (4). Thus a concerted procedure is guaranteed only as far as water is concerned. Existing guidelines for the realization of soil survey programs mainly treat inorganic compounds. As the importance of organic pollutants (including pesticides) has increasingly been recognized, the province of Upper Austria pioneered the inclusion of pesticides, PCBs and PAHs in its own soil survey program (5) in Austria. Hence results from both programs (soil and ground water) are now available for this province and can jointly be interpreted.

The following 17 herbicides were included in the groundwater monitoring program (6): Alachlor, metolachlor, phenoxy carbonic acids (2,4-D; 2,4-DP; MCPA; MCPB; MCPP; 2,4,5-T) and triazines (atrazine, desethylatrazine, desisopropylatrazine, cyanazine, prometryn, propazine, simazine, sebuthylazine, terbuthylazine). This selection of compounds overlaps to a large extent the list of herbicides studied in the soil survey: Atrazine, desethylatrazine, cyanazine, simazine, sebuthylazine, terbuthylazine, methoprotryne, diallate, alachlor, metolachlor, metazachlor.

Apart from a few exceptions however mainly atrazine, desethylatrazine and less frequently desisopropylatrazine could be detected in ground water (6). This fact also applies to soil, as atrazine and desethylatrazine were the herbicides found almost exclusively (5). While desethylatrazine concentrations of water often exceeded those of atrazine (with pronounced regional variations), atrazine contents of soil were consistently much higher than desethylatrazine ones. In regions where maize and cereals represent major crops, higher levels of atrazine and metabolites could be found. The distribution of atrazine contents of soil and water are heavily skewed and there is a significant difference between soil samples from arable land and grassland.

This approach to jointly evaluate soil survey and groundwater monitoring results seems to be a promising starting point for further interpretations and future studies. E.g. soil characteristics may be responsible for regional differences of atrazine/desethylatrazine ratios and will have to be taken into account. On the other hand selection of sites for long term monitoring of soil quality should not be made independent of the ground water situation.

## *Literature*

□      1 GSRW, BGBl. Nr. 502/91, 2147 - 2152.

□      2 WGEV, BGBl. Nr. 338/91, 1631 - 1660.

□      3 Trinkwasser - Pestizidverordnung, BGBl. Nr. 448/91, 2011 - 2013

□      4 O.ö. Bodenschutzgesetz, LGBl. Nr. 115/91, 357 - 374

□      5 Oberöst. Bodenkataster, Bodenzustandsinventur 1993. Amt d. oö. Landesregierung

□      6 Wassergüte in Österreich, Jahresbericht 1993, BMLF

## 4.4.2.     Finnish studies on degradation and leaching

# S. Kurppa[1], P. Laitinen[1],
# R. Mutanen[2] and J.-M. Pihlava[1]

[1]     Agriculture Research Centre, Institute of Plant Protection,
        FIN-31600 Jokioinen
[2]     Plant Protection Inspection Centre, Agricultural Chemistry Department,
        PO Box 83 FIN-01301 Vantaa

In Finland pesticide residues has been studied on river waters and in ground water earlier, but in 1993 outdoor experiments were started on an experimental field in which both drainage and surfage water could be collected plot by plot. The soil type was fine sand with 5 % of clay anf 5 % of organic matter on the surface layer. Water samples were collected automatically, 30 ml sample per each 4,5 l of water flow. MCPA, dichlorprop, propiconazole, dimethoate, pirimicarb and iprodione were studied during the first season the crop being spring barley. Pesticides were sprayed using normal application time and normal dosage recommended in practice. The residues were analysed with gas chromatographic method (eg. SILTANEN and ROSENBERG 1978).

Of the quantity MCPA used on the experimental area 0,016 % was leached in surfage water during the next four months, until the soil frose. Of diclorprop the leaching was 0,018 %, dimethoate 0,04 %, propiconazole 0,01 % and pirimicarb 0,03 %, respectively. The maximum concentrations of the active ingredients were as follows: MCPA 30 µg/l, dichlorprop 70 µg/l, dimethoate 10 µg/l, propiconazole 6 µg/l and pirimicarb 4 µg/l. All the maximums were detected at the first part of August at the latest. Iprodione leached up to 0,08 % of the original amount of application. Iprodione was the only active ingredient found in drainage water, too. The maximum concentration in surfage water was 20 µg/l and in drainage water 0,3 µg/l. 0,23 % of the original quantity of iprodione used in application was detected in plant at harvest and 0,94 % in soil, respectively.

The total leaching during the first summer after application was low. However, the peak concentrations of dimethoate, at least, may lead to some occasional side-effects.

SILTANEN, H. & ROSENBERG, C. 1978. *Bulletin Environmental Contamination and Toxicology* 19:177.

*4.4.3.*    *Environmental Impact - Effect of some Agrochemicals applied to Diester-Rapeseed*

# J.-G. Pierre

French professional organizations working on oilseeds endeavour to promote rapeseed crops aimed at producing diester. This production can take place on energetic set aside, but it implies to take all necessary measures to reduce possible sources of pollution to a minimum. The development of an "Environmental Chart" linked to actions in the field, can lead to important reductions in nuisance caused by nitrates. But as far as agrochemicals are concerned, it is recommended to identify risks first, and if they exist, to define their nature and quantify them.

That is why CETIOM (Centre technique interprofessionnel des Oléagineux metropolitains) and its partners developed a three year programme of work, from 1993 to 1995 included, in two regions where rapeseed is traditionally cultivated : the Centre and Lorraine.

CETIOM's partners are : the "Agences de l'Eau Rhin-Meuse and Loire-Bretagne", the regional and departmental Chambers of Agriculture for Lorraine, the national and local Services of Plant Protection and the Union of Industries for Plant Protection (UIPP). The mixed laboratory of INPL-ENSAIA and CNRS-CPB in NANCY carries out the methodological studies and gives its scientific support. The Laboratory CIRSEE of the "Lyonnaise des Eaux" performs a certain number of analyses on active matters in waters and soils, whereas the "Bureau de Recherches géologiques et minières (BRGM) determines plottings and piezometric studies which are necessary to know the soil in the Centre.

The programme itself deals with complementary aspects in the context specific to both regions : on the departmental level of Lorraine, surveys will define the level of consumption of agrochemicals. In the Centre, surveys at producers will aim at defining cropping practices. Measures of the dispersion of two herbicide molecules will tend to check the quantities and period of presence in the soils and their possible passage into drainage waters (hydrologically isolated plots in Lorraine) or in subterranean waters (small drilling basin in the Centre). A more detailed research programme will be applied in Lorraine from 1994 onwards, which should allow us to draw up a balance of active matters as well as measure the possible effect of run-off on present risks.

The whole action is coordinated by CETIOM on the level of headquarters and local teams. It is globally followed by a National Committee of coordination and by two project groups on the local level.

# SECTION V

# OUTDOOR EXPERIMENTS

Chairman :     S. KURPPA
(Agric. Res. Center, Finland)

# OUTDOOR EXPERIMENTS

## 5.1.     Introductory Presentation

### 5.1.1.    *Predicting Field Behaviour of Pesticides from Laboratory Studies - Theory into Practice*

## D. L. Suett

Horticulture Research International, Wellesbourne, Warwick, CV35 9EF, UK

### Abstract

The extent to which data from laboratory studies done under controlled conditions can be used to predict the behaviour of soil-applied insecticides is examined by evaluating the impact of environmental and treatment variables on soil microbial activity.

### Introduction

For the past half-century, control of a wide range of insect pests, weeds and pathogens has depended largely on the use of soil-applied pesticides. The stability and bioavailability of these pesticides, and hence the level and duration of their biological efficacy, is influenced by a variety of soil, climatic and application variables. In theory, it should be possible to establish and quantify the individual and collective influences of these variables on pesticide behaviour and performance. However, in practice most of these variables are not only comprehensively inter-related but also can differ significantly both spatially and temporally. Using soil insecticides as examples, this paper examines the extent to which data obtained under the singular conditions of a laboratory incubation study can be used to predict the behaviour, and ultimately the performance, of a pesticide in the field.

### Soil Microbial Activity

Although many factors combine to determine the behaviour of an insecticide in soil, it is commonly accepted that the most important influence is that of the soil microbial population (Torstensson, 1987). Large qualitative and quantitative differences in microbial composition and activity occur between soils from different sites (Cook & Greaves, 1987) and these differences may be exacerbated further by seasonal fluctuations and cropping practices. It is known also that the degradative capacities of some soil microbial communities may be enhanced greatly following treatment with insecticides. It is now evident that, in some instances, just a single application of the commercially-recommended dose of an insecticide is sufficient to induce this increased activity and that some soils can retain this modified degradative ability for many years (Suett & Jukes, 1993a).

As long as microbial adaptation does not reduce the efficacy of an insecticide against the target pest, the development and stability of a modified degradative capacity are of little economic or practical adverse significance. On the contrary, in many instances the rapid dissipation of residual insecticide might be considered as environmentally beneficial. However, recent studies at Horticulture Research

International (HRI) have shown that the efficacies of most of the insecticides available for application to European soils are threatened by the accelerated degradation of their residues in previously-treated soils.

## Soil Characteristics

The large inherent variability in the microbial characteristics of soils (Cook & Greaves, 1987) emphasises the importance of exercising the utmost care in all aspects of sampling, storage and treatment of field soils which are transferred into the laboratory. It is essential to avoid exposing soils to prolonged storage, air-drying or freezing, all of which may modify significantly the composition and activity of the microbial population (Walker, 1989). Differences in rates of loss of pesticides in laboratory studies have been correlated closely with variations in microbial biomass (Anderson, 1987) and it has been suggested that, if storage is unavoidable, soils should be maintained in a moist condition at a temperature of 2-4°C for no longer than 3 months (Anderson, 1987). Similarly, it is essential to ensure that soil bioactivity is maintained at an appropriate level throughout an incubation study and it would seem prudent to limit such studies also to no longer than 3 months (Anderson, 1987).

In most instances, the information which can be derived from an incubation study with just a single soil is likely to be limited. Extremes in the behaviour of freshly-applied insecticide may be interpreted relatively confidently. Thus it would be realistic to assume that an initial half-life of a few days or less is unlikely to be associated with reliable long-term efficacy. Similarly, an initial half-life of more than a few weeks would suggest that any reduction of efficacy in the field was unlikely to have occurred as a result of accelerated biodegradation. However, as the majority of studies are likely to fall between these two extremes, comparisons are usually made between the soil under scrutiny and an adjacent and similar but previously-untreated soil. It cannot be emphasised too strongly that selection of the "untreated" soil demands at least as much diligence, and often more, as is used for the previously-treated soil. Although sites might seem, at first glance, to differ only in their insecticide treatment histories, in practice they may have different pH values or cropping histories, have been exposed to different fertiliser regimes, etc, all of which can have a significant impact on the size and activity of the soil microbial population.

Nor should it be assumed that lack of direct treatment with insecticide guarantees the absence of microbial adaptation. Suett & Jukes (1987) reported the accelerated biodegradation of freshly-applied phorate in soil from hedgerows and ditches surrounding carrot fields which had been treated with the insecticide. They also found (Suett & Jukes, 1990) that, at an intensively-treated hop farm, enhanced degradation of the systemic aphicide mephosfolan was almost as rapid in soil from the pathways as in soil from the base of plants. Transfer of rapid-degrading characteristics from treated to previously-untreated soils has also been observed in experimental plots. Suett & Jukes (1993b) reported a gradual decline in the stability of freshly-applied carbofuran in soil from an untreated plot which was sited 50 m from the treated area. In the laboratory, increased degradation rates were induced in previously-untreated soils by incorporating as little as 0.1% of an iprodione-degrading soil (Walker, 1987) or 0.25% of a carbofuran-degrading soil (Harris et al., 1984). This and other evidence of the readiness with which a modified degradative ability can be transferred from one soil to another emphasises the need for care at all stages of soil selection and sampling as well as during the laboratory studies.

With some insecticides, incubation of just a single soil can yield useful evidence of a change in its degradative properties. The degradation of parent aldicarb, disulfoton and phorate in soil to their respective sulphoxides and sulphones is by now well documented (Suett & Jukes, 1987). There is also abundant evidence that the prolonged bioactivity of these insecticides in soil is associated with the formation and stability of these oxidation products, with relatively little contribution from the parent insecticides. It is now apparent that, although the long-established pattern of residue behaviour is still evident in previously-untreated soils, it may be modified greatly in soils which have received previous treatments with these insecticides. Thus Suett & Jukes (1992) found that phorate sulphone accumulated steadily in a previously-untreated soil until, after 12 wk, it comprised > 80% of the initial dose. However, in soil from an adjacent, previously-treated site, the sulphone reached a maximum level equivalent to 10-15% of the applied dose after only 2 wk, subsequently declining very rapidly. A similar marked difference in the behaviour of aldicarb sulphoxide in previously-treated and -untreated soils has also been reported (Smelt et al., 1987; Suett & Jukes, 1988). With these insecticides, therefore, a relatively simple qualitative assessment should reveal the development of accelerated biodegradation.

The singular characteristics of each composite soil sample prepared for a laboratory incubation study will reflect the mean value of what, in the field, can be a wide range of physical and chemical as well as macro- and micro-biological properties. Factors such as nutrient status, organic matter content and soil pH can vary markedly within a field, sometimes over a relatively small area (Walker, 1989). Such differences may be insignificant to the soil chemist (Anderson, 1978) but may be sufficient to determine the extent to which specific microbiological activity, such as adaptation, will occur. Furthermore, while nutrient and organic matter levels exert a direct influence on crop vigour and insecticide availability, the impact of differences in soil pH can be just as, sometimes more, extreme.

In all soils, the diversity of the microbial population is influenced greatly by pH, with even minor changes inducing significant fluctuations in the composition and activity of the microbiological community (Burns, 1976). One of the consequences of these fluctuations is a variation in microbial adaptation to the presence of insecticides. Recent studies have shown that, although adaptation occurs consistently more readily at high than at low pH, the level of optimum response varies with different insecticides. Thus Read (1986) found that accelerated degradation of carbofuran could not be induced in soils with pH <5.8 and that aldicarb was stable in soils of pH 5.6 and below (Read, 1987). Rapid degradation of the organophosphorus insecticide isazophos was associated with pretreatment of soils with pH>6.9 (Somasundaram et al., 1993). Significant correlations between stability, treatment history and soil pH have been reported also for mephosfolan (Suett & Jukes, 1990) and have been observed with chlorfenvinphos and phorate (Suett & Jukes, unpublished data). In practice, therefore, the net effect of within-site variability will depend largely on the characteristics of the insecticide.

It is almost inevitable that soils used for laboratory studies will be root-free. The diversity and activities of microorganisms in soil remote from plant roots are different from those in the rhizosphere (Hill & Wright, 1978) and some of the consequences of this were summarised by Walker (1989). It should be recognised also that relatively large changes in pH can occur locally in soil adjacent to plant roots. Gray et al. (1978) suggested that an increase of one pH unit could occur over a few millimetres around the root. Marschner & Romheld (1983) showed that, in a soil with a bulk pH of 6.0, the pH in the rhizosphere varied from 4.5 to 7.0. Furthermore, the magnitude and limits of this range were influenced also by plant species and nitrogen source.

In the field, the net impact of these macro- and micro-differences in pH on microbial adaptation and activity is reflected by the overall treatment efficacy. The extent to which this might be correlated with a laboratory study with a single composite soil sample is questionable and is complicated further by unavoidable differences in treatment factors.

## Treatment Factors

Laboratory studies of degradation demand that treatment factors are defined and controlled. These studies will therefore comprise a single uniformly-incorporated dose, usually of a liquid formulation, which is maintained at a fixed temperature and moisture level. Such conditions contrast totally with the field situation, where non-uniform doses of granular formulations of insecticides are exposed to fluctuating multi-dimensional gradients of temperature and moisture. The effects of these fluctuations on the induction and expression of microbial adaptation can be extreme.

### *Application rate and treatment uniformity*

It has long been considered that soil microorganisms were affected by pesticides only at doses substantially greater than those encountered in practice (Anderson, 1978). However, it is now evident that adaptation can be induced at, and often below, typical field application rates. It is therefore most important to ensure that the concentration level selected for a laboratory study lies within the range likely to occur in the field. This is hardly difficult to achieve, given that most commercial applications of soil insecticides result in a concentration gradient in the rhizosphere which can range from zero to several hundred mg AI/kg dry soil (Suett & Jukes, 1990)! Behaviour at a single level may therefore bear little resemblance to the extremes of behaviour occurring within this concentration gradient. Nevertheless, a single target treatment level is usually specified for a laboratory study of pesticide effects, although it has been suggested that at least 2 rates should be used (Anderson, 1978).

There is much evidence that, in soil, larger doses of insecticides degrade proportionately more slowly than smaller doses. However, in the presence of adapted microorganisms the influence of insecticide concentration can become even more significant. Thus Suett & Jukes (1990) showed that accelerated degradation of mephosfolan was expressed readily at initial treatment levels of 1 and 10 mg/kg but not at 100 and 1000 mg/kg. Similarly, Read (1987) reported a marked delay in the degradation of aldicarb in previously-treated soils at doses exceeding 750 mg AI/kg whereas carbofuran (Read, 1986) was degraded rapidly at doses up to 5g AI/kg.

In commercial practice the majority of soil insecticides are now applied as localised rather than as dispersed overall treatments. With these localised band or "spot" treatments, mean concentrations of insecticide in the treated zone will commonly be within 5 - 50 mg AI/kg dry soil (Suett & Padbury, 1981). At HRI-W, therefore, all laboratory studies of insecticide behaviour are done using initial treatment doses of 10 - 25 mg AI/kg dry soil.

The limitations of a single level of treatment are slightly less relevant to a field study, where dose can often be based on a commercially-recommended maximum. However, such systems seldom yield sufficient information for establishment of dependable dose-response relationships and there is much to recommend the log-dose evaluation system developed at HRI-W for use with granular formulations (Thompson, 1984). This system provides a comprehensive package of information about the efficacy of granular formulations of insecticides over a wide range of dose and is reliably sensitive to relatively modest shifts in dose response.

Treatment uniformity should be considered within this context of dose rate. Walker (1989) drew attention to the contrast between the uniformity achieved with a laboratory treatment of a homogeneous, debris-free sample of soil and that resulting from the field application of a granular formulation. Even the most meticulous field application is likely to result in large plant-to-plant dosing variabilities. Recent surveys of insecticide doses within a 15 cm dia. of individual field-treated swedes revealed coefficients of variation of 35 - 50% (Suett & Jukes, unpublished data). Furthermore, a survey of 52 granule applicators in commercial use indicated that, in practice, variability was likely to be even greater (Thompson et al., 1984) while Walker (1989) concluded that it would increase further with increasing time after application.

## *Formulation*

Formulation is a major source of difference between field and laboratory studies of insecticide behaviour. In the laboratory, liquid formulations of technical- or analytical-grade materials are used most commonly. They offer total flexibility in selection of dose and, by using appropriate dilutions, optimum uniformity of treatment can be achieved with even the smallest quantity of soil. Although in the field almost all soil insecticides are applied as granular formulations, their use in laboratory studies can be misleading. Suett (1987) found that, after carbofuran was applied as a liquid formulation, there were large differences in its stability in previously-treated and previously-untreated soils. However, after application of a granular formulation, there were no significant differences in rates of loss from the same two soils even though performance against cabbage root fly was reduced greatly in the previously-treated soil. The accelerated degradations of fensulfothion, isofenphos and trimethacarb as well as carbofuran in previously-treated soils also were expressed more readily with an analytical formulation than with a granular formulation (Chapman & Harris, 1990).

The relatively large concentrations of AI in commercial granular formulations (up to 10% w/w) also can cause practical difficulties, since reliably uniform incorporation of a few hundred mg of granules per kg of soil is impossible. In the absence of an appropriate water-miscible liquid formulation, therefore, it may be necessary to use analytical-grade material. In such circumstances the pesticide should be "formulated" on to an inert carrier such as sand (Harris et al., 1984) or clay (Suett & Jukes, 1988) rather than applied in an organic solvent which may modify the soil's microbial composition and activity. However, care must be taken when preparing and handling such "formulations" as some insecticides are susceptible to oxidative transformations during and after formulation of these relatively small concentrations.

## *Soil temperature and moisture*

Irrespective of the degradative capacity of a soil, the stability of a freshly-applied insecticide will be influenced greatly by the temperature and moisture conditions to which it is exposed. It is therefore important that conditions selected for a laboratory incubation do not differ unduly from those encountered in the field, under which appropriate microbial populations have developed. Thus studies with topsoils from temperate regions are usually done at 15-20°C whereas Mediterranean and tropical soils can reasonably be exposed to much greater temperatures. However, it is likely that the effects of storage and transport on the microbial composition of such soils will make at least as much impact on their net degradative activities as would inappropriate selection of incubation temperatures (Hance, 1989).

Although there is abundant evidence that insecticide degradation in soil is correlated positively with both temperature and moisture, the magnitude of the effects of these factors can vary according to soil treatment history. Differences in the stability of insecticides in previously-treated and previously-untreated soils decrease markedly as moisture levels decline (Suett & Jukes, unpublished data) and they eventually become difficult to discern under conditions which would not be uncommon in the rhizosphere under an actively-growing crop in summer.

In practice, the procedure followed in order to maintain a constant moisture level during an incubation study can further exaggerate differences between laboratory and field environments. At HRI-W, initial laboratory studies with insecticides adopted a regime of once- or twice-weekly additions of water. In subsequent studies, placement of the bottles of soils in loosely-covered water-filled containers inside the incubators (Suett & Jukes, 1993a) enabled soil moistures to be maintained at pre-determined levels without intermittent replenishment. By thus eliminating the need to stir each sample regularly, the modified procedure limits another potential discrepancy between laboratory and field as well as reducing the risk of possible cross-contamination.

# Insecticide Behaviour and Biological Efficacy

The above observations illustrate the difficulty of extrapolating from a laboratory study done under controlled conditions in order to predict the behaviour of an insecticide under totally variable field conditions. In terms of further extrapolation to predict biological efficacy - which is the ultimate objective of most of these studies - the task is complicated further by the variability and unpredictability of the biological target. Different species of insect vary considerably in their susceptibilities to insecticides and significant differences can also occur within a single population (Harris, 1972). In practice, pest eradication within a single crop is rarely, if ever, total. Similarly a failed control measure invariably reflects damage to only an unacceptably large proportion of a crop rather than total crop loss. Furthermore, this "damage threshold" can vary with different crops. Wheatley (1973) suggested that, in order to protect the edible parts of crops grown for human consumption, insecticide treatments needed to remain >90% efficient until harvest. In contrast, efficiencies of only 70% for just a few weeks were likely to be sufficient to control damage to the non-edible portions, as long as this indirect damage did not reduce yield (Wheatley & Coaker, 1970). Interpretation of laboratory data also must take the specific crop-pest situation into account. With, for example, the insecticide/nematicide carbofuran, evidence of a reduced initial half-life in a previously-treated soil would probably make little impact on its efficacy against most nematodes, as the nematode population would be reduced immediately after application and would re-establish only slowly. Similarly, control of seedling pests on a crop such as sugar beet might also be adequate whereas longer-term protection of, for example, carrots against carrot fly would be seriously jeopardised.

The problem of specifying an optimum dose is compounded further by inadequate knowledge of the ways in which soil-inhabiting insects accumulate toxic doses of insecticides. The end result is that application methods are rarely optimised in order to meet the demands of specific control problems.

In view of the many complexities discussed above, it is hardly surprising that there is often a reluctance to acknowledge the practical implications of laboratory studies, especially where microbial adaptation is concerned. Nevertheless, both field and laboratory studies will continue to make essential contributions to a better understanding of the principles of crop protection. It is also evident that, as long as appropriate attention is afforded to the relevant aspects, the two systems can complement each other accurately. Recent studies at HRI-W have shown that the field performance of carbofuran against cabbage root fly in previously-treated and -untreated soils could be predicted from laboratory incubation studies (Suett & Jukes, 1993a). The correlation was specific to a single pest on a single crop but there is every likelihood that similar correlations could be established for other crop-pest situations. This will be achieved only by continued efforts to quantify the impact of all the above variables on the complex inter-relationships between soil, pest and insecticide.

## Acknowledgement

The above studies at HRI-W were supported by the Ministry of Agriculture, Fisheries and Food.

## References

□    Anderson, J.P.E. (1987) Handling of soils for pesticide experiments. In: *Pesticide Effects on Soil Microflora*, L. Somerville & M.P. Greaves (Eds), London: Taylor and Francis, pp. 45-60.

□    Anderson, J.R. (1978) Pesticide effects on non-target soil microorganisms. In: *Pesticide Microbiology*, I.R. Hill & S.J. Wright (Eds), London: Academic Press, pp. 313-533.

□    Burns, R.G. (1976) Microbial control of pesticide persistence in soil. In: *The Persistence of Insecticides and Herbicides*, K.I. Beynon (Ed), *BCPC Monograph No. 17*, London: BCPC Publications, pp. 229-239.

□    Chapman, R.A.; Harris, C.R. (1990) Enhanced degradation of insecticides in soil. Factors influencing the development and effects of enhanced microbial activity. In: *Enhanced Biodegradation of Pesticides in the Environment*, K.D. Racke & J.R. Coats (Eds), Washington: American Chemical Society, pp. 82-96.

□    Cook, K.A.; Greaves, M.P. (1987) Natural variability in microbial activities. In: *Pesticide Effects on Soil Microflora*, L. Somerville & M.P. Greaves (Eds), London: Taylor and Francis, pp. 15-43.

□    Gray, T.R.G.; Jones, J.G.; Wright, S.J.L. (1978) Microbiological aspects of the soil, plant, aquatic, air and animal environments. II. The soil and plant environments. In: *Pesticide Microbiology*, I.R. Hill & S.J. Wright (Eds), London: Academic Press, pp. 17-77.

□    Hance, R.J. (1989) Aspects of the physical behaviour of soil applied pesticides that contribute to differences in dissipation rates in the field compared with controlled conditions. *Aspects of Applied Biology*, 21, 147-157.

□  Harris, C.R. (1972) Factors influencing the effectiveness of soil insecticides. *Annual Review of Entomology*, 17, 177-198.

□  Harris, C.R.; Chapman, R.A.; Harris, C.; Tu, C.M. (1984) Biodegradation of pesticides in soil: rapid induction of carbamate degrading factors after carbofuran treatment. *Journal of Environmental Science and Health*, B19, 1-11.

□  Hill, I.R.; Wright, S.J. (1978) The behaviour and fate of pesticides in microbial environments. In: *Pesticide Microbiology*, I.R. Hill & S.J. Wright (Eds), London: Academic Press, pp. 79-136.

□  Marschner, H.; Romheld, V. (1983) In vivo measurement of root-induced pH changes at the soil:root interface: effect of plant species. *Zeitschrift für Pflanzenphysiologie*, 111, 241-251.

□  Read, D.C. (1986) Accelerated microbial breakdown of carbofuran in soil from previously-treated fields. *Agriculture, Ecosystems and Environment*, 15, 51-61.

□  Read, D.C. (1987) Greatly accelerated degradation of aldicarb in re-treated field soil, in flooded soil and in water. *Journal of Economic Entomology*, 80, 156-163.

□  Smelt, J.H.; Crum, S.J.H.; Teunissen, W.; Leistra, M. (1987) Accelerated transformation of aldicarb, oxamyl and ethoprophos after repeated soil treatments. *Crop Protection*, 6, 295-303.

□  Somasundaram, L.; Jayachandran, K. Kruger, E.L.; Racke, K.D.; Moorman, T.B.; Dvorak, T.; Coats, J.R. (1993) Degradation of isazophos in the soil environment. *Journal of Agricultural and Food Chemistry*, 41, 313-318.

□  Suett, D.L. (1987) Influence of treatment of soil with carbofuran on the subsequent performance of insecticides against cabbage root fly and carrot fly. *Crop Protection*, 6, 371-378.

□  Suett, D.L.; Jukes, A.A. (1987) Evidence and implications of accelerated degradation of organophosphorus insecticides in soil. *Toxicological and Environmental Chemistry*, 18, 37-49.

□  Suett, D.L.; Jukes, A.A. (1988) Accelerated degradation of aldicarb and its oxidation products in previously-treated soils. *Crop Protection*, 7, 147-152.

□  Suett, D.L.; Jukes, A.A. (1990) Some factors influencing the accelerated degradation of mephosfolan in soils. *Crop Protection*, 9, 44-51.

□  Suett, D.L.; Jukes, A.A. (1992) Accelerated degradation of phorate: implications for pest control in the United Kingdom. *1992 Brighton Crop Protection Conference - Pests and Diseases*, 3, 1217-1222.

□  Suett, D.L.; Jukes, A.A. (1993a) Stability of accelerated degradation of soil-applied insecticides: laboratory behaviour of aldicarb and carbofuran in relation to their efficacy against cabbage root fly in previously treated field soils. *Crop Protection*, 12, 431-442.

□  Suett, D.L.; Jukes, A.A. (1993b) Accelerated degradation of soil insecticides: comparison of field performance with laboratory behaviour. In: *Fate and Prediction of Environmental Chemicals in Soils, Plants and Aquatic Systems*, M. Mansour (Ed), Boca Raton: Lewis Publishers, pp. 31-41.

□  Suett, D.L.; Padbury, C.E. (1981) Influence of some new application variables on insecticide behaviour and availability in soils. *1981 British Crop Protection Conference - Pests and Diseases*, 1, 157-164.

□  Thompson, A.R. (1984) Use of a log-dose system for evaluating granular insecticide products against cabbage root fly. *Mededelingen van de Faculteit Landbouwwetenschappen Rijksuniversiteit Gent*, 49, 909-918.

□  Thompson, A.R.; Kempton, D.P.H.; Percivall, A.L. (1984) The use and precision of insecticide granule applications for protecting brassicas against cabbage root fly. *1984 British Crop Protection Conference - Pests and Diseases*, 3, 1123-1128.

□    Torstensson, N.T.L. (1987) Microbial decomposition of herbicides in soil. *Progress Pesticide Biochemistry and Physiology*, 6, 249-270.

□    Walker, A. (1987) Further observations on the enhanced degradation of iprodione and vinclozolin in soil. *Pesticide Science*, 21, 219-231.

□    Walker, A. (1989) Factors influencing the variability in pesticide persistence in soils. *Aspects of Applied Biology*, 21, 159-172.

□    Wheatley, G.A. (1973) The effectiveness of insecticides applied to soil. *Proceedings of the 7th British Insecticide and Fungicide Conference*, 3, 991-1003.

□    Wheatley, G.A.; Coaker, T.H. (1970) Pest control objectives in relation to changing practices in agricultural crop production. In: *Technological Economics of Crop Protection and Pest Control, SCI Monograph No. 36*, London: Society of Chemical Industry, pp. 42-55.

# 5.2.    Platform Presentations

## 5.2.1.    *Assessment of Herbicide Persistence in Saskatchewan Field Soils from Laboratory Data*

## A. E. Smith

Agriculture Canada, Research Station,
Box 440, Regina, Saskatchewan, Canada, S4P 3A2

## Summary

The long cold winters and hot dry summers experienced on the Canadian prairies are not conducive to the steady and continuous breakdown of herbicide residues in field soils. It has been demonstrated that, under field conditions, degradation occurs only when temperature and moisture conditions are favourable for microbial activity.

The experimental conditions found in laboratory studies used to determine herbicide persistence at different temperature and moisture levels are not comparable to those observed in the field. However, laboratory studies are still very useful and can indicate whether residues of a particular herbicide are likely to be carried over in the soil to the next crop year.

The prediction of losses of amidosulfuron, asulam, atrazine, cyanazine, metribuzin, simazine, 2,4-D, and 2,4,5-T in clay field plots at the Regina Research Station as derived from laboratory data was reasonably good but, in general, rates of loss were underestimated.

As a result of laboratory and field studies carried out with many herbicides, it has been empirically observed that if the laboratory half-life of a herbicide in soil, incubated at 20°C and 85% of field capacity, is greater than 3-4 weeks, then under prairie field conditions there is potential for the carry-over of spring-applied residues to the next crop year.

## Introduction

Losses of herbicides from treated soils occur as a result of both chemical and biological processes, and also, in the case of certain products, by volatilization. These processes are all affected by soil moisture and temperature conditions. In western Canada, optimum conditions for herbicide breakdown tend to occur during the spring and early summer. In winter, when the ground is frozen, and during the late summer and autumn, when the top soil is dry, losses are minimal (Smith & Hayden, 1981). For these reasons soil persistence data for herbicides are different from those obtained from studies in other parts of the world with a less severe climate. Soil dissipation studies, under prairie field conditions, are therefore required prior to their registration in western Canada.

Considerable efforts are being made toward the development of mathematical models for the prediction of herbicide dissipation under field conditions. One method (Walker & Barnes, 1981; Walker et al., 1983) involves the study of herbicide persistence in soils at different temperature and moisture regimes under carefully controlled laboratory conditions. These data are then used in conjunction with meteorological records in a computer program to simulate the field dissipation pattern.

In Saskatchewan, this approach has been used to predict the field persistence of amidosulfuron (3-(4,6-dimethoxypyrimidin-2-yl)-1-(N-methyl-N-(methylsulfon-yl)amino)sulfonylurea), asulam (methyl 4-aminobenzenesulphonylcarbamate), atrazine (2-chloro-4-ethylamino-6-isopropylamino-1,3,5-triazine), cyanazine (2-chloro-4-(1-cyano-1-methylethylamino)-6-ethylamino-1,3,5-triazine), metribuzin (4-amino-6-tertbutyl-4,5-dihydro-3-methylthio-1,2,4-triazin-5-one), simazine (2-chloro-4,6-bisethylamino-1,3,5-triazine), and 2,4,5-T (2,4,5-trichlophenoxyacetic acid) in clay field plots at the Regina Research Station (Smith & Walker, 1977, 1989; Walker & Smith, 1979; Walker et al., 1983; Smith & Aubin, 1992). This paper discusses these data and also the general applicability of laboratory data for the prediction of field residues under Canadian prairie field conditions.

# Materials and methods

## *Soil*

The composition and physical characteristics of the Dark Brown Chernozem clay used in the various studies are summarised as follows: clay 67%, silt 21%, sand 12%; organic carbon 3%; pH 7.2; C.E.C. 8 meq 100 g$^{-1}$; conductivity 0.4 ms cm$^{-1}$; field capacity 42%; and wilting point 20%.

## *Experimental procedures*

The laboratory and field dissipation studies for asulam, atrazine, cyanazine, metribuzin, simazine, and 2,4,5-T have previously been reported (Smith & Walker, 1977, 1989; Walker & Smith, 1979; Walker et al., 1983). For amidosulfuron, the laboratory degradation studies have been described (Smith & Aubin, 1992). The [14C]amidosulfuron field studies were conducted in small (10 X 10 cm) field plots, similar to those previously reported for studies with [14C]fenoxaprop-ethyl and [14C]fenthiaprop-ethyl (Smith, 1985), with soil extraction and analysis as for the laboratory studies (Smith & Aubin, 1992).

In all studies, the laboratory dissipation of the herbicides approximated to first-order kinetics. The influence of temperature on herbicide degradation rates was characterised using the Arrhenius equation. Moisture effects were evaluated using the empirical equation:

$$H = AM^{-B}$$

where H is the half-life in days at moisture content M (% wt/wt) and A and B are constants. The values derived for the Arrhenius activation energies and the moisture dependence constants A and B for the herbicides under discussion are summarised in Table 1.

| Herbicide | Activation Energy (kj mole$^{-1}$) | Moisture dependence constants A | B |
|---|---|---|---|
| Amidosulfuron | 58 | 145000 | 2.1 |
| Asulam | 40 | 1340 | 1.5 |
| Atrazine | 44 | 7600 | 1.5 |
| Cyanazine | 61 | 2700 | 1.9 |
| Metribuzin | 59 | 60200 | 2.1 |
| Simazine | 45 | 8170 | 1.2 |
| 2,4,5-T | 85 | 80 | 0.7 |

**Table 1 :** Temperature and moisture dependence of herbicides in a Regina clay

## Results and discussion

Of the herbicides studied, asulam, cyanazine and 2,4,5-T were short lived with almost complete breakdown occurring in the field soils within 4 to 8 weeks following treatments made in May or June (Smith & Walker, 1977, 1989; Walker & Smith, 1979). In contrast, residues of amidosulfuron (Table 2), atrazine, metribuzin, and simazine (Smith & Walker, 1989) in late October immediately prior to freeze-up, and therefore destined to be carried over to the next crop year, were significant being over 30% of the spring treatments. In all studies, there was negligible leaching of herbicide from the top 10 cm of field soil.

In the case of asulam, six separate field plot studies were carried out at different times between May and October. In general, there was close correspondence (<25% difference) between observed and predicted residue levels. With applications made in October, when low rainfall was accompanied by low temperatures, the model underestimated the loss (Smith & Walker, 1977).

In five separate field plot studies with 2,4,5-T a greater discrepancy (>25%) was observed between the observed and predicted residues. This was attributed, in part, to the difficulty of obtaining a correct measure of soil moisture for use with the simulation program (Walker & Smith, 1979).

With the triazine herbicides atrazine, cyanazine, metribuzin, and simazine (Walker et al., 1983; Smith & Walker, 1989) there was a reasonable agreement between the observed and predicted levels for atrazine and the short-lived cyanazine. For both metribuzin and simazine, the model underestimated the rate of loss. Thus at the end of the 105-130 day field study periods, the observed amounts of atrazine, metribuzin, and simazine were approximately 32, 31, and 25% of that applied with predicted levels of 38, 64, and 60%, respectively.

For amidosulfuron (Table 2) there was close agreement between the observed and predicted field dissipation.

| Date | Days | Observed amidosulfuron (%) | Predicted amidosulfuron (%) |
|------|------|----------------------------|------------------------------|
| 8 June | 0 | $99 \pm 5^{*}$ | 100 |
| 15 June | 7 | $110 \pm 1$ | 98 |
| 23 June | 15 | $103 \pm 10$ | 97 |
| 14 July | 36 | $86 \pm 2$ | 89 |
| 18 August | 71 | $83 \pm 2$ | 83 |
| 19 October | 133 | $76 \pm 3$ | 70 |

    *  Mean and standard deviation from three replicate plot treatments.
       No leaching observed from the 0-10 cm soil depth.

**Table 2 :** Observed and predicted [$^{14}$C]amidosulfuron recovered with time from top 10 cm of small field plots at Regina following treatment at a rate of 90 g ha$^{-1}$ on 8 June, 1992

Some of the discrepancies in the model predictions may be explained by the fact that laboratory dissipation studies do not accurately reflect dissipation conditions in the field. In the former studies, the soils are maintained under specific moisture and temperature regimes, whereas, in the field, temperature and moisture can vary by the hour. It has also been noted (Anderson, 1987) that the soil biomass under laboratory conditions is very sensitive and prone to loss on soil storage and during incubation. Such losses of microbial biomass can range from 29-76%. As the biomass decreases so can the numbers of herbicide-degrading organisms, thus resulting in increased laboratory half-lives and so increased herbicide prediction rates. For laboratory dissipation studies it has been recommended (Anderson, 1987) that the soil should be freshly collected, excessive drying should be avoided, storage should be carried out at 2 to 4°C for periods of no longer than three months, and that studies should be run for no more than three months or until 50% of the initial biomass has been lost. It has been suggested (Walker et al., 1983) that herbicide losses in the field through volatilisation, photochemical degradation, and wind erosion of the surface soil may also be responsible for some of the discrepancies in model predictions. These and other laboratory/field differences have been summarised (Hance & Führ, 1992).

Although under Saskatchewan conditions there is a tendency for the model to overestimate herbicide residues remaining, it would still be sufficiently accurate to enable most farmers to plan crop rotations.

The persistence of many herbicides in Saskatchewan soils has been studied under field conditions. Laboratory studies in the same soils at 20°C and at 85% of field capacity moisture have also been carried out (Smith, 1982, 1985, 1987, 1989; Walker et al., 1983; Smith & Belyk, 1989; Smith & Walker, 1989; Smith et al., 1990). The results of these investigations are compared in Table 3. In general, if the laboratory half-life is more than 3-4 weeks, then under Saskatchewan field conditions there is the potential for carry-over of spring field treatments to the next crop year.

From the data depicted in Table 3, a plot of field carry-over (% of applied treatment) vs the laboratory half-life (weeks) gives the empirical equation:

$$\text{Carry-over } (\%) = 2.3(\text{Laboratory Half-life}) - 2.3 \ (R^2 = 0.71)$$

Thus for Saskatchewan field soils, the empirically derived carry-over from spring treatments is approximately 2.5 times the half-life (in weeks), expressed as a percent.

| Herbicides | Half-life in weeks | | | % Field carry-over | | |
|---|---|---|---|---|---|---|
| | Clay | Sandy loam | Clay loam | Clay | Sandy loam | Clay loam |
| Asulam | Bromoxynil 2,4-D 2,4-DB Diclorprop Glufosinate MCPA MCPB Mecoprop Thifensulfuron Silvex 2,4,5-T | | | | | |
| | <2 | <2 | <2 | <2 | <2 | <2 |
| Simazine | 14 | -** | - | 40 | - | - |
| Atrazine | 10 | - | - | 30 | - | - |
| Metribuzin | 5 | - | - | 20 | - | - |
| Cyanazine | 2 | - | - | <2 | - | - |
| Diclofop | 5 | 4 | 12 | 10 | 10 | 15 |
| Fenoxaprop | 3 | 2 | - | 5 | 5 | - |
| Fenthiaprop | 8 | 4 | - | 20 | 20 | - |
| Haloxyfop | 6 | 4 | 14 | <5 | <5 | 20 |
| Fluazifop | 4 | 2 | 4 | <5 | <5 | <5 |
| Dicamba | 6 | 3 | 2 | <2 | <2 | <2 |
| Sethoxydim | 4 | 2 | 2 | <5 | <5 | <5 |
| Triallate | 7 | - | - | 25 | - | - |

*   Data from Smith (1982, 1985, 1987, 1989), Walker et al. (1983). Smith & Belyk (1989), Smith & Walker (1989), Smith et al. (1990),
**  Not determined.

**Table 3 :** Comparison of half-lives of herbicides, incubated in Saskatchewan soils at 20°C and 85% of field capacity, with percent of the spring-applied chemical carried over in the field to the next crop year

# Conclusions

1.  Laboratory data are satisfactory for obtaining a qualitative comparison of herbicide persistence in soils at different temperatures and moistures.

2.  With care, laboratory data can be used to predict field persistence, though the field values tend to be underestimated.

3.  By measuring the half-lives of herbicides at 20°C and 85% of field capacity, approximate field carry-over data can be derived for Saskatchewan soils.

# References

□   ANDERSON JPE (1987) Handling and storage of soils for pesticide experiments. In: Somerville L & Greaves MP, eds. *Pesticide Effects on Soil Microflora*, London: Taylor & Francis, pp. 45-60.

□   HANCE RJ & FFHR F (1992) Methods to study fate and behaviour of pesticides in the soil. In: Führ F & Hance RJ, eds. *Lysimeter Studies of the Fate of Pesticides in the Soil, British Crop Protection Council Monograph No 53*, Lavenham: Lavenham Press, pp. 9-18.

□   SMITH AE (1982) Herbicides and the soil environment in Canada. *Canadian Journal of Soil Science* 62, 433-60.

□   SMITH AE (1985) Persistence and transformation of the herbicides [$^{14}$C]fenoxa-prop-ethyl and [$^{14}$C]fenthiaprop-ethyl in two prairie soils under laboratory and field conditions. *Journal of Agricultural and Food Chemistry* 33, 483-88.

□   SMITH AE (1987) On the persistence of herbicides in soils of western Canada with additional comments on the phenoxyalkanoic acids and sulfonylureas. *Expert Committee on Weeds Western Canada Research Report*, Vol 3, pp. 287-90.

□   SMITH AE (1989) Degradation, fate, and persistence of phenoxyalkanoic acid herbicides in soil. *Reviews of Weed Science* 4, 1-24.

□   SMITH AE & AUBIN AJ (1992) Degradation of the sulfonylurea herbicide [$^{14}$C]amidosulfuron (HOE 075032) in Saskatchewan soils under laboratory conditions. *Journal of Agricultural and Food Chemistry* 40, 2500-04.

□   SMITH AE & BELYK MB (1989) Field persistence studies with the herbicide glufosinate-ammonium in Saskatchewan soils. *Journal of Environmental Quality* 18, 475-79.

□   SMITH AE & HAYDEN BJ (1981) On the long-term persistence of 2,4-D and tri-allate in Saskatchewan soils. *Proceedings of the European Weed Research Society Symposium 1981, Theory and Practice of the use of Soil Applied Herbicides*: pp. 156-62.

□   SMITH AE & WALKER A (1977) A quantitative study of asulam persistence in soil *Pesticide Science* 8, 449-56.

□   SMITH AE & WALKER A (1989) Prediction of the persistence of the triazine herbicides atrazine, cyanazine, and metribuzin in Regina Heavy Clay. *Canadian Journal of Soil Science*. 69, 587-95.

□   SMITH AE, SHARMA MP & AUBIN AJ (1990) Soil persistence of thiameturon (DPX M6316) and phytotoxicity of the major degradation product. *Canadian Journal of Soil Science* 70, 485-91.

☐    WALKER A & BARNES A (1981) Simulation of herbicide persistence in soil; a revised
     computer model. *Pesticide Science* 12, 123-32.

☐    WALKER A & SMITH AE (1979) Persistence of 2,4,5-T in a heavy clay soil. *Pesticide Science*
     10, 151-57.

☐    WALKER A, HANCE RJ, ALLEN JG, BRIGGS GG, CHEN Y-L, GAYNOR JD, HOGUE
     EJ, MALQUORI A, MOODY K, MOYER JR, PESTEMER W, RAHMAN A, SMITH AE,
     STREIBIG JC, TORSTENSSON NTL, WIDYANTO LS & ZANDVOORT R (1983) EWRS
     Herbicide-soil working group: Collaborative experiment on simazine persistence in soil. *Weed
     Research* 23, 373-83.

5.2.2.    *The fate of pesticides in soil in colder regions and the regulatory aspects within the European Union (EU) and approval of new pesticides*

## O. M. Eklo & O. Lode

Norwegian Plant Protection Institute, Fellesbygget, N - 1432 Ås

## Summary

During 1988-1993 the behaviour of some selected pesticides were studied in sorption experiments, minilysimeters and field experiments. Results from the field experiments showed that two phenoxy-acids and propiconazole were frequently found in the runoff from these areas until 10 months after application. A monitoring study of a springwater outlet from an aquifer contaminated with atrazine and simazine showed that the pesticides still persist in the water 8 years after treatment was put to an end. These results are discussed in connection with the soil and climatic conditions in the colder regions and the concequences this might have on the harmonization and effort to have common rules for approval of pesticides in Europe.

## Introduction

During several years, harmonization of the approval of new pesticides in the European Union (EU) have been an important topic. By adopting the so-called "uniformal principles" new pesticides approved in one country will be allowed to be distributed to other countries within the EU territory. All the approved pesticides in this area will enter the "positive list" after beeing tested according to special guidlines and criterias.

The sorption - desorption of pesticides in soil is one of the most important processes determining the behaviour of the pesticide applied to the field. Precise instruction for testing these properties are developed by OECD in the Test Guidlines 106 (OECD, 1993) and represent an important criteria for futher testing. Through computerized analyses of maps and soil data 5 representative soil types from the EU territory have been selected (Kuhnt, 1988). Depending on an international ringtest these soil types will truely be selected to serve as reference soils for the approval procedure in the future. Brümmer et al. (1987) tested 12 different soil types from Germany with 7 different pesticides confirming that the selected soil types met the demand of representativity according to the EU soils. The soil fitted well into one of the 3 groups formed by a statistical grouping test that contained at least one of the reference soils. Delmas & Hascoet (1988) confirmed that for France it was possible to find 4 of the 5 selected EU reference soils.

After that time of harmonization, the area with economical agreement with EU has expanded especially to the north, and from 1994 Austria, Finland, Iceland, Norway and Sweden entered into the EEA - agreement (European Economic Area Agreement). For the moment there is exceptions for the positive list, but in the future it is important to know more about the representativity of these 5 soil types in the new areas.

In this communication, result from some field experiments will be presented. These results will serve as an example for the general discussion of the indirect influence of the climatic factor on the sorption processes through the genesis and qualitive content of clay minerals and organic matter.

## Runoff of pesticides from farmland

From three farmland areas at Ullensaker, Norway, pesticides in surface and drainage runoff was monitored during 5 years. During this periode 3 herbicides (chlorsulfuron, MCPA and dichlorprop) 1 fungicide (propiconazole) and 1 insecticide (dimethoate) were analysed depending on what had been used by the farmers. Each area represented small catchment areas which was part of a larger field in conventional use. Barley and wheats were cultivated varying from one year to another. The soil was classified as Orthic Gleysol with the size distribution of sand, silt and clay of 5, 63 and 33% repectively. The other important paramaters for the soil was pH 5.1, Org. C 1.3 % and CEC 20 meq/100 g. the dominating clay minerals were illite and mixed illite-vermiculite. Detailed description is given in Eklo et al. (1994).

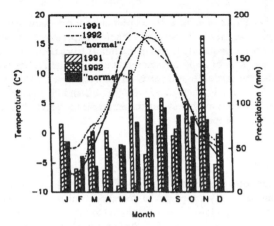

**Figure 1 :** Precipitation and temperature Gardermoen for the years 1991-1992 and the normal periode 1931-1960

The results from these monitoring studies showed that the typical pattern of runoff for three of the pesticides used on these areas was one peak during the autumn with highest concentrations at runoff events close to the application and one peak in spring connected to the snowmelting and when or just after the ground was thawing.

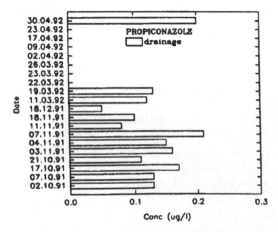

**Figure 2 :** Concentration of propiconazole in drainage water from field 1, Ullensaker 91/92

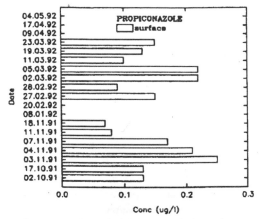

**Figure 3 :** Concentration of propiconazole in surface runoff water
from field 1, Ullensaker, 91/92

Propiconazole appeared in the first runoff both from the surface and the drainage water (Fig. 2 and 3) In the drainage propiconazole was found in the end of April the following season after application. Water samples with no pesticides detected (beyond the detection limit) are marked with a date on the graphs. Distribution coefficient (Kd) for this soil was calculated to 37.2 for the plowing-layer and 25.3 for the subsoil. In sorption experiments with illite the Kd varied from 26,4 - 29.8.

Dichlorprop showed the most extended pattern of runoff and was found frequently in the surface runoff during the whole spring periode untill the beginning of May, 11 months after application (Fig. 4). MCPA followed the same pattern but the concentration of dichlorprop was usually 3 - 5 times higher (Fig. 5). This is perhaps due to the ratio between the concentrations in the commercial product which was a mixture of the two substances.

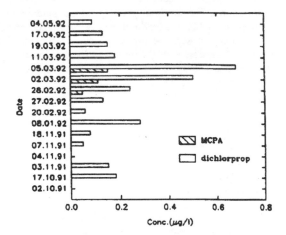

**Figure 4 :** Concentration of MCPA and dichlorprop in surface runoff
water from field 1, Ullensaker, 91/92

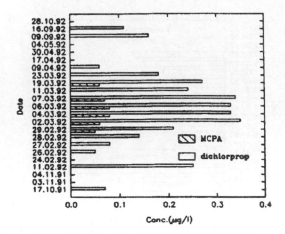

**Figure 5 :** Concentration of MCPA and dichlorprop in surface runoff
water from field 2, Ullensaker, 91/92

Distribution coefficient for both MCPA and dichlorprop was close to 1.0 for this soil (Eklo et al., 1993).

## Contamination of an aquifer with two triazines

In 1986 water from a spring water outlet containing simazine and atrazine caused severe damage for the glasshouseproduction of flowers at Mysen, Norway. After some investigation the pesticides seemed to come from an industrial area within the catchment area of the aquifer for this water outlet. To keep the weeds away from the yards of a cement foundry, this area had been treated with atrazine and simazine for more than 20 years. The dominating soil in the catchment area was loamy sand and sandy soil, material often directly used for the production of cement pipes. More details are given in Lode et al. (1994). After the accident with the flower production the use of the pesticides were immediately stopped, but the two pesticides still remained in the water outlet (fig. 6). As a drinking water source the well was closed and some years later trying to use the water in the glasshouse, the plants soon were starving and the water was not suitable.

## Discussion

Propiconazole is characterized as a weakly basic pesticide with a pKa of 0.8. Under acid conditions it ionizes to a cationic species (Liu & Weber, 1986). As a cationic pesticide sorption to clay minerals is supposed to be important. Propiconazole in Ca-montmorillonite is strongly sorbed with a Kd > 2000 (Liu & Weber, 1986). Some transport of propiconazole throught macropore solved and sorbed to the smallest particle fractions might be an explanation for the findings in the first runoff from the area. From minilysimeter experiment with this soil with other radiolabelled pesticides (Riise et al. 1992) preferential flow in macropores was observed. Transport throught macropores connected to the drainage system was also a possibel explanation for drainage runoff both for fertilizers and other pesticides (Eklo et al., 1994). The findings early and late in the spring may be due to the liberation caused by the freezing and thawing. And specially this year an extremely mild periode in February and and March (Fig. 1) caused situation with surface erosion and runoff of pesticides sorbed to particles. These frequent findings might be a result of the treatment from the last year or it is repeated treatment from several years. However, if it is from the last year of treatment, these findings represent not more than 0.3 % of applied.

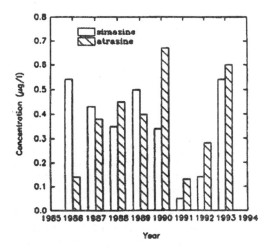

**Figure 6 :** Concentration of simazine and atrazine in a spring water outlet after contamination at Mysen

The content of clay minerals are important for some pesticides but this is differing with the properties of the pesticide (Sanchez - Camanzo & Sanchez - Martin, 1988), (Grice et al. 1973) (Hayes et al. 1974). Soils containing smectite seem to be important (Romero - Taboada et al. 1988) and especially for cationic pesticides. Because of the weathering periode after the last glaciation in Scandinavia is "not more" than 10 000 years, the content of the clay minerals are different from what is the case in the South - Europe. In a project coordinated by FAO, Bergseth (1989) reported that smectite was present in trace amount in 1 and 2 soil samples of 10 in respectively Norway and Finland, while in Portugal 7 of 10 soil samples the clay contained more than 50% smectite (fig. 7).

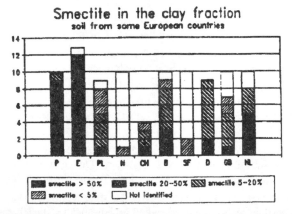

**Figure 7 :** Estimated contents of smectite in the clay fraction of soil samples from some European countries (after Bergseth, 1989)

For the two phenoxyacids MCPA and dichlorprop, it is shown from lysimeter and isotherm studies that these are mobile compounds almost following the waterfront. In some soils rich in organic matter the sorption is found to be quite high and dependent upon pH (Riise et Salbu, 1992). According to the estimated rapid breakdown of this dichlorprop, 90 % within 30 days (Altom & Strizke, 1973) the frequent findings of dichlorprop is surprizing, especially in spring. For pesticides where the organic matter fraction is important, pesticides is sorbed to different fractions of the organic matter according to size, chemical composition (humic acid, fulvic acid or humin) and dependent on time (Barriuso et al, 1991a, 1991b). Generally in acid soils and low temperature the microbial activity is inhibited and containing high amount of poorly humified organic matter.

Atrazine is perhaps one of the most investigated pesticide according to the fate and behaviour in soil. It is a mobile and persistent compound and one of the most frequently found pesticide in Europe (Warren, 1988). In special circumstances and chemical conditions the breakdown of atrazine can be very slow (Boesten, 1993). Generally the prevention of the groundwater has high priority. Using pesticides on areas with sand and coarse material without vegetation to sorb the pesticide, the risks of leaching to groundwater is higher than for cultivated areas. Contamination of the described aquifer may last for many years, because of the low temperature and low microbial activity in such habitats.

These examples have been presented to continue the discussion about the effect of certain parameters, important for the behaviour of pesticides in soil and to ask for more collaboration along a climatic gradient. This seems to be important connected to the harmonization of approval of pesticides in the future. Kuhnt and coworkers have done an excellent and difficult job by the selection of representative Euro - soils, which will be important references in the work of risk assessment. Studying the "Soil map of Europe" (FAO, 1978) most of the agricultural areas of the Scandinavian countries are mapped as Orthic Podzols. This corresponds to Euro - soil 5 selected from Schleswig - Holstein (Kuhnt, 1993) and thereby the selected eurosoils also might be representative for areas in the north. The main question wil be: are the climatic condition so different in these areas that the quality of organic matter and clay, and the degradation rates so different that separate investigations are needed in the approval of new pesticides?

# References

▫  Altom, J. D. & Strizke, J. F. 1973. Degradation of dicamba, picloram, and four phenoxy herbicides in soil. *Weed Sci.* 21, p 556 - 560.

▫  Barriuso, E., Schiavon, M., Andreux, F. & Portal, J. M. 1991a. Localization of atrazine non -extractable (bound) residues in soil size fractions. *Chemosphere*, vol 22, 12, p 1131 - 1140.

▫  Barriuso, E., Schiavon, M., Andreux, F. & Portal, J. M. 1991b. Intérêts et limitations des méthodes de séparation des micropollutants organiques des sols. *Science du sols*, 29, 4, p 301 -320.

▫  Bergseth, H. 1989. Relationships between the degree of weathering and content of Cd in some European soils. *Trends in Trace Elements*, 1, FAO, Regional office of Europe & INRA, Centre de Bordeaux. p 11 - 20.

▫  Boesten, J. J. T. I. & Van der Pas, L. J. T. 1993. Transformation rate of atrazine in water saturated sandy subsoils. 8th EWRS Symposium "Quantitative approaches in weed and herbicide research and their practical application", Braunschweig, p 381 - 389

▫  Brümmer, G., Fränzle, O., Kuhnt, G., Kukowski, H. & Vetter, L. 1987: Fortschreibung der OECD Prüfungslinie "Adsorption/desorption" im Hinblick auf die Übernahme in Anhang der EG - Bereich und Abstufung der Testkonzeption nach Aussagekraft und Kosten. - Umwelt-forschungsplan des Bundesministers für Umwelt, Naturschutz und Reaktorsicherheit: For-schungs.bericht 106 02 045, Kiel.

□    Delmaes, A-B & Hascoet, M. 1988: Recherche de sols équivalents aux standards CEE pour l'étude de la mobilité des pesticides. p 159 - 166.

□    Eklo, O. M., Lode, O. Riise, G. & Aspmo, R. 1994. Mobility of dichlorprop and MCPA in silty clay loam; a comparison of laboratory, lysimeter and field experiments. 8th EWRS Symposium "Quantitative approaches in weed and herbicide research and their practical application", Braunschweig, p 483 - 492.

□    Eklo, O. M., Aspmo, R. & Lode, O. 1994. Runoff and leaching experiments of dichlorprop, MCPA, propiconazole, dimethoate and chlorsulfurone in outdoor lysimeters and catchment areas. *Norwegian Journal of Agricutural Science*. Supplement no. 13, p 53 - 78.

□    FAO/UNESCO 1978. Soil map of Europe, V1/V2.

□    Grice, R. E., Hayes, M. H. B., Lundie, P. R. 1973. Proc. 7 th British insecticide and fungicide conference. Brighton.

□    Hayes, M. H. B., Stacey, M., Toms, B. A. & Quinn C. M. 1972. The different interaction of paraquat and diquat with montmorillonite and vermiculite. *Proc. Xth. Internat. Congr. Soil Sci. Moscow*, 7, p 90 - 96.

□    Kuhnt, G. 1988: Selection of representative soil samples for testing the sorption behaviour of chemicals. In Jamet, P. (ed): *Methodological aspect of the study of pesticide behaviour in soil*. INRA Versailles, June 16-17, 1988, p 151-158.

□    Kuhnt, G. 1993. The Euro - Soil Concept as a Basis for Chemicals Testing and Pesticide Research. In: Mansour M. *Fate and Prediction of Environmental Chemicals in Soils, Plants, and Aquatic Systems*. Lewis Publishers, London, p 83 - 93.

□    Liu, S. L. & Weber, J. B. 1986. Adsorption/desorption , and mobility of propiconazole and prometryn in soils. *Proceedings of Southern Weed Science Society*, p 460 - 471.

□    Lode O., Eklo, O. M., Kraft, P. & Riise, G. 1994.Leaching of simazine and atrazine from an industrial area to source. A long term case study. *Norwegian Journal of Agricutural Science*, Supplement no. 13, p 78 - 88.

□    OECD 1993: OECD Test Guidlines. Paris.

□    Riise, G. & Salbu, B. 1992. Mobility of dichlorprop in the soil - water system as a function of different environmental factors. I. A batch experiment. *The Science of Total Environment*, 123/124.p 399 - 409.

□    Riise, G., Eklo, O. M., Lode, O. & Pettersen, M. N. 1992. Mobility of dichlorprop in the soil-water environment as a function of different environmental factors. II. A lysimeter experiment. *The Science of Total Environment*, 123/124, p411 - 420.

□    Romero - Taboada, E., Guillén - Alfaro, J. A., Sanchez - Rasero, F., & Dios - Cancela, G. 1988. Adsorption of ethirimol by clay minerals, peat and soils. In: Jamet, P. (ed). *Methodological aspect of the study of pesticide behaviour in soil*. INRA Versailles, June 16-17, 1988, p 179 - 184.

□    Sanchez - Camanzo, M. J. & Sanchez - Martin, M. 1988. Influence of organic matter and clay fraction in pesticide adsorption by soils. Importance of pesticide structure. In: Jamet, P. (ed). *Methodological aspect of the study of pesticide behaviour in soil*. INRA Versailles, June 16-17, 1988, p175 - 177.

□    Warren, S. C. 1988. Presentation of the conclusions of the European Institute for Water. May seminar. In/ Final report from seminar on "The EEC - directive 80/778 on the quality of water inteded for human consumption: Pesticides" at Como, Italy. European Institute for Water, Strasbourg, France.

5.2.3.        *Phytotoxic Persistence of Chlorsulfuron, Metsulfuron,*
              *Flurochloridone and Metribuzin in Finnish Soils*

# S. Junnila[1], L-R. Erviö[1] and R. Mutanen[2]

[1]   Institute of Plant Protection, Agricultural Research Centre of Finland,
      FIN-31600 Jokioinen
[2]   Agricultural Chemistry Deparment, Plant Protection Inspection Centre,
      P.O. Box 83, FIN-01301 Vantaa

The phytotoxic persistence and movement of chlorsulfuron, metsulfuron, flurochloridone and metribuzin were studied in field experiments by bioassays in sandy, clay and organic soils at two locations in Finland during 1983-1987.

## Introduction

Soil microbiological activity determines the degradation rate of many herbicides. Factors affecting microbiological activity, such as soil type and weather conditions, are generally significant in their persistence. The growing season in Finland is short and the mean temperature of the summer months is low. Finnish arable soils are usually acid and rich in organic matter (OM) and/or clay. These factors limit the soil microbiological activity and thus the degradation rate of herbicides. On the other hand, sulfonylureas degrade mostly via hydrolysis in acid soils. Therefore, under Finnish soil conditions their adsorption may be higher, water solubility lower and problems possibly related to persistence and leaching less serious than in alkaline, low-humic soils. Many studies testing the persistence of herbicides have been made under laboratory conditions in neutral or alkaline soils with a pH above 7 and/or with doses not used in Finland. The results from those studies cannot directly be applied to Finnish field conditions.

## Materials and Methods

### Field experiments

The field experiments were situated in Jokioinen (60° 49' N, 23° 30' E) in Southwest Finland and in Ruukki (64° 46' N, 25° 05' E). The performance of herbicides in the soil tillage layer was evaluated in sandy and organic soils at both locations and in clay in Jokioinen (Table 1). The experiments were designed as randomized blocks with four replicates and a plot size of 48 m$^2$. The plots were sprayed in June with both the recommended and a double dose (metribuzin) or a threefold dose. Barley plots were sprayed during tillering with chlorsulfuron and metsulfuron and potato plots before emergence with flurochloridone and metribuzin (Table 2). The fields were left unharvested and uncultivated throughout the study period, 2-3 years after spraying, in order to determine the movement in an undisturbed tillage layer.

Soil samples were taken 1 day and 1 month after treatment and, later, at the end and the beginning of the growing seasons until no signs of phytotoxicity were observed on the test Plants grown in a glasshouse. The soil samples were taken from the layers 0-5, 5-15 and 15-25 cm. The pot trials were usually set up immediately after sampling.

| Soil type | OM % | Clay % | pH (H$_2$O) |
|-----------|------|--------|-----------|
| Sandy soils | 3.6 - 8.5 | 2 - 18 | 5.7 - 7.2 |
| Clay soils | 4.7 - 9.7 | 67 - 79 | 5.8 - 6.2 |
| Organic soils | 21 - 67 | | 4.4 - 5.9 |

**Table 1 :** Soil characteristics of field experiments sprayed in 1983 -1985

| Field experiments Herbicide | Dose kg | a.i. ha$^{-1}$ | Spraying year | Bioassays Test plant |
|-----------|------|--------|-----------|-----------|
| Chlorsulfuron | 0.004 | 0.012 | 1983, 1984, 1985 | Allium cepa |
| Metsulfuron | 0.004 | 0.012 | 1984, 1985 | Allium cepa |
| Flurochloridone | 0.75 | 2.25 | 1984, 1985 | Lactuca sativa |
| Metribuzin | 0.70 | 1.40 | 1983, 1984 | Lolium multiflorum |

**Table 2 :** Spraying years of field experiments, doses applied and the test plants used in bioassays

| Year | Length (days) | | Precipitation (mm) | | Sum day degrees (°C) | |
|------|------|------|------|------|------|------|
| | J | R | J | R | J | R |
| 1983 | 185 | 145 | 363 | 308 | 1361 | 1108 |
| 1984 | 186 | 165 | 515 | 282 | 1280 | 1127 |
| 1985 | 170 | 140 | 321 | 349 | 1205 | 1026 |
| 1986 | 191 | 142 | 421 | 313 | 1222 | 997 |
| 1987 | 175 | 176 | 425 | 372 | 994 | 868 |

J=Jokioinen; R=Ruukki

**Table 3 :** Length (days), cumulative precipitation and total sum of effective day degrees (over 5°C) of growing seasons 1983-1987.

## *Bioassays*

The residual effect on test plants was studied in pot trials in the glasshouse. Test plants were grown in the soil samples taken from each plot of the field experiment. Seeds were sown into soil filled 0.5 l plastic pots which were kept at equal moisture during the test period of about 5-10 weeks depending on the test plant. The day/night temperatures were adjusted to 20/14°C, day length to 16 h and light intensity to 20 000 lx. The number of viable plants per pot, the fresh and dry weights of aerial shoots and roots per pot and the number of leaves were recorded.

## *Chemical analyses*

Flurochloridone and metribuzin residues in soil were analyzed chemically from some soil samples by the Plant Protection Inspection Centre employing a gas-chromatographic method.

### Weather

The growing season in Finland is short, 140-180 days. The mean temperature in the summer months varies from 10 to 17°C, and the soil is frozen for 5-7 months. Information about the weather conditions during the growing season in Jokioinen and Ruukki is given in Table 3. There were considerable differences between the growing seasons. Differences in weather conditions between research sites were smaller.

## Results

### Chlorsulfuron and metsulfuron

Chlorsulfuron and metsulfuron persisted and leached to nearly the same extent. Both herbicides were generally found below the 0-5 cm layer. Residues persisted longer in the sandy soils and also leaching was deeper in the sandy soils. The phytotoxicity of the 4 g a.i. ha$^{-1}$ dose persisted in the 0-5 cm layer of all soils mostly for 1 month. The residues tended to leach deeper than 5 cm, and were in some cases recovered still 1 year after application. The phytotoxicity of the 12 g a.i. ha$^{-1}$ dose persisted for at least 1 month and generally for 1 year. Residues of this dose were found in the 0-25 cm layer, in clay mostly in the 0-15 cm layer. Significant growth stimulation was common 1-2 years after application in the 5-15 and 15-25 cm layers of sandy soils.

### Flurochloridone

The phytotoxic effect of flurochloridone was greatest in sandy soils and least in organic soils. The phytotoxicity persisted in the 0-5 cm layer at most for 1 month after application of 0.75 kg a.i. ha$^{-1}$. A dose of 2.25 kg a.i. ha$^{-1}$ caused growth inhibition in the 0-5 cm layer of organic soils only for 1 day, whereas in other soils growth inhibition was observed throughout the growing season, in some cases also in the following summer. In sandy soils phytotoxic effects were found in the 15-25 cm layer at the end of the first or the second growing season. Chemical analysis revealed flurochloridone residues in the 15-25 cm layer in all soils. The residues analysed chemically in the first autumn after application in the very dry conditions in 1985 were about the same as the calculated dose applied to clay soil and 70% of the dose applied to organic soil. Bleaching of plant tissues often persisted about 1 year longer than growth inhibition, being thus a more sensitive method to detect low residues. Bleaching was almost the only symptom of residues in the sandy soil from Ruukki. Growth stimulation of the test plants was observed mostly in deeper soil layers until the second growing season.

### Metribuzin

The phytotoxicity of metribuzin 0.7 kg a.i. ha$^{-1}$ persisted in the surface layer of 0-5 cm for one growing season in sandy and clay soils, but only for some weeks in organic soils. A double dose may have phytotoxic effects on other crops during the next growing season. Phytotoxic residues were seldom detected below the 5 cm layer, although chemical analyses have revealed residues down to 25 cm in all soil types. Growth stimulation in the test plants was observed generally in the 5-25 cm layers of sandy and clay soils from the first to the second autumn after treatment. Residues were detected chemically still 2 years after treatment in the surface layer of sandy soil.

## Discussion

In sandy soils, chlorsulfuron persisted longer in the Ruukki soils than in the Jokioinen soils, probably because the adsorption in the Ruukki soils was stronger due to a higher OM content. In addition, the residues seemed to move more readily into the 5-15 cm layer in the sandy soils of Jokioinen. These herbicides leached below 5 cm in some organic soils sprayed with a threefold dose despite their very high OM content (66%). Under wet soil conditions after spraying, both herbicides leached rapidly, within 1 month, at least into the 15-25 cm layer in the sandy soils. Heavy leaching did not occur under the same conditions in the organic soil, probably because of stronger adsorption due to high OM content. Omitting soil cultivation may have decreased the degradation rate of herbicides leached to the deeper soil layers. When applied at the recommended doses, these herbicides usually disappeared during the first growing season after application. However, there is a risk of phytotoxicity to sensitive crops grown in the following season, which makes restrictions to subsequent crops necessary.

The residual effects of flurochloridone were generally more evident in sandy soils up to the second spring. In the clay soils these effects were smaller than in the sandy soils and smallest in the organic soils. These differences between soil types may be due to the adsorption of flurochloridone being higher in organic soils than in mineral soils. Flurochloridone persisted relatively long in these acid Finnish soils. Its use is therefore prohibited in Finland. A normal dose of flurochloridone seemed to be adsorbed into a form almost unavailable to plants in the soils with an OM content of over 8%, only bleaching of test plants could be seen. At an OM content below 8% growth was greatly inhibited in the surface layer 1 day after application. In some experiments we observed growth inhibition all through the first growing season. The lettuce plants showed well the bleaching caused by flurochloridone residues. Instead of using complicated bioassay measurements, observation of bleaching indicated the presence of residues below the level of growth injuries.

The phytotoxicity of a normal dose of metribuzin may persist for one growing season in the Finnish climate, according to our field experiments. A double dose may have phytotoxic effects on other crops during the next growing season. The differences in persistence were mostly dependent on the soil type or weather, not on the geographical location of the experiments. Ruukki sandy soils of lower pH showed less phytotoxic effects than those of Jokioinen. Phytotoxic residues were seldom detected below the 5-10 cm layer. However, in the Jokioinen sandy and clay soils treated in 1983, growth inhibiting effects were found in deeper layers during the first growing season, perhaps due to the heavy rainfall in June and September. The bioassay method employed was most suitable for sandy soil and poorest in acid organic soil. By measuring the above ground parts of plants the residues were indicated reliably enough.

This research project consists of three original papers. One of the papers has been published in Weed Research, and the other two are in press. These papers include also the effects of herbicide residues on soil micro-organisms, determined by assessment of nitrification and dehydrogenase activities.

The original papers are as follows:

◻ JUNNILA S., HEINONEN-TANSKI H., ERVIÖ L-R. & LAITINEN P. 1993. Phytotoxic persistence and microbiological effects of metribuzin in different soils. *Weed Research* 33, 213-223.

◻ JUNNILA S., HEINONEN-TANSKI H., ERVIÖ L-R., LAITINEN P. & MUTANEN R. 1994. Phytotoxic persistence and microbiological effects of flurochloridone in Finnish soils. *Weed Research* 34, in press.

◻ JUNNILA S., HEINONEN-TANSKI H., ERVIÖ L-R. & LAITINEN P. Phytotoxic persistence and microbiological effects of chlorsulfuron and metsulfuron in Finnish soils. *Weed Research*, in press.

5.2.4.        Mobility and Transformation of Diuron in Railway Embankments

# L. Torstensson

Swedish University of Agricultural Sciences, Department of Microbiology,
Box 7025, S-750 07 Uppsala, Sweden.

## Abstract

Diuron [3-(3,4-dichlorophenyl)-1,1-dimethylurea] has been used for weed control of Swedish railway embankments for many years. There have been several cases of damage to pine trees along the railway tracks and diuron has been found in the dead trees. Therefore, an investigation of the mobility and the transformation of the herbicide in railway embankments was initiated. The $K_d$-values for adsorption in the uppermost part (0-5 cm) of the embankment and at a depth of 0.5 m were 0.56 and 0.29, respectively. The rate of decomposition was low and demethylated diuron accumulated. A front of diuron moving downvard could be followed down to a depth of 1.5 m. It can then be concluded that pine tree roots growing into the railway embankment, at a depth of about 0.5 m, may very well take up toxic amounts of diuron and that diuron and demethylated diuron in the railway embankment cause a great risk for contamination of the ground water.

## Introduction

Weed control on railway embankments is necessary to retain track quality, improve safety and to achieve a good working environment. Chemical weed control on railway embankments has been used in Sweden for many years. To get a good result when spraying herbicides on railway embankments it is necessary to have detailed knowledge of the herbicidal effects on the weeds. From an environmental point of view it is also necessary to be well informed about the appearence and persistence of the herbicide in the railway embankment. For many years the Swedish Railway Company has used diuron [3-(3,4-dichlorophenyl)-1,1-dimethylurea], preparation Karmex 80 DF (DuPont, 80% a.i.) for weed control. Since there have been several cases of damage to pine trees along the railway embankments and diuron has been found in the dead trees (Torstensson 1983), an investigation of mobility and transformation of the herbicide in the embankments was initiated.

## Material and Methods

### The railway embankment

The railway embankment that has been investigated is situated in the middle part of Sweden at Gopshus in the Landscape of Dalarna. Along that embankment a great number of pine trees had been killed and diuron has been found in the dead trees (Torstensson 1983). The embankment crosses here a great area of sandy soils covered with pine trees. Some characteristic data from the embankment, as well as data from some agricultural soils that were also used, are given in Table 1.

### Analysis of diuron

The analyses of diuron and demethylated diuron were carried out mainly as described by Glad et al. (1978). A sample (20 g) and methanol (50 ml) containing internal standard [3-(3,4-dichlorophenyl)-1-methoxy-urea, 100 μg/ml methanol] were placed in a bottle with a screw cap and shaken vigorously for 1 h.

| Material/soil | pH | Organic C (%) | Clay (%) | Texture |
|---|---|---|---|---|
| Railway embankment | | | | |
|    Gopshus, 0-40 cm | 5.4 | 0.2 | - | Gravel, coarse sand |
|    Gopshus, 40-150 cm | 5.6 | 0.1 | - | Fine sand |
| Agricultural soil | | | | |
|    Skärplinge | 5.7 | 1.0 | 4 | Sand |
|    Funbo | 6.8 | 3.3 | 23 | Sandy clay loam |
|    Lövsta | 6.5 | 3.0 | 39 | Clay loam |
|    Skyttorp | 5.8 | 35.7 | - | Organogenic |

**Table 1** : Characteristics of the railway embankment material at Gopshus and of some agricultural soils.

After filtration and evaporation to about 1 ml the herbicide content was determined by reverse-phase liquid chromatography, using Water's chromatographic equipment (Waters Associated inc., Milford, MA). This equipment comprises an automatic sample processor (Model 7108), a solvent delivery system (Model 501), a UV detector (Model 441) adjusted to 254 nm, and a Radial PAK NOVO PAK C18 column (100 by 8 mm). The solvent was methanol:water (55:45). The flow rate was 1.0 ml per min. The peak area was measured with Water's recording integrator (Model 740). The limit of detection was 0.01 µg per g railway embankment material for both diuron and demethylated diuron.

## Determination of adsorption capacity

The adsorption capacity of the railway embankment material and the agricultural soils was determined according to the OECD Guidelines (1981). The data obtained from the adsorption experiments were linearized using the Freundlich adsorption isotherm, and $K_d$- and $K_{oc}$-values were calculated.

## Studies of diuron decomposition

Studies of diuron decomposition in railway embankment material as well as in an agricultural soil have been carried out at the laboratory. Carbonyl-[14]C-labelled diuron (specific activity 4.2 µCi/mg, DuPont, Wilmington, Delaware) was used and the procedure was similar to that described by Torstensson & Stark (1981). Samples of 100 g were moistened (60% of WHC). A dosage of 8 µg diuron per g sample (corresponding to a normal dosage of 0.8 g diuron per $m^2$ mixed to a depth of 5 cm) was applied. Incubation was done in air-tight glass-jars together with KOH at a temperature of 20°C. Four parallells were used. Evolved [14]$CO_2$ was quantified by scintillation counting.

Also 1 kg portions were moistened (60% of WHC) and incubated in glass-jars covered with aluminium foil at 20°C. A dosage of 8 µg non-labelled diuron per g sample was applied. Three parallels were used. Samples of 20 g were removed at different times for chemical analysis of diuron and its metabolites.

## Field experiments

The railway embankment at Gopshus had been sprayed with the normal dosage of Karmex 80, 1 g per $m^2$ (0.8 g diuron per $m^2$), two years before the sampling started. A new normal dosage was applied on a part of the embankment. From the day of the spraying operation sampling started at the earlier sprayed and the newly sprayed embankment. At every sampling depth a stainless steel frame (hight 5 x 10 x 16 cm) was pressed down into the embankment. Every fifth cm was sampled down to 150 cm. The sampling was carried out between the rails and the sleepers. The samples were stored at -20°C until analyzed. At every sampling occasion two randomly chosen spots where used at each of the two embankment parts. From the time of the last application of the herbicide, sampling was carried out at intervals for three years.

# Result

## *Laboratory tests*

As can be seen from Table 2 the $K_d$-values for the railway embankment are low compared to those for the agricultural soils. However the $K_{oc}$-values are at the same level.

| Material/soil | $K_d$ | n | $K_{oc}$ |
|---|---|---|---|
| Railway embankment | | | |
| Gopshus, 0- 5 cm | 0.56 | 0.74 | 280 |
| Gopshus, 25-30 cm | 0.50 | 0.72 | 250 |
| Gopshus, 45-50 cm | 0.29 | 0.70 | 290 |
| Gopshus, 90-95 cm | 0.24 | 0.68 | 240 |
| | | | |
| Agricultural soil | | | |
| Skärplinge | 2.4 | 0.83 | 240 |
| Funbo | 10.6 | 0.70 | 320 |
| Lövsta | 15.3 | 0.99 | 510 |
| Skyttorp | 49.5 | 0.75 | 140 |

**Table 2 :** Adsorption of diuron to railway embankment material
and agricultural soils

Decomposition of carbonyl $^{14}C$-labelled diuron was studied in railway embankment material taken from two depths. As can be seen in Fig. 1 the rate of evolution of $^{14}CO_2$ was slow, especially in material from 45-50 cm depth.

In another experiment with non-labelled diuron, the concentrations of the herbicide as well as demethylated diuron were followed (Fig. 2). Other possible metabolites could not be detected.

## *Field experiment*

The concentrations of diuron and demethylated diuron in the railway embankment at Gopshus were determined. There is a front of diuron moving downward the first 6 months after the application (Table 3). During the following winter with frozen ground, the amounts of diuron in the railway embankment remains constant.

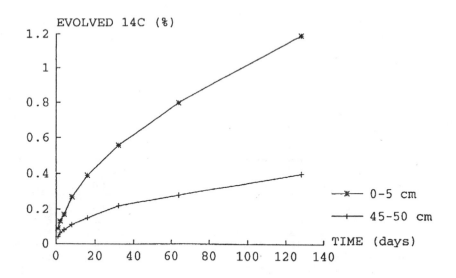

**Figure 1 :** Decomposition of carbonyl-$^{14}$C-labelled diuron in a laboratory experiment with railway embankment material from Gopshus. Mean of 4 parallells.

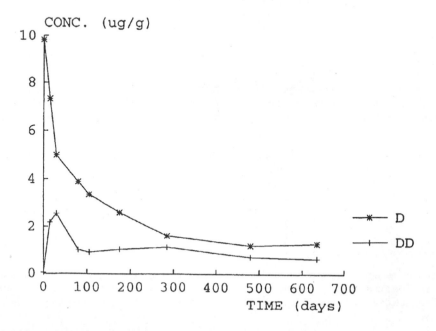

**Figure 2 :** Decomposition of diuron in railway embankment material in a laboratory experiment. Concentrations of diuron (D) and demethylated diuron (DD) after different times. Mean of 3 parallels.

| Depth (cm) | Amounts of D and DD (mg/m2 to the given depths) found at different sampling times (days). | | | | | | | | | | | |
|---|---|---|---|---|---|---|---|---|---|---|---|---|
| **A.** | 730 D | DD | 763 D | DD | 896 D | DD | 1120 D | DD | 1457 D | DD | 1930 D | DD |
| 0- 25 | 99 | 75 | 61 | 48 | 82 | 71 | 88 | 76 | 46 | 58 | 65 | 72 |
| 26- 50 | 51 | 37 | 27 | 11 | 15 | 22 | 10 | 19 | 26 | 28 | 18 | 23 |
| 51- 75 | 104 | 68 | 105 | 77 | 15 | 34 | 6 | 26 | 41 | 32 | 22 | 27 |
| 76-100 | 54 | 27 | 56 | 32 | 72 | 25 | 50 | 31 | 8 | 19 | 4 | 12 |
| 101-125 | 37 | 18 | 31 | 18 | 29 | 11 | 24 | 11 | 16 | 21 | 8 | 16 |
| 126-150 | 36 | 21 | 26 | 12 | 15 | 9 | 17 | 18 | 25 | 19 | 2 | 11 |
|  | 381 | 246 | 306 | 198 | 228 | 172 | 195 | 181 | 162 | 177 | 119 | 161 |
| **B.** | 0 D | DD | 14 D | DD | 33 D | DD | 76 D | DD | 166 D | DD | 390 D | DD |
| 0- 25 | 905 | 65 | 521 | 155 | 125 | 113 | 91 | 53 | 105 | 87 | 181 | 91 |
| 26- 50 | 61 | 45 | 222 | 65 | 150 | 75 | 128 | 90 | 78 | 72 | 93 | 64 |
| 51- 75 | 98 | 58 | 130 | 88 | 306 | 160 | 163 | 141 | 127 | 81 | 157 | 97 |
| 76-100 | 52 | 30 | 68 | 58 | 189 | 58 | 133 | 81 | 112 | 71 | 91 | 40 |
| 101-125 | 39 | 21 | 36 | 27 | 112 | 36 | 121 | 51 | 101 | 40 | 22 | 18 |
| 126-150 | 41 | 19 | 38 | 25 | 46 | 30 | 69 | 33 | 107 | 38 | 38 | 27 |
|  | 1196 | 238 | 1015 | 418 | 928 | 472 | 705 | 449 | 630 | 389 | 582 | 337 |

**Table 3 :**    Amounts of diuron (D) and demethylated diuron (DD) in the railway embankment at Gopshus at different sampling times after the previous treatment (A was sprayed 2 years before the sampling started and B was sprayed at the first day of sampling). Amounts are expressed as mg per m2 at the different depths given. Means of two replicates.

## Discussion

Based on the $K_d$-values and an approximate yearly precipitation of 500 mm the expected mobility of diuron in the embankment can be estimated to 1-2 m according to Mc Call et al.(1981), while for the agricultural soils it is 5-20 cm. According to Table 3 the downward transport of diuron has been > 1.5 m the first year after the application.

The first step at decomposition of diuron is a demethylation and this metabolite accumulates. Other possible metabolites have not been found either in the laboratory or in the field experiment. The rate of decomposition of diuron was very slow in the laboratory experiments. Also in the field experiment the rate of disappearance was very slow. It was not possible to conclude how much of the disappeared herbicide that really was broken down since some of if it just passed deeper down into the ground than the sampling could be done. The half life of diuron plus demethylated diuron in the uppermost 1.5 m of the studied railway embankment is around two years and the time for ninety percent disappearance is about seven years.

It can thus be concluded that pine tree roots growing into the railway embankment, at a depth of about 0.5 m (Torstensson 1983), may very well take up toxic amounts of diuron and that diuron and demethylated diuron in the railway embankment cause a great risk for contamination of the ground water.

# Literature

□ Glad, G., Popoff, T. & Theander, O. 1978. Determination of linuron and its metabolites by GLC and HPLC. *J. Chrom. Sci.* 16, 118-122.

□ Mc Call, P.J., Swann, R.L., Laskowski, D.A., Vrona, S.A., Unger, S.M. & Dishburger, H.J. 1981. Prediction of chemical mobility in soil from sorption coefficients. In: Bransen, D.R. & Dickson, K.L. (Eds.), Aquatic Toxicology and Hazard Assessment: *Fourth Conference,* ASTM STP 737. *Am. Soc. for Testing and Materials,* 49-58.

□ OECD 1981. Guidelines for testing of chemicals 1981. Updated 1984. OECD Environmental Directorate, Paris.

□ Torstensson, L. 1983. Undersökning av diurons rörlighet och nedbrytning i banvallar. Swedish Environmental Protection Agency, SNV PM 1764, 43 pp.

□ Torstensson, L. & Stark, J. 1981. Decomposition of $^{14}$C-labelled glyphosate in Swedish forest soils. Proc. EWRS Symp. *Theory and Practice of the Use of Soil Applied Herbicides,* 1981. 72-79.

## 5.2.5.   *Herbicides and Insecticides Soil Metabolisms in Field Crops: Influences of Organic Fertilizers Treatments*

# J. Rouchaud[1], F. Gustin[1], D. Callens[2], R. Bulcke[2], A. Wauters[3], F. Van De Steene[4], O. Cappellen[5], D. Mouraux[5], F. Benoit[6], N. Ceustermans[6], J. Gillet[7], S. Marchand[7], L. Van Parys[8], G. Vulsteke[8]

1.  IRSIA, Laboratory of Phytopathology, UCL, Louvain-la-Neuve.
2   IRSIA-IWONL, Weed Research Center, RUG, Gent.
3.  Royal Research Institute for Sugar Beet, IRBAB, Tienen.
4   IRSIA-IWONL, Laboratory for Agrozoology, RUG, Gent.
5.  IRSIA, Laboratory of Agriculture Phytotechnology, UCL, Louvain-la-Neuve.
6.  IRSIA-IWONL, Vegetables Research Station, St Katelijne-Waver.
7.  State School for Horticulture, Gembloux.
8.  Provincial Research Station for Agriculture and Horticulture, Rumbeke, Belgium.

The present paper makes the synthesis of the most recent results that we obtained in the study of the influence of the organic fertilizers onto the herbicides and insecticides soil biodegradations in agricultural and horticultural crops.

## Winter wheat crops

The soil metabolism of the herbicide isoxaben has been studied in winter wheat crops. Isoxaben was successively transformed in soil into demethoxy-isoxaben, 5-isoxazolone and 2,6-dimethoxybenza-mide.[1] There was only very low and temporary 5-aminoisoxazole soil concentrations, which is the expected product of the usual amide function hydrolysis. This observation rules out the risk of formation of carcinogenic nitroso compounds in soil during isoxaben metabolism. Isoxaben thus is a good example of soil metabolism of a powerful herbicide without contamination risk for the environment. Professor Bulcke (RUG) observed that demethoxy-isoxaben and 5-aminoisoxazole only have low herbicide activities. On the other hand, the 5-isoxazolone has a great herbicide activity, which extends the one of isoxaben during its soil metabolism. The slowing down of isoxaben soil metabolism in winter wheat crops has been observed when the organic fertilizers pig slurry, cow manure, or green manure had been applied recently; this was observed in the winter wheat trials made in 1990-1991, 1991-1992, and 1992-1993 (Figure 1).[1]

In spite of the reduction of their application rates, isoxaben (applied either at 125 or 65 g ha$^{-1}$ killed practically all the broadleaf weeds, similarly when the organic fertilizer treatments were applied or not; on the other hand, their secondary herbicide action against grass weeds was greater in the plots treated with organic fertilizers than in the plots not treated with these fertilizers. On the other hand, the primary herbicide activity of imazamethabenz (at 500 or 375 g ha$^{-1}$) against grass weeds was greater in the plots treated with organic fertilizers.

## Sugar beet crops

The new soil insecticide imidacloprid has been applied in seed pelleted dressing at the rate of 90 g ha$^{-1}$ in sugar beet.

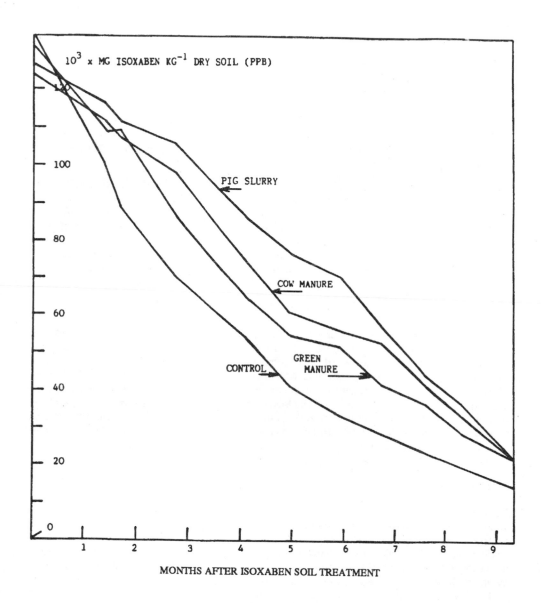

**Figure 1 :** Isoxaben soil concentrations in the winter wheat organic fertilizers trial made in 1990-1991 at Melle, in the soil layer at a 0-10 cm depth.

In spite of this application technique, its soil biodegradation was slowed down in the field plots treated one month before sowing with one of the organic fertilizer cow manure (40 tons ha$^{-1}$), or pig slurry (40 tons ha$^{-1}$), relatively to the control plots not treated with the organic fertilizers (trials at Lubbeek in 1992 and 1993). In another sugar beet trial made at Remicourt in 1992, the imidacloprid soil biodegradation was slower in the field plots whose soil contained a normal concentration of recent soil organic matter (2.4%), than in the plots containing a high concentration (4.3%) of very old soil organic matter, humus. In another trial made in 1993 at Remicourt, the cow manure applied 18 months before sowing on some field plots only had a small effect on the rate of imidacloprid soil biodegradation; the cow manure applied on other field plots either 1 or 6 months before sowing much slowed down, and with similar intensities, the imidacloprid soil biodegradation. These results indicate that it is the recent soil organic matter which, on account of its chemical structure -and at the opposite of the old soil humus-, is able to slow down the pesticides soil biodegradations.[2] On the other hand, to the greater imidacloprid soil concentrations -due to the organic fertilizers treatments- corresponded greater imidacloprid concentrations in the sugar beet leaves. At the rates of 90 or 45 g imidacloprid ha$^{-1}$, in all the field trials about all the aphids were killed up to July 15, as well in the plots treated or not with the organic fertilizers.

One of the insecticides aldicarb or thiofanox were applied at the rate of 1 kg ha$^{-1}$ as granulates in the sowing furrow of other plots of the same sugar beet field trials where imidacloprid was assayed (Lubbeek 1992 and 1993; Remicourt 1992).[3] Similar slowing down effects of the thiofanox and aldicarb soil biodegradations were observed -as with imidacloprid- in the plots treated recently with the organic fertilizers, and relative to the control plots not treated with the organic fertilizers. To the greater aldicarb or thiofanox soil concentrations -due to the recent organic fertilizer treatments- corresponded greater insecticide protection efficiencies of the sugar beet foliage against aphids.

## Maize crops

In maize crops, pig slurry or cow manure applied either 6 months (in November) or 1 month (in March) before sowing, slowed down the atrazine (applied at 1 kg ha$^{-1}$) and metolachlor (2 kg ha$^{-1}$) soil biodegradations, relative to the control plots not treated with these organic fertilizers.[4] The effect of the organic fertilizers was greater with metolachlor than with atrazine. Whereas atrazine killed almost all weeds as well in the plots treated or not with the organic fertilizers, to the greater metolachlor soil concentrations -due to the organic fertilizers treatments- corresponded greater herbicide protection efficiencies. The herbicide rimsulfuron works both by contact and through the soil. In spite of that, rimsulfuron (10 g ha$^{-1}$) has a greater soil persistence, and a greater herbicide efficiency in the plots treated with the organic fertilizers.

## Vegetable crops

In Brussels sprouts and cauliflower crops, the herbicides metazachlor (750 g ha$^{-1}$) and propachlor (4 kg ha$^{-1}$) soil biodegradations were slowed down by the organic fertilizers cow manure, pig slurry, composts and green manure applied either 6 or 1 month before planting (trials at St Katelijne-Waver, the State School for Horticulture at Gembloux, and at Elverdinge). The same was observed with the herbicide chlorbromuron (750 g ha$^{-1}$) in white celery (trial at Esen). The herbicide protection given by these three herbicides was greater in the plots treated with the organic fertilizers. Similar observations were made in carrot, beans and celery trials treated with vegetable originating composts, and organized by the Flemish Office for the Organic Matters and Wastes Management.

The effect of the recent organic fertilizers -to slow down the soil biodegradation of the insecticide chlorpyrifos in cauliflower crops- is not significantly increased when the organic fertilizers treatments are repeated during several years in the past on the same field plots (trials at St Katelijne-Waver from 1990 till 1993).[5,6] In all the assays (at St Katelijne-Waver, Gembloux and Elverdinge), when the rate of 50 mg chlorpyrifos plant$^{-1}$ is applied, practically all the cabbage flies are killed, similarly when there is or not treatment with organic fertilizers. At the rate of 15 mg, the insecticide protection could be as two times better in the plots treated with the organic fertilizers. However, when the infection pressure is abnormally high, the effect of the organic fertilizers is already observed at the rate of 50 mg; at the rate of 15 mg, the attack is almost complete in all the field plots treated or not with organic fertilizers.

## Herbicide and insecticide efficiencies

In all the trials, the effect of the recent organic fertilizers -to slow down the herbicides and insecticides soil biodegradations- was observed during the first and main part of the crop. Thereafter, the rates of pesticides soil biodegradations greatly increased; the herbicides and insecticides soil concentrations progressively became very low, and similar in the plots treated or not with the organic fertilizers.

In about 15 trials made in 1992 and 1993, Mrs Callens, Bulcke, Van de Steene (RUG) and Wauters (IRBAB) measured the biological efficiencies of the herbicides and insecticides. As already told above, to the slowing down of their soil biodegradations -during the main first crop period- due to the recent organic fertilizer treatments, and thus to their greater soil concentrations, generally corresponded greater herbicide and insecticide efficiencies. However, conditions were necessary to make the phenomenon observable : 1) the pesticide rate must be sufficiently low so that there is no levelling off by overrating (situation where practically all insects and weeds are killed, as well in the organic fertilizers treated plots as in the organic fertilizers untreated ones); 2) the herbicide must have an activity against such or such kind of weeds; 3) especially with the cabbage fly, a minimum infection pressure is needed; when the infection pressure is abnormally high, the insecticide rate, at which the organic fertilizer effect is observable, is greater.

## Mechanism of the organic fertilizer effect

The soil microbial activity of the above described trial fields treated with organic fertilizers was measured by incorporating into the freshly sampled soil from the field, low amounts of D,L-[1-$^{14}$C]-glutamic acid, and measuring after a short time of incubation the $^{14}CO_2$ evolved, i.e. the initial rate of glutamic acid mineralization. One of the organic fertilizers green manure, cow manure, pig slurry, or mushroom cultivation compost (at 50 or 100 tons ha$^{-1}$) has been incorporated into soil in the field plots either 1, 3 or 12 months before. On some plots, the same organic fertilizer treatment had been repeated once per year as far as 4 times in the past. In all the organic fertilizer treated plots, the soil microbial activity was 1.2 to 2.4 times greater than in the control plots not treated with organic fertilizers. Repetition in the past of the organic fertilizer treatments did not significantly increase the soil microbial activity, relative to a only one organic fertilizer treatment made within the year. As the soil microbial activity is increased by the organic fertilizer treatments, one should expect that these also should increase the rates of pesticides soil biodegradations. The fact that the reverse is observed in the field (the slowing down by the organic fertilizers of the pesticides soil biodegradations) suggests that the effect is not due to the change of the intensity of the soil microbial activity due to the organic fertilizers treatments.

The way into which the effect of the organic fertilizers was observed (great effect with the recent soil organic matter, low with the old soil organic matter, humus) suggests that this effect is linked to the chemical structure of the soil organic matter. This changes during humification. In a first period the density of the oxygenated chelating chemical functions increases; thereafter, with ageing, the density of the oxygenated chelating functions decreases (coalification). Reversible fixation of the pesticides onto the soil organic matter -by reversible (bio)chemical reaction or adsorption- is the most probable hypothesis.[2] In that way, the pesticides should be protected -in the adsorbed phase- against chemical degradation and the soil microbial activity which metabolizes them. Such a mechanism corresponds to the slow-release effect by the recent soil organic matter.

At the opposite of what is frequently thought, the recent organic fertilizers treatments have a positive effect on the herbicide and insecticide soil treatments. During the first and main crop period (essential for the plant protection, because the young plants are the most sensitive), the organic fertilizer treatments slow down the soil biodegradation (positive kinetic effect) of the herbicides and insecticides, and thus increase their soil concentrations. On the other hand, the low increase of the soil organic matter concentrations -due to the organic fertilizer treatments (which for instance increase from 2 to 2.5% immediately after the organic fertilizer application)-, is insufficient to reduce the pesticide biodisponibility (negative effect due to the increase of pesticide fixation onto the soil organic matter).

Results show that the resultant of these opposite effects -due to the organic fertilizer treatments- generally is positive for the efficiencies of the herbicides and soil insecticides. These fields trials will be pursued. Moreover, laboratory experiments will be made in order to get an experimental knowledge of the mechanism of the organic fertilizers effect.

## Conclusions

As for soil fertility, the recent organic fertilizer treatments thus are beneficial for the herbicide and soil insecticide protections. At crop end, this effect disappears, the herbicides and insecticides residues remaining in soil becoming very low or non-existent. The organic fertilizer treatments thus are a remedy against the accelerated soil biodegradations -to which correspond heavy decreases of the protection efficiencies-, which are observed in some fields. Moreover, when these organic fertilizers are applied at normal rates and at the right periods, they maintain the pesticides at the soil surface, avoiding their leaching to the groundwaters.

## Acknowledgements

Mass spectra were recorded by C. Moulard (Université Libre de Bruxelles, Brussels, Belgium). This work was supported by the Institute for Applied Research in Industry and Agronomy, IRSIA-IWONL, Belgium.

## References

☐       1. Rouchaud, J.; Gustin, F.; Callens, D.; Van Himme, M.; Bulcke, R. Soil metabolism of the herbicide isoxaben in winter wheat crops. *J. Agr. Fd Chem.* 1993, 41, 2142.

☐       2. Honnay, J.P. Scientific Adviser at the Institute for Applied Research in Industry and Agronomy, IRSIA-IWONL, Belgium, 1993.

☐       3. Rouchaud, J.; Gustin, F.; Wauters, A. Effects of organic fertilizer treatments and old humus on thiofanox and aldicarb soil metabolisms in sugar beets. *Arch. Environ. Contam. Toxicol.* 1994, 26, 222.

☐       4. Rouchaud, J.; Gustin, F.; Cappellen, O.; Mouraux, D. Effects of pig slurry and cow manure on the atrazine and metolachlor herbicides soil biodegradations in maize crops. *Bull. Environ. Contam. Toxicol.* 1994, 52, 568.

☐       5. Rouchaud, J.; Gustin, F.; Metsue, M.; Touillaux, R.; Van de Steene, F.; Pelerents, C.; Gillet, J.; Benoit, F.; Ceutermans, N. *Toxicol. Environ. Chem.* 1992, 35, 47.

□    6. Rouchaud, J.; Gustin, F.; Benoit, F.; Ceustermans, N.; Gillet, J.; Van de Steene, F.; Pelerents, C. Influence of cow manure and composts on the effects of chlorfenvinphos on field crops. *Arch. Environ. Contam. Toxicol.*1992, 22, 122.

*5.2.6.      Mass Balance and Fate of $^{14}$C-Terbuthylazine and $^{14}$C-Pendimethalin in Outdoor Lysimeters*

# I. Scheunert[1], U. Dörfler[1], R. Schroll[1] and D. Mourgou[2]

[1]    GSF-Institut für Bodenökologie, Neuherberg D-85764 Oberschleißheim, FRG
[2]    Office of Public Health GR-67100 Xanthi, Greece

## Abstract

The herbicides terbuthylazine and pendimethalin, both $^{14}$C-ringlabelled, were applied to outdoor lysimeters, and agricultural plants were grown. Volatilization of $^{14}$C-residues into the air, mineralization to $^{14}$CO$_2$, uptake by plants, and leaching into percolate water were determined for various time periods. The formation of extractable metabolites and of soil-bound residues was measured also. It could be established that losses into air and percolation water as well as output by plant cropping were only minor pathways for the dissipation of both herbicides from the soil. Transformation to soluble polar metabolites and to soil-bound residues were the main mechanisms for the residue decline of original herbicides in the soil-plant system.

## Introduction

The fate of pesticides in the soil after agricultural use is the result of various physical, physico-chemical, chemical and biological processes. Although these processes may be quantified in the laboratory in dependence on pesticides substance properties, soil properties and environmental conditions, laboratory results cannot be extrapolated quantitatively to outdoor conditions. In order to assess the fate of pesticides under outdoor conditions and the resulting realistic residue levels, lysimeters are a useful tool[1]. They permit to use $^{14}$C-labelled compounds and, thus, to include in the study, in addition to the parent compound, soluble metabolites and unextractable residues, and CO$_2$ resulting from mineralization.

In this contribution, outdoor lysimeter experiments with the $^{14}$C-labelled pesticides terbuthylazine and pendimethalin will be reported.

## Experimental

The round lysimeters (Fig.1) - with a depth of 1 m and a circumference of 2 m (surface area 0.32 m2) - were made from stainless steel. A wirecloth and 3 metal plates which had been punched before, resulting in a "hole area" of about 50% of the total lysimeter area in the bottom of the lysimeters, allowed the leachate to run out of the soil column.

The soils were taken from different sampling sites, and each soil horizon was sampled separately. After transport in plastic bags to the lysimeter station they were put into the lysimeters and pressed to the same density as in the natural soils. In the following, as examples only studies with one soil will be presented. It was a sandy agricultural soil with 85% sand in the upper horizon and nearly 100% in the C-horizon; silt (10%-0%) and clay (5%-0%) decreased from the upper to the deepest horizon. The pH was between 4.9 and 5.5 from the upper to the deepest horizon and the amount of organic matter decreased from 1.77% in the Ap-horizon to 0.99% in the C-horizon.

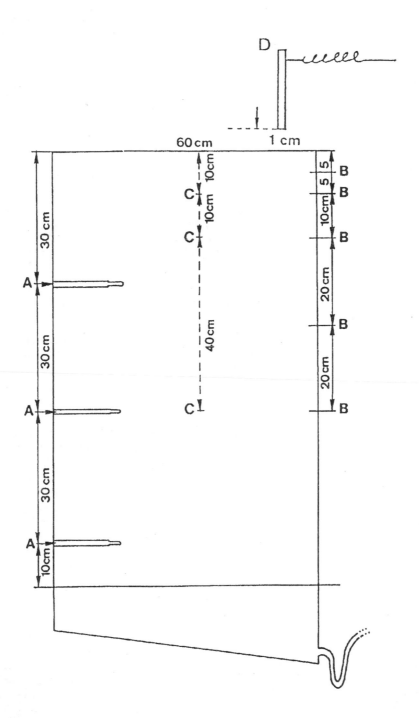

**Fig. 1.** Side-view of a lysimeter
    A. Suction candle
    B. Thermometer
    C. Tensiometer
    D. Thermic anemometer for wind speed measurements

A water reservoir was placed below the soil column in the lysimeter to collect the leachate for at least 4 weeks. The leachate was sucked by means of a vacuum pump, into 5 l glass bottles which were stored in a refrigerator at a temperature of 4°C. Water was collected intermittently every 2 or 4 weeks.

A volatilization chamber was designed to determine $^{14}$C-losses from soil surfaces. The chamber which was made of plexiglas, was 28 cm long, 2 cm high and 8 cm wide. Air was drawn through the chamber at 0.2 m/sec. Wind speed measurements taken 1 cm above the lysimeter soil surface outside the chamber showed that 0.2 m/sec is a good average value. In the inlet funnel of the chamber, 4 metal sieves were placed to produce a nearly laminar air stream in the chamber. The lower part of the chamber consisted of a metal frame which was pressed into the soil surface to make sure that only substances volatilizing from the test area of 20 cm length and 8 cm width were collected in the system. A small part of the total air stream (0.2%) which passed through the volatilization chamber was taken for the determination of the volatile $^{14}$C-labelled compounds. It was passed through a trapping system consisting of two tubes containing ethyleneglycolmonomethylether to remove volatile pesticide residues and two tubes containing a mixture of Carbosorb and Permafluor (3:3;v:v) to trap $^{14}$C-labelled carbon dioxide. At the end of the trapping system there was a charcoal filter, a gas flowmeter and a small membrane pump. The main air stream in the chamber could be determined by a gas flowmeter in front of the large membrane pump.

In the volatilization chamber the microclimate was different from the climate of the open lysimeter. In order to treat the whole lysimeter surface equally, the volatilization chamber was moved several times (after about 30 days each) to different sections of the soil surface within the same lysimeter[1,2].

For the experiments with terbuthylazine, the $^{14}$C-labelled pesticide was applied on the soil surface in a dose as used in agriculture, and maize was grown. For pendimethalin, the crop was winter wheat. After the harvest, plants and soil layers 0-20 cm deep were analyzed for $^{14}$C-labelled compounds. All tests were carried out in triplicates.

$^{14}$C in leachate, in the liquids containing $^{14}$C trapped from the air, and in soil and plant extracts was determined in a liquid scintillation counter. Metabolites in soil and plants were identified, after separation and purification by chromatographic methods, by HPLC with the aid of authentic reference compounds. Unextractable $^{14}$C in soil and plant material was determined by combustion to $^{14}CO_2$ followed by liquid scintillation counting.

# Results and Discussion

The chemical formula of $^{14}$C-labelled herbicides used and of their metabolites identified in this study are shown in Figs. 2 and 3.

From terbuthylazine (Fig.2), only the desethyl-derivative and, in leaching water, traces of the desisobutyl-derivative could be detected by HPLC. None of the other desalkylated and/or hydroxy-lated metabolites analogous to those known from atrazine degradation (3), could be detected in this study. Unchanged terbuthylazine, unidentified highly polar soluble metabolites and soil-bound unextractable residues prevailed, as will be discussed below.

In case of pendimethalin, it was similar. Besides the parent compound, soil-bound residues and highly polar metabolites were most important. A metabolite carrying a hydroxyl group in the side chain was found in soil in trace amounts by TLC (Fig.3). Also in this case, none of the other transformation products reported in literature (4-8) could be detected.

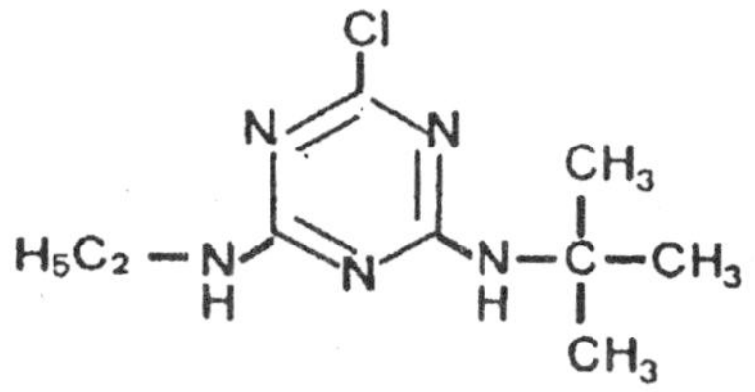

Terbuthylazine

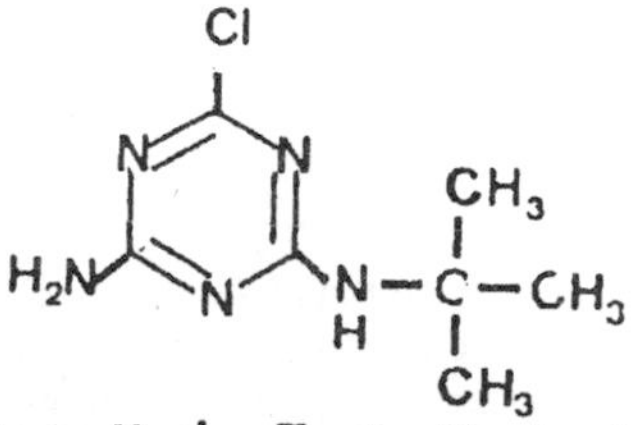

Desethyl — Terbuthylazine

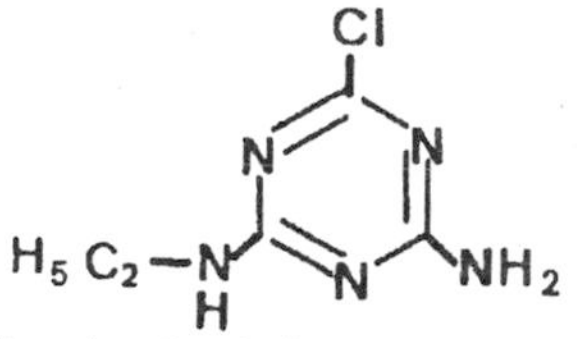

Desisobutyl — Terbuthylazine

**Fig. 2.** Formula of terbuthylazine and its metabolites identified in this study

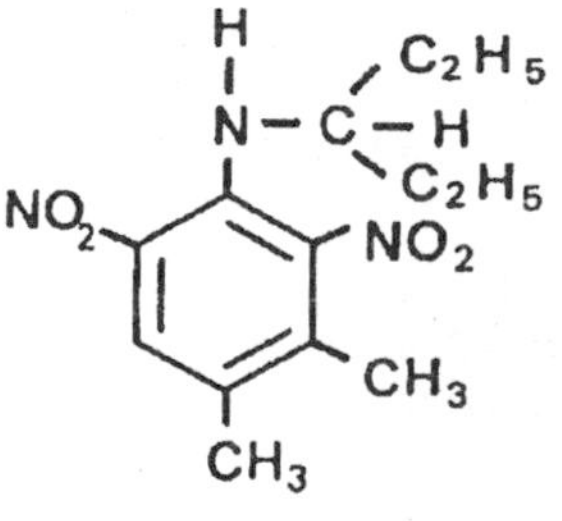

Pendimethalin

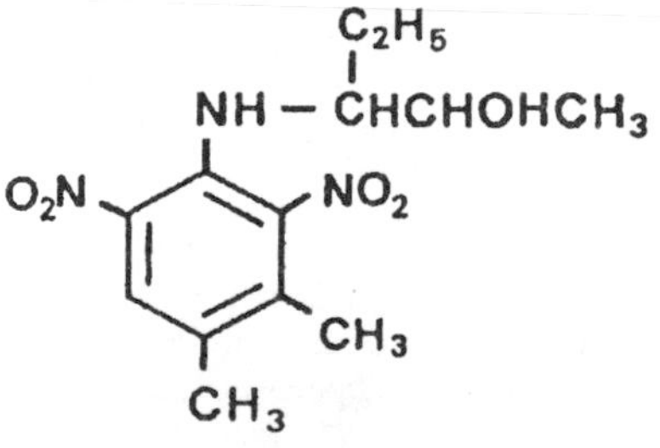

3-(2,6 Dinitro-3,4-xylidino)-2pentanol

**Fig. 3.** Formula of pendimethalin and a metabolite identified by TLC

The transfer of both herbicides and their metabolites from soil into air, plants and leaching water is reported in the following paragraphs.

**Volatilization** of terbuthylazine was only about 0.5% of the applied amount within 67 days. The volatilization of pendimethalin was 6-7 fold. This is in agreement with the difference in vapour pressure (terbuthylazine: 0.15mPa; pendimethalin: 4mPa). However, besides vapour pressure numerous other pesticide properties and environmental conditions are involved in the complex process of volatilization from soil. It is remarkable that the volatilization process slows down strongly or even discontinues after about 20 days; it begins again only after tilling the soil in the following year.

**Mineralization** of terbuthylazine to $^{14}CO_2$, as measured simultaneously during the same time period, was about tenfold as compared to volatilization. This process also discontinued after about 20 days and, also, began again after tilling the soil. Differences between terbuthylazine and pendimethalin were minor.

In the maize **plants** grown in soil treated with [14]C-terbuthylazine, a continuous decrease in [14]C-concentration could be observed with proceeding growth for 89 days (from about 2.3 µg terbuthylazine equivalents per g dry mass on day 28 to 0.3µg/g on day 89). This phenomenon is due to growth dilution and has been reported also for other pesticides, e.g. chlorinated benzenes, and other plants. After the end of growing a slight increase in the concentration of residues follows, due to continuing uptake from soil. The [14]C-redidues were higher in leaves than in cobs. The [14]C in the plants consisted nearly completely of highly polar metabolites; unchanged terbuthylazine and its monodesalkylated metabolite desethylterbuthylazine were found only in trace amounts.

In case of pendimethalin and winter wheat, the [14]C-contents were also higher in straw than in ears. They were significantly lower than the residues of terbuthylazine, which may be partly due to the long growing time of winter wheat. Therefore, [14]C could not be characterized chemically by chromatographic methods.

**Leaching** of [14]C originating from [14]C-terbuthylazine in a 1m-depth amounted to about 0.04% after 67 days. After 459 days, it was 1.5-2%. After extraction of [14]C of the leachate from the terbuthylazine experiments after 459 days, chromatographic separation and identification by HPLC, it turned out that only 1.9% of [14]C in leachate was the unchanged parent herbicide and another, small part were the desalkylated derivatives (desethylterbuthylazine: 3.8%, desisobutylterbuthylazine: 0.3%). The major part were unidentified, highly polar metabolites, most of them probably being bound to higher molecular soluble humic material (DOC).

In case of pendimethalin, the rate of leaching did not differ much from that of terbuthylazine. In the leachate, only highly polar metabolites were present. The dissolved humic material to which the pesticide residues are bound govern the leaching behaviour of [14]C; since they probably do not differ very much between the [14]C from terbuthylazine and that from pendimethalin, leaching in both cases is similar, although the water solubility of the parent herbicides differs by more than one order of magnitude (terbuthylazine: 8.5 mg/l; pendimethalin: 0.3 mg/l).

All the transfer processes discussed do not contribute much to the overall dissipation of the herbicide terbuthylazine from the soil. The major part remains in the soil. As sources for the disappearance of the parent compounds, therefore, from the upper soil layer, only transport to deeper soil layers and conversion to more polar metabolites are important.

The data of **transport in soil** show a steady decrease in [14]C-concentration with soil depth for both herbicides (terbuthylazine: 0.6-0.8µg terbuthylazine-equivalents/g in 0-5 cm depth and <0.1 µg/g in 15-20 cm depth after 242 days; pendimethalin: 1.1-1.4 µg pendimethalin-equivalents/g in 0.5cm depth and <0.1 µg/g in 15-20 cm depth after 288 days).

Exhaustive extraction of [14]C from the soil layers of the experiments with terbuthylazine revealed that only about one third of [14]C present in each layer was extractable. The extractable [14]C consisted, after partition between water and organic solvent, of a minor water-soluble and a major organic-soluble part. In the upper soil layer (0-5 cm depth), unchanged terbuthylazine was the main radioactive component, followed by the desethyl derivative and polar residues. In the deeper layers, the ratio between unchanged terbuthylazine and the metabolites shifted in favour of the metabolites.

In case of pendimethalin, soil samples were taken after two time intervals. 16 days after application of pendimethalin to soil, only the 0-5 cm soil layer was analyzed. The radioactivity, for its major part (more than 3/4), was extractable with organic solvent (methanol). The extracted [14]C consisted mostly of unchanged pendimethalin; the hydroxylated metabolite and other polar metabolites represented a minor fraction. After 288 days, four soil layers down to 20 cm were analyzed separately. The unextractable portion in all layers was higher than after 160 days. The extractable portion contained unchanged pendimethalin only in the upper layer (0-5 cm), where it represented about half of the total

[14]C-residue. The other layers contained only polar metabolites, except for the second layer (5-10 cm), where minimum amounts of the hydroxylated metabolite could be detected.

# Conclusion

These lysimeter studies showed that for terbuthylazine and pendimethalin, the main sink in relation to the persistence of the parent compounds is transformation in soil to highly polar and unextractable soil-bound residues.

# References

□   1. F. Führ and R.J. Hance, Lysimeter Studies of the Fate of Pesticides in the Soil. *BCPC Monograph No. 53,* British Crop Protection Council, Farnham, Surrey, U.K., 1992.

□   2. R. Schroll, T. Langenbach, G. Cao, U. Dörfler, P. Schneider and I. Scheunert, Fate of $[^{14}C]$ terbuthylazine in soil-plant systems. *Sci. Total Environ.* 123/124, 377-389 (1992).

□   3. I. Scheunert, R. Schroll and U. Dörfler, Persistence of herbicides in agricultural soils. In: Proceedings Vol. 1, *International Symposium of the Indian Society of Weed Science,* Hisar, Indien, pp. 107-114 (1993).

□   4. J. Zulalian, CL 92,553- metabolism V. Fate of carbon-14 labelled CL 92,553 (PROWL herbicide) in soil. Progress Report, Project 2-463, American Cyanamid Company, Princeton, NJ, 1973.

□   5. A.S. Barua, J. Saha, S. Chaudhuri, A. Chowdhury, and N. Adityachaudhury, Degradation of pendimethalin by soil fungi. *Pestic. Sci.* 29, 419-425 (1990).

□   6. S.B. Singh and G. Kulshrestha, Microbial degradation of pendimethalin. *J. Environ. Sci. Health* B 26, 309-321 (1991).

□   7. J. Saha, A. Chowdhury and S. Chaudhuri, Stimulation of heterotrophic dinitrogen fixation in barley root association by the herbicide pendimethalin and its metabolic transformation by Azotobacter spp. *Soil Biol. Biochem.* 23, 569-573 (1991).

□   8. G. Kulshrestha and S.B. Singh, Influence of soil moisture and microbial activity on pendimethalin degradation. *Bull. Environ. Contam. Toxicol.* 48, 269-274 (1992).

5.2.7.     *Single application or continuous exposure in ecotoxicity testing*
           *- consequences for interpretation*

# L. Samsøe-Petersen

VKI, Water Quality Institute, Agern Allé 11, 2970 Hørsholm,Denmark

## Introduction

Internationally accepted guidelines for ecotoxicological effects tests are predominantly using aquatic organisms (OECD 1993). These guidelines are used for the registration of industrial chemicals and pesticides. For the registration of pesticides individual countries are in addition using national guidelines for terrestrial organisms such as honey bees and natural enemies of pests (e.g. BBA-guidelines).

Within the OECD coorporation there is a growing interest in developing guidelines for testing of chemicals with terrestrial organisms to supplement the existing guidelines, and the uniform principles for the registration of pesticides in the EEC (Annex VI, Dir. 91/414/EEC) call for internationally accepted guidelines for terrestrial organisms like soil micro-organisms, honey bees, natural enemies of pests and decomposers (in excess of earthworms).

At present the choice of test methods for future guidelines and the use of results in terrestrial effects assessment are being discussed at the European level (e.g. EPPO 1993) as well as in the wider cooperation of the OECD (van Straalen & van Gestel 1993). Part of these discussions is concentrating on the choice of species to be used, part of them on the conditions under which the tests are being performed.

As ecotoxicological testing is undertaken to obtain comparable results, standardisation of the conditions under which the tests are performed is essential. This is necessary whether the sensitivity of species or the toxicity of chemicals are going to be compared. The need for standardisation is especially pertinent when results from different tests are combined for the calculation of the Predicted No Effect Concentration (PNEC) of a chemical in the environment. If the PNEC is calculated from a mixture of results from aquatic and terrestrial tests - as proposed by e.g. Løkke (1994) - the compatibility of data from both types of tests is a prerequisite.

At this point - when tests for more terrestrial organisms are to be included in the standardised guidelines to be used for industrial chemicals as well as for pesticides - several points need consideration. One of them is that there are basic differences between the guidelines for aquatic organisms and the test methods for terrestrial organisms. In order to maximise the usefulness of results from future tests it is appropriate to discuss and decide on the standardisation of several features concerning these differences before more terrestrial tests are developed into guidelines.

## Differences in Experimental Conditions

Some of the differences between tests for aquatic and terrestrial organisms arise from differences in the environments - others from the fact that most terrestrial tests were developed (and used) for the assessment of effects of agricultural pesticides, whereas the aquatic tests were developed to assess effects of industrial chemicals. The pesticides are predominantly applied as a discrete spray resulting in a deposit on either plant surfaces or soil, whereas the industrial chemicals are expected to be released continuously and result in a more or less constant concentration of the chemical in water bodies. These differences are reflected in differences in experimental condition between terrestrial and aquatic tests.

The bioavailability of the chemicals to the test organisms during exposure may be dependent on the presence of sorbing surfaces and/or biological activity metabolising the compounds. In principle the interpretational problems arising from these mechanisms are the same in terrestrial and aquatic testing, but in practise they give rise to different solutions.

The differences in experimental conditions are summarized in table 1.

| | Terrestrial tests | Aquatic tests |
|---|---|---|
| Dosage | Deposit ($\mu g/cm^2$) or "Concentration" (mg/kg) | Concentration (ml/l) |
| Bioavailability | Sand/glass | Water |
| | Soil/leaves | Sediment |
| Exposure concentration | Declining | Constant |

**Table 1 :** Differences in experimental conditions between ecotoxicity tests for terrestrial and aquatic organisms

In the following the first two will be discussed briefly as they are outside the scope of this paper. The difference in exposure concentration raises theoretical as well as practical problems that are discussed in more detail.

## Dosage

In most tests for aquatic organisms exposure is mediated via the water surrounding the experimental organisms, and the dosage can be expressed as a concentration (e.g. ml/l). Terrestrial tests for some soil dwelling species (earthworms, springtails) are based on incorporation of the chemicals into the medium, whereas others are based on application of the chemicals (pesticides) as a spray. In the former the dosage is expressed as a kind of concentration (mg/kg soil). In the latter (for most terrestrial arthropods and soil micro-organisms) the chemical is often administered as a spray resulting in a deposit on a surface ($\mu g/cm^2$). In case the surface is of sand or soil, the dosage may be defined as g/kg soil (or even as the concentration in pore water) assuming that the chemical is distributed homogeneously into a well defined top layer of the sand or soil. In theory a deposit on the surface of an animal may be recalculated to a dose measured in $\mu g/g$ animal after measurements of the surface area and weight of the test organism but this is rarely done.

For aquatic as well as for terrestrial organisms there are other types of tests - e.g. feeding studies, topical application - but the tests most often used (for invertebrates) and the tests considered for guideline development are of the former types.

Testing with animals (invertebrates) living in the canopy implies that exposure is confined to the contact obtained with a deposit of chemical. This contact can be very difficult to define and thus to standardise. Whether the invertebrates are sprayed with the surface or they are placed on it immediately after the spraying. In the first case exposure will be a combination of the deposit on the animal itself and the contact with the deposit on the surface (and possible vapours) during the rest of the test. In the second case exposure is confined to the contact with and possible vapours released from the surface during exposure. In any case exposure is limited and difficult to define in these tests (Ford 1992) compared to the tests for aquatic organisms where the animals are surrounded by (submerged in) the solution of chemical all through the exposure period.

Thus a comparison of the toxicity of chemicals to terrestrial and aquatic organisms is difficult as the dosage is described in different units, and calculations of PNEC cannot combine results from these tests. For soil living organisms this may be overcome by assuming that uptake is via the water phase and calculating the concentration of the chemical in pore water - even though these calculations must be based on several assumptions (Sloof 1992).

For hazard or risk assessment purposes the ratios (Predicted Environmental Concentration) PEC/PNEC and PEDeposit/PNEDeposit can however be compared - even if exposure during the test period is not as well defined in terrestrial (deposit) tests as in aquatic tests. But this implies standardisation of the bioavailability of the chemicals during exposure and/or incorporation of this parameter in the estimation of PEC/PEDeposit.

## Bioavailability

The bioavailability of chemicals during exposure is approached from two sides in aquatic as well as in terrestrial ecotoxicity testing.

In one approach the bioavailability of chemicals is maximised during exposure, and the effects measured reflect the inherent toxicity of the chemicals tested. This approach is used in guidelines for pelagic aquatic species (algae, daphnia, fish) and in test methods for several terrestrial species, including some soil dwelling (Hassan 1992). Hazard and risk assessment based on results from such tests implies that the bioavailability of the chemicals in the environment is incorporated in the estimation of the PEC.

In the other approach the bioavailability of chemicals in the environment is imitated during exposure in the laboratory. This approach is used in aquatic tests for sediment dwelling species (mussels, echinoderms) and in terrestrial tests for some soil dwelling species (earth- worms, springtails). For hazard and risk assessment based on results of this type of tests the elements of bioavailability included in the laboratory tests have to be integrated in the estimation of the PEC.

In terrestrial toxicity tests there is as yet no standard "medium" for the tests at the basic level. Some authors advocate the use of inert surfaces or media (glassplates/sand) in order to measure the inherent toxicity of the chemicals (Hassan 1992) while others give priority to more or less natural media (leaves/soil) to facilitate extrapolation from results of laboratory tests to possible effects in the field (Goats and Edwards 1988, van Gestel 1992). The latter approach is widely accepted and the standardisation of soil(s) is investigated and discussed at present. In the OECD guideline 207 (earthworm toxicity test) an artificial soil is used. This medium has the advantage of being standardised (even though standardisation is not complete (Römbke et al. 1992)) and at the same time resembling natural soils in so far as it consists of sand, clay and organic matter.

The choice of media for tests with aquatic as well as terrestrial organisms is thus important for the possibilities of the comparison and integration of results from different tests. This is especially pertinent for the assessment of side effects of pesticides as QSAR is not immediately applicable for many of these chemicals.

## Exposure Concentration

Aquatic ecotoxicological toxicity tests are designed to allow for the maintenance of a constant (flow thorugh) - or at least refreshed (semi-static) - concentration of chemicals during the entire exposure period. This is due to the "effluents approach" of this testing described earlier, but also conditioned by the relatively simple exposure through renewable water.

Terrestrial toxicity tests that were mostly developed from the "pesticides approach" - and conditioned by the static properties of terrestrial environments - are exclusively static. They are designed to use one application of the compound at the start of the test whereafter the treated surface/medium is left undisturbed during the exposure period. The end-points derived from these tests are thus based on an unknown and probably declining amount of pesticide during the exposure period.

Even in acute (short term) tests this difference may be important, and in chronic studies, the concentration of many chemicals will be reduced during the exposure period. This reduction can be caused by sorption, evaporation, hydrolysis, photodegradation and biological activity like uptake by leaves or biodegradation. In terrestrial effects research only very few studies related the effects recorded to measurements of exposure concentrations during the test period. The importance of declining exposure concentrations in static tests will be illustrated by a few examples.

Even in aquatic tests using only water, the concentration of the pesticide may decrease drastically in static tests. Thus in tests with aquatic invertebrates Stephenson (1982) found a decline in the concentration of cypermethrin of up to 80% after only 24 hours. This was explained by the known propensity of cypermethrin for absorbing onto surfaces; i.e. the surface of the test vials. Stephenson (1982) did not perform tests with constant concentrations of cypermethrin, so the reduction in concentration during the test period could not be related to the effects measured.

In the more complex systems including soil, most of the other causes for declining concentrations are probably active as well - not to mention the bioavailability of the chemicals in soil. Heimbach et al. (1992) investigated - among other parameters - the importance of the age of pesticide residues for the effects on the carabid beetle *Poecilus cupreus*. They exposed adult beetles to residues on sand and two types of natural soil and measured mortality, behaviourial and other effects during a 22 day exposure period. The age of the residues at the initiation of exposure was 0, 24 and 48 h. Gaschromatographic analyses were made on water, acetone-water and n-hexane extracts of soil samples taken at 0, 24 and 48 h after application and four different pesticides were used. Correlations between concentrations of pesticides in the different fractions and effects were not obvious for all pesticides and soil types. With lindane however a reduction of the pesticide was measured on all substrates within the 48 h accompanied by a reduction in mortality - most pronounced on sand. This illustrates the importance of the ageing of residues during the exposure period in terrestrial (soil) tests, but investigations using a "semi-static" approach are needed to compare the effects of declining versus relatively constant pesticide deposits during the exposure period.

Bioassays for persistence are designed to age residues of pesticide deposits on natural media (soil or leaves) under field or simulated field conditions (Samsøe-Petersen 1990). The test organisms are introduced after ageing for various periods (3, 15, 28 days) and effects of the aged residues are measured. Results obtained with such tests show strong reductions in toxicity after only 3 days for several pesticides, and thus support the importance of declining concentrations (e.g. Hassan et al. 1991).

Van Gestel et Al. (1991) also discussed the problem of declining exposure concentration in a comparison of earthworm (pore water concentration) and fish toxicity data. Results from tests with chlorobenzenes and 2,4-dichloroaniline were compared, and chemical analyses showed that the volatile mono- and di-chlorobenzenes were present in at most one third of the nominal concentration in the soil at the start of exposure in the earthworm test. The authors found that fish seemed to be more susceptible to these compounds than earthworms, but concluded that the observed difference was explained by the differences in experimental conditions; static versus semi-static or flow through leading to different exposure concentrations in the terrestrial and aquatic tests. The finding that earthworms seemed to be more sensitive than fish to tetrachlorobenzene and pentachlorobenzene was not discussed, but it could be due to a higher sensitivity. This example illustrates the problems arising from the different exposure regimes in aquatic and terrestrial tests; the seemingly lower sensitivity of earthworms than fish could be due to evaporation, but the results did not enable conclusions as to whether earthworms are actually more sensitive than fish.

Thus it must be emphasized that comparisons between - not to mention integration of - results from tests resulting in different exposure concentrations are most probably seriously biased. For the integration of results from such tests in e.g. the calculation of PNEC, research is needed that verify the exposure concentrations and elucidate the consequences of possible declining concentrations.

# Conclusions

For the standardisation of tests with terrestrial organisms and for the interpretation of results in the context of hazard and risk assessment of chemicals - especially pesticides - several parameters are of importance. And there is a need for decisions to be taken as to the exposure (bioavailability, fate) regime of ecotoxicity tests used for regulatory purpoes as well as for other risk assessment purposes. The comparability of conditions - and thus the interpretation of results - from different tests are reduced by the following parameters:

- In most terrestrial tests the end points are expressed in units of area, while in aquatic (and some soil) tests the units are based on volumes. It should be considered whether it is appropriate to integrate data from terrrestrial and aquatic environments.

- The bioavailability of chemicals is different in tests with inert and "natural" media. The approach chosen for standardised guidelines should be discussed. If the "natural approach" is chosen for the development of terrestrial ecotoxicity guidelines, the choice of substrates and the possibilities of standardisation of leaves and soil types should be considered. For soil types the choice should include the possibilities of investigating fate and behaviour of pesticides/chemicals under standardised conditions (Eurosoils).

- The exposure in terrestrial (pesticide) tests are based on one single application of the chemical resulting in declining exposure concentration during the test. As this feature cannot be changed due to the static properties of the terrestrial environment ways of quantifying the importance of this decrease should be sought. For this purpose the choice of standard soils is again important - and standardisation of the microbial activity in (or sterilisation of) these soils should be discussed from the point of view of effects assessment.

# References

□   EPPO, Decision-making scheme for the environmental risk assessment of plant protection products. *Bulletin OEPP/EPPO Bulletin* 23 (1):1-157, 1993.

□   Ford, M.G. Insecticide exposure, pick-up and pharmacokinetics with target and non-target insects. *Aspects of Applied Biology* 31:29-41, 1992.

□   Goats, G.C. and Edwards, C.A. The prediction of field toxicity of chemicals to earthworms by laboratory methods. In: *Earthworms in Waste and Environmental Management*, edited by Edwards, C.A. & Neuhauser, E.F. The Hague: SPB Academic Publishing, 1988, p. 283-294.

□   Hassan, S.A., Bigler, F., Bogenschütz, H., Boller, E., Brun, J., Calis, J.N.M., Chiverton, P.A., Coremans-Pelsener, J., Duso, C., Lewis, G.B., Mansour, F., Moreth, L., Oomen, P.A., Overmeer, W.P.J., Polgar, L., Rieckmann, W., Samsøe-Petersen, L., Staübli, A., Sterk, G., Tavares, K., Tuset, J.J. & Viggiani, G. Results of the fifth joint pesticide testing programme carried out by the IOBC/WPRS Working Group "Pesticides and Beneficial Organisms". *Entomophaga* 36: 55-67, 1991.

□   Hassan, S.A.(Ed.) Guidelines for testing the effects of pesticides on beneficial organisms: Description of test methods. *IOBC/WPRS Bulletin* XV/3:1-120, 1992.

□   Heimbach, U., Abel, C., Siebers, J. and Wehling, A. Influence of different soils on the effects of pesticides on carabids and spiders. *Aspects of Applied Biology* 31:49-59, 1992.

□   Løkke, H. Ecotoxicological Extrapolation: Tool or Toy? in Donker et al. (Eds): Ecotoxicology of Soil Organisms, Lewis, CRC Press, Florida: 411-426, 1994.

□   OECD, OECD Guidelines for testing of chemicals. *OECD Guidelines*, Paris, 1993.

□   Römbke, J., Vickus, P. and Bauer, C. Experiences and problems with the OECD-earthworm acute test in routine testing. In: *Ecotoxicology of Earthworms*, edited by Greig-Smith, P.W., Becker, H., Edwards, P.J. and Heimbach, F. Andover: Intercept, 1992, p. 209-212.

□   Samsøe-Petersen, L. Sequences of standard methods to test effects of chemicals on terrestrial arthropods. *J.Ecotox.Envir.Safety* 19:310-319, 1990.

□   Sloof, W. *RIVM Guidance Document. Ecotoxicological effect assessment: Deriving Maximum Tolerable Concentrations (MTC) from single-species toxicity data*, 1992. (UnPub)

□   Stephenson, R.R. Aquatic toxicology of cypermethrin. I. Acute toxicity to some freshwater fish and invertebrates in laboratory tests. *Aquat.Toxicol.* 2:175-185, 1982.

□   van Gestel, C.A.M., Ma, W. and Smit, C.E. Development of QSARs in terrestrial ecotoxicology: earthworm toxicity and soil sorption of chlorophenols, chlorobenzenes and dichloroaniline. In: *QSAR in Environmental Toxicology - IV*, edited by Hermens, J.L.M. and Opperhuizen, A. Amsterdam: Elsevier, 1991, p. 589-604.

□   van Gestel, C.A.M. Validation of earthworm toxicity tests by comparison with field studies: A review of benomyl, carbendazim, carbofuran, and carbaryl. *Ecotoxicol.Environ.Saf.* 23:221-236, 1992.

□   van Straalen, N.M. and van Gestel, C.A.M. Ecotoxicological test methods using terrestrial arthropods. *OECD Discussion Paper* 1-63, 1993.

# 5.3.  Poster Presentations

## 5.3.1.   *Field Atrazine Behaviour and Dissipation Kinetics in two Different Soils*

## E. Barriuso, R. Calvet, U. Baer and J.M. Dabadie

I.N.R.A., Unité de Science du Sol, 78850 Thiverval-Grignon, France.

## Abstract

Atrazine dissipation was studied in a hydromorphic soil and a loamy soil under similar climatic conditions. The hydromorphic soil was drained and water samples were collected for atrazine analysis. Dissipation kinetics in the soil surface (0-10 cm) were followed until harvest. A distribution profile of atrazine and its dealkylated derivatives was established one year after application. The results are discussed in relation to the soils characteristics, and are compared to simulations with three mathematical models describing pesticide behaviour in soils (PRZM2, LEACHP3.1 and VARLEACH).

## Introduction

The pesticide dissipation in field conditions includes several phenomena of transformation and transport. The most important processes for most of the pesticides are sorption and biological degradation. Their importance is possible to evaluate by laboratory measurements: calculation of adsorption-desorption parameters, and estimation of degradation by incubations under controlled conditions. Nevertheless, extrapolation of these measurements to prevent the pesticide fate in the field conditions can be problematic. The knowledge of pesticide soil behaviour in terms of basic processes (transport, sorption, degradation) is necessary to obtain parameters to build prediction tools. Numerical modelling of chemical transport coupled with sink/source phenomena can be used to describe the fate of pesticide in the soil. Models may be very useful, because they can accommodate various environmental and hydrological conditions, various chemical properties, management practices, spatial variability of soil and chemical/media interactions. This paper reports atrazine dissipation kinetics under field conditions for two different soils. The field observations are compared to simulation results.

## Materials and Methods

The experimental field sites were located at Rambouillet (Yvelines) on a hydromorphic clay soil (*Aquic eutrochrept*) continuously cropped with corn, and at Grignon (Yvelines) on a loamy soil (*Typic eutrochrept*) alternatively cropped with corn and wheat. The main soil characteristics are given in table 1. Plot sizes at Rambouillet and Grignon were $1000\,m^2$ and $600\,m^2$ respectively. The Rambouillet plot was drained at 80 cm depth and drainage water was sampled at the outlet. Climatic conditions were similar for the two sites. The water balance (rain - potential evapotranspiration) was negative between February and October and no water could be sampled from April to December 1990. After atrazine application climatic conditions were hot and dry, followed by a dry summer with a monthly rain average of 35 mm.

| SOIL SITE | DEPTHS cm | CARBON g.kg$^{-1}$ | pH water | CLAY g.kg$^{-1}$ | SILT g.kg$^{-1}$ | SAND g.kg$^{-1}$ | Atrazine Kd (l kg$^{-1}$) |
|---|---|---|---|---|---|---|---|
| GRIGNON *(Typic eutrochrept)* | 0-30 | 11.1 | 7.3 | 220 | 730 | 50 | 2.04 |
| | 30-60 | 5.9 | 7.6 | 230 | 730 | 30 | 1.46 |
| | 60-90 | 3.5 | 7.9 | 270 | 705 | 25 | 0.61 |
| | > 90 | 3.2 | 8.3 | 205 | 772 | 23 | 0.45 |
| RAMBOUILLET *(Aquic eutrochrept)* | 0-25 | 8.9 | 6.7 | 215 | 351 | 434 | 1.61 |
| | 25-35 | 9.2 | 7.0 | 265 | 329 | 406 | 0.69 |
| | > 35 | 2.4 | 7.2 | 551 | 177 | 272 | 0.5-1.7* |

\* Values from different heterogeneous layers.

**Table 1** : Main soil characteristics of Grignon and Rambouillet sites, and atrazine adsorption coefficients (Kd)

Atrazine was applied on April 30, 1990 at Rambouillet and May 9, 1990 at Grignon. The rate of application on each plot was calculated from amounts of atrazine collected on glass sheets placed on the soil surface during the spraying. The mean amount of applied atrazine was 1091 ± 110 g a.i. ha$^{-1}$ at Rambouillet and 1021 ± 30 g a.i. ha$^{-1}$ at Grignon.

An average sample of the 0-10 cm soil layer was taken 7, 18, 36, 73, 163 and 350 days after application at Rambouillet, and at day 2, 14, 30, 62, 150 and 352 at Grignon. Sampling depth for the soil profiles were 0-10, 10-20, 20-40 and 40-80 cm respectively. Until analysis the moist samples were kept at -20°C in glass containers.

At a soil : solution ratio of 1 : 2 (weight : volume), triplicate samples were shaken for at least 16 hours with methanol at room temperature. The extracts were evaporated to dryness and the residues were redissolved in toluene, which contained metazachlor as an internal standard. Water samples were concentrated under reduced pressure with $C_{18}$-BondElut cartridges after their activation by prewashing with 5 ml of methanol and 5 ml of water. The cartridges were dried with a flow of air for 15 min and eluted with 5 ml of methanol.

Atrazine and its dealkyl metabolites (deethylatrazine and deisopropylatrazine) were measured by gas chromatography with thermoionic detection (Carlo Erba Ins. HRGC 5300). A fused silica capillary column (DB5, 30 m x 0.32 mm i.d., 0.25 μm film thickness, J.&W. Scientific) was used with He carrier gas at 1.7 ml min$^{-1}$.

Kd of atrazine adsorption was measured in batch. 5 g of soil samples were suspended in 10 ml of a solution of 10 mg l$^{-1}$ $^{14}$C-atrazine in 0.01 M $CaCl_2$. After 24 h equilibration by shacking at 25°C, supernatant was recovered by centrifugation and the atrazine content measured.

Dissipation kinetics and distribution in the profile were simulated with three numerical models: LEACHP 3.1 (Hutson and Wagenet, 1992), PRZM 2 (Mullins et al., 1993) and VARLEACH (Walker 1993). The main difference between these models is their description of water and solute transport (Calvet, 1994). An analysis using climatic conditions of Ile de France from 1990-1991 has shown, that the models are highly sensitive to two input parameters: the linear sorption coefficient Kd and the degradation half-life $T_{1/2}$. Accordingly, simulations were run with various values for these two parameters, taking constant the other ones which have been either measured or estimated from literature. To take into account the uncertainty due to spatial variability, Kd values were allowed to vary æ 30% around average Kd values. No laboratory $T_{1/2}$ values were available for the corresponding soil so that a range of values between 30 and 90 days was taken from published data. The extreme results of the different simulation runs and the average result were taken to compare simulations to field observations.

# Results and Discussion

Dissipation kinetics (figure 1) show that field dissipation half-life of atrazine is greater at Rambouillet (104 ± 18 days) than at Grignon (73 ± 22 days). This result could be related to the total microbial biomass which is greater at Grignon than at Rambouillet, respectively 346 and 166 mg of C per kg of dry soil in the upper horizon (Cadot, 1993).

Kinetics of atrazine dealkyl derivatives formation are given in figure 1. At Grignon, residues of these metabolites increased over approximately 14 days after application and then decreased continuously. At Rambouillet, on the contrary, dealkyl metabolites increased during the first 70 days. In both sites the amount of deisopropylatrazine was higher than those of deethylatrazine, mainly at Grignon. In laboratory experiments usually more deethylatrazine than deisopropylatrazine is found because the ethyl chain is lost faster than the isopropyl chain (Nair and Schnoor, 1992; Brambilla et al., 1993). However, some microorganisms dealkylated atrazine with preferential formation of deisopropylatrazine (Behki and Khan, 1986).

The dealkyl derivatives allow to estimate the importance of biological processes in the atrazine dissipation. Dissipation without dealkyl metabolites appearance could mean that degradation processes were not biological. Differences between the concentration of dealkyl derivatives in the two sites could indicate a less important biological degradation process at Rambouillet as compared to Grignon. This may be interpreted in terms of different degradation mechanisms at the two sites.

From the distribution profiles (Fig. 1) it can be seen, that atrazine was two to three times more persistent in the deep layers of Rambouillet soil than at Grignon. The longer persistence can lead to important losses to the water table in times of a positive water balance. The maximum amount of atrazine was found in layer 0-10 cm at Rambouillet, and in layer 10-20 cm at Grignon. The distribution of the dealkyl metabolites in soil profiles was comparable to that of atrazine (Fig. 1). At Grignon, below 40 cm, only deethylatrazine was detected.

Deisopropylatrazine content was higher than deethylatrazine in Rambouillet plot, but it was not detected below 40 cm. Evolution of atrazine metabolites at Grignon, could be interpreted as the result of an important biological degradation with a preferential migration of deethylatrazine. The higher migration of deethylatrazine as compared to that of atrazine may be related to its smaller adsorption in soils (Schiavon, 1988).

In the water samples from Rambouillet atrazine concentration was very variable, with high concentrations at the beginning of the drainage period, probably due to some piston-flow effects as suggested by the transfer characteristics of $^{18}O$ (Coulomb, 1992). Deethylatrazine was found in all samples in similar concentrations as atrazine. Concentrations of deisopropylatrazine were smaller than those of deethylatrazine, which is classically reported (Thurman et al., 1991; Squillace et al., 1993). Deisopropylatrazine was only measurable during the second drainage period. These observations may be related to the adsorption order: deethylatrazine < atrazine < deisopropylatrazine < hydroxyatrazine (Schiavon, 1988).

The models used in this study did not allow to simulate correctly the dissipation and the transport of atrazine. Simulated results appear to be very sensitive to Kd and $T_{1/2}$ input values. Although quite a large range of values was used to simulate atrazine behaviour, simulations were not satisfactory. Also all three models were not able to predict atrazine in the drainage water. A partial explanation could be found in the fact, that these models were not developed to account for macro- and bypass flow. Concerning the poor simulation of dissipation kinetics, no general conclusions can be drawn because the models were run only with one set of climatic data. Several pedoclimatic situations must be studied to get a better view on model performances.

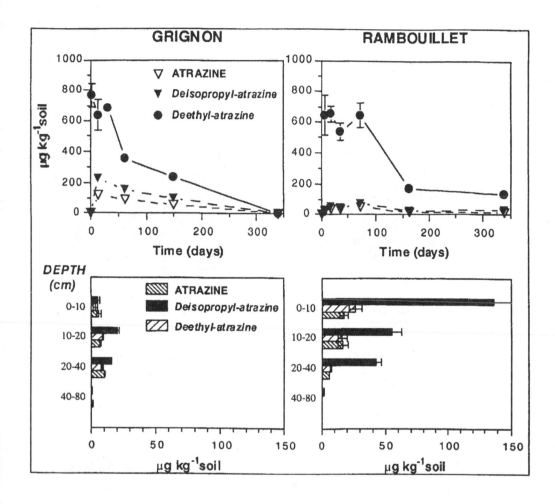

**Figure 1 :** Dissipation kinetics and distribution profiles after 12 months of application of atrazine and their dealkyl derivatives at Grignon and Rambouillet sites

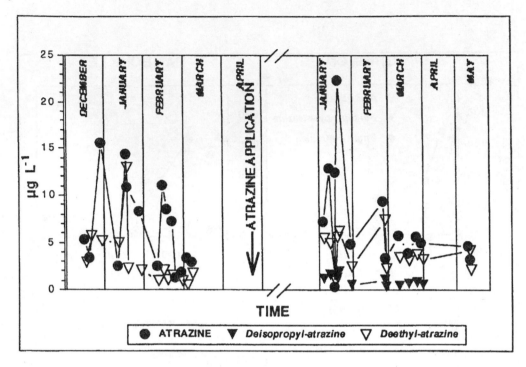

**Figure 2 :** Content of atrazine and their dealkyl derivatives in water
samples of the Rambouillet site

## Acknowledgements

The authors would like to thank Th. Lasnier and J.N. Rampon for their technical assistance. The simulation part of this work was carried out with support from EC-DGXII Environment Research Programme EV5V-CT92-0226.

## References

☐    Behki R. M. and S. U. Khan, 1986. *J. Agric. Food Chem.* 34, 746-749.

☐    Brambilla A., B. Rindone, S. Polesello, S. Galassi and R. Balestrini, 1993. *Sci. Total Environ.* 132, 339-348.

☐    Cadot L., 1993. *Impact de facteurs pédoclimatiques sur la biodégradation de boues résiduaires de station d'épuration dans deux sols. Modélisation et simulation.* Thesis of Institut National Agronomique Paris-Grignon.

☐    Coulomb C., 1992. *Etude de la circulation de l'eau dans un sol argileux drainé. Approches hydrodynamique, isotopique et geochimique.* Thesis of Paris XI Orsay University.

◻ Hutson J. L. and R. J. Wagenet 1992. LEACHM User manual version 3, Dept. of Soil, Crop and Atmospheric Sciences Research Series 92-3, Cornell University, Ithaca NY

◻ Mullins J. A. et al., 1993. PRZM 2 release 1.1 User manual, Environmental Research Laboratory, US EPA, Athens GA 30605-2720

◻ Nair D. R. and J. L. Schnoor, 1992. *Environ. Sci. Technol.* 26, 2298-2300.

◻ Schiavon M., 1988. *Ecotox. Environ. Safety* 15, 55-61.

◻ Squillace P. J., E. M. Thurman and E. T. Furlong, 1993. *Water Resourc. Res.* 29, 1719-1729.

◻ Thurman E.M., D. A. Goolsby, M. T. Meyer and D. W. Kolpin, 1991. *Environ. Sci. Technol.* 25, 1794-1796.

◻ Walker A. 1993. (pers. comm.)

*5.3.2.*        *Leaching of Pesticides in Swedish Soils Measured in Field Lysimeters*

# L. Bergstrom

Swedish University of Agricultural Sciences, P.O.B. 7072, S-75007 Uppsala, Sweden

## Introduction

To characterize leachability of pesticides in soil as a basis for regulatory decisions, results obtained in short-term laboratory leaching tests are commonly used, along with data on physico/chemical properties of pesticides (Tooby & Marsden, 1991). Such leaching tests are mostly performed under saturated steady-state flow conditions and are therefore not typical of natural field situations. In contrast, outdoor lysimeter experiments, which are normally conducted with unsaturated soil over much longer periods, have demonstrated that they can simulate actual field conditions quite well with respect to leaching of pesticides in soil (Kubiak et al., 1988).

In the following presentation, a selection of Swedish pesticide leaching studies carried out in monolith lysimeters are summarized. Attention has been focused on illustrating the advantages and limitations of such studies with regard to mobility in soil and their usefulness for regulatory decisions.

## Materials and Methods

All the results presented here are based on measurements performed in 0.3-m diameter and 1-m deep monoliths, which were collected using a coring technique described by Persson & Bergstrom (1991). With this method a standard PVC sewer pipe is gently pressed into the soil by a steel cylinder, equipped with four cutting teeth, which rotates slowly around the pipe as it penetrates the soil. After collection, the soil cores were transported to a lysimeter station in Uppsala, equipped for collection of leachate.

Treatments have included three soils ranging from loamy sand/sand to clay, "average" and "worst-case" precipitation, and normal and double the normal application rates of spring applied pesticides (see Table 1), i.e. bentazon, chlorsulfuron, clopyralid, dichlorprop, fluroxypyr, and metsulfuron methyl. All lysimeters were cropped with barley, grown according to normal agricultural practices.

## Results and Discussion

### *Influence of physico/chemical properties and degradation on pesticide leaching*

All the compounds included here are considered to be fairly mobile based on information obtained in laboratory tests (Table 2). However, when considering the amounts of the pesticides that actually leached out during periods between 7 and 11 months in the lysimeter studies, all seemed relatively non-leachable, i.e. <0.2 % leached out in percent of applied (Table 2). For example, no detectable concentrations of clopyralid were found in leachate, despite a listed $K_{oc}$-value of 4.6 mL/g (Woodburn & French, 1987) and a half-life of 15-87 days (Swann et al., 1976). Moreover, when comparing adsorption and degradation of bentazon and dichlorprop (Table 2), both tested in the same soils, one would expect the former to leach at larger amounts. However, this was shown not to be the case (Table 2). One explanation to the problems involved in classifying pesticide leaching based only on physico/chemical properties and degradation is the fact that field studies usually take several forms of degradation into account (e.g. leaching, volatilization), whereas in most laboratory studies, losses only occur through degradation and thus overestimate the residence time of a compound in soil.

| Herbicide | Dose (g ai/ha) | Soil | Irrig. treat. | Prec. + Ir- rig. (mm) | Leachate (mm) | Leached mass (g ai/ha) |
|---|---|---|---|---|---|---|
| *Chlorsulfuron* | 4/8 | Loam[a]/sand[b] | A[c] | 555[d] | 156[d] | 0.001/0.013[d] |
| *Metsulfuron me- tyl* | 4/8 | Loam/sand | A | 447[e] | 86[e] | 0.000/0.005[d] |
| *Dichlorprop* | 1600 1600 | Sa loam/sand Clay/clay | A/WC[f] A/WC | 518/603[g] 518/603[g] | 79/142[g] 158/196[g] | 0.48/0.91[g] 3.22/0.26[g] |
| *Bentazon* | 600 600 | Sa loam/sand Clay/clay | A/WC A/WC | 583/619[d] 583/619[d] | 124/250[d] 268/280[d] | 0.00/0.10[d] 0.06/0.44[d] |
| *Clopyralid* | 120/240 | Loam/sand | A | 609[d] | 228[d] | 0.00/0.00[d] |
| *Fluroxypyr* | 188/375 | Loam/sand | A | 609[d] | 229[d] | 0.00/0.00[d] |

[a] Topsoil classification; [b] Subsoil classification; [c] A = "Average" precipitation; [d] 11-month period; [e] 7-month period: [f] WC = "Worst-case" precipitation; [g] 9-month period.

**Table 1 :** Soils, experimental treatments and selected results of some Swedish lysimeter studies

## *Influence of soil properties on leaching*

There are several studies showing that soil physical conditions may have a major impact on pesticide leaching which overshadow compound related properties (e.g. Bergstrom, 1994). In all the studies referred to in this summary, preferential flow processes related to soil structural features considerably influenced pesticide leaching. For example, in the study with bentazon, more than four times as much of the compound leached from the clay soil than the sandy soil (Table 1), which was taken as clear evidence of preferential flow.

An unexpected and consistent effect of irrigation treatment on dichlorprop concentrations in leachate occured in the structured clay soil (Table 1). In the clay monoliths with the "average" irrigation treatment, concentrations varied between 1 and 5 µg/L throughout the 9-month period, resulting in a total load of 3.22 g a.i./ha (Table 1). However, only two samples containing concentrations above the detection limit (>0.5 µg/L) were found in the "worst-case" watered clay monoliths. Again, preferential or bypass flow is the most likely explanation, since the dried-out topsoil of "average" watered monoliths may have allowed water flow in cracks to a much greater extent than in "worst-case" watered monoliths, thus moving some of the pesticides rapidly through the topsoil to the subsoil. Once the compound reched the subsoil, it would be largely protected against degradation and thus stored for later leaching. In laboratory incubation tests using the clay subsoil, no degradation was detected during the course of a 13-day experiment, whereas the dichlorprop half-life in the topsoil was only 4 days (Stenstrom, pers. comm.).

| Herbicide | K$_{oc}$ (mL/g) | DT50 | Water solubility (g/L) | Leached of appl. (%) |
|-----------|-----------------|------|------------------------|----------------------|
| Chlorsulfuron | 21.6-112[a] | 30-120[a] | 27.9 (pH 7)[b] | 0.02-0.16[c] |
| Metsulfuron methyl | 4.6-70[a] | 8-105[a] | 9.5 (pH 6.11)[b] | 0-0.06[c] |
| Dichlorprop | 20-25[d] | 1-4[e] | 0.35[b] | 0.03-0.20[f] |
| Bentazon | 13[g] | 12-20[e] | 0.5[b] | 0-0.07[f] |
| Clopyralid | 4.6[a] | 15-87[a] | 1.0[b] | 0[h] |
| Fluroxypyr | 51-81[a] | 7-55[a] | 0.9[b] 8.0[a] | 0[i] |

[a]Data from the manufacturer; [b]Worthing & Hance (1991); [c]Bergstrom (1990); [d]Ghorayshi & Bergstrom (1991); [e]John Stenstrom (pers. comm.); [f]Bergstrom & Jarvis (1992); [g]Gustafsson (1989); [h]Bergstrom et al. (1991); [i]Bergstrom et al. (1990).

**Table 2 :** Physico/chemical properties, degradation and leaching of some herbicides

## *The significance of analytical detection limits*

Analytical detection limits may have significance for the estimation of leaching loads based on concentrations measured in lysimeter leachate, especially since pesticide concentrations in leachate are commonly close to the detection limit. In the studies represented in this summary, dection limits ranged from 0.01 (metsulfuron methyl) to 1 (fluroxypyr) µg/L. If , for example, the ten times higher analytical detection limit was true for bentazon as for fluroxypyr (i.e. 1 instead of 0.1 µg/L), only one sample would have had a detectable concentration, giving measurable losses only in the clay soil when "worst-case" irrigation was applied. In contrast, if we had taken into account that the analytical detection limit in fact was lowered from 0.1 to 0.05 µg/L for most of the water samples collected later in the season (which was not considered in the estimate listed in Tables 1 and 2), considerably larger losses would have been estimated for bentazon. We must take such considerations into account when regulatory officials determine what is acceptable leaching losses of pesticides.

# *Conclusions*

From the studies discussed above, the following conclusions can be made:

- It is important to carry out environmental risk assessments for pesticides mainly based on results from field studies rather than laboratory tests. The latter are appropriate only as preliminary studies to indicate when further field tests are needed.

- Lysimeters offer an excellent experimental framework for studies of pesticide leaching. The value of lysimeters is that they not only allow the investigator to control all water movements through the soil but also allow manipulation of environmental factors as well.

- Leaching of pesticides in clay soils may be as large or even larger than in sandy soils, due to the occurrence of macropore flow in the former. Therefore, it cannot be safely assumed that measurements of leaching in sandy soils represent worstcase conditions.

- Wet years may not constitute a worst-case scenario if macropore flow exerts a significant influence on leaching.

■    It is very important to correctly evaluate the significance of measured concentrations and loads of pesticides, especially in the context of introducing arbitrary criteria for acceptable leaching losses.

# References

□    Bergstrom, L. (1990). Leaching of chlorsulfuron and metsulfuron methyl in three Swedish soils measured in field lysimeters. *J. Environ. Qual.* 19:701-706.

□    Bergstrom, L.F., McGibbon, A.S., Day, S.R. & Snel, M. 1990. Leaching potential and decomposition of fluroxypyr in Swedish soils under field conditions. *Pestic. Sci.* 29:405-417.

□    Bergstrom, L., McGibbon, A., Day, S. & Snel, M. 1991. Leaching potential and decomposition of clopyralid in Swedish soils under field conditions. *Environ. Toxicol. Chem.* 10:563-571.

□    Bergstrom, L.F. & Jarvis, N.J. 1992. Leaching of dichlorprop, bentazon and [36]Cl in undisturbed field lysimeters of different agricultural soils. *Weed Sci.* 41:251-261.

□    Bergstrom, L. 1994. Leaching of dichlorprop and nitrate in structured soils. *Environ. Poll* (in press).

□    Ghorayshi, M. & Bergstrom, L. 1991. Equilibrium studies of the adsorption of dichlorprop on three Swedish soil profiles. *Swedish J. agric. Res.* 21:157-163.

□    Gustafsson, K. 1989. Bentazone - an ecotoxicological evaluation. *Report from the Swedish National Chemicals Inspectorate*, Solna, Sweden.

□    Kubiak, R., Fuhr, F., Mittelstaedt, W., Hansper, N. & Steffens, W. 1988. Transferability of lysimeter results to actual field situations. *Weed Sci.* 36:514-518.

□    Persson, L. & Bergstrom, L. 1991. A drilling method for collection of undisturbed soil monoliths. *Soil Sci. Soc. Am. J.* 55:285-287.

□    Swann, R.L., Regoli, A.J., Comeaux, L.B. & Laskowski, D.A. 1976. DOWCO 290 (3,6-dichloropicolinic acid) aerobic soil degradation study. *Internal Report GH-C 910*. Dow Chemical Co., Midland, MI.

□    Tooby, T.E. & Marsden, P.K. 1991. Interpretation of environmental fate and behaviour data for regulatory purposes. *In:* Pesticides in Soils and Water. (Ed.) A. Walker. *BCPC Mono.* No. 47, pp. 3-10.

□    Woodburn, K.B. & French, B.W. 1987. A soil sorption/desorption study of clopyralid. *Internal Report GH-C 1873*. Dow Chemical Co., Midland, MI.

□    Worthing, C.R. & Hance, R.J. (Eds) 1991. The Pesticide Manual 9th Ed. Farnham, *The British Crop Protection Council,* 1141 pp.

5.3.3.       *Leaching Potential of Pesticides in Layered Soils in Intensive Plastic Covered Cultivation of the Almeria Region of Spain*

# F. J. Blanco[1], J. V. Cotterill[1], M. Fernandez-Perez[2], M. del Mar Socios-Viciana[2], E. Gonzalez-Pradas[2], R. M. Wilkins[1]

[1]   Department of Agricultural and Environmental Science, Newcastle University, Newcastle upon Tyne NE1 7RU, U.K.

[2]   Departamento de Quimica Inorganica, Universidad de Almeria, Cañada, 03071-Almeria, España

## Introduction

On the southern littoral of Spain there is intensive horticultural production based on plastic greenhouses. This production of high value vegetables has increased rapidly and is now very important to the province of Almeria, where 80% of Spanish greenhouse crops are grown in an area of 16,000 hectares.

Under the conditions prevailing below the plastic canopy insect and mite pests and crop pathogens have become important (Cabello and Canero, 1994) for a wide range of crops. These conditions, which are partway between those of enclosed greenhouses and of the open environment, encourage the pests but suppress natural control. This has led to the use, and perhaps overuse, of pesticides mainly for foliar applications but also applied to the soil. Due to the long growing season operated in the Almeria region (September to July) and the plastic cover pesticides will reach the soil and thus may present a potential leaching hazard considering the amounts of irrigation water used.

The crops are grown in a layered substrate laid upon the original soil surface as the native soil is of poor quality (Agencia de Medio Ambiente, 1984). A typical growing medium may comprise a layer of clay (20cm depth) on the natural soil surface followed by layers of peat (2cm) and then sand (10cm) (Perez de los Cobos, 1960). Surface irrigation with nutrients is employed using metering pumps (Casado and Carlos, 1990)

There is concern that the aquifers supplying irrigation water are becoming depleted and that these are becoming contaminated with pesticides (Bosch et al., 1991). Some evidence exists that pesticides (methiocarb sulphone, methiocarb, methomyl, butocarboxim, carbaryl) occur in the Almeria well waters at 0.01-0.5μg/L (Chiron et al., 1993).

## Objectives

The objective of this study is the preliminary evaluation of the potential to reduce leaching by the use of controlled release formulation technology. These formulations regulate (and often slow) the rate of availability of the pesticide, localizing it in the crop zone and reducing the amount accessible to leaching (Wilkins, 1990). This evaluation included the modelling of the behaviour of a reference pesticide in this unusual layered soil structure. The properties of this reference pesticide (diuron, selected due to availability of suitable formulations) relevant to each of the layers were determined and used in a computer model for layered soils. The prediction for mobility from this model was then supported by the leaching in a laboratory composite soil column representing a typical substrate structure in the greenhouse. An example of a more relevant pesticide (carbofuran) was also modelled using the layered soil computer program.

## Properties of the Individual Layers in the Growing Medium

The individual layers of the soil from the greenhouse were characterized in terms of their physical properties and sorption of diuron. As these materials are added artificially to the original surface each of the layers are individually mixed and can be considered uniform. Of these layers the peat is the most important: high in organic matter (about 70%), low pH and low bulk density, highly sorptive and high water holding capacity. These values were determined, following standard methods, and are shown in Table 1

## Model Used

*Chemical Movement in Layered Soils* (Nofziger D.L. and Hornsby A.G., University of Florida 1987, published as "A microcomputer-based management tool for chemical movement in soil" in Applied Agricultural Research, 1: 50-56, 1986.) The model estimates the location of the peak concentration of non-polar organic chemicals as they move through a soil in response to downward movement of water. The inputs for the model are listed in Tables 1 and 2 for each of the soil layers. The crop chosen was tomato. The water input, as irrigation, and evapotranspiration were taken as for normal agronomic practice in Almeria greenhouses. The sorption coefficients for diuron (and also for the second pesticide, carbofuran) were calculated from experimental data, but the half lives were taken from the literature as supplied in the software program. This assumed the same half life (328 days for diuron and 37 days for carbofuran) for all four layers.

| Layer | pH | % O.C. | Bulk density g/cm$^3$ | Field capacity % v/v | Wilt point % v/v | Water saturation % v/v |
|---|---|---|---|---|---|---|
| Sand | 9.6 | 0.02 | 1.9 | 5.8 | 1.8 | 40 |
| Peat | 3.5 | 41 | 0.2 | 77 | 77 | 128 |
| Clay soil | 8.1 | 0.2 | 1.4 | 38 | 21 | 52 |
| Original soil | 8.6 | 0.2 | 1.2 | 48 | 28 | 51 |

**Table 1 :** Characteristics of the four layers of the greenhouse soil

## Leaching Potential Simulations

The depth of pesticide movement is simulated for the layered soil structure for diuron and carbofuran in Fig 1. Diuron ($K_{oc}$: 380) is clearly held by the peat layer throughout the complete growing season (220 days), whereas carbofuran ($K_{oc}$: 29), with the same irrigation total penetrates the peat barrier (2cm thick).

The relative amounts of pesticide remaining in the soil depend on the half lives. Carbofuran degrades rapidly ($t_{1/2}$ 37 days) and the quantity that penetrates the peat layer is substantially reduced, although metabolites may be present. Continual applications of the pesticides may contribute to higher subsoil levels over the season.

|              | $K_d$ | r     |
|--------------|-------|-------|
| Sand         | 0.05  | 0.990 |
| Peat         | 286   | 0.994 |
| Clay soil    | 0.82  | 0.984 |
| Original soil | 1.4  | 0.996 |

**Table 2 :** Distribution coefficients and correlations
coefficients (r) for the sorption of diuron
on the four layers

| Column level       | Col. 1 | Col. 2 |
|--------------------|--------|--------|
| Sand (10 cm)       | 5.4    | 7.9    |
| Peat (2 cm)        | 90.1   | 86.4   |
| Clay-top (7 cm)    | 0.9    | 1.3    |
| Clay-middle        | 0      | 0      |
| Clay-bottom        | 0      | 0      |
| Original soil (20) | 0      | 0      |
| Total              | 96.4   | 96.1   |

**Table 3 :** Diuron (% of applied) recovered from each
layer of two soil columns, 1 and 2

## Layered Soil Column Experiment

*Objectives*: to compare the simulation with real data obtained using a laboratory soil column reproducing the layered structure. As the layers used in the greenhouse lack any normal soil horizon composition then the behaviour of a pesticide in the corresponding layered column should be a reasonable representation of the field.

*Column design*: The column was made from a PVC tube, 10cm diameter and 60cm in length. The layers were carefully packed according to the typical field depths (see Table 3). Duplicate columns were used. The soil was saturated with water from the bottom and eluted downwards with diuron (7.8 ppm) in water (3L) over 8 days at room temperature. This corresponded to a full season's irrigation and diuron applied at a very high applicance (30kg a.i./ha). Analysis of the eluate and of the soil following methanol extraction was by HPLC.

*Results*: The amounts (% of diuron applied) recovered from each of the layers of the two columns are shown in Table 3.

*Conclusions*: that diuron ($K_{oc}$: 380) is completely sorbed by the peat layer and that no penetration occurred throughout the equivalent of a season's irrigation, thus supporting the simulation.

# Discussion

This preliminary evaluation of the potential for leaching of pesticides for the layered soil conditions prevalent in the greenhouses of the Almeria region of Spain indicated that with diuron there was no leaching potential with a season's irrigation but with carbofuran penetration occurred at 80 days but the simulation indicated substantial degradation.

In comparing the computer model with the layered soil column there was agreement. This provided support for the use of the soil column to represent the field situation. Field studies are planned to provide further evidence for this assertion.

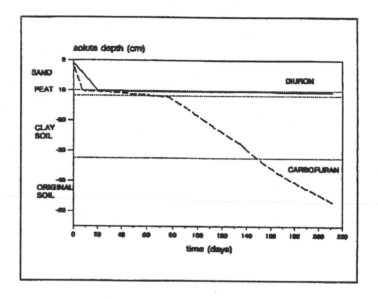

**Figure 1** : Depth of pesticide movement as function of time (simulated)

From evidence provided by residue analysis of groundwater (Chiron et al, 1993) it is confirmed that leaching has occurred. This simulation indicates that only compounds of low $K_{oc}$ values would leach, as expected. Alternatively, leaching of other compounds could be expected if the integrity of the peat layer was disrupted through channelling or root growth or through degradation of the peat. These aspects of the work are the subject of further studies.

## *Acknowledgements*

This work was supported by a grant (No 240 and HB-093) from the Accion Integrada of the British Council, Madrid and Ministero de Educacion y Ciencia, Madrid.

# References

☐ Aguera, A., Contreras, M. (1993) Gas chromatographic analysis of organophosphorus pesticides of horticultural concern. *J. Chromat. A*, 655, 293-300.

☐ Agencia del Medio Ambiente, Catalogo de Suelos de Andalucia, Junta de Andalucia, Sevilla, 1984.

☐ Bosch, A.P, Navarette, F., Molina, L., Martinezvidal, J.L. (1991) Quality and quantity of groundwater in the Campo de Dalias (Almeria, SE Spain). *Water Sci. Technol.*, 24, Ch 30, 87-96.

☐ Cabello, T., Canero, R. (1994) Technical efficiency of plant protection in Spanish greenhouses. *Crop Protection*, 13, 153-159.

☐ Casado, J.P., Carlos, A. (1990) Drip Irrigation demand Campo de Dalias, Almeria. *11th Intern. Cong.: Use of plastics in Agriculture*, Suppl. Vol., Ch 32, 53-59.

☐ Chiron, S., Fernandez Alba, A, Barcelo, D. (1993) Comparison of on-line solid phase disk extraction to liquid-liquid extraction for monitoring selected pesticides in environmental waters. *Environ. Sci. Technol.*, 27, 2352-2359.

☐ Perez de los Cobos, L. (1960) Enarenados en terrenas salinas. Acta IV Congreso Internacional de Riegos y Drenajes. Ministero de Agricultura, Madrid, p. 75.

☐ Wilkins, R.M. (1990) Controlled delivery of crop protection agents, Taylor and Francis, London, 320pp.

## 5.3.4.   Effect of soil N-supply on the fate and phytotoxicity of carbamide herbicides

# K. Bujtás

Research Institute for Soil Science and Agricultural Chemistry of the Hungarian Academy of Sciences
H-1022 Budapest, Herman O. út 15. Hungary

In Hungary and also in other Central Eastern European countries agricultural fields of highly different (low, optimal, or more than optimal) nutrient levels were formed in the past two decades. This was caused by application of build-up fertilization onto a great proportion of the arable land, resulting frequently in an overload of nutrients; while originally poorly supplied soils often remained in this status. An environmentally friendly agriculture demands a harmonized fertilization + pesticide application practice, also considering the specific conditions in the fields. Hence the question was raised: how the fate and effectivity of pesticides, especially of soil-applied herbicides are influenced by nutrient supply of soils of different physical, chemical and biological characteristics?

Altered N levels may directly or indirectly modify:

- soil pH and texture, affecting the adsorption capacity of the soil
- qualitative and quantitative composition of the soil microflora, thus altering biological decomposition
- growth vigour and mineral nutrition of crops, thus making them
  - more fit to withstand harmful effects, and/or in contrast
  - more susceptible to herbicides via enhanced uptake.

On the other hand, many herbicides may affect the N cycle in the soil, changing the soil microflora and biochemical characteristics, influencing not only biological decomposition of herbicides but also the processes involved in the fate of applied N.

A research project was initiated at our Institute to study adsorption, biological and chemical decomposition, effects on soil microbiological processes and phytotoxicity of carbamide herbicides (linuron as representative) with the aim to esteblish how these features depend on soil N supply and soil moisture levels, and also their interrelationships. 3 soils will be used in the experiments, representing soil types characteristic to Central Eastern Europe (Danube basin), and selected from the soil databank of our Institute. In estimating the role of adsorption to soil and of microbiological decomposition in the development of phytotoxic symptoms special emphasis will be placed on the possible interactions between soil N supply and herbicide effectivity.

Specific objectives of the project are:

(i)   to assess decomposition of the herbicide at several soil N supply and soil moisture levels in function of time and in a concentration range from low to toxic application levels, in a specially designed laboratory incubation experiment;

(ii)  to measure amount of microbial biomass and urease activity, and to determine soil toxicity using *Azotobacter chroococcum* as test organism, at different soil N supply;

(iii) to characterize phytotoxicity, in pot experiments, in function of soil N supply.

Further plans involve comparison of the above features on several soils. This should make possible some insight into the role of several soil physical and chemical parameters in the environmental fate of herbicides.

This research is supported by the Hungarian National Scientific Research Fund (OTKA) under grant No.T 006436.

5.3.5.     *Migration of Pesticides in Soil*

# B. Caussade[*], C. Thirriot[*], J.P. Calmon[**], C. Jean[*], G. Mrlina[**], R. Koreta[*] and A. Pinheiro[*]

[*]     INP-ENSSEIHT - Institut de Mécanique des Fluides URA CNRS D005,
        Avenue du Professeur Camille Soula, 31400 Toulouse France.
[**]    INP-ENSAT - Equipe d'Agrochimie, 145, Avenue de Muret,
        31076 Toulouse France.

## Introduction

Contamination of groundwater by chemical compounds due to industrial, agricultural and urban settings, through the unsaturated zone, is a major concern in the management of water resources because these contaminants degrade soils, aquifers and streams. Pesticides and fertilizers used in agriculture represent the most important part of nonpoint-sources pollution. Sometimes, important levels of traces of these substances in water can exceed the existing standard.

The case of herbicides like s-triazines (atrazine, simazine) is representative of such a pollution. Their fate in soil and water is treated under normal use (field doses, unlabelled molecules) with classical analytical methods.

The aim of this work is to present a multidisciplinary approach. Different spatial scales (lysimeter, field, watershed) are studied. Experiments carried out both in the laboratory (undisturbed lysimeters) and in the field, constitute a fondamental mean in the comprehension of the parameters incriminated for the pesticide migration in the soils. The watershed model developed represents today the most efficient tool in the management of non-point source pollution.

## Experimental Methods

### Cultivated field

The experimental site of Poucharramet (INP-ENSAT) is located 35 km from Toulouse, France. This site consists of four drained plots of about 1.1 ha (cultivated maize), and a wheather-measuring centre. S-triazines herbicides (atrazine and simazine) are applied each year as pre-emergent weed control agent on maize (1992 application rate = 0.75 kg/ha). The studied soil is an hydromorphous silty soil (aqualf) whose caracteristics are, belonging to the layer:

| Layer | Depth (cm) | Clay (%) | Silt (%) | Sand (%) |
|-------|-----------|----------|----------|----------|
| Ap | 0-30 | 13.4 | 55.6 | 31.0 |
| $A_2g$ | 35-55 | 22.1 | 51.9 | 26.0 |
| $B_tg$ | 35-55 | 40.5 | 38.8 | 20.7 |

Since 1990, systematic soil sampling at different depths are carried out and herbicide residues (atrazine, simazine and their N-dealkylated derivatives: D.E.A., D.I.A.) are determined. The residues in drained water are evaluated in instant water samples obtained at the end of the drain.

Out come-flows in the drain are continously measured by a neutronic probe. In order to measure the level of the residues in the soil water solution, one of the plots is equipped with six tensiometric canes, placed at 10, 20, 30, 40, 50 and 60 cm, and nine porous ceramic cups, placed at 30, 60 and 90 cm from the soil surface.

### Lysimeters study

Thirty undisturbed lysimeters (20cm in diam. and 75 cm high) have been extracted from the above-mentioned site. An experimental device has been constructed to carry out leaching experiments in laboratory. A spray mechanism simulates different rainfall series. Simulated rainfall and irrigation applied to all lysimeters are equal to the season's rainfall for the region and the year of interest. At the same time, each of the lysimeters received the same quantity of water. One of the lysimeters is equipped with eight capacitive probes placed horizontally at 3, 8, 15, 25, 35, 45, 55 and 67 cm from the soil surface. The gravimetric soil water content is continously recorded using an "Axone" Acquisition device. Thermocouples located at 10, 20, 30, 40, 50 and 60 cm measure the diurnal thermal profile. Leachate samples have been collected at the base of each lysimeter.

### Physical characteristics of the studied soil

During the sectionning of the lysimeters, undisturbed soil samples are taken for physical characterization. The layered soil profile is considered in this study. Soil bulk density, soil porosity and saturated hydraulic conductivity experiments were carried out for each layer. Unsaturated conductivity functions are obtained using Campbell's equation (Campbell, 1974). Soil moisture characteristic/retention curves are developped for each layer with the help of DTE 1000 tensiometer and capacitive probe.

### Pesticide application and analysis

To fit the soil moisture status of the field at pesticide application time, the previous rainfall series application time were simulated for all lysimeters. 10 ml solution of atrazine, simazine and KBr (used as a tracer), was uniformly applied to each lysimeter with a sprayer. The application rate of the pesticide was equivalent to $0.75$ kg ha$^{-1}$ and $1.28$ kg$^{-1}$ for Br. After herbicide application, rainfall and irrigation were simulated for 245 days. Throughout the duration of the study, the lysimeters are sectionned at 5, 10, 15, 20, 25, 30, 40, 50 and 60 cm. Ten soil samples for each lysimeter are then obtained and stored at -20°C until analysis.

Prior to analyses, subsamples of each level have been taken, thawed, air-dried, ground and sieved (2mm). Soil extraction was done with methanol according to Khan and Marriage (1977) and the extract was purified on alumina grade V columns (Ramsteiner et al., 1974). GC analyses were done with a HP5890 chromatograph equipped with a Nitrogen Phosphorus detector and confirmed with a mass detector HP 5971. Residue levels (atrazine, simazine and their N-deal-kylated metabolites) are calculated by the external standard method. The biological activity of the soil was estimated by measurement of the ability of 1g of soil diluted in a buffered medium (Schnürrer and Rosswall, 1982) to hydrolyse fluorosceine diacetate.

The residue analyses on the leachate samples are done on solid phase extracts of the water eluted with methanol.

## Results and discussion

### Cultivated field

The analyse of the soil moisture profiles showed two caracteristic zones: an upper profile (top 30cm) with rapid variation of soil humidity and a deeper zone (65 cm-90cm) with larger and stable soil water content (fig.1.a).

The distribution pattern of atrazine with soil depth (fig.1.b) showed some movement and distribution throughout the soil depth. The excessive season's rainfall (1992) had favoured an important leaching of the molecules to the deepest depth.

The biological activity of the soil microorganisms (fig.1.c) showed strong activity in the upper profile (top 30 cm), and in the intermediary depth (50-60 cm). The presence of DIA showed that biological degradation is one of the most important ways of degradation of these molecules throughout the different levels studied.

## *Lysimeters study*

Soil moisture characteristic/retention curves were determined for each layer (fig.2.a). Soil water content evolution during a rainfall simulation is presented (fig.2.b). A strong evolution is observed at the upper layer (0-35 cm) and nearly saturated conditions at the bottom of the lysimeter. The similarity observed between soil water profiles in the field and in the lysimeter (fig.2.c), proved that hydrous conditions in the field are well restored by the lysimeter.

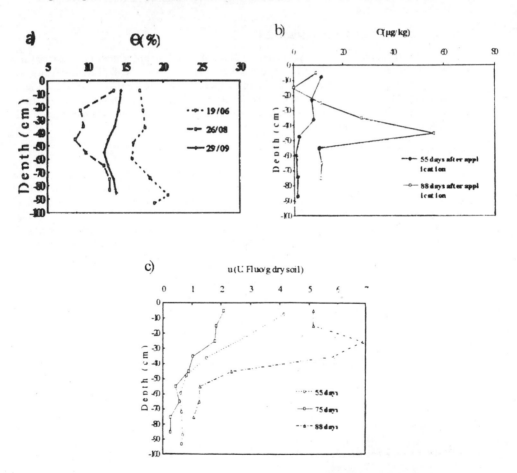

**Figure 1 :** Field measured profiles : a) gravimetric soil water content;
b) atrazine concentration and c) microbial activity of soil microorganisms.

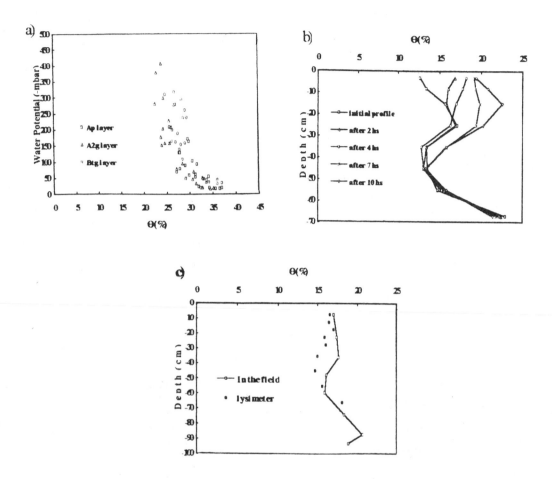

**Figure 2 :** Lysimeter measured profiles : a) tensiometric measurements;
b) evolution of soil water content profile during a rainfall simulation
and c) soil water profiles in the field and in the lysimeter.

After herbicide application, atrazine remained in the upper layer (top 5 cm) and is well conditionned by the rainfall simulated series. After 20 days, half of the quantity applied was found (fig.3.a). Rapid downward movement through macropores is caused by the preferential pathways in the lysimeters. As in the field, traces of DIA (fig.3.b) are found in the rootzone (10-20 cm) and more profoundly (55 cm). Leachate sample analyses will complete the actual results.

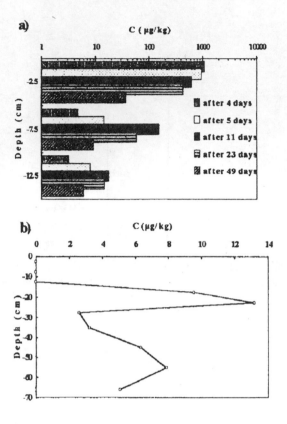

**Figure 3 :** Measured concentration in lysimeter : a) atrazine profiles and b) DIA profile.

## Modelling of Solute Transfer

Two approaches of solute transfer modelling (nitrate, phosphorus, pesticide) in the soil are considered. Firstly, mechanistic model LEACHM (Hutson and Wagenet, 1989) will be used to modelize the transfer of s-triazines herbicides and their metabolites in the lysimeters and in small cultivated field. Secondly, modelization at a large scale (watersheds of different sizes) is carried out based on the CEQUEAU hydrologic model (Morin et al, 1981). It is a conceptual model with distributed parameters which takes into account the variability of land use in the watershed. Solute transfer model, proposed in this study, is composed of three components: a) Hydrologic model (the major component of the model), b) Soil erosion model on agricultural land and c) Solute transfer model (Pinheiro and Caussade, 1994).

Watersheds of different sizes, situated in the south west of France, are studied: experimental site of Poucharramet (4.6 ha); Auradé watershed (322 ha) managed and sampled by the Agronomy Service of Grande Paroisse SA; Ruiné watershed (547 ha) managed and sampled by the Water Quality Division of CEMAGREF-Bordeaux; Save watershed (1130 km$^2$) and Charente watershed (4144 km$^2$) sampled by Agence de l'eau Adour-Garonne.

Tested on Poucharramet (for atrazine) and Auradé (for nitrate), the results given by the developed model are presented (fig.4). For atrazine, the exact time of occurence of the concentration peaks, as well as their magnitudes, are not well reproduced.

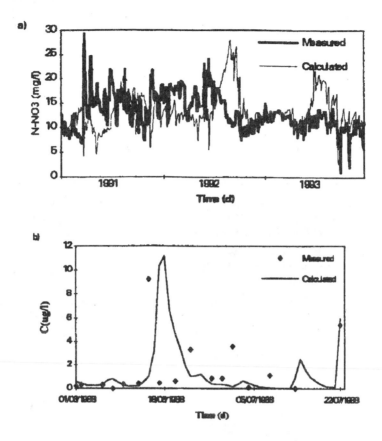

**Figure 4 :** Measured and calculated a) nitrate concentrations (Auradé watershed) and b) atrazine concentrations (Poucharramet).

## Conclusions

Experiments carried out both in the laboratory (undisturbed lysimeters) and in the field are presented. Residue herbicides (atrazine, simazine and their N-dealkylated derivatives: D.E.A. and D.I.A.) have been determined at different soil depths.

Our watershed model for the migration of pesticides and nutrient was tested at two sites. It takes into account satisfactorily solute transfer, but the model is still a developing tool.

In a near future, some conclusions will be drawn from the experimental results, to enrich the numerical models at the local and global scales.

# References

□ CAMPBELL G.S., 1974, A simple method for determining unsaturated conductivity from moisture retention data, *Soil Sci, v. 117, 311-314.*

□ HUTSON J.L. and WAGENET R.J., *1989, LEACHM,* leaching estimation and chemistry model, version 2, *Department of Soil, Crop and Atmospheric Sciences, Cornell University, Ithaca, NY, 148 p.*

□ KHAN S.U. and MARRIAGE P.B. 1977, *J. Agr. Food Chem. vol.25, 1408-1413.*

□ MORIN, G., FORTIN, J.P., LARDEAU, J.P., SOCHANSKA, W. and PAQUETTE, S. (1981), Modèle CEQUEAU: manuel d'utilisation. *INRS-Eau (Université de Québec), Rapport Scientifique n°. 93, 449 p.*

□ PINHEIRO A. and CAUSSADE B., 1994, A model to decribe the fate of agricultural pollutants, *in 1994 National Conference on Hydraulic Engineering, ASCE, Buffalo, NY*

□ RAMSTEINER K., HÖRMANN W.D. and OBERVILLE D.O., 1974, Multiresidue method for the determination of triazine herbicides in field grown agricultural crops, water and soils, *Agrochemicals Division, J. Ass.of Anal. Chem., v. 57, 192-201.*

□ SCHNÜRRER J., ROSSWALL T., 1982. Fluorescein diacetate hydrolysis as a measure of total microbial activity in soil and litter. *Appl. Environ. Microbiology, 43, 1256-1261.*

□ STARR J.L., GLOTFELTY D.E., 1990, Atrazine and bromide movement through a silt loam soil. *J. Environ. Qual. 19: 552-558.*

## 5.3.6.   Study of Atrazine, Simazine and Diuron Residues Mobility in Soil after Long Term Applications

# R. Deleu[1], A. Copin[1], M. Frankinet[2], L. Grévy[2] & R. Bulcke[3]

[1]   CRUPA "Effets secondaires", IRSIA, Fac. Sci. Agr. B-5030 Gembloux, Belgium
[2]   Station de Phytotechnie, C.R.A., B-5030 Gembloux, Belgium
[3]   Laboratory for Weed Research, Univ. Gent, B-9000 Gent, Belgium

Studies involving atrazine, simazine and diuron leaching and long term persistance have been carried out in two Belgian soils : silt loam ( only for atrazine ) and loam sand soil ( for the 3 herbicides ). The caracteristics of the 2 soils are :

|  | Gembloux Silt loam | Melle (Gent) Loam sand |
|---|---|---|
| Granulometry |  |  |
| $< 2 \, \mu m$ | 14.9 | 8.9 |
| 2 to 50 $\mu m$ | 80.6 | 9.9 |
| $> 50 \, \mu m$ | 4.5 | 81.2 |
| pH$_{water}$ | 7.14 | 6.31 |
| pH$_{KCl}$ | 6.47 | 5.39 |
| % C$_{org}$ | 0.79 | 1.27 |
| C.E.C. (meq/100 g) | 10.36 | 9.21 |

In the first soil (silt loam), atrazine is used, since 1981, to control weeds in a silage maize monoculture at the rate of 1 kg a.i./ha. Its residues do not leach below the plough layer (figure 1).

In the soil (loam sand) of an old bush apple plantation, atrazine is applied since 1961 at the rate of 4 kg a.i./ha, simazine 4 kg a.i./ha (since 1958) and diuron 2 kg a.i./ha (since 1960). For atrazine (figure 2) and diuron (figure 3), beyond 50 cm depth, it is not distinctively residues found above the detection limit. For simazine (figure 3), beyond 30 cm, no residue above 0.004 $\mu g/g$ was detected.

# Figure 1

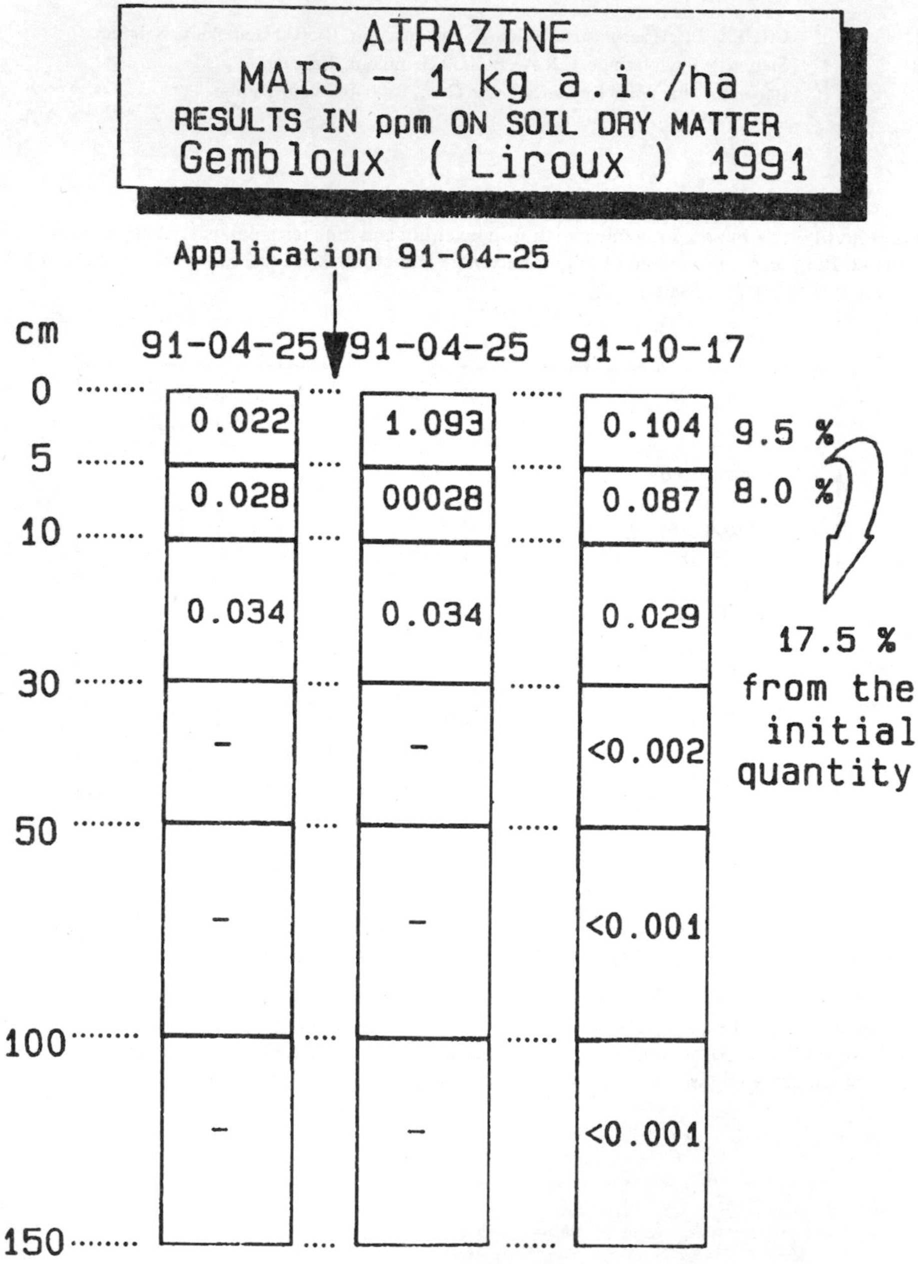

ATRAZINE
MAIS - 1 kg a.i./ha
RESULTS IN ppm ON SOIL DRY MATTER
Gembloux ( Liroux ) 1991

Application 91-04-25

| cm | 91-04-25 | 91-04-25 | 91-10-17 | |
|---|---|---|---|---|
| 0 | 0.022 | 1.093 | 0.104 | 9.5 % |
| 5 | 0.028 | 00028 | 0.087 | 8.0 % |
| 10 | 0.034 | 0.034 | 0.029 | 17.5 % |
| 30 | − | − | <0.002 | from the initial quantity |
| 50 | − | − | <0.001 | |
| 100 | − | − | <0.001 | |
| 150 | | | | |

# Figure 2

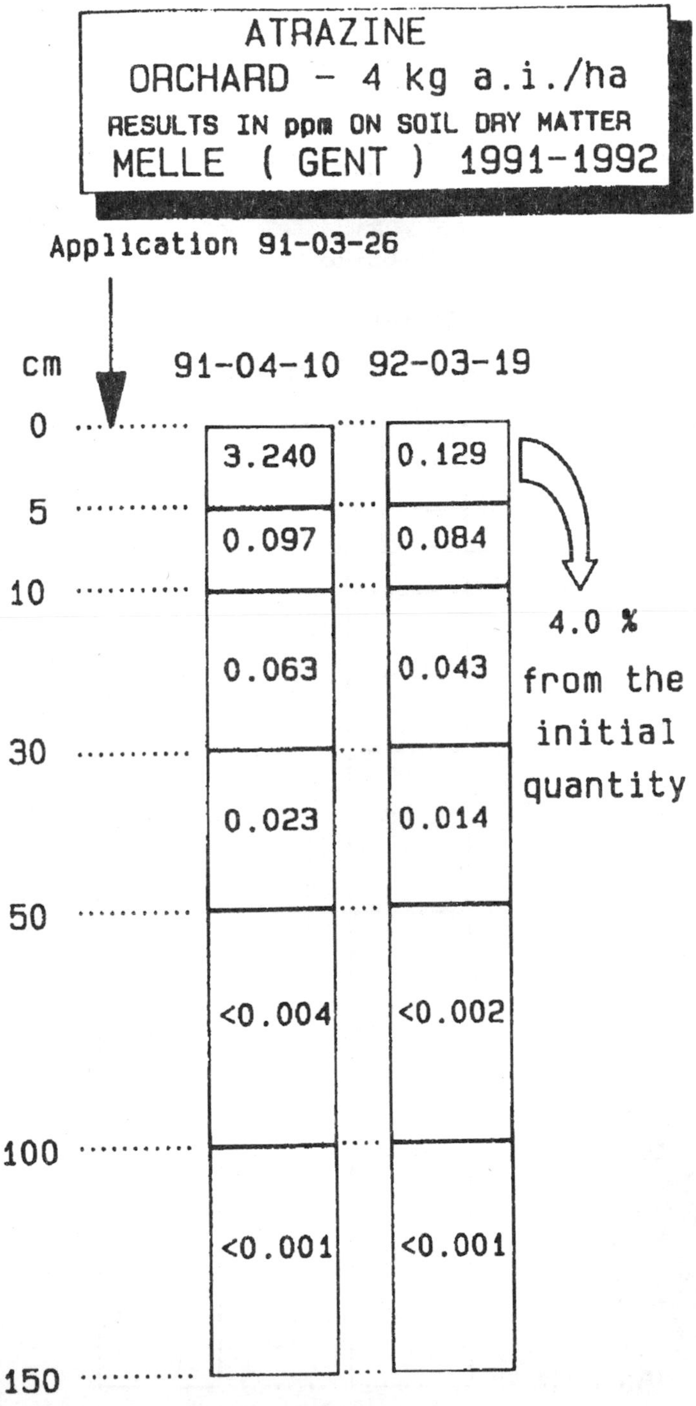

ATRAZINE
ORCHARD – 4 kg a.i./ha
RESULTS IN ppm ON SOIL DRY MATTER
MELLE ( GENT ) 1991–1992

Application 91-03-26

cm          91-04-10  92-03-19

0      3.240      0.129

5      0.097      0.084

10

       0.063      0.043                4.0 %
                                       from the
30                                     initial
                                       quantity
       0.023      0.014

50

       <0.004     <0.002

100

       <0.001     <0.001

150

# Figure 3

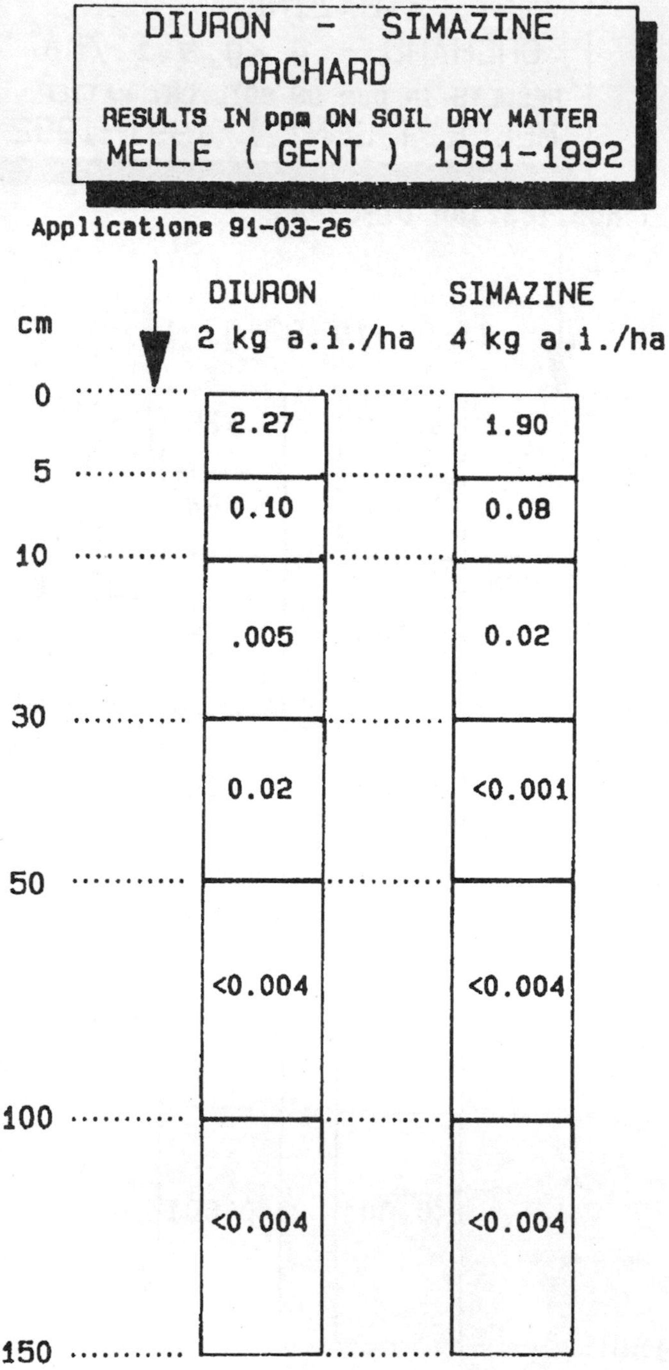

DIURON  -  SIMAZINE
ORCHARD
RESULTS IN ppm ON SOIL DRY MATTER
MELLE ( GENT ) 1991-1992

Applications 91-03-26

| cm | DIURON 2 kg a.i./ha | SIMAZINE 4 kg a.i./ha |
|---|---|---|
| 0 | | |
| | 2.27 | 1.90 |
| 5 | | |
| | 0.10 | 0.08 |
| 10 | | |
| | .005 | 0.02 |
| 30 | | |
| | 0.02 | <0.001 |
| 50 | | |
| | <0.004 | <0.004 |
| 100 | | |
| | <0.004 | <0.004 |
| 150 | | |

5.3.7.     Soil and Water Quality as Affected by Agrochemicals Under
           Different Soil Tillage Systems - Sorption and Transformation
           of Herbicides

## R.-A. Düring, S. Schütz, and H.E. Hummel

Biological and Biotechnical Plant Protection
Justus-Liebig-University Gießen
D-35390 Gießen, GERMANY

## Abstract

An EU-joint research project, started in autumn of 1993, will be presented, which investigates the influence of soil tillage on the behaviour of herbicides in the soil and their potential of surface- and subsurface-water pollution. At different sites in Portugal, Italy, and Germany, six research groups are determining the fate of agrochemicals in soil and runoff-water. Previous investigations, based on long-term tillage experiments, have proven several alterations in soil properties, depending on specific tillage practices (Tebrügge and Dreier, 1994): The amount and concentration of organic matter in the upper soil layer increase with decreasing tillage intensity. The same applies to biological activities. It is a question whether and how far the factors mentioned above affect the behaviour of herbicides in untilled soils compared to the conventionally tilled ones. Moreover, a higher bulk density associated with increased available water capacity or a stabilized pore system with continuous biopores in case of no-tillage have to be taken into account. Based on previous (Düring and Hummel, 1993) and preliminary results illustrated here, the investigations focus on herbicide degradation, adsorption/desorption, distribution in the soil horizon, and, with emphasis, translocation via run-off. Widely used herbicides with active ingredients like isoproturon, terbuthylazine, and metolachlor, including some of their main metabolites are investigated. The obtained data will enable us to determine the main pollution paths of agrochemicals under different tillage systems with respect to different climatic conditions in the temperate and mediterranean areas. Furthermore, the results should be transferred to larger areas to estimate alterations in agricultural practice. The aim is to give recommendations for good agricultural practice regulations with respect to drinking water protection within the EU.

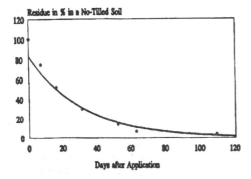 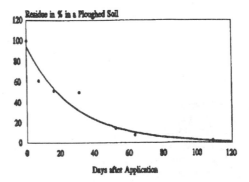

**Figure 1 :** Dissipation of isoproturon in the upper layer of a loamy soil.

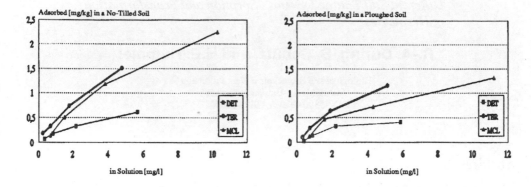

**Figure 2 :** Adsorption of herbicides and metabolite in a sandy soil; DET=deethyl-terbuthylazine, TER= terbuthylazine, MCL= metolachlor.

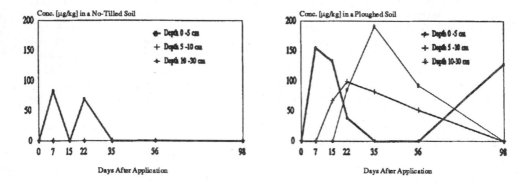

**Figure 3 :** Distribution of hydroxy-terbuthylazine in a sandy soil.

## First Results

The figures depicted above show more or less influence of different tillage practices on herbicide behaviour in the soil. The processes considered are dependent on each other, often in an antagonistic way. If adsorption is high, degradation may be slowed down and may appear to be linked with diminished translocation. On the other hand, high amounts of organic matter, favouring adsorption processes, will lead to accelerated degradation due to higher microbial activity. Thus, in this case, the potential for translocation via runoff or leaching should be reduced. The uncertainty about the role in the leaching process of macropores which are extremely influenced by soil tillage, cannot be clarified with methods used for obtaining the results above. However, interpretation of such data becomes more difficult when considering the various soil properties caused by more or less intensive tillage practices.

# Further investigations

The illustrated preliminary experiments have to be completed with comprehensive field and laboratory trials, concerning horizontal and vertical mobility of the herbicides in soil. In detail, runoff plots in the field with various precipitation intensity and leaching experiments using undisturbed soil columns which elucidate the influence of specific soil properties, will be carried out at the laboratory scale.

# Outlook

The transport parameters for herbicides derived from the field tests, have to be integrated into deterministic model approaches. With the aid of these models it will be possible to reproduce and predict pollution paths of agrochemicals under differentiated tillage systems at contrasting sites and under different climatic conditions within the EU.

# *References*

□  During, R.-A. and H.E. Hummel (1993): Soil Tillage as a Parameter Influencing the Fate of Three Selected Soil Herbicides.- *Med. Fac. Landbouww. Univ. Gent,* 58 (3a), 827-835.

□  Mollenhauer, K. and B. Ortmeier (1992): Zum Einfluß der konservierenden Bodenbearbeitung auf die Bodenerosion.- In: *Einführung von Verfahren der konservierenden Bodenbearbeitung in die Praxis.* FuE-Vorhaben, 4. Zwischenbericht.

□  Tebrügge, F. and M. Dreier (eds.) (1994): Beurteilung von Bodenbearbeitungssystemen hinsichtlich ihrer Arbeitseffekte und deren langfristigen Auswirkungen auf den Boden. *Wissenschaftlicher Fachverlag Dr. Fleck,* Niederkleen, 252 p.

□  Tebrügge, F. et al. (1994, in press): Advantages and Disadvantages of No-Tillage Compared to Conventional Plough Tillage.- In: *Proceedings of the 13. ISTRO Conference at Aalborg,* Denmark, 1994.

## 5.3.8.    Environmental Behaviour of Herbicides
### Atrazine Volatilization Study

# P. Foster, C. Ferrari, S. Turloni, R. Perraud

Groupe de Recherche sur l'Environnement et la Chimie Appliquée.
Université Joseph Fourier
39-41 Boulevard Gambetta, 38000 Grenoble, France

## Introduction

This study is a part of a project on environmental behaviour of atrazine and nitrates in corn culture. It takes place in La Côte Saint-André ( 40 km from Grenoble) in Isère.

This wide project started in 1990 in order to evaluate and understand the risks of agricultural pollutions and adjust cultural practicals. Atrazine is used in corn or sorgho cultures. After spreading, a part of atrazine will be consumed by weeds[1]. The rest will be mineralizated, volatilised, degradated or lixiviated by rain in phreatic underground water[2].

This work is divided in three parts. Volatilization depends on the support nature. Three supports are used : a soil from La Côte Saint-André, a powder of $TiO_2$, and a pyrex glass as a reference support. Atrazine surface concentration influences the kinetics of volatilization : three concentrations are tested on soil.

A comparative study between field measurement and laboratory simulated conditions is made. In this part, Atraphyt, a common agricultural mixture, is used.

## Materials and Methods

**Atrazine:** 97,8 % produced by Cluzeau 33220 Sainte Foix la Grande. **Soil** : from La Côte Saint-André.This soil is an aluminosilicate.

**$TiO_2$ powder (99%):** from Prolabo69102 Vaulx-en-Velin is used. Specific surface: $50 \ m^2/g$.

**Petri dishes** : $66,5 \ cm^2$ surface.

**Methanol** (99,8%) from Prolabo is previously distilled.

**Nebulisation Material** : An atomic absorption nebulisator is used. The nebulisation flow is regulated by an air flow. The nebulisator position is fixed for all the studies.

**Extraction procedure** : A sample of soil or $TiO_2$ (2 g) with atrazine sorbed is mechanically shaken for 15 minutes in a flask with 20 ml of methanol. After, the mixture is vaccum filtered. The solution is put in a 20 ml flask. An aliquot of $0,5 \ \mu l$ to $1 \ \mu l$ is injected in GC or $10 \ \mu l$ in HPLC. The extraction yield is about 98-99 %.

**Analytical Method** : Samples are analysed by Gas Chromatographic (with FID detector), GC/MS and HPLC.

**Irradiation system** : Three lamps of a total power of 100 Watts are used. The spectral field is included between the following wavelenghts: 350 nm to 700 nm. Those lamps are about 60 cm from the Petri dishes containing sorbed atrazine. The temperature is constant during the study ( $23,0 \pm 0,3$ )°C.

## Results and Discussion

### *Support nature influence*

A 2,50 kg/ha surfacic concentration is applied (atrazine in methanol). Three supports are used: soil, TiO$_2$, and pyrex. Atrazine sorbed on different supports is exposed to simulated solar radiations. The evolution of sorbed atrazine is studied during 24 hours (Figure 1). It appears that there are no significant differences between soil and pyrex. Zong Mao Chen, on a pyrex surface, for a 0,67 kg/ha surfacic concentration, has found a $(3,28\pm0,33)\times10^{-6}$ s$^{-1}$ volatilization kinetic constant[3]. For TiO$_2$, the percentage of remaining atrazine decreases quickly. Degradative compounds appears only in the case of TiO$_2$. Those compounds are deethylatrazine and deisopropylatrazine. Those compounds were found by E. Pelizzetti in photocatalytic degradation of atrazine in water with TiO$_2$ particules[4]. Degradation phenomenon represents a small part of total disappearence of atrazine. TiO$_2$ catalytic properties are well known. TiO$_2$ generate hydroxyl radicals when the powder is exposed to wavelenghts inferiors to 370 nm. P. C. Kearney has studied volatility of seven s-triazines on different supports[5]. He has shown a really high atrazine volatilization from metallic surfaces, as nickel for example. Kearney has studied atrazine volatilization on five different soils; he has found a significant volatilization in every cases, but with some differences due to the nature of the soils. Similar results have been found by C. L. Foy[6].

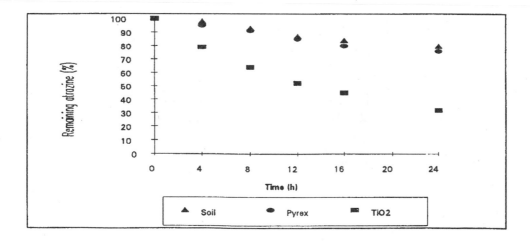

**Figure 1 :** Evolution of sorbed atrazine on different supports in fonction of irradiation time

In order to obtain kinetics constants, the hypothesis of a one order kinetics is verify. Those constants are given in Table 1.

|  | k $(s^{-1})$ | half life (h) |
|---|---|---|
| SOIL | $(2,28 \pm 0,04) \times 10^{-6}$ | $121 \pm 2$ |
| PYREX | $(2,86 \pm 0,04) \times 10^{-6}$ | $97 \pm 2$ |
| TiO$_2$ | $(1,38 \pm 0,04) \times 10^{-5}$ | $21 \pm 2$ |

**Table 1**

## *Influence of atrazine surface concentration*

In this study, three surface concentrations of atrazine are used : 0,83 kg/ha, 1,66 kg/ha and 2,50 kg/ha. Remaining atrazine is followed in fonction of irradiation time on soil (Figure 2).

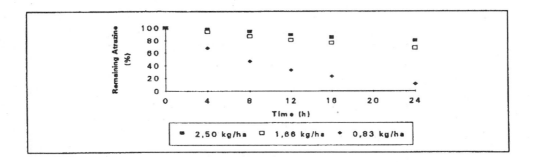

**Figure 2 :** Influence of atrazine surface concentration on volatilization kinetics.
Study on soil.

Kinetic parameters are given in Table 2.

| Surfacic concentration | 0,83 kg/ha | 1,66 kg/ha | 2,50 kg/ha |
|---|---|---|---|
| k $(s^{-1})$ | $(2,60 \pm 0,04) \times 10^{-5}$ | $(4,79 \pm 0,04) \times 10^{-6}$ | $(2,28 \pm 0,04) \times 10^{-6}$ |

**Table 2**

Zong Mao Chen has shown that kinetic parameters are different for different surface concentrations of pesticides. Thirty six pesticides were used. Our results are in good agreement with his results.

## *Comparative study between field measurement and simulated laboratory conditions*

Atraphyt, a common agricultural mixture, is used at 1 kg/ha concentration. In field, a solution made with 1 kg of atrazine diluted in 400 liters of water is spread on soil. Height tanks full of soil are aligned in the field every ten meters. After spreading, one tank is sampled every hour. The comparison between field measurement and laboratory conditions is given in Figure 4. Volatilization phenomenon appears in the first hour following the spreading. In fact, 10 % is volatilized after 1 hour and only 14 % after 7 hours.

Many authors have tried to evaluate atrazine volatility in field. Comparisons between their results are difficult because of the large range of employed methods[7, 8, 9].

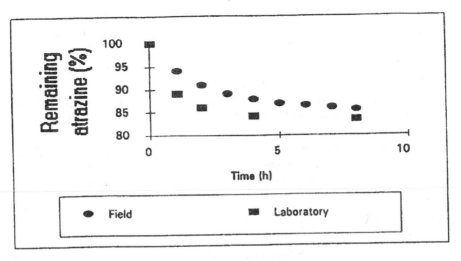

**Figure 3 :** Comparative study between field measurement and
simulated laboratory conditions

To explain this fact, water evaporation from soil after spreading is studied. A 400 l/ha water concentration is spread in the same conditions as before. The remaining water is weighed. A similitude appears between water and atrazine volatilization profiles. Atrazine would be carried up to the atmosphere by a water evaporation phenomenon. Water influence has been studied by several authors[3,8,10,11]. In wet soils, atrazine vapour pressure is higher because of water preferencial adsorption on soil sites. But real interactions have not been clearly defined up to date.

## *Conclusion*

Atrazine volatilization seems to be an important phenomenon, higher than theoretical data.

Volatilization depends on atrazine spreading quantities and soil nature.

To avoid this kind of pollution, new agricultural spreading practices must be investigated : the use of granulated pesticides for example.

# References

□   [1] H. GYSIN. Triazine herbicides; their chemistry, biological properties and mode of action. *Chemistry and industry*. 1962. p 1393.

□   [2] GRAHAM A. Jeffrey. Monitoring groudwater and well water for crop protection chemicals. *Analytical chemistry*. 1991. vol 63, n 11, p 613.

□   [3] CHEN ZONG Mao. Comparative study of thin film photodegradative rates for 36 pesticides. *Industrial english chemistry production residue development*. 1984. vol 23, p 5-11.

□   [4] PELIZZETTI Ezio. Photocatalytic degradation of atrazine and other s-triazine herbicides. *Environmental science and technology*. 1990. Vol 24, p1559.

□   [5] KEARNEY,T.SHEETS, and J.W. SMITH. Volatility of seven s-triazines. *Weeds*.1964. vol 12, p 83.

□   [6] C.L. FOY. Volatility and tracer studies with alkylamino-s-triazines. *Weeds*. 1964. vol 12 , p 103-108.

□   [7] GLOTFELTY E. DWIGHT. Volatilisation and wind erosion of soil surface applied atrazine, simazine, alachlore and toxaphen. *Journal of agricultural food chemistry*. 1989. vol 37, p 546-551.

□   [8] DORFLER et al. A laboratory model system for determining the volatility of pesticides from soil and plant surfaces. *Chemosphere*. 1991. vol 23, n 4, p 485-496.

□   [9] W.F. SPENCER, W.J. FARMER, and M.M. CLIATH. Pesticides volatilization. *Residue reviews*. 1973. vol 49, p 1-47.

□   [10] P. F. SANDERS et al. Measuring pesticide volatilization from small surface areas in field. *Bull. Environ. Contam. Toxicol.* 1985. vol 35, p 569-575.

□   [11] MAJEWSKY S.M. A field measurement of several methods for measuring pesticide evaporation rates from soil. *Environmental science and technology*. 1990. vol 24, p 1490-1497.

*5.3.9.*    *The Usefulness of Monolith Lysimeters for Monitoring Water and Herbicide Leaching*

# S. Issa, M. Ferrari, L. P. Simmonds, D. Barraclough & M. Wood

Department of Soil Science, University of Reading, UK

## Introduction

The risk of ground water contamination by pesticides is mainly determined by the pattern of water flow through the soil, and by the adsorption and degradation of the compound. These depend on environmental conditions (e.g. rainfall and temperature). Lysimeters are widely used to make assessments of the leaching behaviour of pesticides, though there are limitations to the generality of conclusions that can be deduced from the results of a single leaching experiment.

Furthermore, soil is notoriously variable with respect to hydraulic properties (Carter, 1991), which will influence the leaching of pesticides caused by downward water flow through either macropores (preferential flow) or through the soil matrix (Jardine, et al., 1990; Wilson et al., 1990).

The objectives of this study are (i) to illustrate the variability in breakthrough curves obtained from nominally replicate undisturbed lysimeters, and (ii) to present results from studies comparing the leaching of atrazine and isoproturon to demonstrate the usefulness and limitations of monolith lysimeters as devices for studying the environmental fate of applied chemicals.

## Materials and Methods

The 12 replicate monolith lysimeters (50.8 cm i.d. x 100 cm) used in this study were extracted from Rowland series soil at the University of Reading Farm, Sonning, UK in 1974, which is a weakly structured sandy loam, with evidence of continuous earthworm channels when the soil is undisturbed.

**Experiment 1:** 500 ml solution of KCl or KBr (1000 ppm) was applied uniformly onto the surface of each lysimeter. Atrazine or isoproturon was applied at a rate equivalent to 22.5 kg ha$^{-1}$. The volume of drainage water collected from each lysimeter was recorded and then subsamples were taken in glass bottles for the determining the concentrations of Cl, Br and the herbicides. Water inputs were from rainfall alone.

**Experiment 2:** The leaching of bromide and atrazine from two lysimeters was studied using a similar procedure to expt 1, except that irrigation was applied to supplement rainfall, and so provide higher peak infiltration rates.

## Results and Discussion

Figure 1 displays the cumulative precipitation and the cumulative drainage for all lysimeters used in experiment 1. There were small but significant differences between lysimeters in the volume of drainage water collected after each rainfall event, and in the total volume after 500 days (p=0.001). Differences were most apparent during heavy rain events following long, dry periods, implying that the variation was attributable to differences in the amount of drainage by bypass flow. Detailed monitoring of drainage using load-cells to record outflow from these lysimeters in experiment 2 showed that rapid drainage can start within minutes of the onset of intense rainfall or irrigation as reported by Quisenberry and Phillips (1976).

Interpretation of the chloride breakthrough curves (BTCs) is confounded by the high background concentrations present. Even so, there is evident variability between lysimeters in the Cl concentration detected during the initial period. The variations in the early part of the BTC is attributable to movement of Cl in the rapid water pathways (preferential pathways) which in turn suggests that the lysimeters vary significantly in their pore characteristics.

The lysimeters varied in the sharpness of the main peak of the BTC, implying variation in dispersivity caused by differences in pore characteristics. However, the peak was generally close to the nominal value for 1 pore volume of drainage, and most of the applied chloride (after accounting for the background levels) appeared to have been leached in the first 2 pore volumes of drainage. This implies that although bypass flow has occurred, causing some rapid displacement of solute, it is only a minor contributor to drainage.

Examples of the BTC's for bromide, atrazine and isoproturon through individual monolith lysimeters are illustrated in Figure 3. The rapid recovery of small amounts of atrazine and isoproturon in the early stages indicate bypass flow, which might cause a serious risk for groundwater contamination. Although poorly-structured sandy soils are not thought of as being susceptible to bypass flow, these results suggest that the visible earthworm and root channels may be responsible for the rapid movement of water and solutes in undisturbed soil. However, most of the solute leached was via matrix flow.

The two bromide BTCs suggest that the water and solute flow characteristics of the lysimeters used were similar. However, there was a striking contrast between the leaching pattern of atrazine and isoproturon, which has been related to differences in their adsorption and degradation characteristics.

Figure 4 shows the BTC for atrazine in experiment 2, where some of the first 1.8 pore volumes of drainage were generated by high intensity irrigations. The result was that there were small pulses of atrazine leached during the high intensity rain or irrigation periods. The main peak in the BTC (attributable to matrix flow) occurred between 2.2 and 3.8 pore volumes.

Subsequent analysis suggests that the appearance of early pulses of atrazine was associated with water inputs exceeding a threshold of between 1 and 3 mm infiltration per hour. This threshold will commonly be exceeded on occasions in the UK, though will account for only a small proportion of the annual rainfall. The variation between lysimeters in the extent of bypass flow generated by rainfall (Figure 2) might be expected because the threshold rainfall intensity is marginal for bypass flow to be anticipated under rainfed conditions.

# Conclusions

Monolith lysimeters taken from the same field for monitoring water and solute movement can produce apparently different BTC. The early breakthrough of herbicides suggested that preferential flow occurred. For the soil in question, bypass flow was generated by rainfall intensities exceeding 1-3 mm per hour. The intensity of water application was shown to have a dramatic influence on the BTC for halides and pesticides. Differences in the BTC obtained for herbicides added to matched lysimeters under the same conditions (i.e. producing the same BTC for bromide) can be attributed to the properties of the herbicide.

This work is funded by Ministry of Agriculture, Fisheries and Food, UK.

# *References*

□    Carter, A.D. 1991. Methods of monitoring soil water regimes and the interpretation of data
     relevant to pesticides fate and behaviour. p.143-150. In Walker, A. (ed) *Pesticides in soils and
     water: Current perspectives*. British Crop Protection Council, University of Warwick, UK.

□    Jardine, P.M., G.V. Wilson, and R.J. Luxmoore. 1990. Unsaturated solute transport through a
     forest soil during rain storm events. *Geoderma* 46:103-118.

□    Quisenberry, V.L., and R.E. Phillips. 1976. Percolation of surface applied water in the field.
     *Soil Sci. Soc. Am. J.* 40:484-489.

□    Wilson, G.V., P.M. Jardine, R.J. Luxmoore, and J.R. Jones. 1990. Hydrology of a forested
     hillslope during storm events. *Geoderma* 46:119-138.

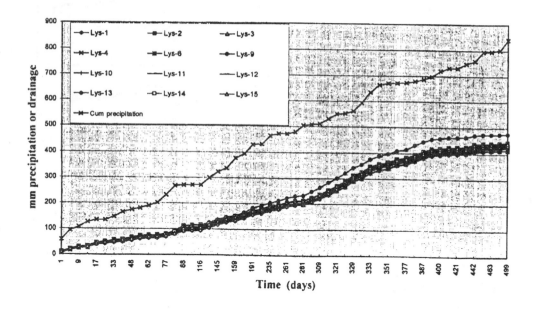

**Figure 1 :** Cumulative drainage water passed through each lysimeter,
and the accumulated measured precipitation

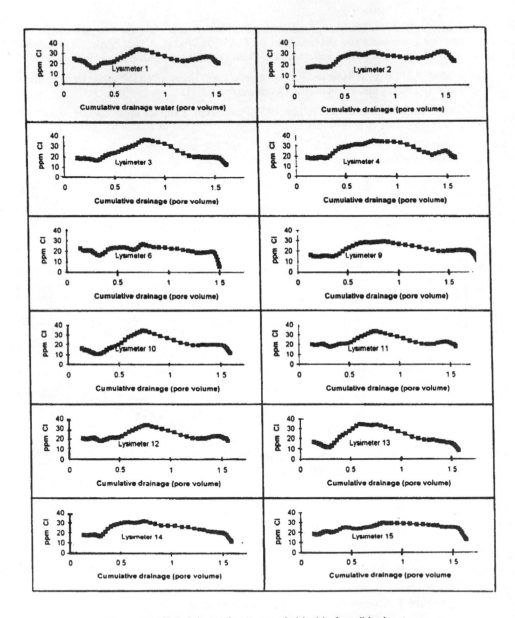

**Figure 2 :** Breakthrough curves of chloride for all lysimeters

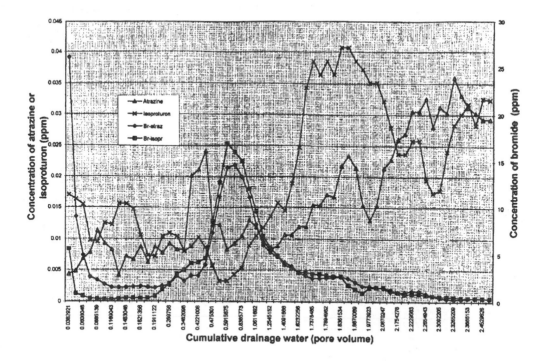

**Figure 3 :** Leaching of bromide, atrazine, and isoproturon
through monolith lysimeters

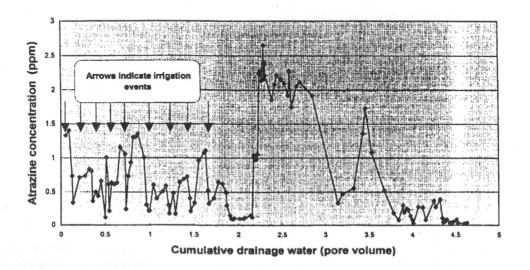

**Figure 4 :** Leaching of atrazine with intensive irrigation

*5.3.10.    The Scarab Project : Pesticide Application Regimes and Changes in Soil Microbial Biomass*

# S. E. Jones & D. B. Johnson

School of Biological Sciences, University of Wales
Bangor, Gwynedd, LL57 2UW, U.K.

## Introduction

The SCARAB project was established by the Ministry of Agriculture, Fisheries and Food (U.K.) in 1990 to investigate the environmental effects of pesticide application regime in different soil types and crop rotations. The duration of the experiment is seven years, incorporating a preliminary year for monitoring baseline faunal and floral differences, followed by one full six-year crop rotation. In relation to the soil microbiology component of the project, particular objectives are (i) to examine the size, activity and composition of the soil microbial biomass within different pesticide application regimes, and (ii) to determine the longevity of pesticide effects with pesticide application regime. This presentation highlights some short- and longer-term effects of pesticide inputs at current commercially-used concentrations at one participating Experimental Husbandry Farm (E.H.F.), High Mowthorpe E.H.F.

## Management Practices

In the first year of the project (1990), the whole field plots were treated identically. Thereafter, pesticides (crop-dependent choice) were applied under two regimes :

(a)    "current farm practice" (CFP), in which pesticides were applied at rates comparable to perceived general usage in similar farming situations;

(b)    "reduced input approach" (RIA), in which minimal levels of fungicide and herbicide were applied (usually half-rate), and zero insecticide application.

Crop rotations in Old Type field (High Mowthorpe) were : winter barley (1990; baseline year), spring beans (1991), winter wheat (1992), winter barley (1993) and oil seed rape (1994).

## Site Description

The experimental site at High Mowthorpe E.H.F. (North Yorkshire, England) was a 32 ha field, containing two sets (A and B) of paired CFP and RIA plots (each 100 x 150m). The soil was a silty clay loam over chalk, pH 7.0 - 7.9, with an organic carbon content of 2.8%. Soil samples were taken (6 samples per plot, 2 - 10 cm. depth) at two monthly intervals during the baseline year, and thereafter at times dictated by differential pesticide applications.

## Experimental Methods

Total soil microbial biomass was quantified using fumigation-extraction (Vance et al., 1987) and direct carbon determination (Ganf & Milburn, 1971) or determination of ninhydrin-reactive nitrogen (Amato & Ladd, 1988). Fungal biomass was estimated microscopically by methylene-blue staining (Sundman & Sivela, 1977). Viable soil bacteria were enumerated using R2A medium (Reasoner & Geldreich, 1985); fungal propagules were isolated and, where possible, identified to genus level, on a range of selective and semi-selective agar plate media.

## Results : Long-Term Effects

Microbial biomass comparisons in set A were not significantly different (P > 0.05) during the initial baseline year (data not presented). Only on one out of six occasions was a significant difference detected in set B, when significant differences were also found in soil moisture contents (data not presented). Comparisons between paired CFP and RIA plots over three successive crop years showed gradual divergences in total soil microbial biomass (Fig. 1, data for set A presented); other microbial parameters measured showed similar trends (data not presented). The greater fluctuation under CFP management became apparent in both sets of plots, and also at two experimental sites in one other E.H.F. (Gleadthorpe E.H.F., Nottinghamshire, England). Mean biomass indicators were generally lower in CFP plots compared with their RIA pairs, although the relative variability of samples (computed as coefficient of variation) remained similar. This diverging pattern in biomass estimates cannot be accounted for by inherent plot or sampling variation, and therefore is deemed to be a consequence of the differential pesticide inputs.

## Results : Short-Term Effects

Short-term effects of individual pesticide applications were detected in transient shifts in microbial community composition (Fig. 2). Following a fungicide (flusilazole + carbendazim + fenpropimorph) application at CFP and RIA rates, the proportion of soil microbial biomass present as fungal biomass declined in the CFP plot, compared with the RIA, plot. The total increment in biomass between pre- and post-sampling remained constant, which was interpreted as a seasonal, crop-related effect. Further evidence for short-term community shifts within CFP soil was shown in viable plate counts, where total bacterial counts rose (P < 0.001) and fungal propagules fell (P < 0.01; Kruskal-Wallis test). No significant differences were found (P > 0.05) in viable populations in pre- and post-fungicide application RIA soils.

## *Conclusions*

Changes in the incidence of particular groups of soil microorganisms were detected in short-term sampling around individual pesticide applications. Information obtained from these types of analyses aided interpretation of concurrent "total" biomass observations. However, the value of individual microbial enumeration for interpretation of long-term pesticide regime effects was curtailed by high inherent variation in seasonal populations. Long-term effects consequent on differential pesticide regimes appeared to be better described and interpreted by the use of "composite" microbial determinations such as estimation of total biomass.

## *References*

□   Amato M. & Ladd J.N. (1988) *Soil Biol, Biochem.* 20:107

□   Ganf G.G. & Milburn T.R. (1971) *Arch. Hydrobiol.* 69:1

□   Reasoner D.J. & Geldreich E.E. (1985) *Appl. Environ. Microbiol.* 49:1

□   Sundman V. & Sivela S. (1977) *Soil Biol. Biochem.* 10:399

□   Vance E.D. et al. (1987) *Soil Biol. Biochem.* 19:703.

## Acknowledgements

This work is funded by the Ministry of Agriculture, Fisheries and Food. The support of M.A.F.F. and A.D.A.S. (Agricultural Development & Advisory Service) personnel is gratefully acknowledged.

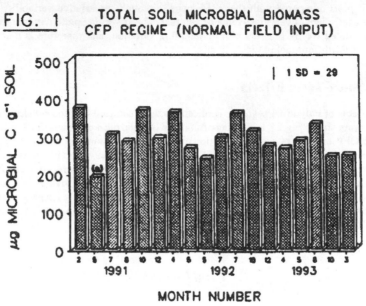

FIG. 1 — TOTAL SOIL MICROBIAL BIOMASS CFP REGIME (NORMAL FIELD INPUT)

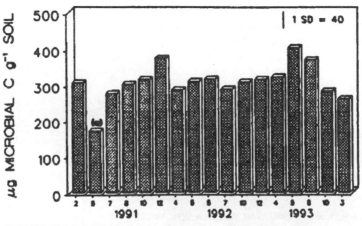

RIA REGIME (HALF RATE)

(a) - REDUCTION IN SOIL CARBON DUE TO HORIZON MIXING DURING DEEP PLOUGHING

## COMPOSITION OF SOIL MICROBIAL BIOMASS
## CFP REGIME (NORMAL FIELD INPUT)

FIG. 2

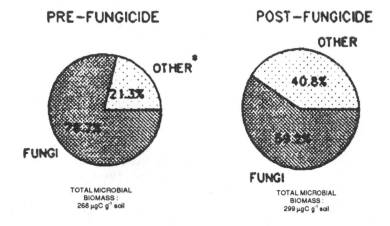

PRE-FUNGICIDE

OTHER*

21.3%

FUNGI

TOTAL MICROBIAL
BIOMASS :
268 µgC g⁻¹ soil

POST-FUNGICIDE

OTHER

40.8%

54.2%

FUNGI

TOTAL MICROBIAL
BIOMASS :
299 µgC g⁻¹ soil

## RIA REGIME (HALF RATE)

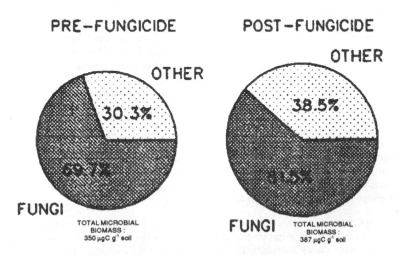

PRE-FUNGICIDE

OTHER

30.3%

69.7%

FUNGI

TOTAL MICROBIAL
BIOMASS :
350 µgC g⁻¹ soil

POST-FUNGICIDE

OTHER

38.5%

61.5%

FUNGI

TOTAL MICROBIAL
BIOMASS :
387 µgC g⁻¹ soil

* - OTHER = BACTERIAL, PROTOZOAL & ALGAL BIOMASS
METHODS : TOTAL MICROBIAL BIOMASS BY FUMIGATION-EXTRACTION
FUNGAL BIOMASS BY MICROSCOPY
OTHERS BY SUBTRACTION

## 5.3.11.    Dynamics of precipitations and evolution of the herbicide isoproturon in the toposoil

# C. Perrin-Ganier[1], M. Schiavon[1], J-M. Portal[2], C. Breuzin[3], M. Babut[3]

[1]    ENSAIA, BP 172 - 54505 Vandoeuvre-lès-Nancy, France.
[2]    CNRS - CPB, BP 5 - 54501 Vandoeuvre-lès-Nancy, France.
[3]    Agence de l'Eau Rhin-Meuse, BP 19 - 57160 Moulins-lès-Metz, France.

## Abstract

After a treatment with isoproturon on outdoor microlysimeters containing the rise level of a loamy soil, the amounts of herbicide recovered in percolates were examined for 6 months and correlations with climatic data were investigated. Simultaneously, microlysimeters were used to determine at intervals : (i) the proportion of free residues (by exhaustive extractions with organic solvents), (ii) the pesticide distribution in the differents compartments of the soil (non humified organic matter after physical separation ; humins, fulvic and humic acids after chemical separation). Particular attention was turned to climatic parameters interfering with pesticide behaviour : quantity and intensity of precipitations, potential evaporation, temperature and humidity.

## Key words

*isoproturon, microlysimeter, non-extractable residues, leaching, climatic parameters*

## Introduction

Isoproturon, a phenylurea herbicide, reacts with soil components. Reactions occur in particular with organic matter (Gaillardon et al., 1978 ; Barthelemy, 1987 ; Pussemier, 1978). Kumar et al. (1987) also brough into relief an adsorption of isoproturon on clay minerals through cations. As a result of adsorption phenomenon, the herbicide is partially kept in the soil. After rains, the other part of residues can also leach and be found in drainage or groundwater (Boesten, 1987). Besides, the distribution of residues in soil, and their concentration in soil-water, is modified with time after an herbicide treatment.

Researches conducted in order to define pollutant distribution in soil compartments have a great interest. Such information can be obtained by experiments using soil columns or lysimeters (Kubiak et al., 1988), radiolabelling technics (Führ, 1985) and outside conditions during several months (Agneessens et al., 1981). Free residues are collected in percolates and the adsorbed part can be determined by extraction proceeding.

Our experiment was carried out with microlysimeters filled with the rise level of an agricultural loamy soil and treated with isoproturon. The evolution of herbicide residues was followed for 6 months, while climatic data were recorded.

# Material and methods

## *Outdoor microlysimeters experiment*

Microlysimeters (5 cm high) were constituted with 120 g of the rise level of a loamy agricultural soil (table 1). The soil was air dried, homogenized and sieved between 1 and 5 mm. The columns were placed on funnels assigned to collect percolates in polyethylen bottles .

Each column was treated with 0.5 mg of isoproturon (N-(isopropyl-4-phenyl-N')N'dimethylu-rea)) in solution. The treatment was equivalent to 1800 g ha$^{-1}$. The solution contained 3.5 μCi of $^{14}$C-ring-labelled isoproturon.

Microlysimeters were then placed in outdoor conditions. Percolates samples were collected after each rainy period. The climatic parameters recorded were : quantity and intensity of precipitations, potential evaporation, temperature and humidity.

| Depth (cm) | Clay (%) | Silt (%) | Sand (%) | pH$_{water}$ | C (%) |
|------------|----------|----------|----------|--------------|-------|
| 0-20 | 20.5 | 58.4 | 19.1 | 6.3 | 1.26 |

**Table 1** : Chemical and physical properties of the loamy soil

## *Extraction of herbicide residues from soil*

Extraction of the soil was performed after 0, 1.5, 3 and 6 months.

### *Extraction of soluble residues*

Easily extractable residues in the soil were determined after rotary shaking of the soil with 150 ml of CaCl$_2$ solution 0.01M for 16 h at 20°C. The soluble components were separated by centrifugation at 5000 g for 20 mn. The procedure was repeated with 100 ml of CH$_3$OH until no more radioactivity was recovered in the supernatant. Finaly, the soil pellet was air-dried.

### *Distribution of residues in soil*

Granulometric fractionation allowed the separation of humified fractions (superior to 50 mm). 5 g of the fraction inferior to 50 mm was shaken 4 times with 50 ml of NaOH-tetrasodium diphosphate solution (0.1N-1%) for 3 h. Alkaline-soluble compounds were separated by centrifugation (5000 g, 20 mn). Pellet residues corresponded to humins. Subsequently the supernatant was acidified to pH 1.5 with HCl 5N. After 12 h at 4°C, fulvic acids (FA) (soluble in acid medium) were separated from humic acids (HA) (coagulated) by centrifugation.

The isoproturon content of the solutions was determined by counting radioactivity. Herbicide residues in solid parts were determined by a combustion method at 950°C.

# Results and discussion

## *Dynamics of leaching*

### *Amounts of residues in leachates*

The first month after the herbicide treatment was particulary dry and no leachate was collected. The first efficient rains led to leachates with high radioactive content. The following rainfall events (up to time 3 months) were comparable to each other but provided leachates with continuously decreasing concentration (Fig. 1). So, up to 3 months, no correlation could be found between rainfall and leaching of residues. On the other hand, between 3 and 6 months, a good correlation (r=0.93) existed between amounts of residues in leachates and quantities of precipitation.

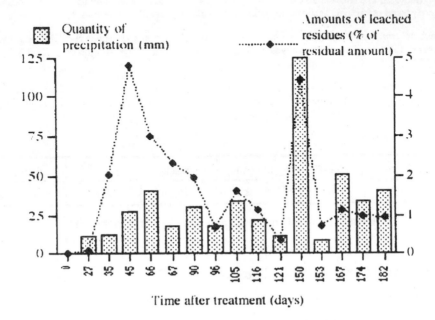

**Figure 1 :** Amounts of residues in leachates and quantities of precipitations during 6 months

*Correlation with climatic parameters*

Correlations were checked by comparing the climatic data with the amounts of $^{14}C$ in leachates. The quantity of precipitations was the only parameter being correlated. Duration or intensity of the rain showed no correlation with leaching.

<u>*Extractable residues*</u>

Water only extracts the residues which are able to be dissolved in soil solution or in leaching water. At the beginning of the experiment (time 0), about 30 % of $^{14}C$ was extracted by $CaCl_2$ (Fig. 2). This rate was 10 % after 1.5 months and was very low after 6 months. The decrease may be due to an increase of immobilization in soil and to the appearance of degradation products better retained in soil. Methanol is able to carry out the less desorbable residues, so the total amount extracted by water and methanol was near 100 % at time 0. The extraction rate decreased rapidly at the beginning, then slowly : extracted residues covered only 20 % of residual radioactivity remaining in soil after 6 months (Fig. 2).

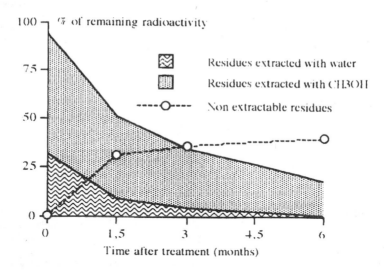

**Figure 2 :** Evolution of isoproturon residues in microlysimeters during 6 months

*Non-extractable residues (NER)*

During the experiment, non-extractable $^{14}$C increased first rapidly then slowly. Fixed residues in soil reached 40 % of the applied quantity after 6 months (Fig 2).

The affinity of every fraction of organic matter for isoprotuorn residues could be evaluated by:

$$\frac{relative^{(*)} \ non \ extractable \ radioactivity \ in \ the \ fraction \ (\%)}{relative^{(*)} \ carbon \ in \ the \ fraction \ (\%)} * 100$$

$^{(*)}$ : relative to the total amount in soil

| | FA | HA | Humins | Non humified |
|---|---|---|---|---|
| % of organic C | 32.4 | 11.6 | 55.3 | 11.3 |

**Table 2 :** Fractions of organic matter in the loamy soil (time 3 month)

| Affinity in :<br>Time | Humic acids<br>NER | A | Fulvic acids<br>NER | A | Humins<br>NER | A | Non-humified<br>NER | A |
|---|---|---|---|---|---|---|---|---|
| 1.5 months | 24.0 | 174 | 29.0 | 84 | 17.4 | 30 | 6.4 | 45 |
| 3 months | 27.9 | 240 | 32.3 | 100 | 21.4 | 39 | 7.1 | 63 |
| 6 months | 24.4 | 197 | 28.9 | 89 | 22.4 | 44 | 8.4 | 48 |

**Table 3 :** Distribution of relative* non-extractable radioactivity (NER) in the fractions
of organic matter and their affinity (A) for isoproturon residues.

Affinity = (relative* NER in the fraction / relative* carbon in the fraction )* 100

(* relative to the total amount in soil, at each time)

The non-humified fraction represented less than 15 % of the total organic carbon in soil (Table 2) but was significantly involved in the formation of bound residues. Unextractable residues regulary increased up to 8 % of the total immobilized $^{14}C$ (Table 3).

Fulvic acids (FA) and humic acids (HA) seemed to be highly reactive materiels and retained 24 to 32% of the non-extractable residues. The distribution between those two fractions was regular through time. The affinity of FA was inferior to that of HA (Table 3) but as they were twice more abundant, FA retained 5 % more residues than HA all along the experiment. Humic substances (FA + HA) took a greater part at time 3 months. The decrease which followed corresponded to an increase in humins.

The evolution of NER in humins was highly similar to the evolution of total NER in the soil. The part of NER increased in humins during 6 months. Although the humins fraction was not very reactive, its abundance in soil gave it an important role : it retained as much as 22 % of bound residues. Those residues can be considered as stable in regard to the stability of the humin compartment.

# *References*

□   Agneessens J.P., Gaspar S., Copin A., Deleu R., Dreze P. étude de la migration des herbicides en colonnes de sol. Proc. *EWRS Symp. Theory and practice of the use of soil applied herbicides*, 1981, 34-41.

□   Barthelemy J-P. Etude de l'adsorption et de la désorption du chlortoluron dans le sol. *Parasitica*, 1987, 43(2), 73-85.

□   Boesten J.J.T.I. Leaching of herbicides to ground water : a revue of important factors and of available measurements. *Proc. British Crop Protection Conference-Weeds*, 1987, 559-568.

□   Führ F. Lysimeters experiments with selected herbicides. *Pollutants and their ecotoxicological significance, Ed. H. W. Nünberg*, 1985, 363-372.

□   Gaillardon P., Calvet R., Rougetet E., Gaudry J.C. étude préliminaire du rôle de la nature des matières organiques dans les phénomènes d'adsorption des herbicides. *Ann. Agro.*, 1978, 3, 243-256.

□   Kubiak R., Führ F., Mittelstaedt W., Hansper M., Steffens W. Transferability of lysimeters results to actual field situations. *Weed Science*, 1988, 36, 514-518.

□   Kumar Y., Ghosh D., Agnihotri A.K. Adsorption of isoproturon on homoionic clays. *J. Indian Soc. Soil Sci.*, 1987, 35, 394-399.

□   Pussemier L. Interaction des pesticides avec la matière organique du sol. *Revue de l'agriculture*, 1978, 3(31), 405-408.

*5.3.12.*      *Reducing the Impact on the Environment of Agricultural Pesticides*

# B. Real[1], J. Masse[1], P. Gaillardon[2], M.P. Arlot[3], J.J Gril[4], V. Gouy[4]

1 : ITCF Institut Technique des Céréales et des Fourrages,
    8 Av du Président Wilson 75116 Paris, France

2 : INRA Institut National de Recherche Agronomique,
    Route de Saint-Cyr, 78026 Versailles, France

3 : CEMAGREF Centre National du Machinisme Agricole, du Génie Rural
    et des Eaux et For+ts, Parc de Tourvoie B.P. 121, 92185 Antony, France

4 : CEMAGREF Centre National du Machinisme Agricole, du Génie Rural
    et des Eaux et For+ts, 3 bis Quai Chauveau, 69336 Lyon, France

## Abstract

An experimental programme is described which aims to evaluate methods of reducing watercourse contamination by agricultural pesticides. Included are studies on pesticide transfer and associated mechanisms, the validation of models and the implementation of management practices. Pesticide transfer is measured by field scale experimentation involving the collection of drainage and runoff samples. In laboratory conditions, pesticide availability in soil solution is studied using glass microfibre filters, at soil moisture levels below field capacity. Potential pesticide mobility in runoff is estimated using rainfall simulation plots, which permit pesticide partition between solid and liquid phases to be followed in dynamic conditions for both active ingredients and formulations. The results of these various experiments are used to validate mathematical models. They are also useful to estimate the environmental impact of preventive methods, in particular improvements in agricultural practices (choice of pesticide, rate, timing, etc) and field management (grass buffer strips).

## Introduction

It is likely that neither restrictions on pesticide use nor the banning of some active ingredients will be sufficient to guarantee drinking water quality in accordance with European Community standards. In order to reconcile the demands of water quality and chemical protection of crops, the latter being essential to obtain produce of good quality, it is necessary to understand the mechanisms of pesticide transfer to surface water.

Within this context ITCF, INRA and CEMAGREF have started a research programme aimed at reducing the impact of pesticides on water quality. This study involves four main lines of research :

- the evaluation of the amounts of agricultural pesticides which may contaminate surface water by runoff or through the drainage system of cultivated land;
- understanding the physical, chemical and biological mechanisms of pesticide transfer, by studying the behaviour of a number of active ingredients in soil, soil solution and runoff;
- the validation and the adaptation of existing mathematical models describing pesticide behaviour in soil and transport towards watercourses, using the results of the experimental programme and which may also generate a data base;
- the adaptation of agricultural practices and testing of the effectiveness of grass buffer strips, to limit the transfer of pesticides in the environment.

All these concerns are brought together at the Water Quality Station of the "La Jaillière" experimental farm (Loire Atlantique/Maine et Loire), where plots provide outdoor experimental facilities at the field scale.

# Evaluation Of Pesticide Transfer To Surface Water

## *Experimental equipment*

The experimental station is located in the domain of the ITCF called "La Jaillière", situated between the cities of Angers (Maine et Loire) and Nantes (Loire Atlantique). The climate (temperate oceanic), the soils (hydromorphic sandy loams over schist) and the agricultural practices are representative of a large part of the north west quarter of France. The station includes 8 plots of 0.5-1 ha. Six are individually equipped with drainage systems and hydrologically isolated (Figure 1). Five plots are fitted with runoff collection facilities (Penel & Lesaffre, 1989). This kind of experiment, with simultaneous runoff and drainage measurement, is unique in France. The runoff collection system has been especially designed to analyse runoff flows and the amount of pesticides transferred. It consists in plastic gutters layed downslope of the plots. Drainage and runoff from each plot are carried through buried pipes to a recording station equipped with automatic samplers. The plots are also equipped with tensiometers and piezometers. Two crop rotations are being studied : 1) winter wheat/catch-crop Italian ryegrass/forage maize, 2) peas/winter wheat.

## *Measurements*

For the plots, agronomic, bioclimatic and hydraulic parameters are recorded. The pesticides contained in runoff and drainage samples are analysed.

— Agronomic records : crop husbandry, soil sections, the yields of crops and the amount of pesticides applied.

— Bioclimatic records : hourly rainfall and soil temperature.

— Hydraulic records : hourly soil water potential, watertable level, drainage and runoff flows (Dutertre, 1993).

— Water quality records : pesticide amounts in soil, drainage and runoff, either on a time basis (every eight hours), or on weekly cumulative samples.

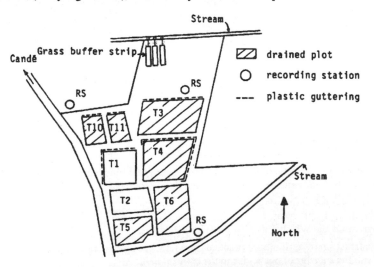

**Figure 1 :** General plan of the Water Quality Station

# Studying Mechanisms

## *Availability of pesticides in soil solution*

Availability of pesticides in soil solution may be evaluated in laboratory conditions (Gaillardon et al., 1991). Dry soil is treated with an aqueous solution of a $^{14}$C-labelled compound. At intervals, soil solution is sampled by means of glass microfibre filters, which retain soil water by capillarity when placed in contact with the soil. Water volume and pesticide amount in filters are determined by weighing and radioactivity counting, and concentration and percentage of pesticide in soil solution are calculated. To ensure that no degradation occurs, the soil is extracted with an appropriate solvent at the end of the experiment. The extract is concentrated and then analysed for AI and breakdown products by tlc. Measurements may also be conducted at low temperature to prevent degradation.

Measuring the amount of pesticide in soil solution allows adsorption and desorption studies of pesticides by unsaturated soils. For example, one day after treatment of a clay loam with 10μg of isoproturon or diuron per g of soil, at 29% soil moisture content, the soil solution contained 28% or 10% respectively of the applied herbicide and these percentages decreased to 23% or 5% respectively two weeks after treatment. Herbicides were rapidely adsorbed by soil initially but adsorption, inducing a decrease in herbicide concentration in soil solution, especially for the most adsorbed herbicide. Current studies with diuron show that adsorption increases when herbicide dose is reduced and might be influenced by soil moisture content. Qualitative data on the soil solution may also be gained and thus changes in pesticide (or metabolite) partition between liquid and solid phases of soil may be observed during degradation (Gaillardon & Sabar, 1993).

## *Pesticide transfer in runoff*

The partition of pesticide between water and suspended particles carried in runoff has been studied by means of a rainfall simulator (Gouy, 1993). An oscillating-nozzle rainfall simulator applied intense rainfall (70 mm/h for one hour) to the surface of a 1 m$^2$ plot filled with 10 cm depth of soil packed into a metal tray and inclined at an angle of 16%. the tray was constructed with troughs to collect runoff and direct it into a glass bottle. Four formulated pesticides were applied together : atrazine (wettable powder), simazine (wettable powder), lindane (soluble concentrate) and methidathion (soluble concentrate), at an application rate for each AI of 70 mg/m$^2$. Pesticides were surface applied on bare soil, 15 hours before rainfall simulation. Three different soils were used : 1) a loamy sand (Ardières, organic matter = 0.2%); 2) a sandy loam (Jaillière, organic matter = 2%) and 3) a calcareous silt loam (Charente, organic matter = 2%). In order to evaluate variation in adsorption coefficient on suspended particles during a storm, runoff was sampled at 5 minute intervals during the first 30 minutes of runoff and at 10 minute intervals thereafter.

A measure of total eroded particle load was made for each sample. Runoff samples were filtered through a 1.2 micron porosity fibre-glass filter. Pesticide content of suspended particles and filtered water were analysed by gas chromatography. Adsorption coefficients, Kd, were calculated for each time period. Kd = Cs/Cw, where Cs is the pesticide concentration in the soil or solid phase (μg/g) and Cw is the pesticide concentration in solution (μg/ml) at equilibrium.

Pesticide transfer into runoff was found to be highly dependant on the soil type. Using the Ardières soil, only 0.2 to 0.5% of total amount applied was carried off in runoff compared to 3 to 8% with the Jaillière soil and 8 to 30% with the soil of Charente. In all three cases, more than 70% of each pesticide was distributed in the water phase. Pesticide concentrations in solution in runoff decreased over time, as did the amount adsorbed per g of suspended particles and the adsorption coefficients Kd varied by several orders of magnitude during a storm. These values of Kd were much greater than values corresponding to standardised methods with pure AI's on soils. This can be seen with the Ardières soil, where mean suspended particle load was lower.

When relating Kd and suspended particle load for each time period, Kd decreased with particle load increase and this relationship could be described using a power function.

## The Validation of Models

Models are required to predict the risk of pesticide transfer to surface water. Most current models require validation and modification to simulate pesticide transfer. The models being studied consider pesticide interaction with soil and pesticide transfer through infiltration (INRA) (PRZM, Carsel et al, 1984), drainage (CEMAGREF, Antony) (SIDRA, Lesaffre & Zimmer, 1988) and runoff (CEMA-GREF, Lyon) (CREAMS, Knisel, 1980). The results of experiments at various scales (laboratory, rainfall simulation plot, lysimeter, field) are used to test and improve the modelling. One of the objectives of this study is to predict the influence of various agricultural pratices on pesticide transfer.

## Agricultural Management to Reduce Pesticide Transfer

At the begining of 1992, a study started examining the methodology necessary to evaluate the effectiveness of grass buffer strips to reduce pesticide transfer in runoff. The experimental system consists on three plots of wheat (5x25 m), located up-slope from a grass buffer strip 12 metres long. Each plot is hydrologically isolated. According to the position of the collection system, the effectiveness of 0, 6 or 12 metre grass buffer strips can be compared. During rain, total runoff on each plot is collected and analysed, measuring the total volume of runoff, the suspended particle load and the pesticide content of both liquid and solid phases.

## Future Work

A campaign of measurement is planned for autumn 1993. It will consist of testing the performance of the gutter system to collect runoff, testing the sampling timing for runoff, validating the methods of pesticide analysis and analysing the residues of atrazine, terbuthylazine, isoproturon, diflufenican, carbofuran, lindane, flusilazole and fluvalinate in runoff and drainage samples. With regard to the grass buffer strips study, a unique event occurred in spring 1993 and isoproturon and diflufenican transport was monitored. However, more information is necessary before conclusions can be drawn.

## *References*

□   Carsel, R.F.; Smith, C.N; Mulkey, L.A.; Dean, J.D.; Jowise, P. (1984) Users Manual for the Pesticide Root Zone Model (PRZM). Release 1, ERL, US EPA, PB85 158913.

□   Dutertre, A. (1993) L'expérimentation "drainage - qualité des eaux" de la Jaillière. *Acte des journées d'étude AFEID, AFGR, SNED "Hydraulique Agricole et Environnement"*, La Rochelle, Juin 1990, 59-64.

□   Gaillardon, P.; Fauconnet, F.; Jamet, P.; Soulas, G.; Calvet, R. (1991) Study of diuron in soil solution by means of a novel simple technique using glass microfibre filters. *Weed Research*, 31, 357-366.

□   Gaillardon, P.; Sabar, M. (1993) Changes in the concentrations of isoproturon and its degradation products in soil and soil solution during incubation at two temperatures. *Weed Research*, in press.

□   Gouy, V. (1993) Contribution de la modelisation a la simulation du transfert des produits phytosanitaires de la parcelle agricole vers les eaux de surface. *Thesis*. ULP de Strasbourg, CEMAGREF de Lyon, 350 pp.

◻     Knisel, W.G (1980) CREAMS, a field scale model for chemicals, runoff and erosion from
       agricultural management systems. Knisel (Ed) USDA. *Conservation Research Report*, 26, 643
       pp.

◻     Lesaffre, B.; Zimmer, D. (1988) Subsurface drainage peak flows in shallow soils. *Journal of
       Irrigation and Drainage Engineering, ASCE*, 114, (3), 387-406.

◻     Penel, M.; Lesaffre, B. (1989) Surface runoff on drained and undrained plots. *ASAE-CSAE
       Summer meeting*. Paper 892104, 11 pp.

**This article has ever been published in BCPC Publications, Weeds 22-25 Nov. 1993, Vol 2, pp
867-872.**

## 5.3.13.   *Persistence of Metribuzin, Pendimethalin and Ethalfluralin in Soil*

# J.L. Tadeo, Y. Lechón, L. Martínez,
# C. Sánchez-Brunete, A. García-Valcárcel

Laboratorio de Productos Fitosanitarios C.I.T.-I.N.I.A.
Apdo 8111, 28080 Madrid, Spain.

## Abstract

Persistence of the herbicides metribuzin, pendimethalin and ethalfluralin in soil from two tomato fields located in Badajoz, Spain, was studied. Soil cores were periodically sampled from the surface layer (0-10 cm) and herbicide levels determined by gas chromatography. Residue analysis showed a rapid loss of metribuzin, mainly in the loamy sand soil and a slower rate of disappearance of the nitroanilines. The influence of soil type and environmental conditions on the persistence of these herbicides is discussed.

## Introduction

Weed control in tomato is generally accomplished by using metribuzin in combination with the nitroanilines pendimethalin and ethalfluralin.

Metribuzin persistence in soil depends on factors like microbial activity, temperature, moisture content and soil type (Sharom and Stephenson 1976; Allen and Walker 1987; Bowman 1991). Dissipation of pendimethalin and ethalfluralin is influenced by cultural practices, environmental conditions and soil type, and differences in persistence have been reported (Zimdahl et al. 1984, Gaynor 1985). The aim of this work was to study the persistence of these herbicides in soil from two tomato fields located in Badajoz, Spain.

## Materials and Methods

The study was carried out in two tomato experimental plots located in Badajoz. Soil characteristics of these fields are shown in Table 1. Rainfall, irrigation and average temperature data during the assay are presented in Table 2. Soil cores were periodically sampled since the date of herbicide application to the harvesting date. Samples were taken from the surface layer (0-10 cm) and maintained at -18°C until analyzed.

Herbicide levels in soil samples were determined by gas chromatography (GC). Soil (30 g) was extracted twice with ethyl acetate (100ml) on an orbit shaker for 30 min each time. The extract was filtered and the filter cake washed with 40 ml of ethyl acetate. The solvent was evaporated to dryness and the residue transferred to a 10ml tube with ethyl acetate. An aliquot was analyzed by GC on a capillary BP-1 column with nitrogen-phosphorous detection.

| Field | % Sand | % Silt | % Clay | % O.M | pH |
|---|---|---|---|---|---|
| A (sandy clay loam) | 59.8 | 24.8 | 15.3 | 0.9 | 5.4 |
| B (loamy sand) | 85.9 | 9.5 | 4.6 | 0.5 | 5.7 |

**Table 1 :** Soil characteristics of the studied fields

| Month | Rainfall (mm) | Irrigation (mm) | Temperature (°C) |
|---|---|---|---|
| May | 50.3 | 28 | 16.5 |
| June | 13.8 | 90 | 22.0 |
| July | 0.0 | 147 | 26.2 |
| August | 0.0 | 147 | 24.8 |
| September | 17.9 | 0 | 19.8 |
| Total | 82.0 | 412 | |

**Table 2 :** Rainfall, irrigation and average temperature data during the experience

## Results and Discussion

Herbicide levels in soil obtained in the studied fields along the season are shown in Table 3 and Table 4. Field half lives attained for metribuzin were 10 and 6 days for the treatment 1 in A and B fields respectively, and 14 days for the treatment 2 in field A. Differences between half lives for both treatments could be explained by the different irrigation system, sprinkling and furrow irrigation for treatment 1 and 2, respectively. These half lives are lower than those reported by different authors in the literature (Hyzak and Zimdahl 1974; Sharom and Stephenson 1976; Walker 1978) but are in good agreement with those obtained by Savage 1977 and Bowman 1991. This behaviour is likely to be due to the conditions in which tomato is grown in this area, with high temperatures under intensive irrigation.

Field half lives for pendimethalin were 116 and 68 days in A and B fields respectively, and 65 and 25 days for ethalfluralin in the same soils. These values are in good agreement with those reported in the literature (Zimdahl et al. 1984; Gaynor 1985).

The rate of herbicide dissipation obtained was remarkably different for metribuzin with respect to the nitroaniline herbicides. Metribuzin residues decreased rapidly to levels below the 10 % of the initial amount by the fifth week in both treatments and both soils, while nitroaniline herbicides were considerably more persistent. The higher adsorption coefficient and the slower degree of degradation reported for these nitroaniline herbicides (Wauchope et al. 1992) may explain their greater persistence in the surface layer of the soil. Ethalfluralin levels were lower than 0.1 mg/Kg at the end of the sampling period. Pendimethalin showed a pattern of decline in which the rate of disappearance reaches a plateau two months after herbicide application. After reaching this plateau, herbicide residues decrease very slowly with time until the end of the growing season. Pendimethalin residues at the end of the sampling period were around 30% to 40% of the initial amount. These values suggest that pendimethalin could persist to the next growing season under determined conditions, e.g. when high herbicide doses are applied or a droughty growing season occurs, and therefore it could present residue carryover problems.

Differences in the persistence of all the herbicides studied were observed between fields A and B. Herbicide half lives were always lower in the loamy sand soil than in the sandy clay loam soil, and the pattern of behaviour described above for pendimethalin was also more evident in the sandy clay loam soil. These differences are likely attributable to differences in texture between the soils. The high sand proportion and low organic matter content of field B could favour mobility of these herbicides and lead to the observed lower half life.

| Days after treatment | Field A[a] | | Field B[a] |
|---|---|---|---|
| | Treatment[b] | | |
| | 1 | 2 | 1 |
| 0 | 0.43 ± 0.08 | | 0.34 ± 0.007 |
| 13 | 0.12 ± 0.03 | | 0.04 ± 0.003 |
| 26 | 0.06 ± 0.01 | | 0.008 ± 0.001 |
| 39 | 0.03 ± 0.01 | | 0.005 ± 0.003 |
| 47 | | 0.42 ± 0.07 | < 0.001 |
| 68 | | 0.05 ± 0.01 | |
| 86 | | 0.02 ± 0.004 | |
| 103 | | 0.01 ± 0.001 | |
| 126 | | 0.007 ± 0.001 | |

[a]    Values are the mean of four replicates±SD.
[b]    Herbicide doses: Treatment 1: 0.21Kg/Ha. Treatment 2: 0.35Kg/Ha applied
       47 days after treatment 1. Sprinkling irrigation was carried out during the first
       period and furrow irrigation after treatment 2.

**Table 3 :** Metribuzin concentration (mg/Kg) in soil

| Days after treatment | Ethalfluralin[a] | | Pendimethalin[a] | |
|---|---|---|---|---|
| | Field | | Field | |
| | A | B | A | B |
| 0 | 0.31 ± 0.06 | 0.14 ± 0.040 | 0.86 ± 0.3 | 0.67 ± 0.22 |
| 30 | 0.18 ± 0.03 | 0.05 ± 0.020 | 0.53 ± 0.2 | 0.39 ± 0.07 |
| 56 | 0.14 ± 0.02 | 0.03 ± 0.001 | 0.37 ± 0.08 | 0.24 ± 0.08 |
| 84 | 0.09 ± 0.02 | 0.02 ± 0.001 | 0.35 ± 0.15 | 0.23 ± 0.02 |
| 119 | 0.07 ± 0.01 | 0.001 ± 0.001 | 0.34 ± 0.09 | 0.19 ± 0.06 |

[a]    Values are the mean of four replicates±SD. Herbicide dose:ethalfluralin
       0.58Kg/Ha, pendimethalin:0.91Kg/Ha. Herbicides were applied preplant
       incorporated.

**Table 4 :** Ethalfluralin and Pendimethalin concentration (mg/Kg) in soil

# Conclusions

The results obtained in the present study show that metribuzin levels in soil decreased rapidly after treatment, while the nitroanilines pendimethalin and ethalfluralin were more persistent. Physical-chemical properties of the studied soils influenced the dissipation of these herbicides, being less persistent in the lighter textured soil. Pendimethalin remained in the soil at noticeable levels at the end of the growing season and residue carryover problems of this herbicide could occur under certain conditions.

# Acknowledgements

The authors wish to thank Esther de Miguel and the staff of Hispareco S.A. for their valuable assistance.

# References

□   Allen, R. and Walker, A. 1987."The Influence of soil Properties on the Rates of Degradation of Metamitron, Metazochlor and Metribuzin". *Pestic. Sci.* 18:95-111.

□   Bowman, B.T. 1991. "Mobility and Dissipation Studies of Metribuzin, Atrazine and their Metabolites in Plainfield Sand Using Field Lysimeters. *Environ. Toxycol. Chem.* 10:573-579.

□   Hyzak D.L. and Zimdahl R.L. 1974. "Rate of Degradation of Metribuzin and Two Analogs in Soil". *Weed Science*, 22(1): 75-79 .

□   Gaynor, J.D. 1985. "Dinitroaniline Herbicide Persistence in Soil in Southwestern Ontario". *Can. J. Soil Sci*; 65:587-592.

□   Savage, K.E. 1977."Metribuzin Persistence in Soil" *Weed Science*, 25(1):55-59.

□   Sharom, M.S. and Stephenson, G.R. 1976. "Behavior and Fate of Metribuzin in Eight Ontario Soils". *Weed Science*, 24(2):153-160.

□   Walker, A. 1978. "Simulation of the Persistence of Eight Soil-Applied Herbicides". *Weed Research*, 18: 305-313.

□   Wauchope, R.D; Buttler, T.M; Hornsby, A.G; Augustijn-Beckers, P.W.M. and Burt J.P.1992. "The SCS/ARS/CES Pesticide Properties Database for Environmental Decision-Making". *Reviews of Environmental Contamination and Toxicology*, 123: 1-155.

□   Webster, G.R.B. and Reimer, G.J. 1976. "Field Degradation of the Herbicide Metribuzin and its Degradation Products in a Manitoba Sandy Loam Soil" *Weed Research*, 16: 191-196.

□   Zimdahl, R.L., Catizone, P. and Butcher, A.C. 1984. "Degradation of Pendimethalin in Soil". *Weed Science*, 32:408-412.

### 5.3.14.   *Volatilization of EPTC, Tri-Allate and Parathion after Spraying onto a Bare Soil Surface*

## F. van den Berg, G. Bor, J.H. Smelt, R.A. Smidt and A.E. van de Peppel-Groen

DLO Winand Staring Centre for Integrated Land, Soil and Water Research,
P.O. Box 125, 6700 AC Wageningen, The Netherlands.

## Abstract

After application of pesticides to soil, a fraction of the dosage of the pesticide volatilizes and disperses in the air. A research project was set up to measure the rate and extent of volatilization of the herbicides EPTC and tri-allate and the insecticide parathion from a loamy soil surface in the field. Because of the differences in the physico-chemical properties of these pesticides, differences in the rate and extent of their volatilization from the soil surface can be expected.

Two methods were used to measure the rate of volatilization of pesticide from the soil into the air (mass volatilized per unit of surface and per unit of time), i.e. a box method (B) and the theoretical profile method (TP). For the B-method, a stainless steel box is placed with its open side onto the sprayed soil surface. An air stream is drawn through the box and the mass of pesticide emitted from the soil into the air is measured by taking samples from the outgoing air. For the TP-method, the concentration of pesticide in the air and the wind speed need to be measured at a specific height above the centre of a circular treated plot. At five time intervals after application, one-hour air samples were taken at a rate of $3 \text{ m}^3 \text{ hr}^{-1}$ with XAD-4 as adsorbent.

The air samples were extracted with ethyl acetate. The extracts were analysed with gas chromatography.

As part of the volatilization rates were too low to be measured with the TP-method, the number of measurements for a comparison of the two methods is limited. Therefore, only a rough indication is obtained from the comparison.

For all three pesticides, the highest rate of volatilization was measured shortly after application. During the first 24 hours after spraying, the volatilization rate decreased sharply. Thereafter, the decrease in volatilization rate was more gradual.

During the first two weeks after application, no rain fell on the treated plot. On the 14th day after the day of application some rain fell. This resulted in a substantial increase in the rate of volatilization of EPTC and tri-allate from the soil into the air. For parathion, the increase was less distinct. The increase in the volatilization rate can be explained by the comparatively strong adsorption of the pesticides onto soil at low moisture contents (below a few percent).

The rate of volatilization of a pesticide is affected by its vapour pressure, its adsorption onto the soil surface and its persistence. The course of the volatilization rate with time is also affected by the weather conditions, e.g. wind speed and air temperature. An attempt is made to relate the measured volatilization rates of EPTC, tri-allate and parathion to their physico-chemical properties and the prevailing weather conditions. The contribution of the fraction volatilized to the material balance of these pesticides is discussed.

# 5.4.     Short Communications

*5.4.1.     A field method to measure the rate of volatilization of fumigants
from soil*

## J.H. Smelt & R.A. Smidt

DLO Winand Staring Centre for Integrated Land, Soil and Water Research
P.O. Box 125, 6700 AC Wageningen, the Netherlands

In the Netherlands the shear injection method is the most common application method for the soil fumigants 1,3-dichloropropene and metham sodium (which forms methyl isothio-cyanate). Recently new application methods have been developed to improve nematode control under wet and cool soil conditions and in soils with fast transformation of the fumigants. Because the new application methods place some of the fumigant near or onto the soil surface, higher volatilization rates can be expected than with injection at a depth of about 0.18 m with shear injection. Field research was set up to quantify the effect of application method on the volatilization rate.

Simultaneous measurements on a comparatively small area provide the best conditions for comparing volatilization rates of fumigants and application methods in the field. Methods which require measurements of concentrations of the compounds in the air above or around the treated fields, like agro-meteorological methods, cannot be used in these conditions, as the concentration plumes interfere. To tackle this problem, the volatilization rates were measured directly at soil surfaces of 1 $m^2$ each during short periods (15 to 60 min.)

Field measuring systems were constructed, which consist of a flat (0.1 m high) stainless steel box with a rectangular form and with an open bottom of 1 $m^2$. This open side with rims is placed on the soil surface and pressed into the soil to prevent air leakage. An air stream of about 0.2 m $s^{-1}$ is drawn over the soil surface by an electric fan. This flow rate results in an air ventilation fold in the box of 6 $min^{-1}$. Guide vanes in the inlet side help to get an even flow and to avoid dead volumes. The air flow rate can be regulated with a restriction valve in the outlet and has to be adjusted and calibrated in the laboratory. The in- and outgoing air is guided by pipes (2 to 4 m high). The relatively great height of the inlet pipe results in an almost zero concentration in the incoming air when the units are placed upwind of small treated areas. The concentration of the fumigant in the in- and outgoing air is measured by subsampling the total air flow with sampling tubes (containing activated carbon) and air sampling pumps with a calibrated air flow. The ratio (1000:1 to 50:1) between the total air flow through the box and the sampling air flow can be changed and depends on the desired detection limit.

The low-cost method has turned out to be easy to handle and quick. In the field, the method gave very little technical problems and it could be used under a wide range of soil and climatic conditions. For the three application methods, the difference between the volatilization rates (mg $m^{-2}$ $h^{-1}$) of the two fumigants could be well measured, as well as the course of the rate with time. The effect of soil temperature and soil moisture conditions on the volatilization rates could be assessed also. The method has shown to be also suitable to measure the volatilization rate of much less volatile pesticides, which are sprayed on bare soil surfaces.

The measured volatilization rates can be used for calibrating and validating computer models, which simulate fumigant emission from soils. The rates can also serve as source strength values for models which calculate the inhalation exposure of people living in the vicinity of treated fields.

# SECTION VI

# MATHEMATICAL MODELS

Chairman :  J. J. T. I. BOESTEN
(Winand Staring Center - The Netherlands)

# MATHEMATICAL MODELS

## 6.1.      Introductory Presentation

*6.1.1.      New Advances in Predicting Pesticide Fate and Transport*

### D. I. Gustafson

Environmental Science Department, Monsanto Agricultural Group
700 Chesterfield Village Parkway North, St. Louis, MO 63198 USA

## Abstract

A variety of field- and laboratory-scale experimental techniques have been developed to describe the environmental fate and transport of agricultural chemicals. Of all the tools available, only computer simulation models can be used in a truly predictive way. Although not sufficiently accurate to be used in the absence of any field data, such computer programs can be used with a high degree of certainty to perform quantitative extrapolations and thereby quantify regional exposure from field-scale monitoring information. New improvements in computer technology have recently made it practical to use Monte Carlo and other probabilistic techniques as a routine tool for estimating human exposure. In addition, geographic information systems (GIS) are currently being employed to develop modeling predictions tied to local soil, weather, and cropping conditions. As the technical basis of such methods improves it will soon be possible, at least in principle, to prepare exposure estimates with known confidence intervals and sufficient statistical validity to be used in the regulatory management of agricultural chemicals. The likely "look" and "feel" of these new predictive methods will be presented, along with some of the technical problems accompanying their development.

## Introduction

Regulators and producers of agricultural chemicals are by necessity interested in the frequency and levels of occurrence of pesticides in various environmental media, especially drinking water. In order to estimate these concentrations more precisely, scientists have further developed mathematical models as a computational tool for predicting pesticide fate and transport in the environment. Some of the newer advances in the further refinement of these models shall be the topic of this paper. The paper begins with a brief overview of the available models. New developments will then be described in the following areas: modeling macropore flow, the integration of GIS ( geographic information systems), accounting for field-scale variability, major new model validation projects, and finally the development of statistically-based probabilistic models. It will be seen that the near-term future (< 5 years) will be accompanied by the introduction of probability-based, GIS modeling tools of sufficient accuracy to be employed in detailed, geographically-specific environmental exposure assessments.

## Available Models

Mathematical and computer models for describing pesticide fate and transport into drinking water supplies may be broadly classified as either screening or simulation models. Screening models typically require no computer, or only a simple spread-sheet type of calculation. They are often empirical or based on a number of severe simplifying assumptions in order to reduce to a bare minimum the amount of computation required. By contrast, simulation models can involve the need for considerable computational resources and generally attempt to describe the behavior of the pesticide as a full function of time and at least one dimension (usually depth in the soil profile). These two classes of models are each widely used and have their own unique place in a comprehensive regulatory control management strategy. Some examples of the currently available models are described below.

### *Screening Models*

Numerous screening models for classifying the drinking water contamination potential of pesticides have been proposed. A few of the more widely used examples are listed here:

#### *Cohen Criteria*

In the early 1980's, Stuart Cohen and others in the United States Environmental Protection Agency published (Cohen et al., 1984) a list of physical properties which they had determined to be associated with pesticides likely to be found in ground water. Seven properties were given, including water solubility (> 30 mg/L), soil/water partition coefficients (Kd < 10 L/kg, Koc < 500 L/kg), and soil degradation half-life (> 2-3 weeks). Other parameters such as volatility and acid dissociation constant were listed, as were some of the hydrogeologic conditions associated with contamination incidents: shallow water tables, sandy soils, and acidic sub-surface conditions.

#### *Jury Screening Models*

Also beginning in the early 1980's, William Jury of the University of California at Riverside published (Jury et al., 1987) a number of papers describing analytical solutions to the convective dispersion equation (CDE) for describing one-dimensional fate and transport of pesticides through soil. The analytical solutions were made possible through the assumption of steady water flows, linear and completely reversible sorption isotherms, first-order and linear pesticide dissipation kinetics, and a number of other simplifying effects. However, a number relatively sophisticated processes were included in some of these models, in particular volatilization and depth-dependent soil degradation rates for the pesticide.

#### *Hornsby Screening Models*

At the University of Florida (Gainesville), Art Hornsby and several co-workers have produced several screening models for describing pesticide leaching behavior (Nofzinger and Hornsby, 1988). These models have the unique feature that they have been linked with a soils database for Florida and with a pesticide physical properties database. Most of the methods (such as CMLS and CHEMRANK) make a rather severe simplifying assumption regarding dispersion of the pesticide through the soil profile -- the pesticide moves as a pulse without spreading.

#### *GUS*

While at Monsanto in the late 1980's, the author proposed a single numerical index for predicting the water contamination potential of pesticides based only on its soil/water partition coefficient (Koc) and the soil half-life (Gustafson, 1989). The index, named Ground water Ubiquity Score or GUS, has been used by the United States Soil Conservation Service as part of its screening technique and by regulatory agencies in Canada and elsewhere to help classify and prioritize pesticides.

## *Simulation Models*

Simulation models are appropriate for those instances in which a considerable amount of computer power and data are available for a specific pesticide use scenario, and when there is a need to have more specific information about the concentrations of the pesticide in adjacent water bodies. Some of the important models are listed here:

### *PRZM*

First released in 1984 by Bob Carsel and others at the US EPA's Environmental Research Laboratory in Athens, Georgia, the Pesticide Root Zone Model continues to be the most widely-used model when a full-fledged simulation of pesticide behavior is required (Carsel et al., 1984). Although a number of technical objections have been raised concerning the assumptions and solution methods utilized by the program, particularly in its handling of hydrology, there is not another model with the breadth of documentation and history of varied uses by a broad range of investigators. The most recent versions of PRZM include an excellent volatilization model and the possibility of modeling up to two degradate products during a single simulation. This last feature is particularly powerful and can be used to simulate non-linear dissipation kinetics for the parent pesticide if so desired.

### *PELMO*

An important variant of the PRZM model has been developed for use in Germany (Klein, 1991). Called PELMO, it utilizes the Freundlich equation for describing pesticide sorption to soil, optional non-linear dissipation kinetics, a different method for estimating evapo-transpiration, and an explicit treatment of the temperature-driven depth-dependence of pesticide degradation rate. Another feature of PELMO that makes it well-suited for certain regulatory applications is that virtually all of the scenario parameters have been hard-coded into the system, reducing the user's input choices to only a handful of very specific properties and dates. To a large extent, this "levels the playing field," and makes it possible to apply the model as a standardized regulatory tool.

### *CALF*

The product of several researchers in the UK, primarily Peter Nicholls and Allan Walker, CALF is the only widely-used simulation model to directly account for the commonly observed slow increase of soil sorption with time (Addiscott and Wagenet, 1985; Walker and Barnes, 1982). A problem in the regulatory use of the CALF model is that there has been no careful version control, and therefore it is difficult to know which version one has when making comparison of runs performed by different parties. CALF is currently being modified to account for macropore or preferential flow processes, which will be a valuable and unique enhancement among available simulation models.

### *GLEAMS*

This program and its predecessor, CREAMS, were both produced by the USDA Agricultural Research Service (Knisel, 1980; Leonard et al., 1987). Among available models, it possesses the most widely-accepted representation of pesticide runoff processes. A considerable amount of field validation work has been reported for the model. It does not have all of the flexibility of PRZM regarding numbers of soil layers or provisions for the use of more sophisticated solution techniques, but these limitations may not necessarily be important for many users or applications of the model.

# New Developments

The advantages of models relative to other techniques for regulatory management of pesticide use have become evident over the past 10 years. Experience with the simulation techniques have demonstrated that they are relatively inexpensive when compared with experimental work and monitoring studies, that they allow the full information value of available data to be extracted, and that they make possible the establishment of critically-required standardized registration criteria. However, a great deal of work is ongoing in a number of areas to further refine and increase the validity of such methods. Some of the these are discussed below.

## *Modeling Macropore Flow*

Today's "work-horse" models, such as PRZM and GLEAMS, assume piston-flow of water through soil, which is probably a reasonable assumption for unstructured sands. However, many structured soils and especially those with a large degree of clay content can exhibit preferential or macropore flow. Proper prediction of water and pesticide transport through such soils requires a new modeling approach in which runoff water is shunted through a second transport phase, bypassing much of the soil matrix. Two of the new models that have recently been developed to account for these processes are VARLEACH (Walker et al.) and MACRO (Jarvis et al.), both of which are discussed in greater detail elsewhere in this symposium. The results thus far are very encouraging for these models, although a good deal of work remains to demonstrate their wide applicability and methods for parameterizing the models based on available soil property information.

## *Integration of GIS*

GIS (geographic information system) techniques have exploded onto the environmental exposure scene in recent years with the advent of much lower-cost, PC-based technologies and the increased availability of data layers of the kind needed for regional and smaller-scale exposure assessments. Two particularly affordable new GIS tools include ArcView II (ESRI) and MapLand (Software Illustrated). ArcView II opens ArcInfo-type files (the de facto standard for GIS data) and contains a sophisticated object-oriented programming language for the rapid development of on-screen dynamic exposure assessment tools. MapLand is a much simpler tool that simply allow Microsoft Excel to graph geographic data onto map-style charts. Its utility lies in the wide acceptance and use of spreadsheet tools, and the reasonably seamless way in which it allows a user to go directly to fairly sophisticated maps straight from a spreadsheet of data on a PC.

Data layers for driving GIS tools are available both in Europe and in the US. In Europe, the CORINE data base contains the most complete set of data. In the US, both the USGS and the USDA have several useful soil and land use data sets, and a private company (Hydrosphere) has recently made meteorological data available in an ArcInfo format. For regional scale assessments at the 1:250,000 or 1:1,000,000 scale, the data layers are available. Those wishing to conduct assessments at finer scale will s till need to wait for the true integration of remotely sensed data into the GIS data layers, but this day will surely come within the next 5-10 years.

## *Accounting for Field-Scale Variability*

Besides accounting for regional-scale variability via GIS, models are being enhanced in the area of dealing with field-scale (or even finer) variability. In most of today's models, for instance, soil half-lives, partition coefficients, and other soil properties are entered as single point values, with no explicit acknowledgment of the true variability and heterogeneity of any real soil system. Such heterogeneities are known to have very significant field-scale effects. For instance, spatial variability in the first-order degradation rate constant can be shown theoretically to result in nonlinear pesticide dissipation curves (Gustafson and Holden, 1990). This nonlinear behavior obviously affects long-term predictions such as carry-over and leaching (Gustafson, 1994). Variabilities in the transport process parameters introduce further complexity into models, but are sometimes necessary in order to improve the accurate depiction of real behavior. This topic will also be explored in greater detail elsewhere in the symposium (Vanclooster et al.).

## *Major Model Validation Projects*

A major complaint on the part of "non-believers" in computer modeling is that the models themselves have never been validated. While it is true that no model of natural systems can ever be shown to be perfectly accurate, it is also true that useful approximations of natural system behavior can be developed. For instance, the planets were thought by Copernicus to orbit in circles, Kepler brought us ellipses, and Einstein's Theory of General Relativity brought us yet another curve through a 4-dimensional time-space continuum. But for the purposes of calculating how many days are in a year, the assumption of a circle is probably adequate. Similarly, for the purpose of simply achieving order-of-magnitude accuracy in our environmental exposure assessments, it is debatable how much further refinement or testing of PESTLA or PRZM is required.

But work is needed to establish the practical accuracy levels of today's models, and whether there are relevant combinations of pesticide, soil, and climatic conditions under which they fail to provide useful approximations of observed behavior. To this end, some major validation projects have recently been undertaken, both in the US and Europe. In the United States, the EPA, NACA, and USDA are about to embark on a major validation of the two most widely-used fate models, PRZM and GLEAMS. In Europe, there are several major efforts underway, including the PESTLA validation project (the subject of the next symposium paper), and broad comparisons of several leaching models by both Allan Walker (UK) and a group of Danish researchers. It is therefore certain that much more complete information about the practical accuracy of the important environmental fate models will be become available within the next 2-3 years.

## *Probabilistic Modeling*

The direction in which all of the modeling advances appear to be headed is a sound, probabilistic approach in which model predictions are presented as distributions having known accuracies and statistically-valid confidence intervals. The approach will explicitly consider uncertainties concerning the physical properties of the pesticides, climatic conditions, and soil and land-use information. Monte-Carlo or other, more efficient sampling technique will then be used to conduct simulations for a range of input parameter values. The predictioned environmental concentrations (PEC's) are then presented as a distribution, and decisions can be based on the concept of picking a percentile-level of the exposure curve and comparing it against a percentile-level of the hazard, whether to man or the eco-system. A quantitative risk-assessment can then be made upon which to base a regulatory decision. This will become a routine reality within five years.

# *Conclusions*

Computers are dangerously powerful machines, and are becoming more powerful with an ever-quickening pace. To a certain extent, the sophistication of computer modeling approaches for describing the drinking water contamination potential of pesticides has increased over the past several years in a similar manner. In most instances, however, the theoretical advances have significantly lagged behind the technological strides, placing rapid computational power in the hands of those regulatory personnel managing and hoping to control pesticide use. As discussed in this paper, the models are being improved in a number of areas. Thus, despite the many potential pitfalls in the use computer models, it seems clear that none of the other methods for regulatory control of pesticide usage have the same power, cost-effectiveness, and accuracy. Care must be taken in interpreting modeling results. This is especially true today because of the limited amount of true validation work that has been performed on the existing models, and due to the absence of good modeling practices for performing and documenting results. There is reason to be optimistic that the future will bring with it not only faster simulation models, but more accurate ones too. This circumstance will make it easier and more appropriate for regulatory agencies to more widely embrace computer models as the final, discriminating technique for establishing sound regulatory policies on pesticide use.

# References

◻    Addiscott, T.M.; Wagenet, R.J. (1985) Concepts of solute leaching in soils: a review of modelling approaches. *Soil Science*, 36, 411-424.

◻    Carsel, R.F.; Smith, C.N.; Mulkey, L.A., Dean, J.D.; Jowise, P. (1984) *Users Manual for the Pesticide Root Zone Model (PRZM), Release 1*, ERL, US EPA, PB85 158913.

◻    Cohen, S.Z.; Creeger, S.M.; Carsel, R.F.; Enfield, C.G. (1984) Potential for pesticide contamination of groundwater resulting from agricultural uses. In R.F. Kruger and J.N. Seiber, eds., *Treatment and Disposal of Pesticide Wastes*, ACS Symposium Series No. 259, Ameircan Chemical Society, Washington, DC, pp. 297-325.

◻    Gustafson, D.I. (1989) Ground water ubiquity score: a simple method for assessing pesticide leachability. *Journal of Environmental Toxicology & Chemistry*, 8, 339-357.

◻    Gustafson, D.I., Holden, L.R. (1990) Nonlinear pesticide dissipation in soil: a new model based on spatial variability. *Environmental Science & Technology*, 24, 1032-1038.

◻    Gustafson, D.I. (1994) Effect of nonlinear dissipation kinetics on the predictions of environmental fate models. *Journal of Environmental Toxicology & Chemistry*, in press.

◻    Jury, W.A.; Focht, D.D.; Farmer, W.J. (1987) Evaluation of pesticide groundwater pollution potential from standard indices of soil-chemical adsorption and biodegradation. *Journal of Environmental Quality*, 16, 422-428.

◻    Klein, M. (1991) *PELMO*, Fraunhofer Institut, Scmallenberg, Germany.

◻    Knisel, W.G. ed. (1980) CREAMS: *A Field-Scale Model for Chemicals, Runoff, and Erosion from Agricultural Systems*, USDA Cons. Rep No. 26, 640 pp.

◻    Leonard, R.A.; Knisel, W.G.; Still, D.A. (1987) GLEAMS, ground water loading effects of agricultural management systems, *Transactions of the ASAE*, 30, 1403-1418.

◻    Nofzinger, D.L.; Hornsby, A.G. (1988) *Chemical Movement in Layered Soils: User's Manual*, University of Florida, Gainesville, FL, 44 pp.

◻    Walker, A.; Barnes, A. (1981) Simulation of herbicide persistence in soil: a revised computer model. *Pesticide Science*, 12, 123-132.

# 6.2.　　　Platform Presentations

## 6.2.1.　　　*Field test of the PESTLA model for ethoprophos on a water-repellent sandy soil*

## J.J.T.I. Boesten, L.J.T. Van der Pas, J.H. Smelt & H. Van den Bosch

DLO Winand Staring Centre for Integrated Land, Soil and Water Research,
P.O. Box 125, 6700 AC Wageningen, Netherlands

## Abstract

Field tests of models for pesticide leaching to groundwater are urgently needed to justify use of such models in pesticide registration procedures. The nematicide ethoprophos was applied to a water-repellent sandy soil (5% organic matter) in autumn. In the field, rainfall, soil temperature and groundwater level were continuously recorded. Concentration profiles in soil were measured at selected sampling times. In the laboratory, the transformation rate and the adsorption of ethoprophos were measured using soil sampled in the field before application of ethoprophos. The results were used to test the PESTLA model which is based on the convection/dispersion equation and which assumes equilibrium sorption and first-order transformation kinetics. Measured and calculated persistence in soil corresponded well between 22 and 103 days but after 474 days the calculated amount remaining was much higher than the measured amount. Measured and calculated concentration profiles corresponded reasonably well after 103 days. However, after 474 days the model overestimated both the average penetration depth and the penetration depth of the leading edge of the concentration profile.

## Introduction

Leaching of pesticides to groundwater and their persistence in the plough layer are important environmental aspects of pesticide use. In recent years mathematical models have become important tools in the pesticide registration process in some Western European countries. It is expected that models will also play an important role in the EU pesticide registration procedure. Research aimed at testing such models under realistic field conditions is urgent.

The Dutch authorities have been using the PESTLA model since 1989 for assessment of the leaching potential of pesticides. The present paper deals with a field test of the PESTLA model (Boesten and Van der Linden, 1991) in which penetration of a sorbing and quite persistent pesticide is measured with a detection limit in soil close to the EC drinking water limit of 0.1 μg/L. The study was done for a water-repellent sandy soil. Most of the agricultural sandy soils in the Netherlands are water repellent (Dekker, 1988). Ritsema et al. (1993) showed that preferential flow of water may be significant in such soils. This could lead to faster pesticide leaching than expected from the PESTLA model which ignores preferential flow of water. In a previous paper (Boesten et al., 1993) we presented results up to 278 days after application whereas this paper describes the results up to the end of the study after 474 days.

Sensitivity analysis of the model has shown that leaching is very sensitive to transformation and sorption parameters (Boesten, 1991). Therefore we measured the transformation rate and the sorption isotherm of the pesticide using soil from the experimental field.

## Description of the PESTLA model

The PESTLA model consists of a submodel describing water flow in soil and a submodel describing pesticide behaviour. The submodel for water flow is SWACROP (De Jong and Kabat, 1990). The submodel for pesticide behaviour has been described in detail by Boesten and Van der Linden (1991) and a brief description is given here. Version 2.3 of PESTLA was used.

Pesticide transport is described by the convection/dispersion equation. Pesticide sorption onto the solid phase in soil is described by a Freundlich-type equation:

$$X = m_{OM} K_{OM} c_{REF} (c / c_{REF})^N \tag{1}$$

in which $X$ is the content of pesticide sorbed, $m_{OM}$ is the mass fraction of organic matter, $K_{OM}$ is the organic-matter/water distribution coefficient, $c$ is the mass concentration of pesticide in the liquid phase, $c_{REF}$ is a reference value of $c$ ($c_{REF} = 1$ mg/dm$^3$) and $N$ is the Freundlich exponent. The first-order transformation rate coefficient, $k$, is calculated from the rate coefficient at reference conditions, $k_{REF}$, accounting by factors $f$ for the influence of temperature ($f_T$), soil moisture ($f_\theta$) and depth in soil ($f_Z$):

$$k = k_{REF} f_T f_\theta f_Z \tag{2}$$

The factor $f_T$ is described by:

$$f_T = \exp[\gamma (T - T_{REF})] \tag{3}$$

in which $\gamma$ is a parameter, $T$ is the prevailing temperature in soil and $T_{REF}$ is the temperature at reference conditions (20 °C). The soil moisture factor $f_\theta$ is described by:

$$f_\theta = (\theta/\theta_{REF})^B \tag{4}$$

in which $\theta$ is the volume fraction of liquid in the soil, $\theta_{REF}$ is $\theta$ at reference conditions (i.e. at a matric potential of -100 hPa) and $B$ is a parameter. The factor $f_Z$ was set to unity in the present study.

Plant uptake of the pesticide was described as a passive process using the concept of the transpiration stream concentration factor.

## Experimental methods

The field experiment was done at experimental farm "Vredepeel" in the south-eastern part of the Netherlands. The experimental field was 54 x 80 m. At 22 November 1990 ethoprophos was sprayed on the soil surface at a dose of approximately 3 kg active ingredient per ha, formulated as an emulsifiable concentrate. A few hours before spraying winter wheat was sown on the plot. The wheat crop was harvested in August 1991. In September 1991 mustard was sown on the plot which was incorporated into the soil in November 1991. At the experimental field soil temperature at 2.5 and 100 cm depth was recorded continuously and rainfall was recorded with a rain gauge whose rim was flush with the soil surface. The groundwater level was continuously recorded at one spot in the field. The field was divided into 16 sections (each 10 x 27 m). After 1, 22, 103 and 474 days 16 soil columns were taken (one from each section) and sliced into layers. Samples from corresponding layers of four sections were combined and mixed. Samples from sections 1-4, 5-8, 9-12 and 13-16 were combined. In the laboratory, subsamples of the four mixed samples of each layer were extracted with hexane. The extracts were analyzed for ethoprophos using capillary gas chromatography.

Before spraying ethoprophos, the soil was sampled to 30 cm depth by taking approximately 50 cores. The sample was mixed and it was used for the laboratory experiments on transformation rate and adsorption. Its organic matter content was 4.9% and its pH-KCl was 5.3. It contained 3% clay and

6% silt. The transformation rate was measured at 5, 15 and 25 °C at an initial content of ethoprophos of 5 mg/kg. Sorption was measured at a solid/liquid ratio of 1 kg/dm$^3$ using an equilibration time of 24 h.

## Model parameters

The parameter $m_{OM}$ was 4.9% in the 0-30 cm layer, 0.4% in the 30-50 cm layer and 0.2% below. Daily averages of rainfall, potential evapotranspiration, soil temperature and groundwater level were used as input. Cumulative rainfall was 0.77 m after 474 days. Cumulative evapotranspiration calculated by the submodel for water flow was 0.51 m after 474 days. The groundwater level varied between 0.6 and 1.7 m depth. The pesticide dose was 1.3 kg of active ingredient per ha (based on the amount measured after 22 days; see Results and discussion). The $K_{OM}$ was 79 dm$^3$/kg and the Freundlich exponent, $N$, was 0.87 (both derived from sorption experiments). The rate coefficient $k_{REF}$ was 0.009 d$^{-1}$ and $\gamma$ was 0.09 K$^{-1}$ (both derived from the transformation experiments). The parameter $B$ was set at 0.7 which is the average value found for a range of soils and pesticides (see the review by Boesten, 1986). The dispersion length parameter was set at 3 cm. The relationship between the matric pressure and θ (pF-curve) and that between the hydraulic conductivity and θ were calibrated using field-measured moisture profiles. The transpiration stream concentration factor was set at 0.5.

## Results and discussion

After one day the average areic mass of ethoprophos recovered was 2.7 kg/ha and the standard deviation from this average was 0.1 kg/ha. After 22 days this average was 1.3 ± 0.1 kg/ha. This corresponds with a decrease of about 50% in the first 22 days. However, the model calculated a decrease of only 4% in this period. In the first 17 days 7 mm of rain fell distributed over 7 days. At days 18 to 20 in total 31 mm of rain fell on the site. A possible explanation is that the rapid loss in the field was the result of volatilization (which is not included in the model) during the first 17 days. Leistra (1979) used a model that included volatilization and calculated a loss of 2% due to volatilization assuming that ethoprophos was incorporated in the top 2 cm in spring. In the first 17 days of our experiment ethoprophos was still in the top few millimetres of the soil which may have lead to a considerably higher loss than the 2% calculated by Leistra (1979). All further model calculations were made assuming a dose of 1.3 kg/ha. After applying this correction the calculated areic mass remaining after 103 days corresponded well with the measured one: 1.1 kg/ha calculated versus 1.2 ± 0.05 kg/ha measured. After 474 days the measured areic mass was about 0.02 kg/ha whereas the calculated mass was about 0.3 kg/ha. So persistence was satisfactorily simulated between days 22 and 103 whereas between days 103 and 474 the decline in the field proceeded more rapidly than calculated by the model. Possibly the growth of the wheat crop resulted in a more active microbial population which may have caused the transformation rate to be higher in the field than in the laboratory.

Figure 1a shows that the model reasonably explained the movement in the soil after 103 days. However, Figure 1b shows that the model overestimated movement after 474 days. Both the average penetration depth and the penetration of the leading edge of the concentration profile were overestimated. Note that in none of the layers below 15 cm depth ethoprophos was detected using a the detection limit in soil which was close to 0.1 μg/L (see Fig. 1b).

Leistra and Smelt (1981) also found that a similar model simulated somewhat greater movement than that measured for ethoprophos in soil columns. Boesten (1987) showed that including long-term sorption kinetics into the model eliminated the discrepancies found by Leistra and Smelt to a large extent. Therefore it seems worthwhile to include long-term sorption kinetics in the model. This requires additional measurements on long-term sorption kinetics in the laboratory for this pesticide/soil system.

Van der Zee and Boesten (1991) analyzed theoretically the effect of spatial variability in transformation rate and sorption coefficient on pesticide leaching. They calculated more leaching when spatial variability was taken into account. PESTLA as applied in the present study ignores spatial variability.

Nevertheless we measured less leaching than was calculated with PESTLA. Evidently, long-term sorption kinetics is more significant under these field conditions than spatial variability of pesticide/soil properties.

## Acknowledgements

We thank W. Hamminga, C.J. Ritsema and L.W. Dekker from the Department of Soil Physical Transport Phenomena of DLO Winand Staring Centre for their co-operation. We thank also M. Leistra for his critical comments on the manuscript.

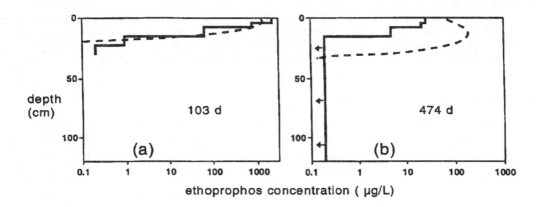

**Figure 1 :**     Comparison of calculated and measured concentration profiles of ethoprophos in soil after 103 days (part a) and 474 days (part b). Solid lines are averages of measured concentrations and dashed lines are calculated. The arrows in part b indicate that the concentration was below the detection limit indicated by the solid line.

## References

◻    Boesten, J.J.T.I. (1986). Behaviour of herbicides in soil: simulation and experimental assessment. *Doctoral thesis*, Institute for Pesticide Research, Wageningen, Netherlands, 263 pp.

◻    Boesten, J.J.T.I. (1987). Modelling pesticide transport with a three-site sorption sub-model: a field test. *Netherlands Journal of Agricultural Science* 35, 315-324.

◻    Boesten, J.J.T.I. (1991). Sensitivity analysis of a mathematical model for pesticide leaching to groundwater. *Pestic. Sci.* 31, 375-388.

◻    Boesten, J.J.T.I. and Linden, A.M.A. van der (1991). Modeling the influence of sorption and transformation on pesticide leaching and persistence. *J. Environ. Qual.* 20, 425-435.

◻    Boesten, J.J.T.I., Pas, L.J.T. van der and Smelt, J.H. (1993). Field test of the PESTLA model for ethoprophos on a sandy soil. In: H.J.P. Eijsackers & T. Hamers (Eds), *Integrated soil and sediment research: a basis for proper protection*, p. 241-245. Kluwer Academic Publishers, Dordrecht, Netherlands, 763 pp.

□   Dekker, L.W. (1988). Distribution, causes, consequences and possibilities for reclamation of water-repellent soils in the Netherlands. Report 2046 [in Dutch], Soil Survey Institute (STI-BOKA), Wageningen, Netherlands, 54 pp.

□   Jong, R. de and Kabat, P. (1990). Modeling water balance and grass production. *Soil Sci. Soc. Am. J.* 54, 1725-1732.

□   Leistra, M. (1979). Computing the movement of ethoprophos in soil after application in spring. *Soil Sci.* 128, 303-311.

□   Leistra, M. and Smelt, J.H. (1981). Movement and conversion of ethoprophos in soil in winter: 2. computer simulation. *Soil Sci.* 131, 296-302.

□   Ritsema, C.J., Dekker, L.W., Hendrickx, J.H.M. and Hamminga, W. (1993). Preferential flow mechanism in a water repellent sandy soil. *Water Resour. Res.* 29, 2183-2193.

□   Zee, S.E.A.T.M. van der and Boesten, J.J.T.I (1991). Effects of soil heterogeneity on pesticide leaching to groundwater. *Water Resour. Res.* 27, 3051-3063.

6.2.2.      *The implications of preferential flow for the use of simulation*
             *models in the registration process*

# N. J. Jarvis

Department of Soil Sciences, Swedish University of Agricultural Sciences,
Box 7072, 750 07 Uppsala, Sweden

## Introduction

Leaching of pesticides to groundwater has long been considered a potentially serious environmental problem. Thus, criteria have been introduced to regulate standards for drinking water quality with regard to acceptable pesticide concentrations (e.g. EEC directive 80/778/EEC). Together with this concern regarding the occurrence of pesticides in groundwater, effective means of assessing the risks of pesticide leaching need to be developed for use in screening and registration programs. In the past, it has been common practice to estimate pesticide mobility from simple short-term laboratory tests or from easily measurable physico-chemical properties of the compound (e.g. sorption constants, water solubilities, degradation rates in soil incubation studies). Simple steady-state screening models with limited data requirements have also been developed (Jury et al. 1987). Such approaches can be useful for classifying pesticides into different groups based on expected mobility in the field, but should not be used to predict likely concentrations reaching groundwater.

Reliable information on pesticide fate and mobility can be obtained from field or lysimeter experiments. However, such experiments are time-consuming, site-specific and expensive. For these reasons, simulation models are now being increasingly used in pesticide registration programs as potentially effective and inexpensive screening tools (Russell and Layton, 1992). With this increasing use of simulation models, it is vital that confidence can be placed in model outputs. Thus, it is important that the models used should be technically sound, incorporating treatments of all processes which are known to significantly affect fate and mobility, and with model descriptions of these processes which reflect current understanding (Russell and Layton, 1992). Preferential flow is perhaps one of the most significant processes which is currently not included in pesticide leaching models being used in registration programs. This paper presents the results of a simulation study performed to investigate the likely consequences of accounting for preferential flow in models used for registration purposes. The dual-porosity MACRO model (Jarvis, 1991 ; Jarvis, 1994a) is used to simulate the leaching of 15 herbicide compounds in 11 hypothetical soils representing the main U.S.D.A. textural classes, to gain insight into interactions between soil and pesticide properties as they affect leaching. For each soil/pesticide combination, simulations are performed with and without accounting for macropore flow. It should be stressed that simulation results are presented for the selected compounds to illustrate likely general trends and that the predictions of concentrations for any individual compound may not be reliable, since the model is run without calibration. Furthermore, the objective of this study is not to discuss whether preferential flow should be accounted for in registration, but rather to illustrate the likely consequences of doing so. In the following sections, the MACRO model is first briefly presented. This is followed by a short description of how the model was parameterized in this study. Finally, the results of the model simulations are discussed.

# The MACRO model

MACRO is a physically-based model which accounts for preferential flow in an explicit way, with the soil porosity divided into two flow systems or domains (macropores and micropores) each characterized by a flow rate and solute concentration. Richards' equation and the convection-dispersion equation are used to model soil water flow and solute transport in the soil micropores, while a simplified capacitance type-approach is used to calculate fluxes in the macropores. Exchange between the flow domains is calculated using approximate, physically-based, expressions based on an effective aggregate half-width (Gerke and van Genuchten, 1993). Additional model assumptions include first-order kinetics for degradation in each of four "pools" of pesticide in the soil (micro- and macropores, solid/liquid phases), together with an instantaneous sorption equilibrium and a linear sorption isotherm in each pore domain.

MACRO has been successfully tested in several recent field and lysimeter studies using a number of different pesticide compounds including dichlorprop and bentazone (Jarvis, et al., 1994), simazin, methabenzthiazuron and metamitron (Jarvis, 1994b) and dichlorprop, MCPA and 2,4-D (Miljøstyrelsen, 1994).

# Model scenarios and parameter estimation

In this study, the MACRO model is used to predict mean pesticide concentrations in water exiting the root zone at a depth of 1 m for 11 soil profiles representing the major U.S.D.A. soil textural classes. The soil profile was divided into ten layers, each 0.1 m in thickness, and a unit hydraulic gradient was assumed at the base. It should be noted that, in terms of receiving water-bodies, the precise destination of the pesticide predicted to leach past 1 m depth is not specified. However, it is likely that for the fine-textured soils, local soil and geohydrological conditions would be such that most leached pesticide would be routed to surface waters rather than groundwater, via field drainage systems.

Simulations were run using weather data for Mellby, south Sweden (56° 29'N, 13° 00'E) for a ten-year period (1st January 1983 to 31st December 1992). Driving variables for the simulations consisted of daily rainfall totals input at an intensity of 2 mm $h^{-1}$, daily maximum and minimum air temperatures and daily potential evapotranspiration calculated by the Penman-Monteith equation. An identical spring-sown crop was "grown" for each year of the simulation, with emergence on 1st May and harvest on 28th August. Mean predicted leachate concentrations were calculated by dividing the accumulated leaching load of pesticide by the total water seepage for the ten-year period. Based on the predicted leachate concentrations expressed in $\mu g\ l^{-1}$, each simulation was classified into one of three risk categories : low risk 0.001, medium risk < 0.001 -1, and high risk > 1.

Fifteen herbicide compounds were selected covering a wide range in sorption and degradation properties (Table 1). MACRO cannot easily account for volatilization, so that all the chosen compounds (with possibly one exception) are non-volatile, or only slightly volatile. In order to simplify the analysis, all compounds were assumed to be applied at the same rate (2 kg $ha^{-1}$) and on the same day each year (10th May). The sorption and degradation rate constants listed in Table 1 were obtained from a variety of literature sources, but primarily from the database presented by Wauchope et al. (1992). The values shown in Table 1 were used as input to the model for the topsoil layers to 30 cm depth. $K_{oc}$ values were converted to sorption constants ($K_d$ values) assuming an organic carbon content of 1% in the topsoil to 30 cm and 0.25% in deeper layers, in all soils. Similarly, the degradation rate coefficients (= 0.69/half-life) were reduced fourfold for layers below 30 cm depth. Degradation rates were assumed not to be affected by soil water conditions. In two-domain simulations, the fraction of sorption sites in the macropore region was set to 0.05, giving approximately equal retardation factors in micro- and macropores.

| Compound | Half-life (days) | $K_{oc}$ (cm$^3$ g$^{-1}$) | Henry's constant (Pa.m$^3$ mol$^{-1}$) | Normal dose (kg ha$^{-1}$) |
|---|---|---|---|---|
| Bentazone | 60 | 95 | - | 1.0 -2.0 |
| Butylate | 13 | 400 | 0.56 | 3.0 - 4.0 |
| Chloridazone | 42 | 66 | - | 1.3 - 3.3 |
| 2,4-D | 10 | 20 | 0 | 0.3 - 3.0 |
| Desmedipham | 30 | 1500 | 0 | 0.8 - 1.0 |
| Dicamba | 14 | 2 | 0 | 0.1 - 11.0 |
| Dinoseb | 20 | 63 | 0.13 | < 2.5 |
| Diuron | 90 | 480 | 0 | 0.6 - 4.8 |
| Ethofumesate | 30 | 340 | 0 | 0.5 - 3.0 |
| MCPA | 25 | 20 | - | 0.3 - 2.0 |
| Oryzalin | 20 | 600 | 0 | 1.0 - 2.0 |
| Prometryn | 60 | 400 | 0 | 0.5 - 1.5 |
| Propanil | 1 | 149 | 0 | 1.5 - 1.8 |
| Simazine | 60 | 130 | 0 | 1.0 - 4.0 |
| Triallate | 82 | 2400 | 1.02 | 1.1 - 1.7 |

**Table 1** : Properties of the selected herbicides

Soil properties were estimated from a combination of literature sources and general knowledge/experience. The parameters describing the soil water retention curve for each U.S.D.A. soil textural class were taken from the extensive database of U.S. soils published by Rawls et al. (1982). They reported mean saturated water contents $\theta_s$ for each textural class ranging from 0.4 to 0.5 m$^3$ m$^{-3}$. In this study, $\theta_s$ has been set to 0.45 and 0.4 m$^3$ m$^{-3}$ in topsoil and subsoil layers respectively in each soil type, since the variation within soil texture classes is often as large, or even larger, than the variation between classes (Rawls et al., 1982). Similarly, $\theta_b$ was set to 0.4 and 0.38 m$^3$ m$^{-3}$ in topsoil and subsoil respectively in all soils, since macroporosity does not depend on texture in any predictable way. Also, saturated hydraulic conductivity $K_s$ was set to 10 and 1 cm h$^{-1}$ in the topsoil and subsoil layers respectively for all soil types. Although it is known that $K_s$ may vary considerably from soil to soil, such variations depend more on soil structure than on soil texture. For this reason, well-structured clays usually have somewhat larger $K_s$ values than unstructured sands (Jarvis and Messing, 1994). Nevertheless, for the sake of simplicity, $K_s$ values are here assumed the same for all soils. This assumption should not greatly affect the pesticide leachate concentrations predicted by the model, since $K_s$ is not a particularly sensitive parameter (Jarvis, 1991). In contrast, the saturated hydraulic conductivity of the micropores $K_b$ should be strongly and consistently dependent on soil texture and is also a more critical model parameter (Jarvis, 1991), since in two-domain simulations, $K_b$ controls the partitioning of fluxes between the two pore systems. Thus, values of $K_b$ for each soil were estimated from the soil water retention curve :

$$K_b = C_{\theta_r}^{\theta_b} \int \frac{d\theta}{d\psi^2} \qquad (1)$$

where $\theta$, $\theta_b$ and $\theta_r$ are the soil water content, the saturated micropore water content and the residual water content respectively, $\psi$ is the soil water pressure head and $C$ is an empirical matching constant. Using the Brooks-Corey (1964) model for q(y) and evaluating the integral in equation 1 gives :

$$K_b = C . \left( \frac{\theta_b - \theta_r}{\psi_b^2} \right) . \left( \frac{\lambda}{\lambda + 2} \right) \qquad (2)$$

where $\lambda$ is the pore size distribution index and $\psi_b$ is the soil water pressure corresponding to $\theta_b$. An estimate of the matching factor $C$ (= 130 cm$^3$ h$^{-1}$) was derived by comparing predicted $K_b$ values with the field measurements reported by Jarvis and Messing (1994) for six tilled soils of contrasting texture. The resulting estimates of $K_b$ range from 0.002 cm h$^{-1}$ in the clay soil to 0.2 cm h$^{-1}$ in the sand. The effective aggregate half-width was set to 4 cm in all soils. All remaining solute transport parameters were set to the default values supplied with the model.

| Soil | With macropore flow | | | Without macropore flow | | |
|---|---|---|---|---|---|---|
| | Low risk | Medium risk | High risk | Low risk | Medium risk | High risk |
| Sand | 2 | 4 | 9 | 3 | 3 | 9 |
| Loamy sand | 2 | 4 | 9 | 4 | 2 | 9 |
| Sandy loam | 3 | 3 | 9 | 4 | 2 | 9 |
| Loam | 3 | 3 | 9 | 4 | 2 | 9 |
| Silt loam | 0 | 5 | 10 | 4 | 2 | 9 |
| Sandy clay loam | 0 | 5 | 10 | 4 | 3 | 8 |
| Clay loam | 0 | 5 | 10 | 4 | 3 | 8 |
| Silty clay loam | 0 | 4 | 11 | 4 | 3 | 8 |
| Sandy clay | 0 | 4 | 11 | 4 | 2 | 9 |
| Silty clay | 0 | 3 | 12 | 4 | 2 | 9 |
| Clay | 0 | 3 | 12 | 4 | 2 | 9 |

**Table 2 :** Classification of the selected herbicides into risk categories for leaching

## Results and Discussion

As expected, without macropore flow, the sand soil proved to be the "worst-case" scenario for all compounds, although the differences in predicted leachate concentrations between soils were not large. For example, three compounds were classified in the "low risk" category for the sand soil (Table 2). Compared to the "best-case" soils (sandy clay loam, clay loam, silty clay loam), only two compounds were placed in higher risk categories for the sand soil (one moves from medium to high, one from low to medium). Of course, differences in leachability between soil types may be larger in reality. For example, although not accounted for here, sand soils may have shallower root depths and smaller organic carbon contents than finer-textured soils, although this is not necessarily always the case. Pesticide leachability becomes markedly more sensitive to soil type when macropore flow is accounted for in the model, such that interactions between compound properties and soil properties become important. Overall, the clay and silty clay soils constitute "worst-case" soil types, while the loam and sandy loam soils represents the "best-case" (Table 2). If the predictions for the sand soil in one domain (which is a typical model scenario used for registration purposes) are compared with the clay soil in two domains, it can be seen that ten of the fifteen compounds remain in the same risk category, while two move from low to medium risk, two from medium to high and one from low to high. In the structured soils, none of the selected compounds are classified as low risk when macropore flow is included (Table 2).

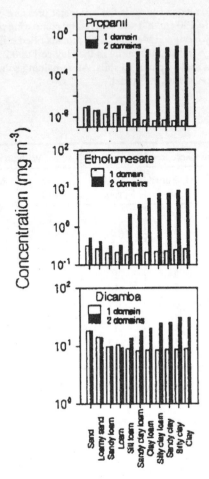

**Figure 1 :** Predicted leachate concentrations for three
selected herbicides

Figure 1 shows predicted leachate concentrations for three of the compounds (Propanil, Ethofumesate and Dicamba), selected as examples of low, medium and high risk compounds when the model is run in one flow domain. Figure 1 demonstrates that the effects of macropore flow may be most dramatic for "non-leachable" compounds. Thus, the predicted leachate concentrations for Propanil, with an assumed half-life of 1 day and a $K_{oc}$ value of 149 cm$^3$ g$^{-1}$, increase by seven orders of magnitude for finer-textured soils, when macropore flow is taken into account. Although not shown here, this dramatic increase results almost entirely from leaching in only one year during the ten-year simulation, when very heavy rainfall occurred soon after application of the compound in the model (108 mm in the month after application, $c$. 3 times the long-term average). In contrast, predicted leachate concentrations are much less affected by assumptions concerning macropore flow for compounds that are readily leachable. For example, leachate concentrations for Dicamba (half-life 14 days, $K_{oc}$ value 2 cm$^3$ g$^{-1}$) only increase by a factor two for the structured soils when macropore flow is taken into account (Fig. 1). An intermediate sensitivity to macropore flow is found for the moderately leachable compound (Ethofumesate). As expected, Fig. 1 shows that macropore flow is much less significant for the coarser-textured soils, with only marginal effects on leachate concentrations predicted by the model. As also noted above, Fig.1 suggests that sandy loam and loamy soils may represent the best protection against pesticide leaching, although this conclusion must be considered tentative until confirmed by experiment.

## Acknowledgements

I am grateful to Jenny Kreuger and Martin Larsson (Dept. Soil Sciences, Swedish University of Agricultural Sciences) for assistance in obtaining information on herbicide properties.

## References

◻   Brooks, R.H. & Corey, A.T. 1964. Hydraulic properties of porous media. *Hydrology Paper no. 3*, Colorado State Univ., Ft. Collins, CO, 27 p.

◻   Gerke, H.H. & van Genuchten, M.T. 1993. Evaluation of a first-order water transfer term for variably saturated dual-porosity flow models. *Water Resour. Res.*, 29, 1225-1238.

◻   Jarvis, N.J. 1991. MACRO - A model of water movement and solute transport in macroporous soils. *Reports & Dissertations* 9, Dept. Soil Sci., Swed. Univ. Agric. Sci., Uppsala, 58 p.

◻   Jarvis, N.J., Stähli, M., Bergström, L. & Johnsson, H. 1994. Simulation of dichlorprop and bentazon leaching in soils of contrasting texture using the MACRO model. *J. Environ. Sci. & Health*, in press.

◻   Jarvis, N.J. 1994a. The MACRO model (Version 3.1). Technical description and sample simulations. *Reports & Dissertations* 19, Dept. Soil Sci., Swed. Univ. Agric. Sci., in press.

◻   Jarvis, N.J. 1994b. Simulation of soil water dynamics and herbicide persistence in a siltloam soil using the MACRO model. *Model. Geo-Bio. Proc.*, in press.

◻   Jarvis, N.J. & Messing, I. 1994. Near-saturated hydraulic conductivity in soils of contrasting texture as measured by tension infiltrometers. *Soil Sci. Soc. Amer. J.*, in press.

◻   Jury, W.A., Focht, D.D. & Farmer, W.J. 1987. Evaluation of pesticide groundwater pollution potential from standard indices of soil-chemical adsorption and biodegradation. *J. Environ. Qual.*, 16, 422-428.

◻   Miljøstyrelsen. 1994. Pesticide modelling and models. *Pesticid-forskning fra Miljøstyrelsen nr.1994*, National Agency of Environmental Protection (Miljøstyrelsen), Denmark, 85 pp.

◻   Rawls, W.J., Brakensiek, D.L. & Saxton, K.E. 1982. Estimation of soil water properties. *Trans. ASAE*, 25, 1316-1320, 1328.

◻   Russell, M.H. & Layton, R.J. 1992. Models and modelling in a regulatory setting: considerations, applications, and problems. *Weed Techn.*, 6, 673-676.

◻   Wauchope, R.D., Buttler, T.M., Hornsby, A.G., Augustijn-Beckers, P.W.M., & Burt, J.P. 1992. The SCS/ARS/CES pesticide properties database for environmental decision-making. *Rev. Env. Contam. Toxic.* 123, 1-155.

6.2.3.        *Characterizing the Impact of Uncertain Soil Properties on*
              *Simulated Pesticide Leaching*

# M. Vanclooster, J. Diels, D. Mallants, J. Feyen[1], M. Dust[2] and H. Vereecken[3]

1)   Institute for Land and Water Management
     Katholieke Universiteit Leuven, Vital Decosterstraat 102
     B-3000 Leuven, Belgium
2)   Institute for Radioagronomy
3)   Department of Chemistry and Dyanmics of the Geosphere
     ICG-4, Forschungszentrum Jülich, KFA, Germany

## Abstract

The paper characterizes the impact of soil variability on the fate and transfer of three pesticides (atrazine, simazine and lindane) in the plow layer of a loamy field. The pesticide leaching flux, which integrates transport, sorption and transformation processes, was calculated using a holistic process-based modelling approach. The within field variability of sensitive model parameters, such as the soil hydraulic properties, the pesticide half-life time and the pesticide distribution constant, were estimated using field experimental and literature data. The impact of the soil variability on the pesticide leaching was assessed using a Monte Carlo simulation procedure.

## Introduction

All over Europe increasing levels of pesticides in surface waters have been reported (e.g. Leistra en Boesten, 1989). It is generally accepted that the study of the pesticide leaching process is cumbersome as it integrates complex mechanisms controlling the fate and transfer of pesticides in the soil environment. Mathematical modelling is believed to help to quantify the different processes governing the fate of pesticides in the soil environment. Since soils are a variable medium in space and time, and subject to a high degree of uncertainty in its characterization, it is to be expected that the model output often deviates from measured data. Model users should be aware of the degree of uncertainty in the model prediction, which is caused not necessarily by the model, but by the uncertainty and the variability in model input and parameters.

There are many ways to tackle the problem of parameter uncertainty in mathematical models. In a first approach, stochastic theory can be applied to develop an analytical solution of the model output probability density function (PDF) in terms of the characterized PDF of the model input. This approach appeals on simple analytical model representations and is not appropriate for holistic mechanistic models. Alternatively, one can use first order uncertainty analysis, which is obtained by truncating a Taylor serie expansion about the mean of functionally related variables (Cornell, 1972). First order uncertainty analysis appeals on continuity and differentiability of the model. In addition, the coefficient of variation of the input parameters needs to be less to 20 %, which is generally not the case for soil properties (Jury, 1986). Hence, for complex numerical models, Monte Carlo simulation is a better alternative. When carrying out Monte Carlo simulations, one designs the PDF of the input parameters. In a next step a sample is drawn from the constructed PDF for which the model output is calculated. The iteration continues until inferences about the model output can be made. This paper present the results of a Monte Carlo approach used for the assessment of the effect of uncertainty on model input and parameters on the output of a holistic mechanistic model, simulating the fate and transfer of pesticides in the rootzone of soils. The simulation model WAVE, describing Darcian water flow, convective solute flow, linear equilibrium sorption and first order decay, was used to simulate the fate of atrazine, simazine and lindane in the plow layer of a loamy soil.

## Materials and Methods

### *The model*

The model used in this study is the WAVE-model (Water and Agrochemicals in soil and Vadose Environment, (Vanclooster et al., 1993)). WAVE is a deterministic mechanistic model describing the fate of water and solutes in a layered soil subjected to different unsteady boundary conditions. The model is a revised and extended version of the SWATNIT-model (Vereecken et al., 1991), integrating a water flow module (SWATRER (Dierckx et al., 1986)), a solute transport module, a solute transformation module, a heat flow module and a crop growth module (SUCROS87, (Spitters et al., 1988)). Initially, the model was designed to study water and nitrogen flow in the crop continuum (see Vanclooster et al., 1994, amongst others). The current version of the model was extended in order to account for pesticide sorption and transformation processes (Vereecken et al., 1994; Dust et al.,1994). Water flow in the model is described using the Richards equation. Parametric models are used in this study for describing the moisture retention and the hydraulic conductivity relationships. Solute transport is described using a physical non-equilibrium convection dispersion equation. Heat transport is described using the heat flow equation. In the present study only convective flow and no crop growth were considered. Additionally, it was assummed that pesticide sorption follows an equilibrium linear isotherm. The distribution coefficient is estimated from the fraction of the organic carbon in the soil and the $K_{oc}$-value. First order decay of the pesticide is considered both in the soil water and the sorbed phase. Reduction of the decay rates in terms of moisture and temperature is described according to Boesten and van der Linden (1991).

### *Uncertainty analysis*

An uncertainty analysis aims at quantifying the effect of input and parameter uncertainty on model output. In the framework of pesticide modelling, the system is poorly defined because many unknown uncertainty factors influence to an unknown extent the model prediction. In this context, a second objective of an uncertainty analysis is to define the input uncertainty which dominates the overall model output uncertainty. The selected model output in this paper is the fraction of pesticide leached form the top soil layer of a loamy field (10 cm thickness). The simulation period covers one climatic year, starting the first of May, the day that the pesticide was applied.

A Monte Carlo approach was developed in which 500 realizations were sampled from the constructed PDF's of the soil hydraulic, sorption and transformation properties. In order to construct these field scale PDF's, it was assumed that the soil physical parameters ($q_r$, $q_s$, a, n , m, $K_{sat}$, B, N, Bd, and %C (see legend Table 1)) were correlated with each other, but uncorrelated with the first order decay rates, $K_{decay}$. The statistical properties of the physical model input were derived from a transect study in the plow layer of a Typic Hapludalf (Diels, 1986). Thirty undisturbed soil samples (0.20 m height, 0.20 m ID) were collected at a regular grid. The unsaturated hydraulic conductivity was determined using the crust method (Bouma et al., 1971). The measurements of the crust method were combined with measurements from the hot-air method applied to samples taken within the larger soil cores (Arya et al., 1975). In addition, undisturbed soil cores (100 $cm^3$) were collected in the samples used for the crust method for the determination of the bulk density and the moisture retention characteristic. The measured hydraulic conductivity and moisture retention data were fitted to parametric models using the RETC-code (van Genuchten et al., 1991). Disturbed samples were used to determine the organic carbon content. Thirty observations for the physical properties were available. The Shapiro-Wilk test for normality on the measured frequency distributions (SAS Institute,1987) indicated that $\theta_r$, $\theta_s$, N and Bd were normally distributed. The remaining parameters ($\alpha$, n , m $K_{sat}$, B, and %

C) were transformed in order to obtain a normal distribution similar to Carsel and Parrish (1988). After normalization of all parameters, a 9-dimensional variance-covariance matrix $\Sigma$, was calculated. Given P being a vector of 9 independent univariate standard normal variates, a 9-dimensional multivariate normal distribution Q can be generated by transforming P through (Johnson et al., 1984):

$$Q = LP + \mu \qquad (1)$$

where L is the lower triangular Choleski decomposition of $\Sigma$, and $\mu$ the mean vector of the multivariate normal distribution. In this study we obtained 500 realizations of transformed physical parameters by generating 9*500 realizations of independent standard normals and using (1). Both L and m are estimated from the experimentally determined mean and variance-covariance structure. The obtained multivariate normal distribution was transformed back to obtain the PDF's of the soil hydraulic properties, organic carbon and bulk density. The values for the pesticide distribution coefficient were obtained by multiplying the realizations of the generated organic carbon content with the pesticide specific $K_{oc}$-value. The $K_{oc}$-values for atrazine, simazine and lindane used were 160, 140 and 1300 l$kg^{-1}$, respectively (Jury and Ghodrati, 1989). The PDF of the pesticide half-life time was estimated from literature data. It was assumed that half-life times of the three considered pesticides were uniformly distibuted between 47-110, 11-94 and 26-630 $day^{-1}$, respectively (Nash, 1980; quoted by Jury and Ghodrati, 1989). The generated half-life times were next transformed to obtain a uniform PDF for the first order decay rates. A summary of the statistical parameters of the generated model input is given in Table 1.

| Parameter | Mean | St.Dev. | Skewness |
|---|---|---|---|
| $\theta_s^{(1)}$ | 0.3978 | 0.016081 | -0.07274 |
| $\theta_r^{(1)}$ | 0.05 | 0. | - |
| $\alpha^{(1)}$ | 0.00369 | 0.001385 | -0.74334 |
| $n^{(1)}$ | 1.406253 | 0.553934 | 0.702228 |
| $m^{(1)}$ | 0.515501 | 0.307306 | 1.287961 |
| $K_{sat}^{(2)}$ | 28.39529 | 237.7812 | 16.66243 |
| $B^{(2)}$ | 0.201305 | 1.063947 | 13.55623 |
| $N^{(2)}$ | 1.913188 | 0.303796 | -0.11123 |
| $Bd^{(3)}$ | 1.526475 | 0.072204 | 0.032968 |
| $\%C^{(4)}$ | 2.2657 | 0.115066 | 0.704063 |
| $K_{decay}(atrazine)^{(5)}$ | 0.00957 | 0.002468 | 0.450481 |
| $K_{decay}(simazine)^{(5)}$ | 0.016419 | 0.010427 | 1.84296 |
| $K_{decay}(lindane)^{(5)}$ | 0.003794 | 0.004083 | 2.766597 |

Legend:

(1) the parameters of the moisture retention characteristic function ($\theta_{(h)} = \theta_r + ((\theta_s - \theta_r)/(1+\alpha h^n)^m)$, where h is the soil water pressure head (cm));
(2) the parameters of the hydraulic conductivity function ($K_{(h)} = K_{sat}/(1+B\,h^N)$);
(3) the soil bulk density (kg $l^{-1}$);
(4) the soil organic carbon (%);
(5) the potential first order decay coefficient ($day^{-1}$).

**Table 1 :** Statistical properties of the 500 generated parameters as used in the Monte Carlo simulation

# Results and Discussion

Summary statistics of the calculated frequency distributions for atrazine, simazine and lindane are given in Table 2. The analysis reveals that the mean of the calculated $PDF_{flea}$'s is proportional to the mobility of the pesticide (here flea refers to the fraction of the applied pesticide, which leaches out of the first soil compartiment). When comparing the means of the $PDF_{flea}$ generated for atrazine and simazine, it can be concluded that atrazine is more mobile, in spite of a higher $K_{oc}$ constant (160 vs. 140 l kg$^{-1}$). The apparent higher mobility for atrazine as compared to simazine is a result of the interaction between the sorption and decay behaviour. A low sorption capacity, as expressed by a low Koc-value, enhances the mobility of the pesticide, though the positive effect can be easily counter-balanced by the decomposition rate. A high decomposition rate minimizes the persistence of the pesticide in the soil environment and hence minimizes the probability of being leached. The counterbalancing effect of sorption and decay on pesticide mobility is not realized, looking at the fate of lindane. In this case, the high sorption capacity dominates the low decay rate which results in a low mean fraction of lindane leached. Given the dynamic nature of the considered system, one should be careful when using screening models to rank different pesticides in terms of leaching vulnerability.

From Table 2 it can be concluded that considerable uncertainty in terms of variance, standard deviation, and range is associated with the predicted leaching flux. Given the many assumptions and simplifications which were established when developing the uncertainty analysis (no crop, only piston flow, lineair equilibrium sorption phenomena, first order decay, deterministic climatic conditions), already large coefficients of variation are generated (22 %, 43 %, 24 % for atrazine, simazine and lindane, respectively). Hence, when using models for pesticide regulatory purposes, the impact of model input uncertainty on model output can not be neglected. When developing risk analysis strategies, attention is merely focused on the right hand tails of the PDF's rather than the expected mean. The value of the fraction leached corresponding to the one-side 95 % exceedance level was 29.65 %, 30.55 % and 7.0 % for atrazine, simazine and lindane, respectively. Increasing the probability to 99 % shifts these critical levels to 34.5 %, 33.65 % and 7.9 %. From a risk analysis point of view, simazine and atrazine are behaving similarly, unlike their different expected mean values.

| Statistic parameter | Atrazine | Simazine | Lindane |
|---|---|---|---|
| Mean | 21.53 | 18.09 | 5.30 |
| Variance | 24.28 | 62.73 | 1.72 |
| Standard Deviation | 4.92 | 7.92 | 1.31 |
| Minimum | 9.2 | 1.3 | 0.9 |
| Maximum | 46.1 | 47.7 | 9.8 |
| Range | 36.9 | 46.4 | 8.9 |
| Skewness | -0.52 | 0.04 | 0.81 |
| Kurtosis | 0.92 | -0.47 | 1.37 |
| Shapiro-Wilk value | 0.98 | 0.97 | 0.94 |

**Table 2 :** Summary statistics of the probability density of the percentage pesticide leached out of the 10 cm soil layer (PDF$_{flea}$)

Analyzing the shape parameters of the three generated PDF's (skewness, kurtosis, normality statistic), it is concluded that the three investigated pesticides behave considerably different, notwithstanding the apparent normality of the generated distributions (Shapiro-Wilk value close to 1.0). The PDF$_{flea}$(atrazine) is skewed left, the PDF$_{flea}$(lindane) is skewed right, while the PDF$_{flea}$(simazine) is bell-shaped. The kurtosis values indicate that the PDF$_{flea}$(simazine) is more uniform, while the PDF$_{flea}$(atrazine) and PDF$_{flea}$(lindane) are more narrow as compared to a normal distribution. It is believed that including additional uncertainty terms will influence the shape parameters of the PDF's.

# Conclusions

This paper reviews the results of a study conducted to assess the impact of variable soil properties on predicted pesticide leaching. A mechanistic model, describing convective pesticide transport based on the Darcian approach, with first-order decay and linear equilibrium adsorption, was integrated in a stochastic framework using a Monte Carlo approach, yielding the relation between input and output uncertainty. The procedure was applied to predict the fraction of atrazine, simazine and lindane leached from the first 10 cm of a loamy soil. Probability density functions of the hydraulic properties, the pesticide distribution coefficients and first order decay rates were derived from both measured and literature values. For the soil hydraulic and sorption parameters, a multivariate approach was adopted.

The predicted mean fraction of pesticide leached is influenced by the interaction between different strong non-linear transport and transformation processes. Therefore, results from simplified screening models that do not account for the dynamic and stochastic nature of the pesticide transport and transformation processes, need a careful interpretation when used for regulatory purposes.

The uncertainty on the model prediction is considerable, given the many simplifications which were adopted when carrying out the study. It is expected that the inclusion of other processes in the uncertainty analysis will effect the prediction uncertainty in a different way. In terms of risk analysis, one should consider the uncertainty on the model prediction.

# References

☐   Arya, L.M., D.A. Farell and G.R. Blake, 1975. A field study of soil water depletion patterns in the presence of growing soybean roots. 1: Determination of hydraulic properties of the soil. *Soil Science Society of America J.*, 39:424-430.,

☐   Boesten, J.J.T.I and A.M.A. van der Linden, 1991. Modeling the influence of sorption and transformation on pesticide leaching and persistence. *J. of Environmental Quality*, 20:425-435.

☐   Bouma, J., D.I. Hillel, F.D. Hole and C.R. Amerman, 1971. Field measurement of unsaturated hydraulic conductivity by infiltration through artificial crusts. *Soil Science Society of America J.*, 35: 362-364.

☐   Carsel, R.F and R.S. Parrish, 1988. Developing joint probability distributions of soil water retention characteristics. *Water Res.*, 24:755-769.

☐   Cornell, 1972. First order analysis of model and parameter uncertainty. In: *Proc. Int. Symp. on Uncertainties in Hydrologic and Water Res. Systems.* Tucson, Arizona, USA, pp. 1245-1274.

☐   Dierckx, J., C. Belmans and P. Pauwels, 1986. SWATRER. A computer package for modelling the field water balance. *Reference Manual*, Soil Water Engng. Lab., KULeuven, Belgium, 114pp.

☐   Diels, J., 1986. De ruimtelijke variabiliteit van bodemkenmerken in de bouwvoor van een Aba bodemtype. *Eindwerk Faculteit Landbouwwetenschappen*, KULeuven, 106 pp.

☐   Dust M., H. Vereecken, M. Vanclooster and F. Führ, 1994. Comparison of model calculations and lysimeter results on dissipation and leaching behaviour of ($^{14}$C) clopyralid. *Submitted to the eight IUPAC congress of pesticide chemistry*, Washington D.C, July 4-9, 1994.

☐   Johnson, M.E., C. Wang and J. Ramberg, 1984. Generation of multivariate distributions for statistical applications. *Am. J. of Mathematical and Management Sciences*, 4(3-4):225-248.

□ Jury, W.A., 1986. Spatial variability of soil properties. In : S.C. Hern and S.M. Melancon (eds.) *Vadose zone modeling of organic pollutants*. Lewis, Chelsea, MI, USA. pp.245-269.

□ Jury, W.A. and M. Ghodrati, 1989. Overview of organic chemical environmental fate and transport modeling approaches. In : *Reactions and movement of organic chemicals in soils*. SSSA Special Publication, 22:271-304.

□ Leistra, M and J.J.J.T.I. Boesten, 1989. Pesticide contamination of groundwater in Western Europe. *Agricultural Ecosystems and Environment*, 26:369-389.

□ SAS, Institute Inc., 1987. SAS/STAT, guide for personal computers. Version 6, Edition Cary, NC, USA.

□ Spitters, C.J.T., H. van Keulen and D.W.G. van Kraailingen, 1988. A simple and universal crop growth simulation model, SUCROS87. In: R. Rabbinge, H. van Laar and S. Ward (eds.). *Simulation and systems management in crop protection*. Simulation Monographs, PUDOC, Wageningen, the Netherlands.

□ Vanclooster, M., P. Viaene, J. Diels and J. Feyen, 1993. The WAVE-modules: theory and input requirement. *Internal Report 10*, Institute for Land and Water Management, KULeuven, Belgium. 85pp.

□ Vanclooster, M., P. Viaene, J. Diels and J. Feyen, 1994. A deterministic evaluation procedure applied to an integrated soil-crop model. *Modelling Geo-Biosphere Processes*. (In press).

□ van Genuchten, M. Th., F.J. Leij and S.R. Yates, 1991. The RETC-code for quantifying the hydraulic functions of unsaturated soils. EPA/600/2-91/065. 83 pp.

□ Vereecken, H., M. Vanclooster, M. Swerts and J. Diels, 1991. Simulating nitrogen behaviour in a soil cropped with winter wheat. *Fertilizer Research*.

□ Vereecken H., M. Dust, M. Vanclooster and F. Führ, 1994. Modelling the fate of methatbenzthiazuron in arable soil using a two year lysimeter study. *Submitted to the eight IUPAC congress of pesticide chemistry*. Washington D.C., July 4-9, 1994.

*6.2.4.*      *Criteria of existing pesticide models evaluation*

# A.A.M. Del Re, M. Trevisan, E. Capri, S.P. Evans and E. Brasa[*]

Istituto di Chimica Agraria ed Ambientale, Facoltà di Agraria, Università Cattolica del Sacro Cuore,
Via Emilia Parmense 84, I-29100 Piacenza, Italia.
[*] Zero Computing srl, Milano, Italia

## Introduction

Mathematical models forecasting pesticide groundwater pollution are increasingly used both for scientific and normative reasons. Model evolution is continuous and evaluation of model performance is an important issue. The aim of this paper is to propose possible solutions towards establishing criteria for evaluation of pesticide soil transport models.

Many researchers have worked, and are currently working in the field of model evaluation (Wagenet & Rao 1990; Wagenet 1993; Del Re & Trevisan 1993a; Vanclooster et al. 1994; Yon et al. 1994). So far no consensus has developed towards the definition of a glossary of terms for all operations connected to models, to avoid that the same things be identified by different terms: for instance validation, testing or evaluation are terms frequently used to identify the same thing.

A range of aspects must be taken into consideration prior to model use, bearing in mind that no one model has been shown to be a reliable predictor of all field cases. Considerations may range from: the purpose for which a model was originally developed; how a model may supply information for the achievement of a given objective; the limitations of each model; the availability of model-required inputs; the uncertainty range of input data; limited current understanding of basic processes; how this is reflected in their quantitative solution in models; how much uncertainty is present in model output, and so on.

In the present context model evaluation is intended as a means of providing the user with a scheme which should be as complete and detailed as possible concerning the model, so that all the questions, doubts and problems which arise may be solved in a coherent and reasonable fashion. Evaluation will supply the user with basic information on how a model may be applied and validated, and what degree of precision to expect from simulations in a given context.

Evaluation procedures of model performance may be broken down into three components: model classification, model analysis and validation procedures.

## Model classification

In an ideal world model classification should be carried out by the user, based on the individual's specific knowledge. In reality users have different degrees of understanding of what models are. Classificatory schemes have been proposed to help the user select the most suitable model given the specific context of use and the outputs required.

| | |
|---|---|
| What is the purpose of the model? | research |
| What does the model predict? | soil pesticide leachate |
| At what scale does the model work? | field |
| Programming language: | FORTRAN |
| Minimum system size: | PC |
| Type of model: | deterministic |
| Author name: | Walker |
| Year of publication: | 1993 |

**Table 1** : Example of model classification for the VARLEACH
model release 2.0 (Walker, pers. comm., 1993)

Model classifications should therefore guide the user, for instance as shown in table 1, helping to answer basic question such as:

(a)     *What is the purpose of the model?* Models have been developed for research, screening, management, regulatory purposes, etc. Using a model outside the prescribed scope of use for which it was developed may be detrimental.

(b)     *What does the model do?* Is the model a leaching, runoff etc. model?

(c)     *At what scale does the model work?* Models have been developed to operate at different spatial scales: plot, lysimeter, field, watershed, etc.;

(d)     *How does the model work? How did the developer approach and solve basic processes and how is this reflected in model structure?* Informations such as the name of the author(s), what language the programme is written in (i.e. FORTRAN, BASIC, C etc.); what is the minimum system size required (PC, mini, mainframe) etc. fall within this category. The developers view of the world and his understanding of how basic processes work is also reflected in model structure: models may be stochastic or deterministic, stochastic if incorporating uncertainty in the model, deterministic if the system's response will be univocal given a set of assumed conditions; mixed models are also available.

(e)     *What data are necessary to make the model work?* Data may be divided into state and site parameters: state parameters defines inputs required to run basic processes, site inputs define specific characteristics of the location for which the model is run. Each model should be provided with a table identifying which input parameters are required, the degree of precision requested, etc.

For illustrative purposes, Jarvis' scheme (1991) of input parameters required by the VARLEACH model (Walker, comm. pers. 1993) is summarized in Table 2. The scheme indicates which input parameters are required and if these have to be measured directly or whether they may be general knowledge, literature values or calibrated values. With this classification it is easier to understand the importance of a given output and more importantly, what scientific information are required in order to make the model work. It is self-evident that a measured input is more time-consuming to obtain than one which may be found in the literature. For instance prior to VARLEACH implementation it is necessary to measure the half-life of a given pesticide in soil at different temperatures and moisture levels, otherwise the model will not function correctly.

| parameter | estimation method |
|---|---|
| Field capacity | 1,2 |
| Water at 200 KPa | 1,2 |
| Initial top soil moisture | 1 |
| Bulk density | 1,2 |
| Sorption coefficient | 1,2 |
| Increment of sorption coefficient | 1,2,3 |
| Half-life and correlated constant | 1 |
| Factor changing FC/Kd/HF with depth | 1,2,3 |
| Water solubility | 1,2 |
| Application dose | 1 |
| Application date | 1 |
| Total soil depth | 1 |
| Horizon | 1 |
| Weather data | 1,2 |
| Latitude, altitude | 1 |

1 = directly measured
2 = general knowledge or literature values
3 = calibrated

**Table 2 :** Input parameters required by the VARLEACH model release 2.0
(Walker, pers. comm., 1993) and estimation method proposed by Jarvis (1991)

(f)     *What type of processes are simulated?* The identification of which basic processes are
        simulated allows a good classification of models. In this phase it is necessary to remember
        both hydrological processes (evapotranspiration, runoff, water flow, etc.), pesticide processes
        (degradation, sorption, volatilization, plant uptake, etc.) and parameters connected to the
        numerical techniques adopted to solve equations (for instance dispersion). Prior knowledge of
        this is important since it highlights which processes are not simulated; this is of some
        consequence also for model validation.

(g)     *How is the model programmed?* In general all models are structured (Del Re & Trevisan 1994)
        with a main programme which reads the input data, initializes the variables, and then launches
        temporal and spatial loops within which transport and pesticide transformation processes are
        simulated at different temporal scales, to obtain a series of outputs. In an evaluation procedure
        this step is fundamental: if the user does not know how a model 'thinks' it is impossible to
        carry on with the successive evaluation steps. Consequently it is necessary either to create or
        to have the author send the flow-chart of model codes, in order to highlight within the different
        subroutines the path taken by each variable. Figure 1 shows a part of the flow diagram of the
        VARLEACH model, which should the path taken by the variable K (Kd). Frequently the
        absence of this sort of information in sensitivity tests or validation procedures lead to variables
        being modified in an incorrect way, resulting in unrealistic results.

(h)     *What types of outputs are produced?* Which outputs are selected for successive visualiza-
        tion/print and which are ignored? Different types of information are required according to the
        purpose of model implementation: pesticide mass in soil, centre of mass, peak concentration,
        mass in leachate etc. It is necessary to know which are calculated but not addressed to output,
        and viceversa. Few models provide this sort of information directly in graphical form.

(i)     *Is the model supported by exhaustive documentation, both theoretical and in terms of software?*
        The model must not be a container for a series of black boxes which cannot be opened or
        validated.

(j)     *Is the programme user friendly, and what level of expertise, both in terms of computing and hard science, is required?* The term 'user-friendly' is intended to cover all aspects concerning the degree of interactivity between user and programme, ease in input file preparation, manageable and suitable outputs, well-documented source codes, a complete user manual, well documented theories, presence of a literature-derived data base of input data, and so on.

(k)     *What algorithms are used to solve differential equations?* In deterministic models what is the resolution technique of these equations? What degree of precision/numerical dispersion is associated with their solution? For example the most widely used method for solving Richard's equations is the Crank-Nicolson, which overcomes the limitations on the size of the time sub-interval compared to the size of the spatial sub-interval, but sometimes other solutions may be adopted.

In conclusion the general the rule should be: the more detailed the information, the quicker the user will be able to focus in onto the model and determine what is necessary to make the model run.

## Analysis of model ruggedness

Analysis of model ruggedness is an evaluation procedure of model stability in which intrinsic control parameters are varied and extreme values of input variables are used.

Ruggedness serves different purposes, for instance:

- what are the extreme ranges beyond which the model ceases to function;

- assess model fragility in extreme conditions in order to reveal weaknesses, contradictions or voids in simulated basic processes;

- if pre-determined and not normally modifiable parameters (i.e. layer size, time loop size) may be changes, what degree of difficulty this entails, and whether the model still forecasts reasonably. This is often carried out by model developers, but remains either confidential information or is never published;

- help to determine if the model may be used by primary or advanced users;

- improve user-friendly interfaces with, for instance 'passive suggestions' made to an inexpert user inputting extreme values.

## Sensitivity analysis

Sensitivity analysis aims at evaluating the variations induced by the uncertainty of input parameters and their relative influence on outputs (Boesten 1991). Sensitivity analysis may be carried out using different techniques: amongst these are linear models using the Plackett-Burman screening design (Plackett & Burman 1946), the FAST method (Fontaine et al. 1992), the response surface technique (Del Re & Trevisan 1993b), bootstrapping (....) etc. The main goals of sensitivity analysis are to: rank input parameters on the basis of their effect in output; determine the level of accuracy required for each input; define the relative importance of the different basic processes in the models; assess the propagation of error.

## Model validation

Validation is the final step in a model evaluation evaluation procedure, and answers questions such as *"How close are my estimates to real world conditions?"* Validation is the term to describe an assessment of goodness-of-fit between observed and simulated data in a fixed scenario.

In this phase it is important to bear in mind the uncertainty existing both in experimental and in simulated data.

How is validation carried out? It is necessary to define a minimum data set for the model and an appropriate experimental protocol for the collection on required inputs. A minimum data set for each model identifies which state and site parameters must be measured for each validation exercise.

In the experimental design it is also necessary to take spatial variability into account (Rao & Wagenet 1985). It is also essential that a large number of replicates or data points be collected in order to allow a statistical evaluation of data.

When field data and model simulations are compared it is necessary to take into account what is defined as uncertainty, which in models derives from inputs and from the way basic processes are represented. To estimate uncertainty in a deterministic model, which produces only one forecast, this must be run several times varying input parameters within a pre-defined range. In first approximation, runs should be as numerous as possible, since variations linking input to output are unknown; in stochastic models this is obtained, for instance, with a Monte Carlo simulator. It is important that only inputs which previous sensitivity tests have identified as being those with the greatest effect on model outputs be varied; often these are half-life, Kd, hydraulic properties and rainfall amount.

Output goodness-of-fit with observed data may be assessed qualitatively and quantitatively using graphical methods and statistical indices, respectively. Graphical methods highlight the superimposition between simulated and observed data for a qualitative evaluation of similarity: for instance the absolute differences between simulated and observed highlight over- and under-estimation; the comparison between the mean, the $1^{st}$ standard deviation and 95% confidence intervals of simulated and observed data (Trevisan et al. 1994). are also useful graphical techniques for an improved visual assessment.

A large number of indices are available which may be used to highlight differences between observed and simulated data (Loague & Green 1991); amongst these are the maximum error, modelling efficiency and the correlation coefficient. We however consider them as having limited statistical validity since requiring a numerous data set; in general these become useful when the number of observations is greater than the number of variables used by the model. It should be borne in mind that if simulated values are below the detection limit, these should be set to zero as occurs for observed data.

In conclusion criteria for assessing model performance may be carried out using graphical methods, for a qualitative assessment in the presence of small data sets, and indices and statistics, for a quantitative evaluation of larger sets.

# *Conclusions*

It is important to emphasize that there are no definitive conclusions. Rather this is an open-ended paper, targeted to stimulate, in the short- to medium term, a lively discussion aimed at the production of criteria or guidelines for model evaluation. In future, models will play a central role both for registration of new products and for environmental assessment purposes. It is felt that criteria for model evaluation may also be extended to become a tool for scientific and legal purposes; this therefore becomes a key issue which needs to be tackled quickly, given the vast increase in model users.

# *Acknowledgements*

Research was carried out as part of EC-DG XII Environment 1990-1994 programme, project EVCV-CT92-0226.

# Bibliography

▫ Boesten J.J.T.I. (1991). Sensitivity analysis of a mathematical model for pesticide leaching to groundwater. *Pesticide Science* 31: 375-388.

▫ Del Re A.A.M. & Trevisan M. (1993a). La modélisation des déplacements et de l'evolution des pesticides dans la zone insaturée du sol. 23$^{\text{ème}}$ Congrès Groupe Fran'ais des Pesticides, 'Dégradation des pesticides dans l'environnement' 25-27 May 1993, *in press.*

▫ Del Re A.A.M. & Trevisan M. (1993b). Testing models of the unsaturated zone. In: Proceedings IX$^{\text{th}}$ Symposium Pesticide Chemistry, Del Re et al., (eds), pp. 5-31, Piacenza 11-13 October 1993, Edizioni Biagini, Lucca, Italia.

▫ Del Re A.AA.M. & Trevisan M. (1994). Selection criteria of xenobiotic dissipation models in soil. ESA International Workshop 'Modelling the fate of agrochemicals and fertilizers in the environment', Venice (Italy) 3-5 March, 1994: in press.

▫ Fontaine D.D., Havens P.L., Blau G.E. & Tillotson P.M. (1992). The role of sensitivity analysis in groundwater risk modeling of pesticides. *Weed Technology* 6: 716-724.

▫ Jarvis N. (1991). MACRO - A model of water movement and solute transport in macroporous soils. Swedish University of Agricultural Sciences, Department of Soil Sciences, Reports and Dissertations 9, 58 pp.

▫ Loague K. & Green R.E. (1990). Modeling pesticide fate in soil. In: Cheng H.H. et al. (eds). Pesticides in the Soil Environment, pp. 351-399, SSSA Book Series, no. 2, Soil Science Society of America, Madison, WI, USA.

▫ Plackett R.L. & Burman J.P. (1946). The design of optimum multifactorial experiments. *Biometrika* 33: 305-325.

▫ Rao P.S.C. & Wagenet R.J. (1985). Spatial variability of pesticides in field soils: methods for data analysis and consequences. *Weed Science* 33: 18-24.

▫ Trevisan M., Capri E., Evans S.P. & Del Re A.A.M. (1994). Validation and comparison of pesticide soil transport models for field dissipation of metamitron. Proceedings 5$^{\text{th}}$ International Workshop on Environmental Behaviour of Pesticides and Regulatory Aspects, *this volume.*

▫ Vanclooster M., Diels J., Mallants D. & Feyen J. Analyzing the effect of variable soil properties on predicted pesticide leaching. Proceedings 5$^{\text{th}}$ International Workshop on Environmental Behaviour of Pesticides and Regulatory Aspects, *this volume.*

▫ Wagenet R.J. & Rao P.S.C. (1990). Description of nitrogen movement in the presence of spatially variable soil hydraulic properties. *Agric. Water Mgt.* 6: 227-242.

▫ Wagenet R.J. (1993). A review of pesticide leaching models and their application to field and laboratory data. In: Proceedings IX$^{\text{th}}$ Symposium Pesticide Chemistry, Del Re et al., (eds), pp. 33-62, Piacenza 11-13 October 1993, Edizioni Biagini, Lucca, Italia.

▫ Yon D., Boesten J.J.T.I. & Travis K. The use of pesticide leaching models in regulatory setting. Proceedings 5$^{\text{th}}$ International Workshop on Environmental Behaviour of Pesticides and Regulatory Aspects, *this volume.*

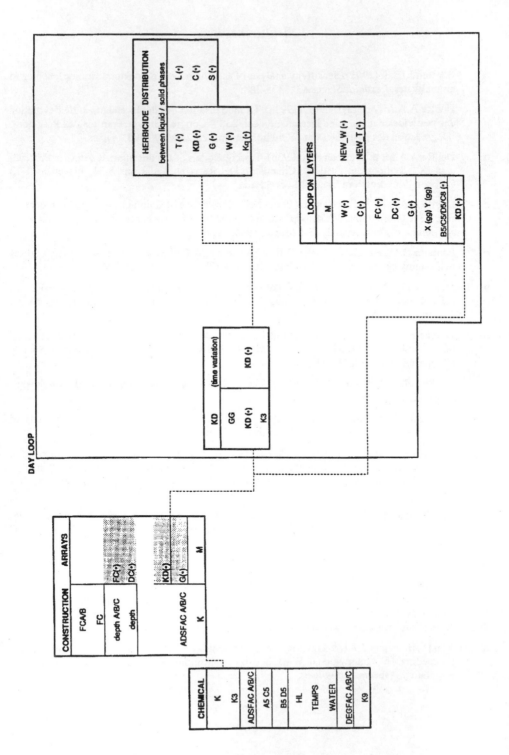

**Figure 1 :** VARLEACH model: flow chart of variable (Kd)

6.2.5.    Use of USEPA Computer Models to Predict Mancozeb
          Concentration in Surface Waters Following its Use in Potato Fields

# L. Keller[1], J. Carbone[1], J. Hamilton[1], S. Hurt[1], N. Gocha[2], J. Ollinger[3]

Rohm and Haas Company

[1]   Ecotoxicology and Environmental Risk Assessment, Toxicology Department,
      Research Division, 727 Norristown Road, Spring House PA 19477;
[2]   European Agricultural Regulatory Department, 185, Rue de Bercy,
      75579 Paris Cedex 12,
[3]   Agricultural Chemicals Regulatory Department, Independence Mall West,
      Philadelphia PA 19105

## Abstract

Three USEPA computer simulation models were used to predict mancozeb concentration entering ponds in Northern European potato-growing regions due to spray drift or runoff. Assuming a worst case of 5% spray drift, the Tier I pseudo first order decay model EFATE predicted that mancozeb would not exceed typical USEPA levels of concern for nontarget aquatic organisms from spray drift. PRZM2 and EXAMS runoff modeling used the same application pattern to estimate the risk from runoff in the field. This scenario demonstrated negligible risk to nontarget aquatic organisms from mancozeb exposure due to the low runoff potential and rapid dissipation of mancozeb in surface water adjacent to treated fields. The presentation discusses how modeling is used in the USEPA risk assessment process.

## Introduction

An important aspect of the U.S. Environmental Protection Agency (USEPA) registration process is to assess potential effects of agricultural chemicals on nontarget terrestrial and aquatic wildlife that may come into contact with agricultural chemicals during approved patterns of use. Ecological risk assessment is performed using a tiered approach to decision-making in which predicted or estimated environmental concentrations (EECs) of an agricultural chemical are compared to toxicity thresholds for nontarget organisms representing major taxonomic groups of environmental concern. To assess Tier I aquatic risk, the "worst case" EECs are typically calculated for ponds of standard size and depth using the maximum label application rate. In Tier II and higher tiers, the EECs are refined by additional laboratory or field testing, or by computer simulation modeling, because Tier II refinement provides more realistic estimates of EECs. The USEPA framework (Norton et al. 1992) comprises three major phases; *problem formulation, analysis (characterization of exposure, characterization of ecological effects), and risk characterization.* Computer simulation models may be used to predict spatial and temporal exposure to evaluate the likelihood of adverse effects on nontarget organisms from agricultural chemical exposure during routine use. Effects on nontarget organisms are measured using established indicator species to determine acute and chronic toxicity values. Once the ecological risk assessment is completed, *risk management* considers, on a case-by-case basis, the results of the risk assessment as well as economic, political, social and legal issues to determine whether to register agricultural chemicals for commercial use and whether restrictions on use patterns should be imposed to maximize efficacy while maintaining adequate safety margins for nontarget organisms. In this example, tiered analysis was used to predict environmental residue concentrations resulting from the application of the ethylene bisdithiocarbamate (EBDC) fungicide mancozeb to potato fields using three USEPA computer simulation models.

# Ecological Risk Assessment for Mancozeb Application on Potatoes

The conceptual model for mancozeb application identified spray drift and runoff as possible routes of exposure to aquatic organisms. Three computer simulation models, EFATE (Pesticide Residue Fate Simulation Computer Program (R. Lee, unpublished)), PRZM2 (Pesticide Root Zone Model (Mullins et al. 1993)), and EXAMS (Exposure Analysis Modeling System (Burns et al. 1982)) were used to develop spray drift or runoff exposure scenarios using a reasonable worst case application pattern of multiple applications at three different use rates (Table 1). EECs entering surface waters as spray drift or runoff following ground spray application were compared to aquatic organism ecotoxicity data to assess potential risks to nontarget organisms inhabiting ponds adjacent to agricultural fields.

The effects of mancozeb were characterized by comparing EECs to toxicity thresholds for the aquatic organisms most sensitive to short-term (rainbow trout 48 hr EC50 = 1.9 mg/L) and chronic (fathead minnow early life stage test NOEC = 0.0052 mg/L) exposure. Within the USEPA assessment scheme, the level of concern for mancozeb to acute exposure would be defined as 1/10th of the lowest acute value (0.190 mg/L) and the lowest NOEC for chronic exposure (0.0052 mg/L).

*Tier I Example of Modeling Using EFATE: EECs Of Mancozeb In Surface Waters From Spray Drift.* Prediction of mancozeb entering surface water as spray drift was assessed by the Tier 1 USEPA model EFATE, a pseudo-first order residue fate model. EFATE calculates the dissipation rate of an agricultural chemical and predicts daily and average residue concentrations assuming first-order decay of the compound in a single environmental medium (e.g., water or soil). This model tends to overestimate EECs because it cannot account for compartmentalization of the compound in different environmental media. EFATE inputs are estimated or measured initial residue concentration, application pattern (use rate and frequency) and half-life of the compound appropriate for the matrix being modeled.

The estimate of initial EECs in surface waters from spray drift immediately after each mancozeb application was calculated according to the method described in the USEPA manual *Standard Evaluation Procedure, Ecological Risk Assessment* (USEPA 1986). The mancozeb application pattern used in this example required four separate EFATE runs (Table 1) that were later linked using Microsoft Excel to calculate seasonal average residue concentration. The use pattern selected for the modeling included three different application rates, and the intervals between applications varied between 7 and 14 days. A single EFATE run can calculate EECs for single or multiple applications; however, more than one EFATE run is required if either the application rate or the number of days between applications varies within the treatment scenario.

The amount of mancozeb entering surface water as spray drift from each ground spray application was assumed to be the worst case USEPA default concentration of 5% of the applied mancozeb (USEPA 1986; WWF 1992). The initial pond residue concentration of 0.014 mg/L was calculated for a 30 cm deep pond having a surface area of 1 ha. An aerobic aquatic half-life of 24 hours was assumed for modeling the fate of mancozeb dissipation in water and sediments. Mancozeb has been shown to be completely degraded in water and sediments within one day (Mueller-Kallert 1994), therefore 24 hours was considered to be highly conservative for modelling. The initial residue concentration, the aerobic aquatic half-life, and the first of the four application patterns listed in Table 1 were entered into EFATE to perform EFATE Run Number 1. For EFATE Runs No. 2, 3, and 4 (Table 1), Day 1 residue concentration was the sum of the calculated initial residue concentration for that application rate, and the residue concentration estimated for the last day of the preceding EFATE run.

The EFATE program predicts maximum, average, and daily residue concentration for each day of the time interval specified in the input data. The overall predicted seasonal average mancozeb residue concentration was 0.0041 mg/L. Mancozeb dissipates rapidly in water following each application (Figure 1A). Because the dissipation half-life used in the modeling was more conservative than the experimentally determined dissipation half-life (Mueller-Kallert 1994), actual mancozeb dissipation would be more rapid than predicted by EFATE modeling. EFATE also overestimates the dissolved concentration of mancozeb in the water column because adsorption to sediments, an important removal pathway for mancozeb in water, cannot be included in this model.

Assuming a worst case of 5% spray drift, the maximum mancozeb concentration was calculated to be 0.033 mg/L, with the seasonal average residue concentration equal to 0.0041 mg/L. The maximum seasonal concentration does not exceed the USEPA level of concern for acute exposure (1/10th the EC50 of the most sensitive species, the rainbow trout (0.190 mg/L)) nor does the seasonal average concentration exceed the chronic NOEC for the most sensitive species, the fathead minnow (0.0052 mg/L). Although ecological risk assessment uses both acute and chronic exposure to assess risks to nontarget species, chronic exposure of mancozeb to aquatic organisms would not occur because mancozeb dissipates in water within one day (Mueller-Kallert 1994).

In order to refine EFATE estimates, the percent spray drift should be defined as realistically as possible. The default value of 5% is considered to be a worst case for ground spray application. In the United States, the National Agricultural Chemicals Association (NACA) and USEPA have jointly formed a Spray Drift Task Force to define guidelines for estmating spray drift. Several European publications have addressed the issue of estimating of spray drift (EPPO 1992; BBA 1993; Lloyd and Bell 1983; de Heer et al. 1985). For example, BBA (1993) recognizes 1.5% spray drift from groundspray application on potatoes and similar crops at sprayer heights of 50 cm and above.

*Advantages and Disadvantages of the EFATE Model.* The EFATE model is useful as for Tier I analysis in the risk assessment process because it requires minimal input data and provides a rapid means for calculating EECs in any environmental matrix. EFATE may be used to predict a reasonable worst case EECs, and if the EECs for the worst case do not exceed nontarget species levels of concern, there is no need to proceed with more complex Tier II modeling due to the highly conservative nature of this approach. EFATE cannot interface with higher tier models such as EXAMS. The EFATE model cannot account for the many important factors that influence the distribution of agricultural chemicals in various environmental media, such as multiple and simultaneous physical (e.g., photolysis, hydrolysis, adsorption) and biological (biodegradation) removal pathways, which limits the ability to extrapolate behavior in the real environment. The predicted worst case EFATE EECs overestimate potential adverse effects on nontarget organisms if unrealistic assumptions are used, and EFATE must be carefully interpreted when used to make risk management decisions.

*Tier II Modeling Example Using PRZM2 and EXAMS: EECs of Mancozeb Entering Surface Waters As Runoff.* Mancozeb entry into water bodies adjacent to potato fields as runoff was modeled as a second potential exposure pathway for nontarget organisms in the conceptual model. Mancozeb concentration in surface water due to runoff was derived using the Tier II USEPA models PRZM2 (Mullins, et al. 1993; Carsel et al. 1985) and EXAMS (Burns et al. 1982). PRZM2 can calculate both agricultural chemical leaching and runoff for parent compounds and metabolites. To estimate worst case mancozeb runoff concentration, the runoff component of PRZM2 was used. Mancozeb loading was estimated by PRZM2 as both the amount dissolved in water (kg/ha) and adsorbed to soil particles (erosion) (kg/ha). The PRZM2 output was interfaced with EXAMS to estimate the maximum and yearly average concentration of mancozeb in the water column and benthic sediments.

PRZM2 and EXAMS were used to calculate a more refined estimate of runoff using the same potato application pattern used for the EFATE spray drift example (Table 1). For modeling runoff, USEPA defines the typical pond as being a one acre and six feet deep (for this example, a 1 ha, 2 m deep pond was assumed), having no outflow. Runoff into the pond was assumed to drain from a 10 ha watershed.

Files containing the soil and meterological information used in PRZM2 modeling were obtained from PIRANHA (Pesticide and Industrial Chemical Risk ANalysis and Hazard Assessment (USEPA 1991)), a U.S. meteorological and soils database which also serves as the "shell" software for running the PRZM and EXAMS software modules. For PRZM2 modeling, Eastern Pennsylvania chosen as a suitable climatic surrogate (e.g., for rainfall and average temperature) that was representative of reasonable worst case climate conditions and runoff potential for potato-growing regions in Northern Europe. Northern European meterological files from the European database PELMO were not used because PELMO files could not be interfaced with PRZM2, nor with EXAMS. A silty clay loam soil of hydrologic group D, having a maximum potential for runoff, was also selected using the PIRANHA database.

PRZM2 input data requirements have been described in detail elsewhere (Carsel et al. 1985; OECD 1989). Selected parameters used in the PRZM2 modeling for potato application are listed in Table 2. 1982 was selected as the 90th percentile worst case year for PRZM2 rainfall simulation. PRZM2 integrated temporal events (typical dates of crop planting, first emergence, and harvest; shortest frequency intervals for mancozeb applications; rainfall dates), meteorological data (amount of rainfall per rainstorm), mancozeb use rates (kg/ha applied during each application) and soil characteristics to generate output as daily and total kg/ha of dissolved and bound mancozeb in runoff (mancozeb in runoff was zero on days with no rainfall). A worst case of no foliage interception (100% deposition of mancozeb on the soil surface) was assumed. This is a highly conservative assumption, since in actual practice, crop foliage interception during ground spray application has been observed to account for 62 ($\pm$27)% of the applied agricultural chemical (Willis and MacDowell 1987). For the application pattern in Table 2, PRZM2 estimated that a total of 1.8E-04 kg of mancozeb, or 1.6E-03 percent of the total yearly application rate, would enter a 1 ha, 2 m deep pond immediately adjacent to a treated 10 ha field when all runoff events in a worst case year were combined.

To calculate mancozeb concentrations in the pond water column and sediments following each rainfall, the PRZM2 output data file was input into EXAMS. Other EXAMS input is listed in Table 2. EXAMS estimated total and daily mancozeb loading into the water column and top 5 cm of sediment (benthic zone) for each rainfall in 1982. The predicted maximum concentration of mancozeb in the water column was 7.90E-05 mg/L, with an average concentration over the course of the year of 4.69E-07 mg/L. Using EXAMS, the role of important removal pathways such as hydrolysis, adsorption and biodegradation are incorporated into the modeling scenario to provide a more complete picture of mancozeb fate in surface water. The mancozeb concentration in surface water following multiple applications as predicted by EXAMS was also represented graphically (Figure 1B), and illustrated that the amount of mancozeb loading into the water column was minimal and the mancozeb loaded by worst case rainfall would be rapidly degraded via hydrolysis and biodegradation.

When EXAMS output data were compared to ecotoxicological data, the predicted maximum pond concentration of 7.90E-05 mg/L was orders of magnitude less than 1/10th the acute EC50 for the rainbow trout (48 hour EC50 = 1.9 mg/L), and the predicted average concentration 4.69E-07 mg/L was orders of magnitude less than the NOEC for most sensitive species, the fathead minnow (NOEC = 0.0052 mg/L). Based upon the predictions of PRZM2 AND EXAMS modeling, no unacceptable risks to aquatic organisms from mancozeb exposure were predicted to occur from runoff of mancozeb following a reasonable worst case pattern of application on potatoes.

*Advantages and Disadvantages of Tier II Modeling in Risk Characterization and Risk Management.* A number of computer models have been developed to better understand the fate and persistence of chemicals in water and soil (ECETOC 1992). Tier II models can be used to closely simulate environmental fate and exposure of agricultural chemicals in various environmental media. The creation of realistic environments requires a number of assumptions to be made about factors such soil characteristics, geographic location, and climatology, each of which can significantly affect the distribution and persistence of chemicals in the environment. The complex and potentially subjective process of selecting environmental parameters can lead to significant variations in modeling outcome when one or more of these factors is modified. When environmental data for a specific location are unavailable, as in the example presented here, the model user must prepare a composite of surrogate data files to construct a modeling scenario which reflects as closely as possible the actual location and environmental exposures being simulated. While realistic simulations can result from such modeling exercises, advanced expertise is required to run these programs and to properly interpret the program output.

Validation trials have shown that these models can generate realistic depictions of actual environmental fate processes (Kolset and Heiberg 1988; Mueller et al. 1992; Reinert and Rodgers 1986). Computer simulation modeling is an invaluable tool in the risk characterization and risk management processes, especially in instances where simplistic models such as EFATE cannot account for important fate and removal processes that must be understood in order to make appropriate risk management decisions. Tier III assessment, or field monitoring trials, may be conducted to support modeling assumptions or ecological risk assessment if it is perceived that EECs or dissipation pathways are not adequately addressed by computer models.

## Summary

The tiered approach used by USEPA to estimate environmental concentrations of agricultural chemicals was illustrated by modeling application of mancozeb to potato fields using a reasonable worst case application pattern. The Tier I EFATE model, a USEPA pseudo-first order decay program, was used to estimate the amount of mancozeb that would enter a 30 cm deep pond of 1 ha surface area as spray drift following repeated applications. Using 5% as a worst case for spray drift, and a conservative estimate of the aerobic aquatic half-life of mancozeb of 24 hours, EFATE predicted that mancozeb concentrations due to spray drift would not exceed levels of concern for nontarget aquatic species. Within the USEPA assessment scheme, there would be no need to refine the EECs using higher tier modeling. The Tier II models PRZM2 and EXAMS predicted that concentrations of mancozeb in the water column and in sediments entering a pond as runoff in a year of high rainfall would be low (4.09E-07). and that predicted mancozeb concentrations did not exceed any typical USEPA levels of concern for nontarget aquatic species.

Computer simulation modeling for ecological risk assessment is becoming increasingly important in the regulatory decision-making process. Simulation models can be used to help define use patterns for agricultural chemicals that are both efficacious and safe for use in the environment. However, the process of defining appropriate modeling scenarios is complex, and is only one facet of a larger process which must, on a case-by-case basis, consider experimental data, economic, political, and social factors to determine appropriate use patterns for registration of agricultural chemicals.

## *Literature Cited*

□    BBA 1993. Criteria for assessment of plant protections products. Berlin. Heft 285.

□    Burns, L.A., D.M.Cline, R.R. Lassiter 1982. USEPA-ERL, EPA-600/3-82-023.

□    Carsel R.F., L.A. Mulkey, M.N.Lorber, L.B. Baskin 1985. *Ecol Modeling* 30:49-69.

□    de Heer, H.C. Schut, H. Porskamp, L.M. Lumkes 1985. *Gewasbescherming* 16:195-197.

□    ECETOC 1992. *Technical Report* No. 50.

□    EPPO 1993. *EPPO Bulletin.* 23:1-151.

□    Kolset, K. and A. Heiberg 1988. *Wat Sci Tech* 20:1-12.

□    Lee, R. (unpublished). EFATE. Pesticide Residue Fate Simulation Model. USEPA Environmental Effects Branch, Washington, D.C.

□    Lloyd, G., G.J. Bell 1983. Hydraulic nozzles: Comparative spray drift study. MAFF/BBA.

□    Mueller, T.C., R.E. Jones, P.B. Bush, P. Banks 1992. *Env Toxicol Chem* 11:427-436.

□    Mueller-Kallert, H.-M. 1994. [14]C mancozeb: Degradation in aquatic systems. Rohm and Haas Co: RCC project No. 317970.

□    Mullins, J.A., R.F Carsel, J.E. Scarbrough, A.M. Ivery 1993. PRZM-2 Release 2.0. USEPA-ERL, Office of Research and Development, Athens, GA.

□    Norton, S.B., D.J. Rodier, D. H. Gentile, W.H. van der Schalie, W.P. Wood, M. W. Slimak 1992. *Env Toxicol Chem* 11:1663-1672.

▫   OECD 1989. OECD Environment Monographs No. 27.

▫   PIRANHA Version 2.0. 1991 USEPA-ERL Office of Research and Development, Athens, GA.

▫   Reinert, K.H. and J. H. Rodgers, Jr. 1986. . *Env Toxicol Chem* 5:449-461.

▫   USEPA 1986 Standard Evaluation Procedure- Ecological Risk Assessment. Hazard Evaluation Division, Office of Pesticide Programs. June 1986.

▫   Willis, G.H. and L.L. MacDowell 1987. *Rev Env Contam Toxicol* 100:23-73.

▫   World Wildlife Fund 1992. RESOLVE, World Wildlife Fund, Washington, D.C.

| Mancozeb use rate (kg/ha) | Number of applications | Application interval (days) | Number of days in EFATE run | EFATE Simulation sequence |
|---|---|---|---|---|
| 0.85 | 4 | 7 | 28 | EFATE run 1 |
| 1.80 | 4 | 10 | 40 | EFATE run 2 |
| 0.85 | 1 | 10 | 10 | EFATE run 3 |
| 2.0 | 1 | 14 (preharvest) | 14 | EFATE run 4 |

**Table 1 :** Inputs for EFATE modeling to predict mancozeb concentration in surface water due to 5% spray drift. The application pattern is a reasonable worst case example for potato application in Northern Europe.

| Parameter | Input value | Input into PRZM2 | EXAMS |
|---|---|---|---|
| Molecular weight | 271.24 | | X |
| Vapor pressure | > 1.0E-07 torr | | X |
| Henry's Law Constant (dimensionless) | 9.39E-08 | X | |
| Henry's Law Constant | 2.03E-09 atm m$^3$/mol | | X |
| Solubility in water | 15.6 mg/L | | X |
| Kow | 21.4 | | X |
| Koc (clay loam soil) | 675 | | X |
| Kd [1] | 7.83; 1.19 | X | |
| Soil curve number | 91 | X | |
| Neutral aqueous hydrolysis (half-life) | 1.27E-02/hr | | X |
| Aerobic soil metabolism (half-life) | 0.693/day | X | |
| Aerobic aquatic metabolism | 24 hr | | X |
| Second order biolysis [2] | 2.8 X $10^{-6}$ cfu/ml/hr | | X |
| Meterological profile | Eastern Pennsylvania | X | |
| Potato culture | | | |
| Date of planting | | X | |
| Date of first emergence | | X | |
| Date of harvest | | X | |
| Mancozeb applications | Table 1 | | |
| Date of first application | | X | |
| Soil Series | D (Dunning) | X | |

[1]  Kd = (Koc X OC/100). Kd was calculated for each soil horizon in Dunning soil.

[2]  A first order rate constant, K, was derived using the aerobic aquatic half-life of 24 hours: K = ln2/24 hr = 2.8 X $10^{-2}$ h$^{-1}$. K was converted to a second order rate constant by dividing by 1 X $10^5$ colony forming units (cfu) per gram of benthic sediment or per ml of water in the water column, representing 5% of the total cfu in the benthic sediment and limnetic zone in the EXAMS AERL pond environment data set.

**Table 2 :** Physical chemistry, environmental fate, soil and plant culture inputs for PRZM2 and EXAMS simulation modeling of mancozeb runoff

A.

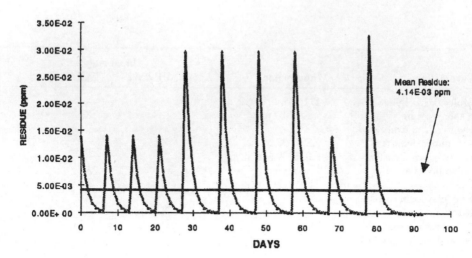

B.

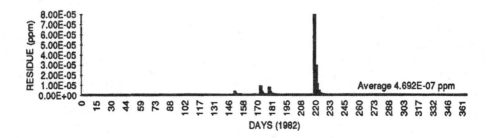

**Figure 1 :**

A.    Seasonal mancozeb concentrations due to 5% spray drift into a 30 cm pond of 1 ha surface
       area, predicted using EFATE.

B.    EXAMS output illustrating the frequency and magnitude of mancozeb runoff loading from a
       10 ha watershed into a 1 ha, 2 m deep pond.

6.2.6.       The Use of Pesticide Leaching Models in a Regulatory Setting

# D. Yon, J.J.T.I. Boesten & K. Travis

for FOCUS

The new European Community (EC) procedure for pesticide registration is now being implemented. The use of models to obtain predicted environmental concentrations for pesticides in soil, ground-water, surface water and air is central to the EC risk assessment procedures. The FOrum for international Co-ordination of pesticide fate models and their USe (FOCUS) has been set up to develop guidance and consensus on such issues as the state of the art, selection and validation of models etc. In particular, the Regulatory Modelling Workgroup of FOCUS was formed in 1993, made up of regulators, industry and other pesticide fate modellers, and has worked specifically in the field of leaching models. Three goals of this group have been

- To develop practical guidance on the use leaching models to address EC registration needs
- To define current modelling deficiencies, recommend appropriate conditions for using selected models, and suggest improvements
- To develop general recommendations for validation and calibration of leaching models

Progress towards achieving these goals has involved several steps : agreeing terminology in the modelling process (especially the term "validation"); defining a procedure for assessing the suitability of a model; describing available models and their validation status with respect to use in the EC registration process; the relationship between modelling and GLP; establishing EC soil, weather and cropping scenarios and using these in test exercises with selected models. Results of the work of the group will be presented.

Modelling is a relatively new area which is open to a large variety of approaches, and both regulators and industry see the importance of guidance on the appropriate use of models in a regulatory setting. Once the short-term need for guidance on leaching models has been addressed, it is hoped that medium-term issues such as model improvements and validation can also make progress by interna-tional co-ordination and co-operation.

# 6.3.　　　Poster Presentations

*6.3.1.*　　*Using the Creams Pesticides Transfer Sub-Model at a Rainfall Simulation Scale*

## V. Gouy and R. Bélamie

CEMAGREF - Centre National du Machinisme Agricole, du Génie Rural, des Eaux et Forêts
3 bis Quai Chauveau, 69 336 Lyon, France

## Abstract

The pesticide sub-model of CREAMS was tested to describe pesticides transfer into runoff generated by a rainfall simulation. Adsorption coefficient was found to be a sensitive parameter of the model. So, we developed an experimental method in order to measure Kd in a dynamic way. We studied four pesticides (atrazine, simazine, lindane, methidathion) and three soils (loam sand, sand loam, calcareous silt loam). The values of Kd obtained were compared to the values measured on the soils by the standardized method of batches and to the values of the literature. Mean Kd measured in the runoff of a rainfall simulation were considerably greater, especially for low suspended particles loads. We ran the model with these different data for Kd and we found that predictions were more accurate with Kd obtained by rainfall simulation particularly for lindane and simazine. We improved the modelisation for atrazine and methidathion. Nevertheless, the relative error between observed and predicted concentrations was not the same according to the pesticide, which made prediction of risk dependent on pesticides classes of behaviour.

## Keywords

CREAMS; modeling, rainfall simulation; insecticide, herbicide, partition coefficient.

## Introduction

The increasing use of pesticides in agriculture was responsible for significant background contamination. The aim of this study is to find a method which permits to evaluate molecules potential mobility into runoff. We have decided to test the ability of the CREAMS model to reach this goal at a 1 square metre scale using rainfall simulation.

## Pesticide-Sub-Model of Creams

The CREAMS was developed in 1980 by the United States Department of Agriculture in order to evaluate the impact of agricultural practices on pesticides transfer into runoff from the land to the river. It consists of three major components : hydrology, erosion/sedimentation, chemistry which includes nutrients and pesticides. Each sub-model can be used separately owing to the appropriate pass file. We focused on the pesticide sub-model. It was elaborated to estimate concentrations in runoff (including water and sediment) and total masses transferred during a storm. In this study, we intended to use the pesticide sub-model of CREAMS at a 1 square metre scale. Owing to the erosion pass file we calculated pesticides concentrations in runoff using hydrologic and erosion results of a rainfall simulation.

A sensitivity analysis, with the Tomovic method, showed that the adsorption coefficient of pesticides on sediment , Kd, was a sensitive parameter, especially for pesticides with a high Kd.

Since this parameter is difficult to estimate as it is very dependent on molecule and soil characteristics, and environmental conditions (humidity, pH, temperature, composition of the solution...), we decided to compare modelisation results using values of Kd obtained by different ways.

Kd is defined by the relation : Kd = Cs/Cw where Cs is the pesticide concentration in the soil or solid phase (mg/g) and Cw is the pesticide concentration in solution (mg/ml) at the equilibrium of adsorption. Three sources of estimation of Kd were used :

- the values available in the tables of handbooks,
- the values measured on soils with pure $^{14}$C labelled molecules within a standardized method in batches,
- the values measured in runoff during a rainfall simulation with formulated molecules (Gouy, 1993).

The third source of estimation of Kd was developed by the authors in order to propose a method which permits to approach natural conditions of runoff.

## Rainfall Simulation Experimentation

### *Apparatus*

An oscillating-nozzle rainfall simulator, applied an intense rainfall to the surface of a 1 square metre plot filled with 10 cm deep soil packed in a metallic tray and inclined at an angle of 16% (GOUY, 1993). The tray was constructed with troughs that collected runoff and infiltration, and directed them into two glass bottles. The figure 1 presents the experimental system. The average rainfall intensity was 70 mm/h during an hour.

### *Chemicals Application*

Four formulated pesticides were applied : atrazine (Triazine, solubility = 30 ppm, Kd = 2.9 in the literature, wettable powder), simazine (Triazine, solubility = 3.5 ppm, Kd = 2.3 in the literature, wettable powder), lindane (Organochlorine, solubility = 7 ppm, Kd = 21.9 in the literature, concentrated solution) and methidathion (Organophosphate, solubility = 240 ppm, Kd = 2.2 in the literature, concentrated solution). The application rate was 70 mg/m$^2$. They were surface applied on bare soil, 15 hours before rainfall simulation. We present the results obtained using three different soils: (1) a loam sand (Ardières, organic matter = 0.2%), (2) a sand loam (Jaillière, organic matter = 2%) and (3) a calcareous silt loam (Charente, organic matter = 2%).

### *Experimental Method*

In order to evaluate adsorption coefficient evolution during a storm, we sampled runoff at a 5 minutes time step during the first 30 minutes of runoff and at a 10 minutes time step thereafter.

### *Analysis of Runoff Samples*

Total eroded particles load was measured for each sample. Then, runoff samples were filtered through a 1,2 microns porosity fibre-glass filter. Suspended particles and filtered water pesticides contents were analysed by gas chromatography. The extracted solvents were methylene chloride and methanol (2/1). Extraction efficiencies were 107% for atrazine, 67% for simazine, 96% for lindane and 98% for methidathion in water; 102% for atrazine, 99% for simazine, 92% for lindane, 116% for methidathion adsorbed on Ardières suspended particles; 107% for atrazine, 109% for simazine, 85% for lindane and 112% for methidathion adsorbed on Jaillière suspended particles; 99% for atrazine, 98% for simazine, 80% for lindane and 109% for methidathion adsorbed on Charente suspended particles. Adsorption coefficient, Kd, was calculated for each soil and each time step.

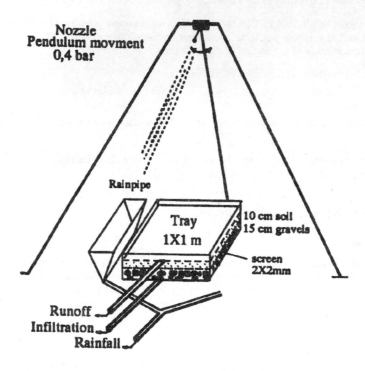

**Figure 1 :** Experimental rainfall simulation apparatus

## Results and Discussion concerning the Adsorption Coefficient

We observed that total pesticides transfer into runoff was highly dependant on hydrodynamic behaviour of soils. Using the soil of Ardières, only 0.2 to 0.5% of total amount applied was carried out into runoff instead of 3 to 8% with the soil of Jaillière and 8 to 30% with the soil of Charente. Mean suspended sediment loads were 0.2 g/l for Ardières, 5.7 g/l for Jaillière and 8.3 g/l for Charente. But, in both cases, more than 70% of each pesticide were distributed into water phase. We noted that pesticides concentrations in solution decreased during runoff as well as adsorbed concentrations and the adsorption coefficient Kd, varied within several orders of magnitude during a storm. Thus, it was not always possible to adjust linear isotherms to describe the adsorption. Means values are given in table 1 compared with literature data and batches experimentation results.

| Molecule | Kd of literature | Kd of batches Ardières | Kd of batches Jaillière | Kd of batches Charente | Kd of rainfall simulation Ardières | Kd of rainfall simulation Jaillière | Kd of rainfall simulation Charente |
|----------|------------------|------------------------|-------------------------|------------------------|-----------------------------------|-------------------------------------|------------------------------------|
| Atrazine | 2,9[*] | - | 2,8 | 2,1 | 36 | 7 | 4 |
| Simazine | 2,3[*] | - | - | - | 228 | 34 | 48 |
| Lindane | 21,9[*] | 38,4 | - | - | 157 | 41 | 37 |
| Methidathion | 2,2[**] | 3,3 | 6,7 | - | 60 | 8 | 8 |

[*]  USDA, Knisel, 1980.
[**]  calculated from Koc (organic carbon/water partitioning) given by the USDA.

**Table 1 :** Comparison of Mean Kd Values (ml/g) Obtained By Different Methods

Batches results were not significantly different from values of the literature. On the contrary, we concluded that Kd measured in runoff during a rainfall simulation with formulated pesticides can be much greater than values corresponding to standardized methods with the pure molecules on soils. It was obvious for the soil of Ardières where mean suspended particles load was lower. When relating Kd and suspended particles loads for each time step, it appeared that Kd decreased when particles loads increased and for the soil of Ardières, we could adjust a power function. We tested the CREAMS pesticide sub-model with the values of Kd obtained on soils in closed systems (literature, batches), or on suspended particles in a dynamic way (simulation).

## Results of Modelisation

It appeared that predictions of the model, using values of Kd measured on soils, were bad. Especially, the distribution of pesticides between suspended particles and water was incorrect. However, the Kd values of rainfall simulations permitted to obtain better predictions. Nevertheless, the quality of the adjustment depended of the class of pesticides : thus, simazine and lindane showed mean relative errors of 50 and 30% for water concentrations (respectively 70 and 40% for adsorbed concentrations) instead of 123 and 100% for atrazine and methidathion concentrations in water (respectively 195 and 108% for adsorbed concentrations). So, we tried to improve the modelisation by dividing the calculation in two steps : (1) We used the Kd measured with soils in batches or the Kd of the literature if not available, in order to calculate pesticides leaching, (2) we used the mean Kd obtained through a rainfall simulation to evaluate pesticides partition between water and suspended particles in runoff. Owing to this distinction in the modelisation processes, the mean relative errors for atrazine and methidathion decreased to respectively 32 and 60% regarding to water concentrations (50 and 80%, regarding to adsorbed concentrations). But, when using this method, simazine concentrations were badly reproduced for the simulation with the soil of Ardières (fig. 2). This result may to be a consequence of the filtering behaviour of this soil. Concentrations of lindane were under-estimated by a factor 2 with the soil of Charente.

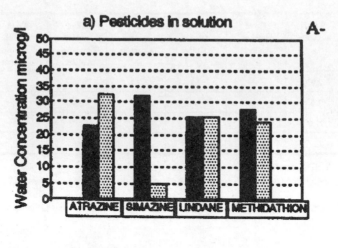

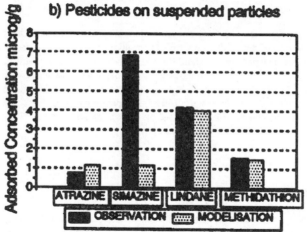

**Figure 2 A :** Results of modelisation using two steps of calculation
with to two different values of Kd.
With the soil of Ardières.

In conclusion we can note that the determination of Kd by rainfall simulation studies was useful to improve the modelisation of runoff concentrations of pesticides. However, it appeared that the relative error of predictions was not the same according to pesticides. Especially, lindane which is the most adsorbed showed the best fits (except for Charente soil), atrazine and methidathion which are more soluble in water showed greater relative error. And finally, the modelisation of simazine, which has a low solubility, needed to be improved for soils characterised by high infiltration during a rainfall simulation.

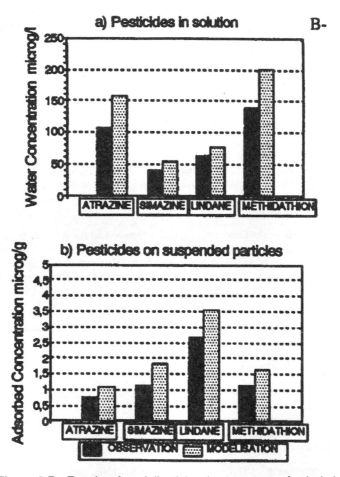

**Figure 2 B :** Results of modelisation using two steps of calculation
with to two different values of Kd.
With the soil of Jaillière.

## References

☐  Gouy V., (1993). Contribution de la modelisation a la simulation du transfert des produits
phytosanitaires de la parcelle agricole vers les eaux de surface. *Thesis.* ULP de Strasbourg.
CEMAGREF de Lyon 3 bis Quai Chauveau, 69336 Lyon France. 350 p.

☐  Knisel W.G. (1980). CREAMS, a field scale model for chemicals, runoff and erosion from
agricultural management systems, Knisel Editor USDA. *Conservation Research Report* n°26,
643p.

This article has ever been published in Wat. Sci. Tech. vol. 28, n°3-5, pp. 679-683, 1993.

## 6.3.2.      *Fate Modelling of EC Formulated Chlorfenvinphos in Water*

# Z. Mulinski, J. Piasecki and M. Swarcewicz

Agricultural University, Department of Chemistry, 71-434 Szczecin,Poland

## Introduction

Pesticides are almost always formulations.For example,an emulsifiable concentrate [EC] contains the active ingredient in a definite concentration and additional adjuvants like emulsifiers, inerts,organic solvents, thickeners as well as a great multitude of specific ingredients, necessary for the distinct purpose of the whole preparation.

The purpose of the present work was to use a general kinetic model for the mass transfer mechanism accompanied by a first-order irreversible chemical reaction which allows the estimation of the pesticide concentration both in oil droplets and water phase. It was also possible to compute the distribution coefficient of pesticides between oil and water phase and the pseudo first-order rate constant of pesticide decay in the water phase.

For evaluation of the model used in this work to assess the pesticide fate in water, the results from two experiments with chlorfenvinphos [Mulinski,1989 ; Piasecki and Swarcewicz,1980] were taken.

## Materials and Methods

The insecticide chlorfenvinphos was used in two commercial formulations: ENOLOFOS 50 EC [Polish-manufacturing licence of Shell Chemicals] and BIRLANE 24 EC [Shell Chemicals].The initial concentration of chlorfenvinphos $Co(0)$ was 1.389 mol/l in the commercial formulation of ENOLOFOS, and that of insecticide $CT(0)$ was $1.39*10^{-4}$ mol/l in aqueous suspension. The water used came from Miedwie Lake ,pH 8.7.

The initial concentration of chlorfenvinphos in BIRLANE was $Co(0)=0.6505$ mol/l and $CT(0)=1.0*10^{-5}$ mol/l. In this experiment distilled water, pH 4,was used. The insecticide residues in all water samples were measured by GLC.

## Results and Discussion

The model B was described in detail by Mulinski [1994]. The integrations were performed using the Runge-Kutta method variable time step.This model relatively well described the hydrolysis of chlorfenvinphos [ENOLOFOS] after 7 days of the experiment, the same was noticed for BIRLANE.

During the initial period of the study [up to 7 days] rates of degradation obtained from model B were greater than from the experimental data. It is possible that on this could have been caused by the variability of the overall mass-transfer coefficient K, and no accurate initial parameters : volumetric fraction of adjuvants Ya and the rate change of adjuvants $r_3$ [table 1].

The value of volumetric mass-transfer coefficient ,KA/Vo,for BIRLANE was $6.8*10^{-3}$ day$^{-1}$,and $54.4*10^{-3}$ day$^{-1}$ for ENOLOFOS ,respectively (from model A). The both commercial formulations were similar and this difference in values was caused probably by change in boundary layer in consequence of the pH differences of water.

| Time day | ENOLOFOS | | | | BIRLANE | | | |
|---|---|---|---|---|---|---|---|---|
|  | $Y_e$ | $Y_B$ | Co | $C_w 10^8$ | $Y_e$ | $Y_B$ | Co | $C_w 10^8$ |
| 3 | 0.98 | 0.86 | 1.03 | 1560 | 0.95 | 0.86 | 0.47 | 12 |
| 6 |  | 0.86 | 0.96 | 2568 | 0.95 | 0.85 | 0.47 | 18 |
| 7 | 0.87 | 0.86 | 0.94 | 2817 |  | 0.85 | 0.47 | 19 |
| 14 | 0.80 | 0.84 | 0.85 | 3830 |  | 0.85 | 0.47 | 24 |
| 28 | 0.80 | 0.80 | 0.76 | 4235 |  | 0.85 | 0.46 | 24 |
| 30 |  | 0.79 | 0.76 | 4232 | 0.82 | 0.85 | 0.46 | 24 |
| 60 | 0.68 | 0.70 | 0.68 | 3893 |  | 0.84 | 0.46 | 24 |
| 90 |  |  |  |  | 0.77 | 0.84 | 0.46 | 24 |

$Y_e$ - experimental data

$Y_B$ - simulated data from model B

Co , Cw - concentration of pesticide in oil and water phase, respectively,mol/l

**Table 1 :** Kinetics of chlorfenvinphos simulated by model B
compared to a measurement in water

The estimated parameters for predicting the pesticide decay with model B are shown in table 2.

The mathematical model B is the first step for predicting the behaviour of EC formulated pesticide in water emulsion.

| Parameter | ENOLOFOS | BIRLANE |
|---|---|---|
| $k_2$, day$^{-1}$ [a] | 0.0104 | 0.0065 |
| m [a] | 16082.5 | 1837121 |
| $r_3$, day$^{-1}$ | 130 | 130 |
| Ya(0) | 0.1 | 0.1 |
| Ya | 0.2 | 0.25 |

[a] - parameter obtained from model A

**Table 2 :** The parameters used for model B

## Conclusions

1.  The model B has gives positive results for describtion the decay of chlorfenvinphos in the second part of experiments from 7 days in waters.

2.  As to the first part of insecticide decay, a good estimation was not obtained because it was assumed that both K and partition coefficient were constant and because a correct value of parameters : Ya (0), Ya and $r_3$ was not used.

3.    The influence of pH on the boundary layer structure in the oil - water interface was noticed
      because the obtained value of volumetric mass-transfer coefficient was greater for emulsion
      at pH 8.7 than at pH 4, instead a lesser value of the partition coefficient was found . Probably
      the boundary layer had another structure at pH 4, therefore the velocity of release of pesticide
      from oil droplets was slower.

# *References*

□    1. Mulinski Z.,Physical-chemical aspects of persistence of some insecticides in model water-
     bottom sediment systems. *Doctoral Thesis*, Agric. Univ. Szczecin (1989).

□    2. Piasecki J. and M.Swarcewicz , Influence of temperature on the decay rate of bromfenvin-
     phos and chlorfenvinphos in water. *Zesz.Nauk.AR* (1980),no.84,197.

□    3.Mulinski Z.,Modelling partition between water and formulation,in print (1994).

*6.3.3.*     *Validation and Comparison of Pesticide Soil Transport Models*
            *for Field Dissipation of Metamitron*

# M. Trevisan, E. Capri, S.P. Evans and A.A.M. Del Re

Istituto di Chimica Agraria ed Ambientale, Università Cattolica del Sacro Cuore,
Via Emilia Parmense 84, I-29100 Piacenza - Italy.

## Summary

Pesticide soil transport models PRZM-2, Varleach, LEACHMP 3.1 were run simulating the field dissipation of metamitron in the top unsaturated soil zone. Using a field data set model outputs were validated using graphical and numerical techniques and assessing the goodness of fit when uncertainty of input parameters on ouput is taken into account. Results show good agreement between observed and simulated data. Using different Kd and Kdeg values, representative of the variability observed in field conditions and reported in the literature, the range of simulated data overlaps with the observed one.

## Introduction

Numerous pesticide soil transport models models are currently available; some have been selected for evaluation in view of future use for registration in Europe. The aim of this paper is to validate and compare three widely known models which have been tested in a range of pedoclimatic settings as well as in Italian conditions (Trevisan et al., 1993): LEACHMP-3.1 (Hutson and Wagenet, 1992), VarLeach 2.0 (Walker, pers. comm.) and PRZM-2.release 1.1 (Mullins et al., 1993). When field data and model simulations are compared it is necessary to take into account uncertainty, which in models derives from inputs and from the way basic processes are represented. To estimate uncertainty in a determistic model, which produces only one forecast, this must be run several times varying input parameters within a predefined range. In first approximation runs should be as numerous as possible since variations linking input to output are unknown. In this paper model comparison is carried out with a fixed scenario and with input parameter variations accounting for an uncertainty range observed in field conditions and reported in the literature. For illustrative purposes, sensitive parameters known to influence outputs have been modified. Field data derives from a study on the dissipation pathways of metamitron conducted during 1992 on a representative site farm of the Po Valley (Northern Italy) (Capri et al., 1994).

## Materials and Methods

### *Field data*

The site is located in Piacenza (45°N; 51 m) and has 1 ‰ slope and with good soil permeability; consequently simulated runoff and erosion will not be considered. Soil characteristics were measured for 2 horizons (0-10 and 10-30 cm): soil is a clay silty loam with average 1.3% organic carbon. Daily temperature (max, min) and precipitation were recorded by a meteo station located at the study site. Metamitron dissipation and degradation were measured as described in Capri et al. (1994); soil samples for residue determination were collected randomly from each of 3 monitoring plots at 3 depths (0-5; 5-10; 10-20 cm) located along a longitudinal transect, (220•14 m), perpendicular to crop rows. Laboratory half-lives were measured at different temperatures and soil moisture in a 60-day period (Capri et al., 1994).

## Models

Models have been run to predict pesticide mobility in two soil horizons (0-10 cm; 10-20 cm) using 1 cm-thick compartments (2.5 cm for LEACHMP) for the March to June 1992 period. Total amounts of rainfall are the same for all models (77 mm). Some data were calibrated to obtain similar forecasted amounts of water lost by evapotraspiration (67 mm) and water percolating at 30 cm depth (6 mm).

**Estimated and measured input.** The same parameters have been used for all models: water solubility (1700 mg/l); Kdeg ($0.018$ day$^{-1}$); Koc (71 l/kg); organic carbon content (1.28 and 1.24 for the two horizons respectively); metamitron application rate and date (1.54 Kg/ha; 03.03.1992); surface application (1 cm); bulk density (1.35 and 1.60 Kg/l for the two horizons respectively); field capacity (26%, except 29% for PRZM-2); lower boundary conditions (free-draining profile); pan factor (1.0); pesticide plant uptake (1.0); crop information (sowing, germination, emergence, maturity and harvest dates); crop cover (0.8%), maximum root depth (30 cm); total depth (30 cm); initial moisture (15.3%); molecular diffusion coefficient in air ($4.3 \times 10^3$ mm/day). *Specific VARLEACH inputs* are: water content at 2 bar (15.6%); factor changing Kd with time (10% of Kd value); 1 cm layer thickness is pre-determined by the model. *Specific LEACHMP inputs* are: dispersivity (10 mm); maximum ratio of actual to potential transpiration (1.05); root flow resistance (1.05); maximum water flux per time step (0.01); largest time interval per day (0.05); Richards equation (yes); correction for temperature (yes); saturated hydraulic conductivity (92.6 mm/day); vapour density ($8.6 \times 10^{-8}$); layer thickness is 2.5 cm. *Specific PRZM-2 inputs* are: minimum depth from which evaporation is extracted (30 cm); maximum interception storage of the crop (0.15 cm); maximum canopy height at maturation date (40.0 cm); enthalpy of vaporization of metamitron (20 Kcal/mole); Henry constant ($5.9 \times 10^{-7}$); hydrodynamic dispersion (0); Method Of Characteristics (Yes); wilting point (21.7 %); layer thickness is 1.0 cm for 0-5 cm; 2.5 for 5-10 cm and then 5 cm for successive layers.

**Calibrated input.** Minimum depth at which evaporation is extracted, field capacity (PRZM-2), saturated hydraulic conductivity and maximum ratio of actual to potential transpiration (LEACHMP) are calibrated to obtain the same evapotranspiration and water percolation.

**Uncertainty.** Previous sensitivity tests (Del Re and Trevisan, 1993) have identified Kdeg, Kd and bulk density (BD) as parameters with the greatest effect on model outputs. Input values used reflect uncertainty ranges associated with each parameter: for KDeg e BD these are in the min-max range of field measurements. Estimated values of KD have been used: range and standard deviations are from Smith et al. (1987).

## Evaluation of model outputs

Output goodness-of-fit with observed data were assessed qualitatively and quantitatively using graphical methods and indices, respectively. Indices evaluate different properties of the difference between predicted and simulated data. The following have been used for each run and depth: data **maximum error** (ME) and **modelling efficiency** (EF) (Loague and Green, 1990); **amplitude of the maximum error** (AME), which expresses the amplitude in the run between observed and simulated data; the **correlation coefficient** (R) between observed and simulated data. Complete agreement between observed and simulated data is 1 for R and EF and 0 for ME and AME. All simulated values < detection limit (0.032 µg/kg) are set to 0.

To take into account single model output variations all runs were averaged to produce a mean ($\overline{X}$), standard deviation (SD), and a confidence interval ($\alpha$=0.05). These were graphically compared to the observed X±SD and confidence interval.

## Results and Discussion

Table 1 gives values of indices for depth 0-5 cm; at lower depths no indices are available since simulated values are below detection limit. Indices show variations in values, with the exception of R, which is the least sensitive of the indices used.

On average VARLEACH simulations provide the best fit with observed data (R=0.89-0.90): indices remain constant and values of ME and AME show maximum reduction of error range, with the sole exception of A and B constants, indicating greater sensitivity to inputted values. EF values indicate stability of VARLEACH and noticeable variations for PRZM-2 and LEACHMP.

Ranking of simulations indicate that values vary > with Kdeg>KD>BD; EF values show that VARLEACH remains stable with high values of Kdeg.

Figure 1 indicates that VARLEACH presents a better fit, with a reduced difference between (observed-simulated) also for depths > 5 cm, at which all models tend to show poor fit. PRZM-2 and LEACHMP overestimate at 5-10 cm and underestimate at depths > 10 cm. Bearing in mind the variability of field data, all models fit into the $\alpha=0.05$ confidence interval of observed data.

Simulated dissipation at 0-5 cm are validated by field data. At greater depths validation is not considered possible due to limited temporal sampling; indicatively all models seem to fail correct estimation. Model ranking indicates a best fit for VARLEACH>LEACHMP≥PRZM-2.

With the present data set, graphical techniques appear to be sufficient for validation purposes. With small data sets indices provide only limited indications, which appear to be more suitable for larger sets (i.e. Del Re et al., 1994).

## Acknowledgements

Research was carried out as part of EC-DG XII-Environment 1990-1994 project. EVCV-CT92-0226.

## References

□   Capri E., Ghebbioni C. and Trevisan M. (1994). Metamitron and chloridazon dissipation in a silty clay loam soil. *Journal of Agriculture and Food Chemistry* (submitted).

□   Del Re A.A.M. and Trevisan M. (1993). Testing models of the unsaturated zone. *Proceedings IX Symposium Pesticide Chemistry*, Piacenza (Italy), (ed. Del Re et al.), pp. 5-31.

□   Del Re A.A.M., Trevisan M., Capri E., Evans S.P. and Brasa E. (1994). Criteria for existing pesticide models evaluation. *This volume.*

□   Hutson J.L. and Wagenet R.J. (1992). LEACHM: Leaching Estimation and Chemistry Model. A process-based model of water and solute movement, transformations, plant uptake and chemical reactions in the unsaturated zone. Version 3. Res. Series 92-3, Cornell University, Ithaca, New York.

□   Loague K. and Green R.E. (1990). Criteria for evaluating pesticide leaching models. Field-Scale Water and Solute Flux in Soils (ed. Roth et al.), pp. 175-208.

□   Mullins J.A., Carsel R.F., Scarbrough J.E. and A.M. Ivery (1993). PRZM-2, a model for predicting pesticide fate in the crop root and unsaturated soil zones: users manual for release 2.0. Users Manual for release 2.0. USEPA, Athens, GA 30605-2720.

□   Smith C.N., Parrish R.S. and Carsel R.F. (1987). Estimating sample requirements for field evaluations of pesticide leaching. *Environmental Toxicology and Chemistry* 6: 343-357.

□   Trevisan M., Capri E., Del Re A.A.M. (1993). Pesticide soil transport model: model comparison and field evaluation. *Toxicological Environment and Chemistry* 40: 71-81.

| DEPTH : 0-5 CM | | | | | | | | | | | | | |
| Value | PRZM-2 ME | AME | EF | R | | LEACHMP ME | AME | EF | R | VARLEACH ME | AME | EF | R |
|---|---|---|---|---|---|---|---|---|---|---|---|---|---|
| reference | 24.94 | 46.54 | 0.63 | 0.89 | | 29.90 | 57.79 | 0.42 | 0.88 | 28.20 | 50.59 | 0.49 | 0.90 |
| mean | 21.35 | 52.70 | 0.48 | 0.89 | | 29.11 | 61.63 | 0.31 | 0.88 | 25.62 | 52.63 | 0.54 | 0.90 |
| **Kdeg** 0.0149 | 24.94 | 46.54 | 0.63 | 0.89 | | 20.30 | 52.69 | 0.48 | 0.89 | 27.79 | 47.09 | 0.54 | 0.90 |
| 0.0238 | 21.76 | 55.56 | 0.42 | 0.88 | | 29.30 | 64.79 | 0.25 | 0.88 | 26.26 | 56.41 | 0.47 | 0.90 |
| 0.0282 | 20.20 | 57.91 | 0.30 | 0.87 | | 28.80 | 68.19 | 0.12 | 0.87 | 24.57 | 62.38 | 0.33 | 0.89 |
| 0.0370 | 17.28 | 60.22 | 0.30 | 0.86 | | 27.80 | 72.19 | -0.10 | 0.86 | 26.26 | 56.39 | 0.47 | 0.89 |
| 0.0414 | 15.85 | 60.50 | -0.07 | 0.85 | | 27.30 | 73.29 | -0.20 | 0.85 | 27.98 | 45.59 | 0.54 | 0.90 |
| 0.0924 | 5.20 | 55.13 | | 0.78 | | 21.91 | 72.24 | | 0.78 | 28.63 | 39.75 | 0.52 | 0.90 |
| **BD** 1.01 | 23.44 | 50.76 | 0.57 | 0.89 | | 29.90 | 60.69 | 0.35 | 0.89 | 28.50 | 52.39 | 0.47 | 0.89 |
| 1.02 | 23.44 | 50.78 | 0.57 | 0.89 | | 29.90 | 60.69 | 0.35 | 0.89 | 28.40 | 52.29 | 0.47 | 0.89 |
| 1.07 | 23.49 | 50.96 | 0.57 | 0.89 | | 29.90 | 60.29 | 0.36 | 0.89 | 28.40 | 51.89 | 0.47 | 0.90 |
| 1.27 | 23.72 | 50.55 | 0.57 | 0.89 | | 29.90 | 58.59 | 0.32 | 0.88 | 28.30 | 50.89 | 0.49 | 0.90 |
| 1.37 | 23.84 | 50.41 | 0.58 | 0.89 | | 29.90 | 59.09 | 0.39 | 0.88 | 28.23 | 50.58 | 0.50 | 0.91 |
| 1.40 | 23.84 | 50.30 | 0.58 | 0.89 | | 29.09 | 57.39 | 0.42 | 0.89 | 28.21 | 50.48 | 0.50 | 0.90 |
| **Koc/KD[*]** 10 | 19.75 | 33.91 | | 0.83 | 0.128 | 29.90 | 65.19 | 0.10 | 0.87 | 28.20 | 58.29 | 0.37 | 0.89 |
| 23 | 21.86 | 63.03 | | 0.85 | 0.294 | 29.90 | 62.69 | 0.24 | 0.88 | 28.20 | 54.99 | 0.45 | 0.89 |
| 47 | 23.27 | 50.76 | 0.56 | 0.89 | 0.602 | 29.90 | 59.69 | 0.36 | 0.88 | 28.20 | 51.99 | 0.48 | 0.90 |
| 95 | 24.09 | 49.18 | 0.60 | 0.92 | 1.216 | 29.90 | 56.49 | 0.44 | 0.89 | 28.20 | 49.68 | 0.51 | 0.90 |
| 119 | 24.29 | 48.28 | 0.61 | 0.89 | 1.523 | 29.90 | 55.59 | 0.46 | 0.89 | 28.20 | 49.09 | 0.51 | 0.90 |
| 140 | 24.41 | 47.75 | 0.62 | 0.89 | 1.792 | 29.90 | 54.99 | 0.47 | 0.89 | 28.20 | 48.69 | 0.51 | 0.90 |
| **A & B** 69 & 0.458 | | | | | | | | | | 17.30 | 65.24 | -0.09 | 0.86 |
| 146 & 0.750 | | | | | | | | | | 19.00 | 66.69 | -0.04 | 0.87 |
| 730 & 1.000 | | | | | | | | | | 27.40 | 53.09 | 0.50 | 0.90 |
| 2300 & 1.750 | | | | | | | | | | 25.40 | 70.28 | 0.04 | 0.88 |
| 3300 & 2.150 | | | | | | | | | | 21.20 | 67.50 | | 0.85 |
| 6000 & 3.000 | | | | | | | | | | 2.34 | 37.64 | | 0.82 |

[*] Koc for PRZM and KD for LEACHMP and VARLEACH

**Table 1 :**   INDICES. Complete agreement between observed and simulated data is 1 for R and EF and 0 for ME and AME. All simulated values < detection limit (0.032 µg/kg) are set to 0; selected indices are consequently not calculated.

**Figures :**   Comparison between observed [obs] and simulated [simul] mean data for all depths and sampling intervals : (A) difference between means; (B) obs and simul mean; (C) obs and simul mean ± standard deviation; (D) obs and simul mean ± confidence interval ($\alpha = 0.05$).

**(A)**

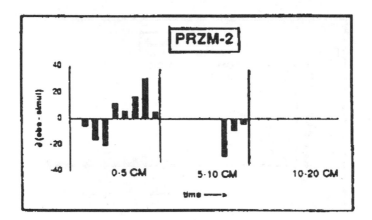

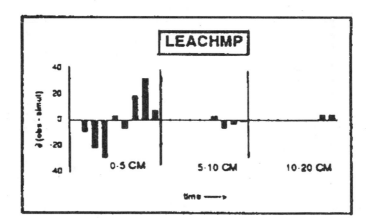

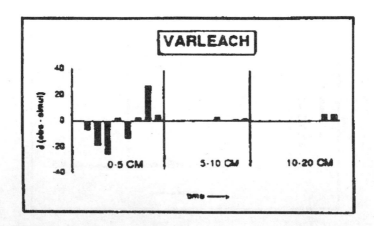

**(B)**

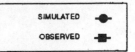

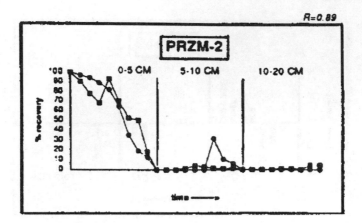

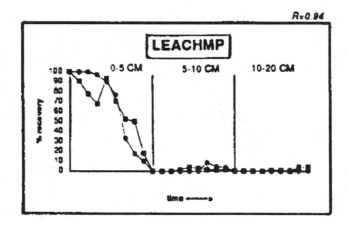

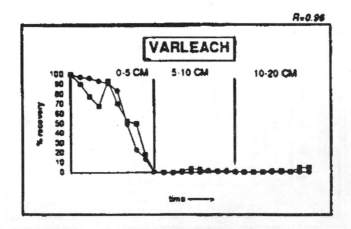

(C)

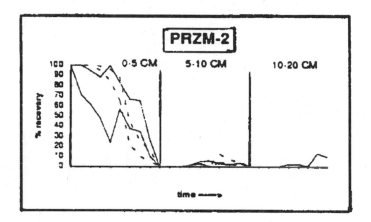

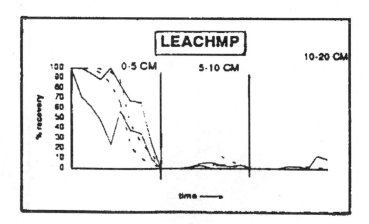

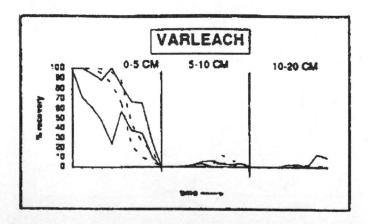

**(D)**

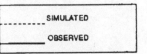

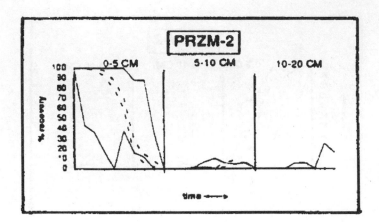

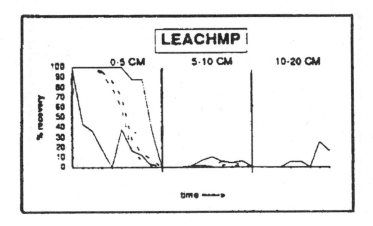

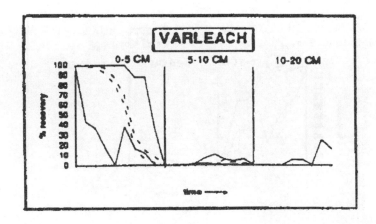

## 6.3.4. Modelling the Microbial Breakdown of Pesticides in Soil Using a Parameter Estimation Technique

# J.P.M. Vink[1], P. Nörtersheuser[2], O. Richter[3], B. Diekkrüger[3] & K. P. Groen[1]

[1]   Ministry of Transport, Public Works and Water Management, Directorate
      Flevoland, P.O. box 600, 8200 AP Lelystad, Netherlands.
[2]   Agricultural Research station BASF, P.O. box 220, W-6703 Limburgershof,
      Germany.
[3]   Technical University Braunschweig, Inst. Geography and Geo-Ecology,
      Langer Kamp 19C, 3300 Braunschweig, Germany.

In those few papers that focus on pesticide transport behaviour in soil, and apply non-linear kinetics, transport phenomena are often totally neglected. It is challenging to study the spatio temporal patterns produced by mathematical models which comprise both biological processes with their inherent non-linearities and physical transport processes.

Descriptions of microbial activity in soil is a pre-set condition to compute transformation rates of pesticides as dependent variables of concentration level and temperature. In general, the differential equations resulting from non-linear degradation laws cannot be solved analytically, and solutions cannot be written in forms of known exponential functions. The actual transformation rate of biologically transformed pesticides reflects both microbial activity and population density. To couple microbial population development and consumption of the compound in time and depth, use is made of an iterative parameter estimation technique. The empirical O'Neill temperature equation was inhibited in the model and coupled with a water transport model. Field data of the nematicide 1,3-dichloropropene (1,3-D) from an experimental plot were used in combination with laboratory sorption and incubation experiments to calibrate transport simulation in time and depth.

A multi-layered approach and the assignment of realistic concentrations to various soils layers will improve approximations of microbial inhibition or growth in depth. The resultant kinetics are in general non-linear and first order kinetics are only a crude approximation valid at low concentrations. The analysis of non-linear transformation and the establishment of a mathematical model necessitates experimental designs involving series of runs under a variety of initial conditions and the application of advanced statistical techniques for parameter estimation in systems of non-linear equations.

This population dynamic model coupled with pesticide transformation is able to produce the dynamic patterns characteristic for non-linear transformation, like time lags and a dependence of these patterns on the initial concentrations. Furthermore, it is demonstrated that parameter estimation is feasible even if analytical solutions do not exist. The application of a linear, chemical Arrhenius law is defective at temperatures higher than 20°C and at high pesticide concentrations: biological reactions appear to have a different temperature sensitivity than pure chemical reactions.

# 6.4.        Short Communications

## 6.4.1.        *Application of CSTR Model : Simulation of Pesticide Movement in Soil*

## T. A. Albanis and P. J. Pomonis

Department of Chemistry, University of Ioannina, Ioannina 45110, Greece

It is rather common practice that analyses of pesticides residue in the soil are carried out by using conventional methods in analysing samples of soil taken from a certain depth. Sometimes the sampling may take place from successive soil depths and analyses then show the extent of penetration of pollutants though the soil. These successive soil layers can be considered as a cascade of continuously stirred tank reactor vessels connected in series (CSTR's) and the problem may probably be treated according the standard theory of chemical reactor design.

The mass balance of the i soil layer, or the i tank-in-series will be :

Inflow - Outflow - Reaction = Accumulation

$$\frac{n_{i-1}\,(Q_{i-1}/V_{i-1})}{V_{i-1}} - \frac{n_i\cdot(Q_i/V_i)}{V_i} - \frac{k.n_i}{V_i} = \frac{dn_i}{V_i.dt} \qquad (1)$$

where i is the total number of vessels (soil layers), $n_{i-1}/V_{i-1}$ the concentration $C_{i-1}$ of n compounds in the i-1 vessel, $n_i/V_i$ the concentration in the i vessel, $Q_{i-1}/V_{i-1}/t_{i-1}$ where $t_{i-1}$ the mean resident time of the compounds in the compartment i-1, k the rate constant of decomposition, $Q_{i-1}$ is the volumetric inflow and -1 the volume of the vessel. Assuming that there are several parallel rates of decomposition, like chemical, biochemical and photochemical which all obey first order kinetics we can write for the total rate :

$$R_{total} = (k_{chem} + k_{biochem} + k_{photochem}) . C = k^1 . C \qquad (2)$$

Solutions of the equation (1) proved useful in predicting the maxima of three pesticide concentrations, namely methyl parathion, lindane and atrazine in different soil layer layer[1]. The corrections that were necessary for this model were the water solubility of each molecule and the adsorption of these pesticides onto the soil. So we attempted a kind of simulation of the experimental results for the three pesticides noticed above according to the set of equations provided by the CSTR's well known theroy[2,3]. The kinetic results of disappearance of pesticides metolaclor, atrazine and cyanazine from soil are better described by a second order approximation[4].

# References

□ 1. T. Albanis, P. Pomonis and A. Sdoukos, *Water Air and Soil Pollution* 39, 293 (1988).

□ 2. T. Albanis, P. Pomonis and A. Sdoukos, "Model of pesticide movement in soil", In *Modelling in Environmental Chemistry*, Ed. S.E. Jorgensen, Elsevier, Amsterdam 1991, pp. 381-414.

□ 3. T. Albanis, P. Pomonis, *Toxicol. and Environ. Chemistry* 38, 109 (1993).

□ 4. P.J. Pomonis, T.A. Albanis and C.S. Skordilis, "kinetic description of disappearance of pesticide residues from soil based on the tank-in-series model", Submitted in *Chemosphere*, 1994.

## 6.4.2.    *Modelling Partition between Water and Formulation*

# Z. Mulinski

Agricultural University, Department of Chemistry, 71-434 Szczecin, Poland

The purpose of the present work was an analysis of mass transfer of the pesticide in water emulsion. The model A was proposed to estimate all parameters which are necessary to describe the pesticide mass transfer from oil to water phase. In the following development a rigorous model (the model B) is then presented to compute the concentration of pesticide both in oil and water phase.

## The model A

It was assumed that : (i) the volume of phases is constant, (ii) the partition coefficient , m, is constant. The material balance for pesticide can then be described by the following equations system:

$$dCo / dt = -KA / Vo ( Co - mCw ) + Gexp(r_2 t)$$

$$dCw /dt = KA / Vo (Vo / Vw ) ( Co - mCw ) - k_2 Cw + Hexp (r_2 t)$$

where: $Co$ , $Cw$ - intrinsic volume-averaged concentration of pesticide in oil and water phase , respectively , mol / l; $Vo$ , $Vw$ - volume of the oil and water phase , respectively , l; $k_2$ - rate constant of pesticide decay , $day^{-1}$ ; $K$ - overall mass transfer coefficient , dm /day; $A$ - interfacial area , $dm^2$ ; $G$ , $H$ - correction for initial change both phase volume and the hydrodynamics.

These equations can be solved with the initial conditions :                          **(1)**

$$Co = Co (0), Cw = Cw(0) =0 \text{ at } t=0 \text{ and } H = -G \, Vo/ Vw$$

to yield : $Co, Cw = f ( KA /Vo , Co(0) , m , G , k_2 , Vo , Vw , t )$ .          **(2)**

In practice only the total concentration of pesticide in dispersion can be measured:

$$CT = CoVo / V + ( 1- Vo / V ) Cw \qquad \text{where} : V = Vo + Vw. \qquad \textbf{(3)}$$

The rate of decay of pesticide in both phases is expresed by :

$$Y(t) = CT / CT(0) = Co(t) / Co(0) + ( Cw \, Vw ) / (Vo \, Co(0)) \qquad \textbf{(4)}$$

and finally : $Y(t) = J \exp(r1 \, t) + ( 1- J ) \exp(r2 \, t) + G \, k2 \, t / ( Co(0)( r2 - r1 )) \exp(r2 \, t)$   **(5)**

$$\text{where} : J = r2 / ( r2 - r_1 ) + G \, k2 / ( Co(0)( r2 - r_1 )^2 )$$

$r_1$ , $r_2$ - the roots of characteristic equation of NDE system (1).

All parameters : $KA /Vo$ , m and $k_2$ may be estimated by nonlineral regression method.

## The model B

The presence of surface-active agents greatly reduces the tendency to develop circulation and tend to make drops more rigid. On the other hand the mass-transfer rate is changed with drop size. It is of interest to estimate the possible importance of these effects on absorption from an emulsion drop. Let $Vo(t)$ represent the particle volume , then from mass balance of inerts :

$$Vo(t) = Vo(0) \ ( \ 1- Y_p - Y_a \ (0)) \ / \ (1- Y_p \ (t) - Y_a - ( \ Y_a(0) - Y_a) \ exp( \ -r_3t))) \qquad (6)$$

where : $Y_p = Co(t) \, M_p \, / \, d$ - volume fraction of pesticide ; $M_p$ - molecular mass of pesticide; d - density of pesticide , g/l ;$Y_a(0)$ , $Y_a$ - volume fraction of adjuvants at t=0 and when equlibrium at the interface is estabilised,respectively ; $r_3$ - rate change of adjuvants ,$day^{-1}$ .

The surface area of oil particle is proportional to $Vo^{0.666}$ , so the mass-transfer rate equations are :

$$dCo/dt = -KA/Vo(Vo(t) \, / \, Vo(0))^{-0.333} \ (Co - mCw) - (Co(t) \, /Vo(t))(d(Vo(t)/dt))$$

$$(7)$$

$$dCw/dt = KA/Vo(Vo(t)/Vo(0))^{0.666} \ (Vo(0)/Vw(0))(Co - mCw) - k_2Cw.$$

The solution of these differential equations (7) on a computer involves numerical integration, f.e., fourth-order Runge-Kutta method with assumption of the $Y_a(0)$ ,$Y_a$ and $r_3$ values.

# Conclusion

A general method is presented for description of the behaviour of pesticide in aqueous emulsion. This model describes mass-transfer through an ageing interface in the presence of surface-active agents. Since both the overall mass transfer coefficient and partition coefficient may not be constant extention of the model appears possible and needed.

ORIENTALISTE, KLEIN DALENSTRAAT 42, B-3020 HERENT